Process Safety Spotlights 2019

Held at the 2019 AIChE Spring Meeting and
15th Global Congress on Process Safety

New Orleans, Louisiana, USA
31 March - 4 April 2019

ISBN: 978-1-5108-8691-9

Printed from e-media with permission by:

Curran Associates, Inc.
57 Morehouse Lane
Red Hook, NY 12571

CURRAN ASSOCIATES INC.
proceedings
.com

Some format issues inherent in the e-media version may also appear in this print version.

Printed by Curran Associates, Inc. (2019)

For permission requests, please contact AIChE
at the address below.

AIChE
120 Wall Street, FL 23
New York, NY 10005-4020

Phone: (800) 242-4363
Fax: (203) 775-5177

www.aiche.org

Additional copies of this publication are available from:

Curran Associates, Inc.
57 Morehouse Lane
Red Hook, NY 12571 USA
Phone: 845-758-0400
Fax: 845-758-2633
Email: curran@proceedings.com
Web: www.proceedings.com

TABLE OF CONTENTS

19 AIChE Spring Meeting
15TH GLOBAL CONGRESS ON PROCESS SAFETY
NEW ORLEANS • 2019

Relief Systems Anonymous: Answers to Questions You Don't Want to Ask

Justin Phillips, P.E.
Provenance Consulting, LLC
Houston, TX
Justin.Phillips@ProvPSM.com

Prepared for Presentation at
American Institute of Chemical Engineers
2019 Spring Meeting and 15th Global Congress on Process Safety
New Orleans, LA
March 31 – April 3, 2019

Relief Systems Anonymous: Answers to Questions You Don't Want to Ask

Justin Phillips, P.E.
Provenance Consulting, LLC
Houston, TX
Justin.Phillips@ProvPSM.com

Keywords: PSV, PRD, pressure relief design

Abstract

This paper shares questions submitted from numerous engineers when asked, "What do you want to know about pressure relief systems but were too afraid to ask?" It then answers them using the latest RAGAGEP and industry direction. This multi-Q&A paper is filled with straight-forward information to help the audience better understand basic, intermediate and complex pressure relief system problems and situations faced by engineers today. It is written for an intermediate level audience that may have regular exposure to pressure relief systems and their design.

1 Introduction

I'm Justin Phillips, and I'm a relief systems engineer. In my experience, many industry professionals avoid asking "dumb" technical questions openly for reasons of fear or pride. What engineers need are answers to the technical pressure relief systems questions they just can't find a way to ask.

This paper shares questions submitted from numerous engineers when asked, "What do you want to know about pressure relief systems but were too afraid to ask?" It then answers them using the latest RAGAGEP and industry direction. This multi-Q&A paper is filled with straight-forward information to help the audience better understand basic, intermediate and complex pressure relief system problems and situations faced by engineers today. It is intended for an intermediate level audience that may have regular exposure to pressure relief systems and their design.

2 Choosing API 526 Make/Model

Why would someone pick one API 526 make/model over another?

- *Anonymous Employee, Major Chemical Company*

Optimization is why!

Let's ignore price, lead time, and preferred manufacturer arrangements for a moment. Also, as a reminder: API 526 established standard dimensions for PSVs with flanges, allowing facilities to swap one manufacturer's model for another without needing to re-pipe. However, the same thinking extends to non-API 526 valves (portable valves).

We can perform preliminary PRD sizing with API 520 Part I using generic discharge coefficients and correction factors (i.e., K_d, K_b/K_w, K_c). Let us use the generic sizing equation in Figure 1 as a reference point.

$$A = \frac{0.04W}{K_d K_b K_c K_v G}$$

Figure 1. API 520 Pt. 1 Eqn. C.9[1]

After preliminary sizing and selection (or if the device already exists and we are revalidating it), it behooves us to use the actual areas, coefficients, and correction factors.

Does it make a difference? Yes! Let's assume we have a single 2J3 in a vapor service (k = 1.4) set at 100 psig with a bellows and a rupture disk on the inlet. The rupture disk/PSV combination has a certified factor for Manufacturer 1 and 2, but not 3.

Table 1. Final Sizing and Selection Comparison Using API 520 Pt. 1

	API Prelim	Mfr 1	Mfr 2	Mfr 3
A, Actual Area (in^2), *normalized to ASME basis	1.43* (1.287 / 0.9)	1.453	1.43	1.496
K_d Discharge Coefficient, *normalized to ASME basis	0.878* (0.975 × 0.9)	0.87	0.858	0.855
K_b, Backpressure Correction Factor **@ 40% backpressure	0.868	0.953	1.000	1.000
K_c, PSV-Rupture Disk Combination Correction Factor (Mfr 4)	0.900	0.986	0.982	0.900
$\Pi(K)$	0.686	0.818	0.843	0.770
$W \propto A \times K$	0.981	1.189	1.205	1.152

Values obtained from NB-18 [2] for anonymized makes and models.
*Never use these values when sizing. They are for illustrative purposes only.
**Depends on the calculated backpressure for a given capacity

If we want the highest guaranteed capacity for this service, we might choose Manufacturer 2. We can also see that the capacity is significantly higher than the preliminary/generic sizing dictates, which pushes us to recalculate pressure drop.

There are many additional considerations:

- Suitability of a device for multi-phase service. I.e., some valves are certified for both liquid and vapor service without a derating factor that penalizes the liquid capacity.
- Set pressure ranges (not all vendors offer something for our desired set pressure)
- Bellows backpressure limitations (discussed later)
- Blowdown compared to operating pressures
- Intangibles: customer service, reputability, ease of model naming conventions, familiarity

3 Low Temp Valves & API 526

Are the low temperature carbon steel valves considered to meet API STD 526?

- Anonymous Employee, Major Chemical Company

Yes, but probably not at the intended temperature.

Most of us think "-20 °F" when we think of Carbon Steel's lower temperature limit. Low Temperature Carbon Steel is designed for even lower temperatures (not as low as Stainless Steel).

API 526 limits Carbon Steel to -20 °F and does not discriminate.[3] My experience leads me to believe that if the material and temperature range are not explicit, then the valve is not API 526 compliant. Let's say that we have LTCS body operating and relieving around -30 °F but which is otherwise completely in line with API 526 (dimensions, spring material, set pressure, flange ratings, so forth). If you have a performance guarantee from the manufacturer, API 526 compliance is not a practical concern. What *is* concerning is implying or advertising that a valve is in strict compliance with API 526 when it isn't.

4 Bellows Ratings & API 526

Is a bellows rated for a lower maximum pressure than those stated within API STD 526 still considered an API 526 valve?

- API SCPRS 526/527 Task Force Technical Inquiry

No.

This was a very good question posed to and answered by API a few years ago. Basically, API 526 provides the mechanical backpressure limits for bellows (because a very high backpressure can destroy or deform the bellows). So, for example, a 3K4 might have a bellows rating of 150 psig per API 526. However, a certain manufacturer might advertise an API-526 compliant valve which has a backpressure limit of 65 psig. Caveat emptor.

5 Multiple PSVs on Same Inlet/Outlet

Why did my facility install two large, equally sized PSVs on the same inlet line and outlet line? The pressure drops are absurdly high.

- Anonymous Employee, Vertically-Integrated Supermajor

I have seen two basic flavors of this:

1) The inadvertent whoops. We just forgot to double-check the design before we built it.

2) The unintended whoops. We never intended to use the spare PSV.

The inadvertent situation occurs during the early stages of new design when the required relief rates are determined, the preliminary device selections are made, the pressure drops are calculated. The early design P&IDs show independent inlet and outlet piping, but thenthe project optimizes the design. Ex: Two 8T10 PSVs are chosen for a very large, low pressure (65 psig MAWP) vessel. The vessel has not been constructed yet, and the project desires to cut down on the cost by using a single nozzle for the PSVs with common inlet piping. Even more money is saved by manifolding the outlet together before it goes to the subheader.

Eureka! The project team saves a million dollars and celebrates over lunch. Meanwhile, the change never makes its way back to the process design engineers so the new pressure drops using common piping are unknown for a few years until the owner starts poking around the calculations. Whoops.

The unintended situation occurs when a once-spared PSV comes into active service. This could be the result of discovering a new overpressure scenario requiring more area than provided by the online PSV. Or, it could also happen because a LOPA needed one more IPL and determined that the additional IPL could be achieved by putting the spare into full-time service. Whoops.

6 Resources for Engineers

Aside from an industry professional, what is the best resource for new engineers to educate themselves on the Do's and Don't's of pressure relief systems?

- *u/hotcheetosandtakis, Reddit*

This answer will not surprise anyone: AIChE (www.aiche.org) is a pretty good online resource. There are several free (sponsored) webinars (even to non-members) on this topic. For more hands-on training, a number of relief valve manufacturers as well as specialty consultants will offer training courses that cover the Do's and Don't's. As far as handbooks go, maybe Marc Hellemans *The Safety Relief Valve Handbook.*[4] A free resource is the Emerson *Pressure Relief Valve Engineering Handbook.*[5] And, of course, API 520 parts 1 & 2[1,6], and API 521[7].

7 RS Design Considerations & Process Fluid

What are the different pressure relief system design considerations comparing the process or utility fluid being relieved? E.g., we aren't venting phosgene but we will vent steam. Are there LD50, flammability, corrosion considerations, etc.?

- *u/hotcheetosandtakis, Reddit*

The short answer is yes. It's all totally dependent on a matrix of rules/ considerations/ situations. How you vent your Methane outside of Pampa, TX will be different than the way you vent it in Contra Costa County, CA.

Let's look at the following specific considerations mentioned:

- LD_{50} – The dose of something that will kill 50% of the exposed population is not a consideration for pressure relief system design because the risk associated with something *that* dangerous is too high to vent it anywhere except a closed system. In other words, you would not consider anything except closed disposal because you meet your risk tolerance threshold well before you get to LD_{50}.

What might be a better, more arguable consideration is whether the relief effluent contains HAPs (hazardous air pollutants found in the Clean Air Act). Many facilities will use HAPs as one decision point for where discharge location. E.g., the existence of any appreciable concentration of HAPs may be cause to route to a closed system.

- Flammability – Assuming that we have a flammable gas that is otherwise safe to discharge to atmosphere, most facilities will review the lower flammability limits (LFLs) for a number of weather conditions to predict how high and how far a plume may travel (Figure 2). Comparing the plumes to surrounding buildings, equipment, or places where people may be will help guide the design. We might extend our tailpipe, for instance, if we see that our plume dips into a breathable zone.

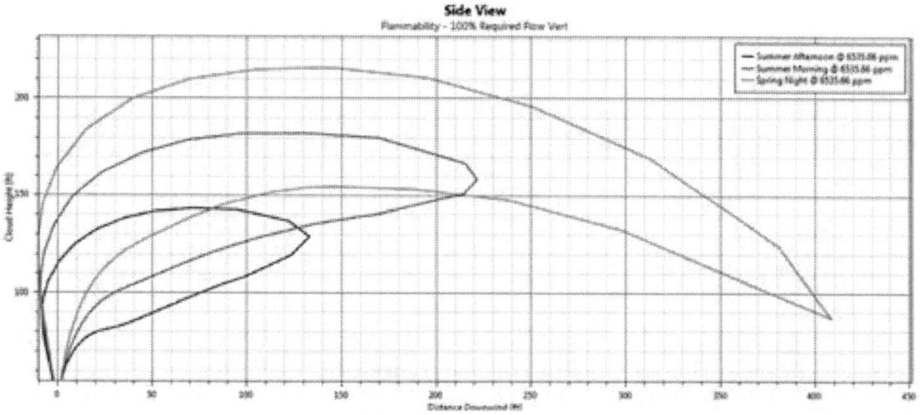

Figure 2. Example Flammability Plumes for Varying Atmospheric Conditions

- Corrosion – Acidic services tend to garner the most attention and are often addressed with resilient metallurgy. In the case of HF Alkylation units, the HF content of the process usually warrants a segregated flare header made of resilient metallurgy to handle the relief effluent (not to mention resilient PSVs). Aside from the relief effluent, corrosion may also occur from external sources: under the insulation where rain or moisture get stuck, or in liquid pockets where microbiologically induced corrosion can fester.

Many utility fluids (air, steam, water, nitrogen) can be discharged safely to the atmosphere. The key consideration is safety. Always discharge away from breathing zones and away from where people will be working. 10 ft above any standing surface and to the outside is a good rule of thumb.

8 Preventing Tail Pipe Liquid Accumulation

What is the best way to keep the tail pipe free of liquid accumulation including rain water (opened to atmosphere)?

- u/cheme1, Reddit

At a minimum, I would recommend a 3/8" hole drilled at the base of the elbow in the discharge to allow for any accumulated water to drain out (the minimum size is ¼"). Sometimes this doesn't work well enough (bugs/freezing rain/sleet/hard-to-check-or-clear on rounds) so a yellow, orange, or other colored cap (usually made of a rubberized plastic) can be placed over the top. They're snug enough to keep wasps/birds out, but loose enough to blow off. It's also a good way of knowing the PRD lifted if it's not otherwise obvious.

Can you go bigger than 3/8"?

Yes, but I'm not aware of an upper limit. Good judgment should dictate an appropriate size.

Isn't the easy way out of this is to design the PSV to relieve at a higher back pressure? Let's say I'm concern with rain filling up the rail pipe, should I just size the PSV to make sure it can displace the required flow rate with a certain orifice size at that given back pressure?

Most PSVs would not have a problem blasting accumulated water, wasp nests, or bird nests out of the way. And, you can even do the math to know what settings you would need to mathematically work your way out of the issue. However, the "easy way out" that you suggest violates a couple of rules and standards and would not be considered safe. Aside from the citation risk (and the rules that tell you to not do this), the physical risk that can be created with accumulated liquid (like water) on the discharge is:

1) It can create corrosion products or harbor biological products (slime etc) that block or impede the discharge piping or impede the opening and closing of the device.

2) It can freeze solid.

What rules does this violate? I agree with your first point but I believe the second point only applies to plants in cold climates.

The applicable industry RAGAGEP would be:

- API 520 Part II Section 5.3 provides a should statement regarding self-draining

- API 520 Part II Section 11.1 provides a shall statement re the same but specifically for PSVs
- ASME VIII UG-135(f)[9] provides a shall statement re the same but generalized to all PRDs
- On top of these you might find manufacturers requiring the same to guarantee performance.

9 Modulating PSVs

Why aren't all PSVs modulating, especially for new design? There seems to only be a few models that exhibit modulating characteristics. Are there a drawback to them?

- u/JermaineCharlemaine, Reddit

Generally speaking, it's difficult to get modulating behavior from a conventional spring-loaded device in a vapor service because the vapor tends to expand rapidly and push the valve all the way open as it moves from the pressurized system and into the valve body and to the downstream pressure.

In fact, this type of behavior is desirable because the idea is to guarantee the valve is 100% open at or before 10% overpressure. For liquids, the idea is the same but the discs are shaped differently because liquids tend to bounces off of surfaces instead of expanding like a vapor. There is some theory that the liquid conventional valves might tend to exhibit modulating characteristics (opening proportional to the system pressure), but this is not always true and some manufacturers flat out say that their valves will not modulate (but instead surge open).

The go-to for modulating behavior continues to be modulating pilot-operated valves, which don't care about the fluid behavior on the disc but only care about the system pressure. While these types of devices can mitigate inlet or outlet pressure drop concerns, they tend to be more expensive, have more parts, and be more tedious to install and maintain properly. Additionally, they can't solve all situations have some limitations on service (dirty fluid or extremely high temperatures, e.g.).

For one reason or another, some facilities absolutely hate dealing with them, whereas other facilities prefer nothing but (e.g., offshore platforms where a big pilot not only saves on installation weight compared to a conventional but also helps out on pressure drop because of some ridiculous piping layout).

10 Coke Drum Overpressure Scenarios

Do you have a specific method for evaluating a loss of power coupled with a loss of feed for a coke drum? Having to relieve the entire normal overhead seems excessive.

- Anonymous Employee, Major Refining Company

A detailed response gets into the realm of consulting, so let's keep this answer generalized. For background, the delayed coking process employees a furnace to heat up vacuum column bottoms (or similar heavy hydrocarbon) to the point that it thermally cracks into more volatile, lighter hydrocarbon components. These lighter components rise upward in a coke drum and are sent to a fractionator for separation, leaving behind heavy, carbon-rich coke in the drum.

Considering that a power failure, or loss of feed, or some other scenario may block the normal outlet of the coke drum, it is not *too* far-fetched to assume that you need to relieve the normal overhead vapor flow. Remember, the coking reaction is still occurring because the drums, coke, and feed still contain substantial amounts of heat, even with a loss of feed. Moreover, emergency steam may be injected into the furnace to keep stagnant feed from coking in the furnace tubes, and that emergency steam may need to be relieved along with the normal overhead.

I would recommend speaking with the delayed coker technologist or vendor (it is often a specialized black box design) regarding required relief rates during relief.

11 Uncommon Flange Ratings

Do PSVs come with 300# × 300#? I've never seen that, but a client asked to change to 300# and I want to make sure before I say they don't come like that.

- Fellow Employee, Provenance Consulting

No, not normally, since most PSVs discharge to a much lower system. An inlet-outlet combination like that is not found in API 526, so it's unlikely to be kept on the shelf. In a high-pressure discharge (back to process, e.g.), you might need flanges like this.

12 Fire on Rotating Equipment

A client asked me to add fire as applicable in a compressor because the service is flammable. I know we don't do fire for rotating equipment and in this case it is vapor full, but, I want to give her some references. Is there anything in API that says fire is not applicable for compressors?

- Fellow Employee, Provenance Consulting

Unfortunately, this is not as explicit as we would expect, so a good RAGAGEP source is hard to come by.

I was always raised to exclude fire from rotating equipment and piping. It is so embedded in my mind that I swore that API 521 explicitly excluded consideration of fire for rotating equipment. However, this is not the case—API 521 does not provide this guidance. From my point of view, the non-consideration of fire case for these types of equipment are rooted in their *exclusion* in the relevant RAGAGEP.

Recall the ASME VIII and API 521 (for pressure vessels), as well as API 2000[8] (for tanks) are very explicit when it comes to fire. Piping (e.g., B31.3 process piping) is excluded, although we often consider some additional wetted surface area because of connected piping. Pumps and compressors? Well, we might look to their design standards, but maybe it's as simple as this: Pump seals will fail in a fire and the whole of the pump is like a piece of deformed piping, anyway. Compressors are vapor-filled and we often don't consider fire on vapor-filled equipment because their metal walls may fail before the internal pressure reaches relief pressure.

13 Conclusions

Our industry develops its pressure relief design standards in a number of consensus groups, but these groups and their members are not always accessible to the general engineering public. The specialized nature of pressure relief design may make the subject matter even less accessible. To help bridge this gap in accessibility, I always encourage questions, especially anonymous questions, from the engineering public. This approach is beneficial to all parties: it answers an uncertainty and it also elucidates gaps in our standards or gaps in general industry knowledge.

14 References

[1] American Petroleum Institute, “Sizing, Selection, and Installation of Pressure-Relieving Devices, Part I — Sizing and Selection,” Standard 520, Part I, 9th ed., API, Washington, DC (July 2014).

[2] National Board of Pressure Vessel Inspectors, “National Board Pressure Relief Device Certifications,” NB-18, Columbus, OH, www.nationalboard.org/Index.aspx?pageID=64, (accessed Feb. 2016).

[3] American Petroleum Institute, “Flanged Steel Pressure-Relief Valves,” Standard 526, 7th ed., API, Washington, DC (Sept. 2017).

[4] Marc Hellemans, *The Safety Relief Valve Handbook: Design and Use of Process Safety Valves to ASME and International Codes and Standards*, Butterworth-Heinemann/IChemE, Oxford, UK (2009).

[5] Emerson, “Pressure Relief Valve Engineering Handbook,” Technical Publication No. TP-V300 (2012).

[6] American Petroleum Institute, “Sizing, Selection, and Installation of Pressure-Relieving Devices, Part II — Installation,” Standard 520, Part II, 6th ed., API, Washington, DC, (March 2015).

[7] American Petroleum Institute, “Pressure-Relieving and Depressuring Systems,” Standard 2000, 7th ed., API, Washington, DC (Mar. 2014).

[8] American Petroleum Institute, “Venting Atmospheric and Low-pressure Storage Tanks,” Standard 521, 6th ed., API, Washington, DC (Jan. 2014).

[9] American Society of Mechanical Engineers, “Rules for Construction of Pressure Vessels,” BPVC Section VIII Division 1, New York, NY (2017).

[10] U.S. Occupational Safety & Health Administration, “Process Safety Management of Highly Hazardous Chemicals,” 29 CFR 1910.119, U.S. Dept. of Labor, Washington, DC, www.osha.gov/pls/oshaweb/owadisp.show_document?p_table=STANDARDS&p_id=9760, (accessed Jan. 2019).

Modeling Shell Side Pressure Profiles for Heat Exchanger Tube Rupture Scenarios

Ahmed Harhara
Texas A&M University
ahmed.harhara@tamu.edu

M. Sam Mannan
Texas A&M University
mannan@tamu.edu

M. M. Faruque Hasan
Texas A&M University
hasan@tamu.edu

Prepared for Presentation at
American Institute of Chemical Engineers
2019 Spring Meeting and 15th Global Congress on Process Safety
New Orleans, LA
March 31 – April 3, 2019

Modeling Shell Side Pressure Profiles for Heat Exchanger Tube Rupture Scenarios

Ahmed Harhara
Texas A&M University
Address
ahmed.harhara@tamu.edu

M. Sam Mannan
Texas A&M University
mannan@tamu.edu

M. M. Faruque Hasan
Texas A&M University
hasan@tamu.edu

Keywords: Compliance with Standards, Safe Design and Venting & Emergency Relief

Abstract

Shell and tube heat exchangers are commonly used in the oil, gas, chemical, and nuclear industries. One fault that may occur with heat exchangers is a tube rupture, an overpressure scenario in which high pressure fluid flows into the low pressure region. This overpressure event may compromise the mechanical integrity of the exchanger and can lead to the equipment's failure. Significant property damage associated with tube ruptures have been documented in the past. To protect against such scenarios, overpressure protection measures such as relief devices are installed on the low pressure side of the exchanger. However, care has to be taken to accurately model the shell side pressure profile in order to avoid exceeding allowable design limits. To determine the size of the relief device that must be installed, API 521 recommends a dynamic analysis be performed when there are large differences in pressure between the shell and tube side. Using a time step-size of one millisecond, a dynamic analysis is performed on a liquid-liquid system and liquid-vapor system. The transient effects of the relief valve are taken into account. The challenges encountered in modeling these scenarios are addressed. This includes appropriate assumptions that can be made with minimal loss in accuracy. The effects of the pressure ratio and shell volume on the liquid-liquid and liquid-vapor systems pressure profiles are mentioned. Lastly, this paper also discusses ways of mitigating tube rupture consequences.

1 Introduction

Overpressure protection measures are critical to a process plant's safety. Pressure relief standards set by API 521 help guide engineers on how to tackle various overpressure scenarios. Little guidance however is given for modeling heat exchanger tube ruptures, a phenomenon in which a high pressure tube ruptures and overpressures the shell side of an exchanger (API Standard 521, 2014).

One technique on estimating the relief rate during a tube rupture is to apply a steady state calculation. This is done by first assuming a complete tube rupture has occurred (referred to in the literature as a guillotine rupture). A standard hydraulic orifice calculation is then performed to determine what is the rate of fluid entering the shell side. This number is then doubled due to the fact that fluid is entering from both ends of the ruptured tube. Finally, this rate is then imputed into a relief valve calculation in order to determine what size valve is required for adequate protection of the heat exchanger.

There are some issues that exist with this method however. The first is the concern that the shell side of the exchanger can fail before the relief valve can open. The reasoning behind this is that tube rupture scenarios result in fast-acting transient shockwaves which can reach peak pressure in milliseconds (Integrity, 2002). This is especially the case for high pressure differences between the shell and tube side.

Another concern with this technique is that the orifice calculation method ignores the shell side throughout the calculation. This is a problem because the tube rupture failure scenario is entirely focused on the shell side failing since it is the low pressure side of the exchanger. Orifice-style calculations overlook how the shell side pressure profile develops throughout the tube failure.

Lastly, orifice calculations are more appropriate for tube rupture calculations that do not have a subcooled liquid on the tube side. Subcooled liquids will flash once entering the low-pressure shell side, further complicating the calculation of the required relief rate.

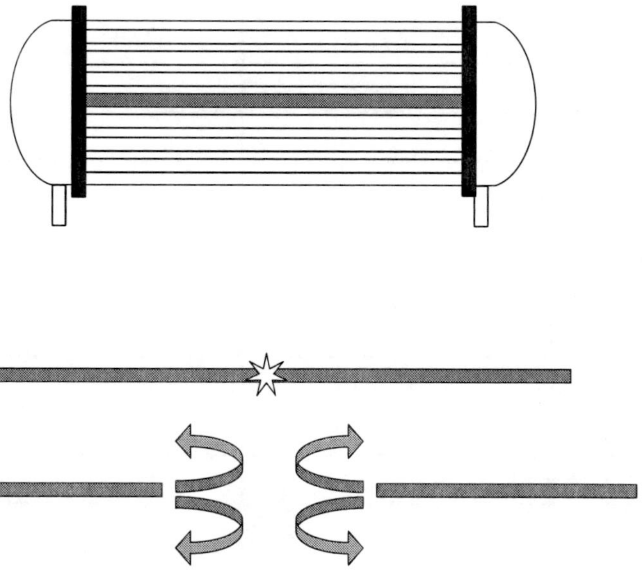

Figure 1. Sketch of Heat Exchanger Tube Rupture

For heat exchanger tube ruptures, the goal is to prevent the shell side of the exchanger from failing. Thus, a dynamic simulation is a more appropriate way to model this phenomenon. With a dynamic simulation, the pressure profile of the shell side of the exchanger can be obtained.

2 Relevance of Tube Ruptures in Industry

Before modeling the tube rupture phenomena, it is helpful to understand its relevance to industry. Knowing where tube ruptures are more common allows engineers to scrutinize one set of systems over others. Reliability data in industry is hard to come by, although there are efforts to counter this such as the Mary Kay O'Connor Process Safety Center's Instrument Reliability Network (Ahammad et al, 2016).

Because of this, it is often difficult to find statistics on the frequency of tube ruptures. Due to tube ruptures not being an everyday occurrence, few statistics are available. Moreover, some failure rate databases may record a tube rupture failure under a different name. This may end up categorizing a tube rupture under corrosion, making it even harder to estimate the likelihood of this event. Fortunately, there are some databases that give us a clue as to where tube ruptures occur. The United Kingdom's Hydrocarbon Release Database collects data on offshore incidents (Health & Safety Executive, 2016). By filtering out exchanger related releases, Table 1 is an estimate of which areas in a plant experienced a tube rupture.

From the table, it is evident that most tube ruptures occur in gas compression systems. Gas compression systems contain a series of intercoolers, with high pressure gases on the tube side, and low pressure cooling water on the shell side.

System	Severity			
	Major	Significant	Minor	Total
Flare/HP		1		1
Gas compression	4	21	13	38
Processing/gas/dehydration		3	5	8
Processing/gas/LPG/condensate		1	2	3
Processing/gas/sour gas		1		1
Processing/oil/oil treatment		1		1
Separation/gas production	1			1
Utilities/gas/fuel gas			4	4
Utilities/oil/heat transfer oil			1	1
Total	**5**	**28**	**25**	**58**

Table 1. Recorded Offshore Systems Experiencing Tube Rupture Incidents

3 Adequacy of Relief Valve Opening Times

For choosing how to protect a system from overpressure scenarios, plants have the option to use either rupture disks or relief valves. Rupture disks have the advantage of quicker response times compared to relief valves. Historically, rupture disks were used on exchangers. However, plants tend to prefer relief valves due to them not upsetting a process. Whereas relief valves close upon relieving the required pressure, rupture disks must be replaced, causing a section of the plant to temporarily halt operations for maintenance. Moreover, depending on the fluid in service, rupture disks can cause environmental damage when opened.

For most overpressure scenarios, relief valves are able to open in time for pressure increases. With heat exchanger tube ruptures however, there is some concern that tube ruptures (which can generate peak pressures in milliseconds) can overpressure the shell side of an exchanger before a pressure relief valve can take effect. A 2002 study prepared by Pipeline Simulation and Integrity Ltd for the United Kingdom's Health and Safety Executive tested the response times of pressure relief valves and rupture disks (Integrity, 2002). The study found that fast-acting relief valves do exist.

4 Modeling Strategy

For engineers who are performing dynamic tube rupture calculations, a flowchart illustrating what steps must be performed (and in what sequence) is helpful. Figure 2 lists the major steps in generating a tube rupture pressure profile. The first step in modeling a tube rupture is collecting the appropriate data about the system. This includes equipment diagrams, process simulation parameters (pressure, temperature, density, etc), and relief valve datasheets.

The second step is to obtain the necessary thermodynamic properties. The best way to achieve this is by creating a table of constant entropy flashes from the tube side conditions to the shell side conditions. Table 2 lists the constant entropy flashes for the liquid-liquid system solved in Section 6. These properties then allow the engineer to determine the mass flux which can be used to calculate the mass flow rate entering the shell side.

Once the above information has been obtained, the equations relating the mass flux and pressurization of the shell can be used to calculate the shell side pressure. Since numerical techniques will be used in solving the differential equation, an appropriate time-step needs to be used. A time-step in the milliseconds is recommended. Too large of a time-step and the resolution will not be detailed enough. If this occurs, the pressure vs. time profile obtained may not display the true peak of the tube rupture phenomena. However, too small of a time-step increases the number of iterations needed to achieve a solution.

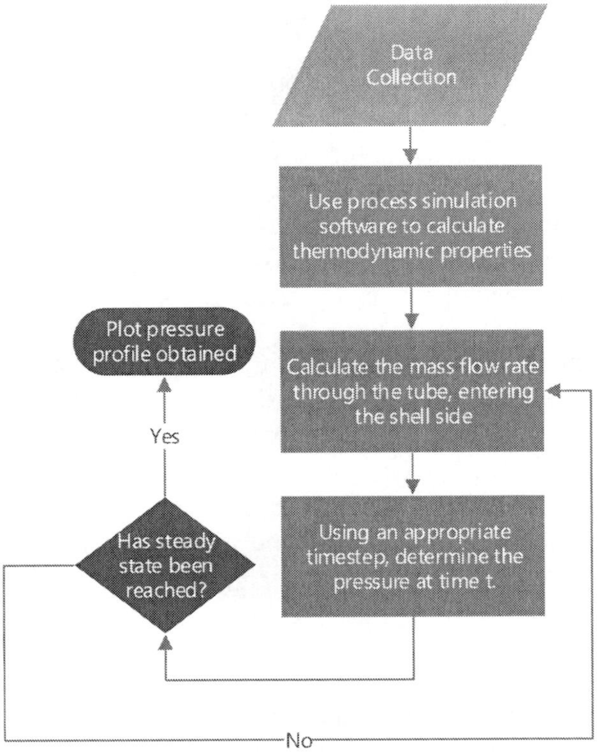

Figure 1. Steps needed to calculate tube rupture

5 Governing Tube Rupture Equations

Rate of tube side fluid entering shell side

$G^2 = \frac{h\left(v_1+v_n+2\times\sum_{j=2}^{n-1} v_j\right)}{{v_t}^2}$ ....Tube side mass flux

$A_{tube} = \pi {r_{tube}}^2$ ....Area of a single tube

$\dot{m}_{tl} = 2GA_{tube}$ ....Mass flow rate entering shell side

Rate of shell side fluid exiting PSV

If $P(t) > Pset$, then

$\dot{m}_{psv} = A_{psv}C_o\sqrt{2\rho_{sl}g_cP(t)}$ ....Mass flow rate through PSV

Else

$\dot{m}_{psv} = 0$

Shell Pressurization Equation for Liquid-Liquid System (Ennis et al., 2011)

$$\frac{dP}{dt} = \frac{\frac{\dot{m}_{tl}}{\rho_{tl}} - \frac{\dot{m}_{psv}}{\rho_{sl}}}{\frac{V_{tl}}{B_{tl}} + \frac{V_{sl}}{B_{sl}} + \frac{V_{shell}}{K_{shell}}}$$ ....Transient shell side pressure

$$B = -\frac{P1 - P0}{\frac{V1 - V0}{V0}}$$ ....Bulk modulus of elasticity

6 Illustrative Example for a Liquid-Liquid System

Suppose the following problem. A carbon steel shell and tube exchanger has ethylene glycol in the tube side and cooling water in the shell side. The exchanger contains 150 2m tubes with an inner diameter of 15mm. The shell's volume is 7.5 m³. The high and low side operating pressures are 10 bar and 1 bar, respectively. The high and low side operating temperatures are 100 °C and 20 °C, respectively. The shell side has a PSV set pressure of 1.2 bar (equal to its design pressure) and a hydrotest pressure of 1.8 bar. Determine the required PSV size in order to adequately protect this system.

Step 1: Collect the physical properties of the ethylene-glycol water system.

Using an appropriate equation of state,

- At 1 bar and 20 °C, the density of water is 1011.0825kg/m3.
- At 10 bar and 100 °C, the density of ethylene glycol is 1054.5917kg/m3.

We then determine the bulk modulus for each fluid. It is determined that

- Water has a bulk modulus of 3.44931×10^9 Pa
- Ethylene Glycol has a bulk modulus of 0.89769×10^9 Pa

The bulk modulus of common metals can be found in engineering reference manuals. The bulk modulus for carbon steel (K_{shell}) is 159×10^9 Pa.

Step 2: Use process simulation software to calculate thermodynamic properties.

A property table is created for the tube side fluid. Constant entropy flash calculations (in increments of 1 bar) are performed for ethylene glycol from the tube side pressure (10 bar) to the shell side pressure (1 bar). The results are listed in the table below.

Pressure [bar]	Molar Entropy [kJ/kgmole-C]	Mass Density [kg/m3]	Specific Volume [m3/kg]
10	-130.64	1054.59	0.0009482
9	-130.64	1054.48	0.0009483
8	-130.64	1054.37	0.0009484
7	-130.64	1054.26	0.0009485
6	-130.64	1054.15	0.0009486
5	-130.64	1054.04	0.0009487
4	-130.64	1053.93	0.0009488
3	-130.64	1053.82	0.0009489
2	-130.64	1053.71	0.0009490
1	-130.64	1053.61	0.0009491

Table 2. Properties of ethylene-glycol for constant entropy flashes

Step 3: Calculate the mass flow rate through the tube, entering the shell side.

Using the mass flux equation and constant entropy table, the mass flux can be calculated.

Pressure [bar]	Specific Volume [m3/kg]	Integral Term [m2/s2]	Mass Flux [kg/s-m2]
10	0.0009482	-	-
9	0.0009483	94.8	14522.1
8	0.0009484	189.7	20535.8
7	0.0009485	284.5	25149.1
6	0.0009486	379.4	29037.4
5	0.0009487	474.2	32462.2
4	0.0009488	569.1	35557.8
3	0.0009489	664.0	38403.8
2	0.0009490	758.9	41052.1
1	0.0009491	853.8	43538.9

Table 3. Ethylene-glycol water system mass flux at different pressure intervals

Once the mass flux for each pressure is determined, the data is then fitted to a quadratic equation in order to obtain the mass flux as a function of pressure. For the ethylene-glycol water system described above, the following mass flux equation is generated. Note that the P in this equation refers to the shell side pressure. At a pressure of 1 bar and 10 bar, this equation yields the maximum and minimum tube side flow rates.

$$G = -434.4P(t)^2 + 526.4P(t) + 41854.5 \text{..........tube side mass flux}$$

Step 4: Using an appropriate timestep, determine the pressure at time t.

At time is equal to zero, the shell side pressure is set to 1 bar. The following equations are then simultaneously solved using a tim-estep of 1 millisecond. Note the calculations presented shall assume a zero second response time for relief valves.

$$\frac{dP}{dt} = \frac{\frac{\dot{m}_{tl}}{\rho_{tl}} - \frac{\dot{m}_{psv}}{\rho_{sl}}}{\frac{V_{tl}}{B_{tl}} + \frac{V_{sl}}{B_{sl}} + \frac{V_{shell}}{K_{shell}}} \text{..........(1)}$$

$$G = -434.4P(t)^2 + 526.4P(t) + 41854.5 \text{..........(2)}$$

$$A_{tube} = \pi r_{tube}{}^2 \text{..........(3)}$$

$$\dot{m}_{tl} = 2GA_{tube} \text{..........(4)}$$

$$\frac{dV_{tl}}{dt} = \frac{\dot{m}_{tl}}{\rho_{tl}} \text{..........(5)}$$

$$\dot{m}_{psv} = \begin{cases} A_{psv}C_o\sqrt{2\rho_{sl}g_cP(t)} & P(t) > Pset \\ 0 & \text{otherwise} \end{cases} \text{..........(6)}$$

r_{tube}	0.0075 m
ρ_{tl}	1054.5917 kg/m^3
B_{tl}	0.89769×10^9 Pa
ρ_{sl}	1011.0825 kg/m^3
B_{sl}	3.44931×10^9 Pa
V_{shell}	7.5 m^3
V_{sl}	7.5 m^3
K_{shell}	159×10^9 Pa

Table 4. Parameters used in ethylene-glycol water system

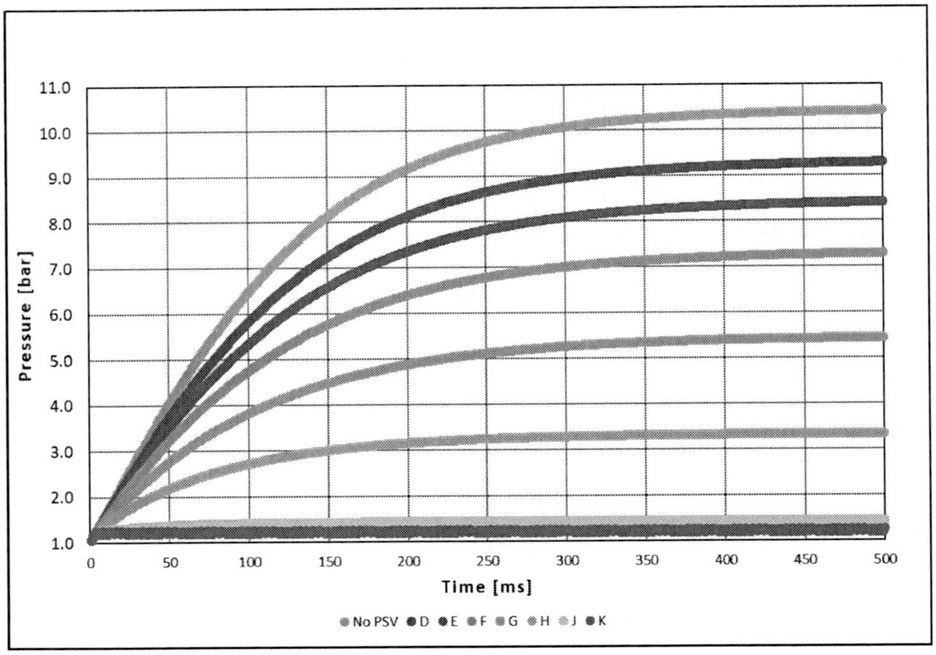

Figure 2. Pressure profiles for ethylene-glycol water systems for different PSV sizes

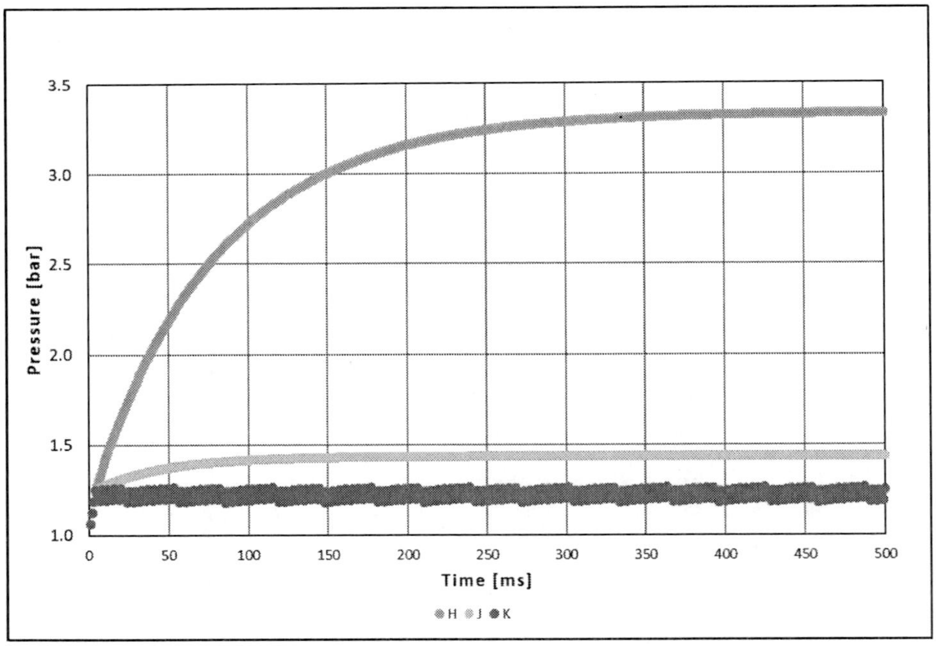

Figure 3. Pressure profiles for ethylene-glycol water system for PSV sizes H, J, and K

6.1 Mitigating Tube Rupture Scenarios

Depending on how far along in the lifecycle of an exchanger a plant is, mitigating the severity of tube rupture incidents can be done through one of several ways. First, a plant should flag exchangers that will have large differences in pressure. These exchangers are most susceptible to failure in the event of a tube rupture. For exchangers that are still in the design phase, increasing the strength of the exchanger is one way of mitigating the effects of a tube rupture. This design change does increase the investment cost of the exchanger. If the exchanger is upgraded to comply with the two-thirds rule, then a pressure relief valve would not be required for that system.

Additionally, plants can select an appropriately sized relief valve to handle a tube rupture scenario. For the ethylene-glycol water example in Section 6, without a pressure relief valve, the shell side would be pressurized from 1 bar to 10 bar in 300 ms. However, by installing either relief valve sizes J or K, the shell side would exhibit peak pressures of 1.43 and 1.27 bar, respectively. Moreover, while relief size K exhibits chattering during this tube rupture, relief size J does not.

7 Conclusion

An overview of the challenges in modeling tube ruptures was presented. The importance of more accurately modeling tube rupture scenarios increases with large differences in pressure between the shell and tube side. By switching from an orifice-style calculation for tube ruptures into a more rigorous dynamic simulation, plants can better predict the effects of a tube rupture. The main steps in generating a pressure profile were discussed, and a step-by-step calculation was performed for an ethylene-glycol water system, using relief sizes D, E, F, G, H, J, K.

8 References

[1] API Standard 521. (2014). Pressure-Relieving and Depressuring Systems.

[2] API Standard 520 Part 1. (2014) Sizing and Selection of Pressure-relieving Devices.

[3] Ahammad, M., Hatanaka, L., Jin, H., Lapeyrouse, K., & Roche, E. (2016). Learnings from Mary Kay O'Connor Process Safety Center (MKOPSC) *Instrument Reliability Network's Project on Pressure Transmitter Maintenance Data Collection* (19th Annual International Symposium).

[4] Health & Safety Executive. (2016) *Offshore Hydrocarbon Releases.*

[5] Integrity, L. T. D. (2002). Testing and Analysis of Relief Device Opening Times, Health & Safety Executive. *Offshore Technology Report, 23*, 20-23.

[6] Ennis, C. J., Botros, K. K., & Patel, C. (2011). Dynamic model for a heat exchanger tube rupture discharging a high-pressure flashing liquid into a low-pressure liquid-filled shell. *Journal of Loss Prevention in the Process Industries*, *24*(1), 111-121.

19 AIChE Spring Meeting
15TH GLOBAL CONGRESS ON PROCESS SAFETY
NEW ORLEANS • 2019

Azmi M. Shariff
CO_2 Research Center, Institute of Contaminant Management, Chemical Engineering Department, Universiti Teknologi PETRONAS, Bandar Seri Iskandar, Tronoh, 32610, Perak, Malaysia
azmish@utp.edu.my

Abdullah Al-Mubarak Md Jalil, Khairul Rostani, Nurzatil Aqmar Othman, M Faudzi Mat Isa,
PETRONAS Research Sdn Bhd, Lot 3288 & 3289, Off Jalan Ayer Itam, Kawasan Institusi Bangi, 43000 Kajang, Malaysia
almubarak@petronas.com, khairul_rostani@petronas.com, nurzatil.othman@petronas.com, faudzimi@petronas.com

Behzad Partoon, Kok Keong Lau
CO_2 Research Center, Institute of Contaminant Management, Chemical Engineering Department, Universiti Teknologi PETRONAS, Bandar Seri Iskandar, Tronoh, 32610, Perak, Malaysia
Behzad.partoon@utp.edu.my, Laukokkeong@utp.edu.my

Prepared for Presentation at
American Institute of Chemical Engineers
2019 Spring Meeting and 15th Global Congress on Process Safety
New Orleans, LA
March 31 – April 3, 2019

High CO_2-Natural gas Depressurization via CO_2 Flow Loop Facility: Design & Engineering Implementation

Azmi M. Shariff
CO_2 Research Center, Institute of Contaminant Management, Chemical Engineering Department, Universiti Teknologi PETRONAS, Bandar Seri Iskandar, Tronoh, 32610, Perak, Malaysia
azmish@utp.edu.my

Abdullah Al-Mubarak Md Jalil, Khairul Rostani, Nurzatil Aqmar Othman, M Faudzi Mat Isa,
PETRONAS Research Sdn Bhd, Lot 3288 & 3289, Off Jalan Ayer Itam, Kawasan Institusi Bangi, 43000 Kajang, Malaysia

Behzad Partoon, Kok Keong Lau
CO_2 Research Center, Institute of Contaminant Management, Chemical Engineering Department, Universiti Teknologi PETRONAS, Bandar Seri Iskandar, Tronoh, 32610, Perak, Malaysia

Keywords: high CO2, flow loop, emergency depressurization, cryogenic

Abstract

Sour natural gas resources contain high carbon dioxide content and require complex separation technologies to be developed. Cryogenic technologies such as cryogenic distillation and supersonic nozzle are among the identified technologies for natural gas conditioning, especially for offshore refinement. One of the challenges with offshore application of cryogenic technology is the absence of specific guideline in mitigating the blowdown philosophy which includes handling of CO_2 solids. Depressurization using a CO_2 closed flow loop facility was performed to further understand CO_2 solid behavior under rapid depressurization. It is shown that CO_2 solids formed in the pipelines and design of the flow loop is viable in order to handle the solids without any Health, Safety and Environmental (HSE) issues.

1 Introduction

As conventional gas resources around the world are depleting; unconventional, high Carbon Dioxide (CO_2)-content gas fields are thrust into the spotlight to become new candidates for field development [1]. However, the presence of substantial amount of CO_2

in the gas reservoir (can be up to 80% volumetric) introduced a huge technical and economic challenges towards the field exploitation [2]. Over the last few years, several studies have been conducted on cryogenic technologies such as cryogenic distillation and supersonic nozzle in CO_2 separation for fields containing more than 40% of CO_2. Based on the studies, these new cryogenic technologies have shown to have high potential in separating CO_2 from natural gas offshore to be utilized under carbon, capture, utilization, and storage (CCUS) project.

The new cryogenic technologies are currently being tested for the proof of concept. In common technology development practices, a pilot plant, which is a scaled down version of the technology, is normally developed to generate the design data and other process information. One of the major challenges faced during the pilot plant testing is the emergency depressurization philosophy as the process involves CO_2 solids handling which is uncommon to the industry standard. Depressurization of high CO_2 fluid at cryogenic temperature would lead to possibility of solid CO_2 formation, hence potential blockage of process equipment and venting line. This has been identified in the industry as American Petroleum Institute (API) guideline has clearly mentioned this issue in its guideline on emergency depressurization API 521 [3] and a recommended practice has also been developed to address the need for guidance on how to mitigate the risks and uncertainties related to CO_2 in pipelines [4]. However, there is no specific experimental studies conducted to specifically measure CO_2 solid formation during blowdown [5].

PETRONAS has designed and built a CO_2 flow loop with a capacity of 2.5 MMSCFD, 45 bar and capable of handling exceptionally low temperature to test cryogenic technologies. A depressurization study using natural gas mixtures containing 40 mol% CO_2 was conducted using the flow loop and the results are presented in this paper.

2 Facility description and test procedure

The flow loop is a closed system where the products from the main equipment, i.e. V-10, is recycled back as the feed to this equipment. Figure 1 illustrated a simple block diagram of this setup. The flow loop can be divided in cold (blue) and warm or moderate (black) sections and two cold and hot utilities as shown in Figure 1. Simplified Block Diagram of CO2 Flow Loop. The temperatures around the CO_2 solid handling vessel (V-10) are in the range of -65°C to -50°C and liquid CO_2 (around 300 liter) is accumulated in this cold section. The upper part of the block diagram shows the warm section with temperatures in the range of 10 to 80°C. The vast majority of medium in this section, namely in V-20 and V-25, is the vapor. Buffer vessel V-20 is at the low-pressure side of the system that run at 16 to 17 bara at nominal conditions (T=15°C). This is approximately the same pressure as in the V-10. Buffer vessel V-25 is at the high-pressure side of the system, which is around 40 bara at nominal conditions (T=40°C).

The warm section has three locations for depressurizing the system, as per control philosophy of the loop, during an emergency shutdown (ESD), the high- and low-pressure sides of warm section as well as compressor (K-20) skid and cold section are isolated by automatic shutdown valves. PCV-225 which controls the pressure in the high-pressure side of the loop (V-25), is also depressurized this section during an ESD. The low-pressure side

(V-20) is depressurized through blowdown valve BDV-220. During the ESD, the compressor skid will be isolated and depressurized through BDV-204.This prevents the back flow of gas through the compressor during an ESD event. The cold section will be mainly depressurized through BDV-110, while small amount of gas and liquid that traps between cold box and V-10 will depressurized by BDV-100.

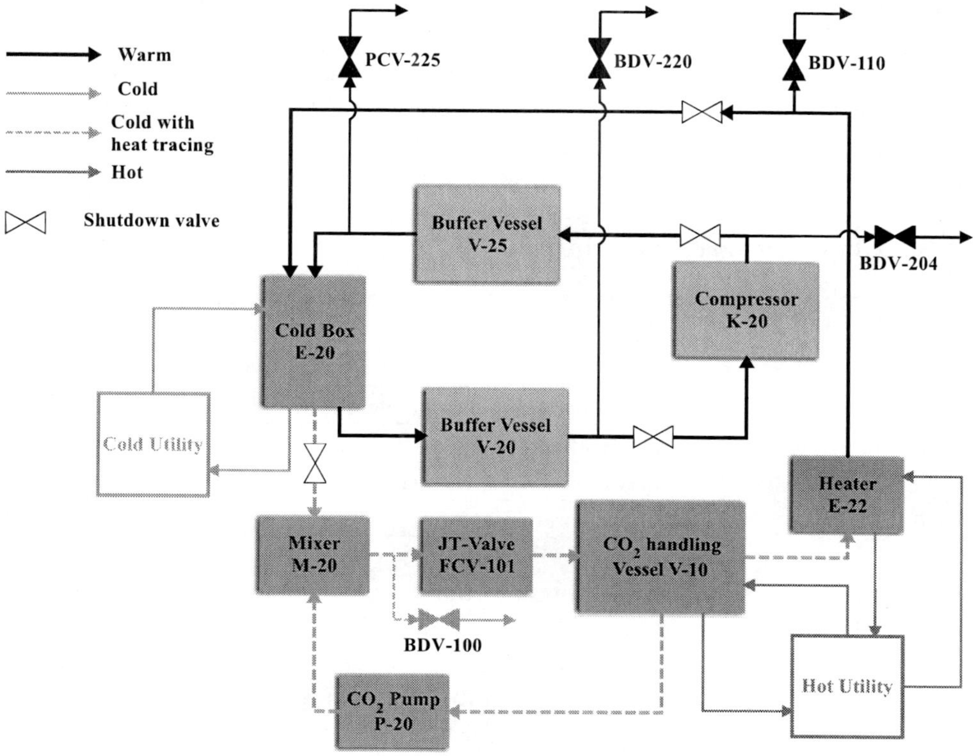

Figure 1. Simplified Block Diagram of CO2 Flow Loop

During a blowdown event, the test loop is divided into a warm and a cold section by shutdown valves and each section has its own vent line. The reason for this division is that nearly all flammable gas is situated in two vessels (V-20 and V-25) which are at higher temperatures where solids cannot form (warm section). The CO_2 liquids in V-10 on the other hand, are at temperatures where there is a substantial risk of solid formation and blockage of piping (cold section). The division ensures that flammable gases in the warm section can be released without any risk of blockages. For the cold section, measures are proposed to depressurize safely. After each blowdown valve, a restrictive orifice is placed to control the flow and pressure of venting system. The size of orifice is calculated based on maximum flow rate during various blowdown scenarios as explained in next section.

For the experimental blowdown test, an ESD is simulated during the normal operation of flow loop at the maximum capacity of 2.5 MMSCFD and the operating pressure of 35 bara at high-pressure section of the loop and 17 bara at the low-pressure section. The inlet stream of CO_2 handling vessel, V-10, had a 2-phase flow of natural gas with an average molecular weight of 28 kg/kmol and 40 mol% of CO_2.

3 Simulation

Before the actual blow down calculations can be done, the content of the system should be known. Therefore, the volume of the system is estimated based on the line list and equipment data sheets. The volume data of the (sectionalized) system is used in the *UniSim® Design* (USD) simulation to determine the contents of the blocked-in system at nominal conditions. There are 4 scenarios considered for the blowdown study as shown in Table 1. The target in these scenarios is to reduce the pressure in each vessel to the back pressure of 1.5 bara in the venting systems at 15 minutes or less.

Table 1. Blowdown scenarios description

No	Scenario	Description	Initial condition (V-10)	
			T (°C)	P (bara)
1	V-10 Operating	Blowdown from nominal operating conditions	-58	17
2	V-10 Settling out	Blowdown after pressure equalization	-51	27.8
3	V-10 design pressure	Blowdown from design pressure	15	46
4	V-10 fire	Blowdown during fire	-7	46

Figure 2 shows the setup in USD to prepare the streams for each scenario for the CO_2 handling vessel (V-10). The stream 'V-10 Operating' in Figure 2.*a* is the result of a combination of streams (103, 203 and 206) around V-10, in such a way that it represents (pressure, temperature, composition, volume) the contents of V-10 during operation. The VS modules are the virtual stream generator for this purpose.

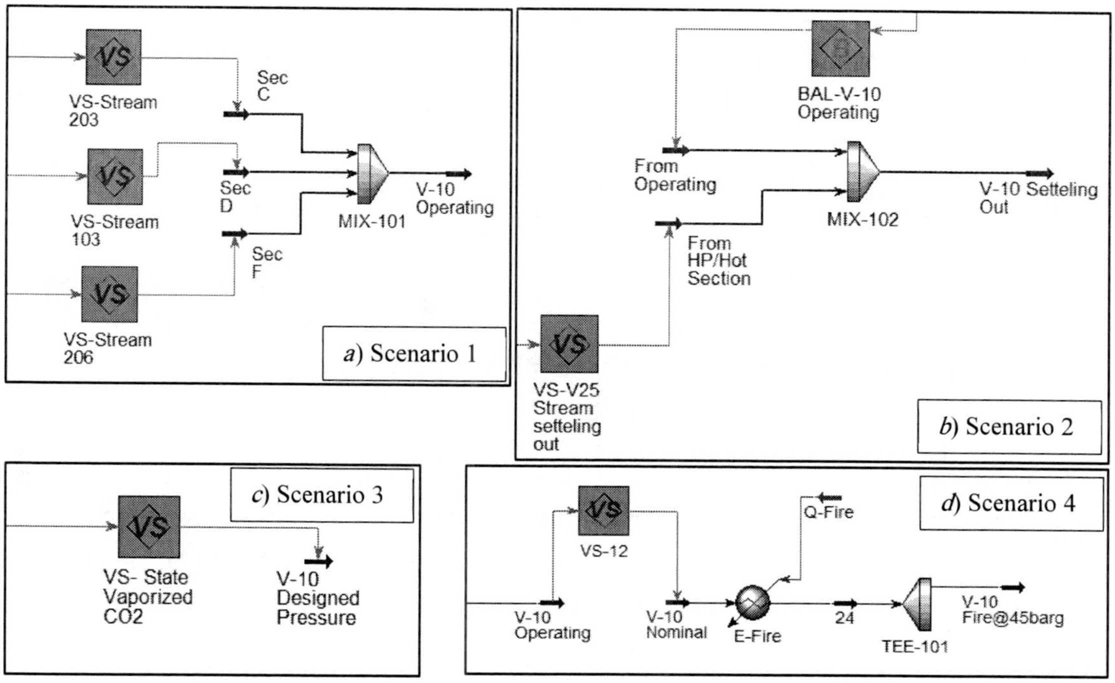

Figure 2. Setup in UniSim Design software to calculate V-10 scenario streams

The 'V-10 Settling out' stream in Figure 2.*b* is constructed by adding gas from the warm section until the pressure equals the settling out pressure (28 bar). 'V-10 design pressure' in Figure 2.*c* originates from a stream in which the liquid CO_2 in V-10, present during operation, is vaporized and the pressure has increased to the design pressure (46 bara). For the fire case, as shown in Figure 2.*d*, stream 'V-10 fire @45 barg' is created by adding heat to the operating case until the pressure is equal to the design pressure (46 bara) for the same volume. The four streams are used for initialization of the blow down utility for each scenario.

Similarly, for the hot section for each scenario, a model is generated in the USD simulation. In each scenario, two streams are simulated to represents vessels V-20 and V-25, as the majority of mass are accumulated in these 2 vessels. These vessels hold only gas with the same composition that leaves V-10 (molecular weight of 24 kg/kmol, ~20 mol% CO_2) in the vapor phase for the scenarios 1 (Operating), 2 (Settling out) and 4 (Fire). The third scenario (Design pressure) have a different composition with a higher CO_2 content due to evaporation of (all liquid) CO_2 in the loop.

Figure 3 shows the setup in USD to calculate the combined relief load of V-20 and V-25 in order to determine the required size of the downstream piping, including the vent header. This results in initial piping sizes that should be verified in a later stage, when other relief loads (namely from pressure relief valves) have become available.

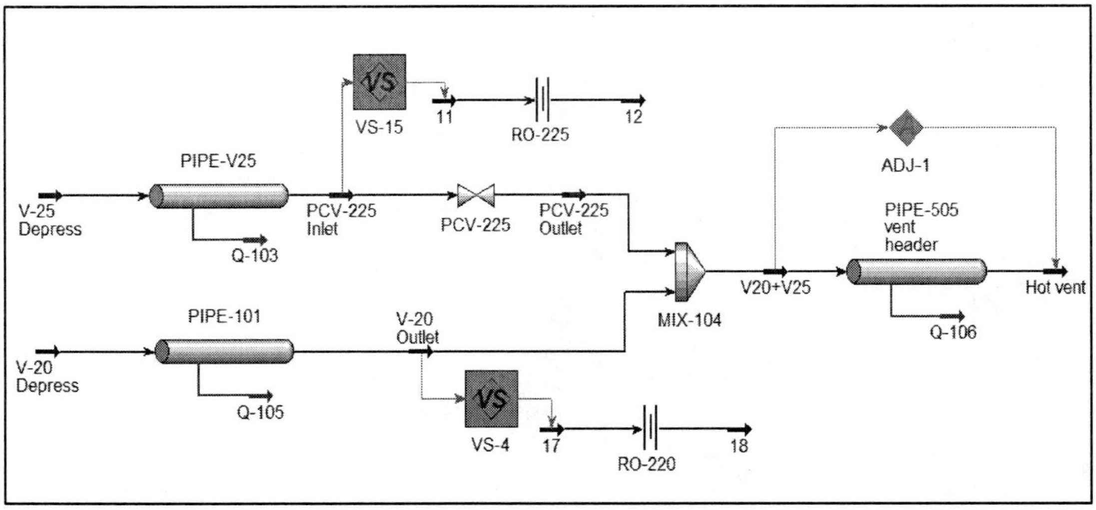

Figure 3. Setup in UniSim Design software to calculate combined relief load of V-20 and V-25 for vent header

The maximum relief flow rate for the warm section occurs for the scenario 3 in which all liquid CO_2 has vaporized, increasing the pressure of the complete loop to 46 bara. The combined maximum relief flow equals 3791 kg/h.

The fire scenario for the cold section shows a very high relief load of 15078 kg/h through its blow down valve/orifice combination. This is due to the large orifice size required for the boiling CO_2 in V-10 after the pressure has been reduced to 8 bara. As this high flow rate determines the orifice size (230 mm^2), the depressurizing times for the other (3) scenarios are very short. After 15 minutes, the pressure in the section is close to the back pressure (1.5 bara).

4 Results & Discussion

During the blowdown test, following the first scenario of simulations, pressure and temperature profile of the vessels at both cold and warm sections were monitored and recorded. Figure 4 shows the pressure profile of V-25 and V-20 at the warm section and V-10 at the cold section.

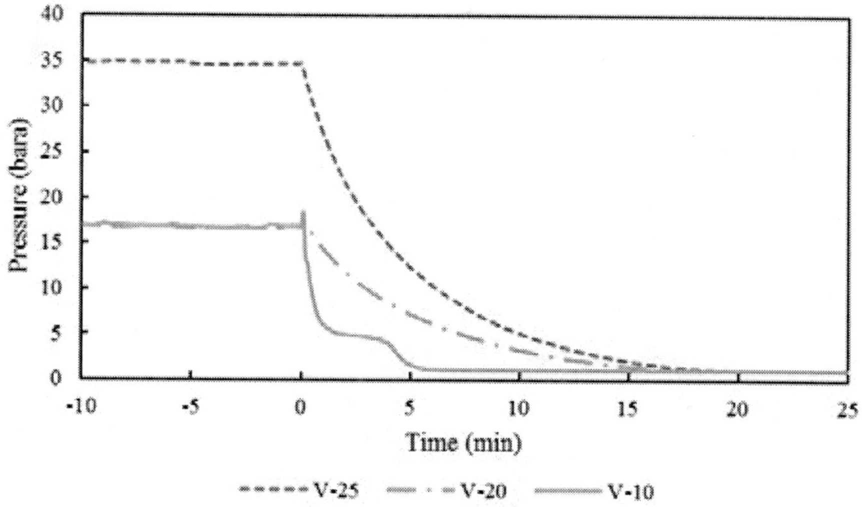

Figure 4. Pressure profile during the blow down

As shown in Figure 4, as soon as the emergency depressurization was initiated the pressure of V-20 and V-25 dropped from 17 and 35 bara, respectively, and reached the safe pressure level of 7 bara within 8 minutes. Pressure of V-10 initially increased from 17 bara to 19 bara upon depressurization and later reduced to 5 bara within the next 2 minutes. The initial peak is thought to be due to immediate CO_2 solid formation in the vessel and piping leading to the cold vent but later displaced due to the high shear stress of the flow acted on the deposited CO_2 solid at the wall. The pressure then maintained at 5 bara for around 3 minutes, and this is due to blockage of solid CO_2 formed after the BDV leading to cold vent stack. The corresponding plot of temperature profile inside V-10 (Figure 6) shows no indication of temperature drop at the same time frame, hence the blockage is most likely happened in the piping section and not inside the vessel. For better illustration of solid formation in the loop, the temperature profile in various location of the loop is presented in Figure 5 and 6. Following the control philosophy of flow loop, upon receiving and ESD signal, the controller will send the stop signal to both hot and cold utilities. Therefore, any reduction in the temperature is a sign of freezing liquid CO_2 and forming solid CO_2 in the pipes/vessels.

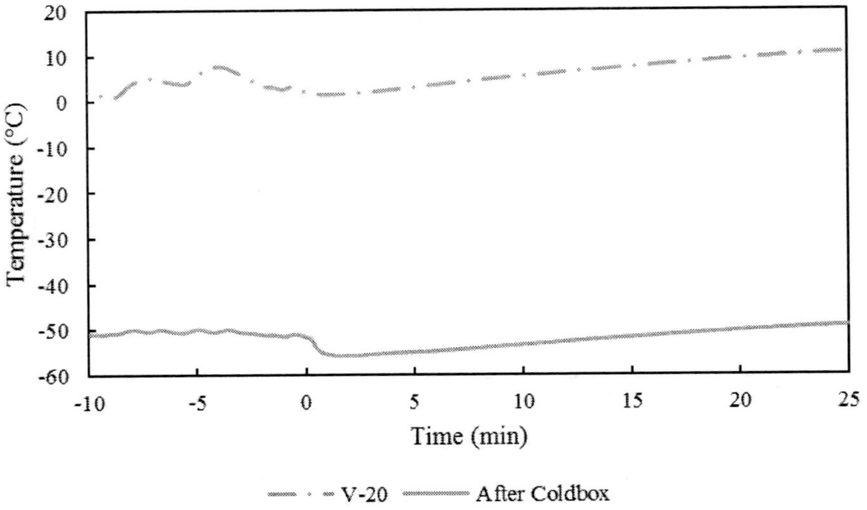

Figure 5. Loop temperature profile during the blow down

Figure 5 shows the temperature profile of V-20 during the blow down practice. As the V-25 vessel did not equipped with a thermometer, the temperature reading after the cold box, which is included in V-25 depressurization, is shown in Figure 5. As illustrated in this figure, after a sudden and small drop at the beginning, the temperature is increased steadily for the next 20 minutes as it balances out with the ambient temperature. The initial temperature drop could be attributed to the small Joule-Thompson effect upon depressurization and it is more prominent at higher pressure, i.e. V-25.

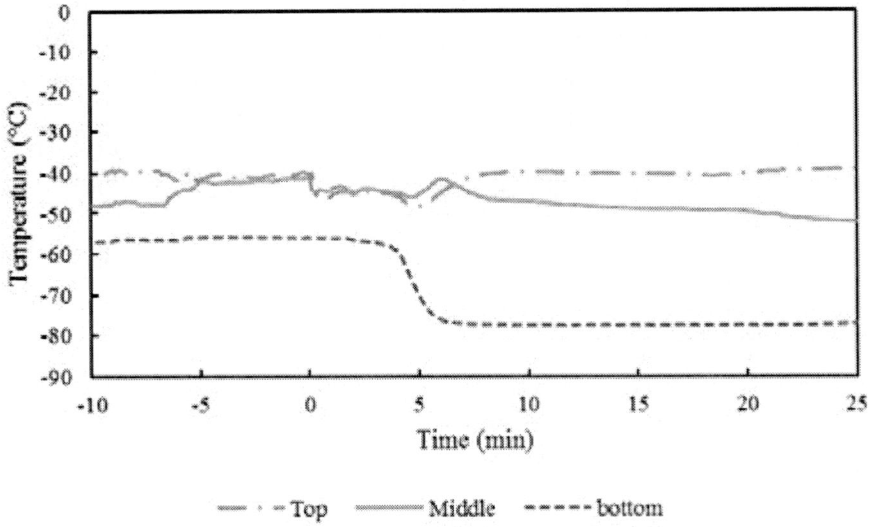

Figure 6. Temperature profile of the liquid compartment of crystallizer during the blow down

A different profile was observed inside V-10. This vessel held around 300 l of liquid CO_2 at the start of blowdown test. Decreasing the pressure in the vessel is changed the phase equilibria of the vessel from liquid CO_2 and vapor phases to solid CO_2 and vapor phases. This was resulted in different temperature profiles in the vessel, as illustrated in Figure 6. It should be noted that the are several thermometers installed at the top, middle and bottom sections of V-10 that enabled us to generate the temperature profiles presented in Figure 6.

The temperature profiles in top and middle sections showed an immediate temperature drop upon blowdown and for the first 5 minutes that follow the trend shown in Figure 5 and could be the result of Joule-Thompson effect and reduction of temperature due to the reduction of pressure. However, this reduction is higher than what observed for the rest of the loop, as the J-T coefficient increases by increasing the CO_2 content. After the first 5 minutes, the temperature profiles started to show different behavior. For the top section, where the area was mostly occupied by vapor phase, the temperature become steady and maintained at -40°C for the next 20 minutes. The middle section temperature started to drop further to below -50°C, showing a possibility of CO_2 solid formation at this level. The bottom section that held the liquid CO_2 and experienced a small temperature drop as the beginning of depressurization, showed a sudden temperature drop to -80 °C when the pressure dropped to 5 bara after 4 minutes. The pressure and temperature profile after this huge temperature drop clearly indicate of solid formation at the bottom of V-10.

Majority of this solid formed inside V-10 where CO_2 was in liquid form before blowdown took place. The solid formed in the piping leading to the cold vent stack caused a blockage as indicated in the pressure profile but the force acted on the solid adhered to the wall was enough to detach the solid off the pipe wall. It is also worth to mention here that the heat tracing system was turned on all the time during operation to help subliming CO_2 solid formed during blowdown. There was no indication of blockage that would cause overpressure in the vessel and piping. The pressure of V-10 also dropped to the safe level of 7 bara within the recommended time frame of 15 minutes.

5 Conclusion

In this study, natural gas mixtures containing high CO_2 content has successfully been depressurized using CO_2 flow loop facility with a capacity of 2.5 MMSCFD. Our strategy of segmenting the flow loop into warm and cold sections is proved to be viable for the cryogenic CO_2 depressurization. It was found that solid CO_2 did form in the vessel and piping but there is no indication that it caused blockage and overpressure in all sections and the test run smoothly without any HSE issues. All flammable gases are safely evacuated within the recommended timeframe of 15 minutes. Future plan includes installing a CO_2 solid detection system to further understand the behavior of CO2 solid in terms of formation time and morphology.

6 References

[1] M. F. M. I. Jalil, A.M., N. A. Othman, Z. Kassim, R. S. Salihuddin, A. A. Seman, F. H. Rahman, "Phase Behaviour Study for Carbon Dioxide Rich Natural Gas : Carbon Dioxide Solids and Hydrate Formation," in *255th American Chemical*

Society National Meeting & Exposition, 2018.

[2] R. S. S. H. Hassan, N. F. N. Omar, A. A. M. Jalil, "Gearing Towards CCUS for CO_2 Reduction in Malaysia," in *Offshore Technology Conference Asia*, 2018.

[3] American Petroluem Institute, "Pressure Relieving & Depressuring Systems" 6th Edition, January 2014.

[4] DNV, "Recommended Practice DNV-RP-J202 : Design and Operation of CO_2 Pipelines." 2010.

[5] F. H. Rahman, F. M. Isa, R. S. Salihuddin, M. Y. J. Jalani, and F. Karim, "High CO_2 Blowdown During Emergency Depressurization EDP at Offshore Environment" in proceeding of Offshore Technology Conference Asia 2018, Vol. 3, 1702-1711, 2018.

Pressure Relief Valves - Restricting Capacity to Mitigate Oversizing

Bill Atkins
Emerson Automation Solutions
Houston, Texas
Bill.Atkins@emerson.com

Chris Buxton
Emerson Automation Solutions
Houston, Texas
Chris.Buxton@emerson.com

Freeman Self
Bechtel Oil, Gas and Chemicals, Inc
Houston, Texas
feself@bechtel.com

UNPUBLISHED

Prepared for Presentation at
American Institute of Chemical Engineers
2019 Spring Meeting and 15th Global Congress on Process Safety
New Orleans, LA
March 31 – April 3, 2019

Pressure Relief Valves - Restricting Capacity to Mitigate Oversizing

Bill Atkins
Emerson Automation Solutions
Houston, Texas
Bill.Atkins@emerson.com

Chris Buxton
Emerson Automation Solutions
Houston, Texas
Chris.Buxton@emerson.com

Freeman Self
Bechtel Oil, Gas and Chemicals, Inc
Houston, Texas
feself@bechtel.com
Keywords: pressure relief valve, safety valve, capacity, restricted lift

Abstract

Pressure relief valves are frequently oversized above the required relief rate. Although oversizing is conservative, it can lead to other difficulties. Therefore, it may be advantageous to limit the capacity to only the rate required. Physically restricting the lift is an inexpensive and simple option.

Oversizing of pressure relief valves is frequently exacerbated because relief valves are normally available in standard sizes, with an increase of 21% to 78% from one orifice area size to the next standard size. Oversizing results in excessive flow through the valve, and may incur high inlet pressure, high backpressure and choking in the body bowl. The resulting high velocities increase erosion and intensify pipe vibration induced by flow or acoustic pressure waves. The relief valve operation may become unstable. Personnel may be affected by noise levels. Piping and engineering costs can increase.

Many solutions are not appealing, including revising the process and equipment or utilizing safety instrumented systems to reduce the required relief rate or relief probability. Purchasing a custom relief valve may be expensive.

However, one successful method is to mechanically restrict the opening - or the lift of the disk - to flow only the required rate. The option adheres to applicable ASME Boiler Codes and Code Cases, plus API-520 and API-526. Most vendors provide the option for both spring operated and pilot operated pressure relief valves. The additional cost is minimal per relief valve with no delay in schedule.

1 Introduction

1.1 Overview of Restricted Lift

One method to limit the flow through the pressure relief valve is to mechanically restrict the opening by restraining the lift of the disk.

The simple schematic illustrates how the lift may be restricted using spacers or washers. There are several other methods for the restraining device which are described below.

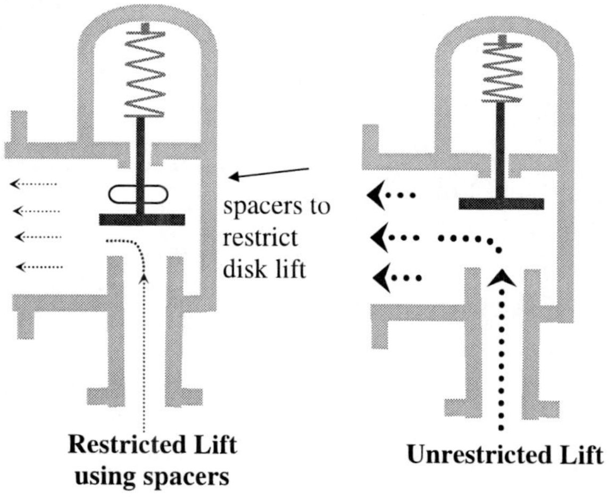

Figure 1 Restricted Lift Overview

Lift restrictions have been allowed by ASME Code since the issue of Code Case 1923 (Section I power boilers) prior to 1984. The final version 1923-3 was issued in 1995. In 1984, Code Case 1945 (Section VIII pressure vessels) was issued with the last version 1945-3 in 1994. They were initially primarily used for boiler safety valves due to their extremely high lift characteristic at set pressure. These Code Cases have now been incorporated into both the ASME Section I and VIII Code.

1.2 Why Restrict Lift?

Pressure relief valves are normally available in standard sizes with each given a letter designation. The difference in orifice areas between each size ranges from 21% to 78%. Refer to Table 1. Selection of a large orifice size may create a disproportionately over-sized relief valve and rated flow [2]. Oversizing results in excessive flow through the valve which may incur high inlet piping non-recoverable pressure loss and high discharge piping backpressure.

Table 1 Differences in Orifice Areas

Designation [7]	Orifice Area [7] (in^2)	Increase in Orifice Area
D	0.110	0%
E	0.196	78%
F	0.307	57%
G	0.503	64%
H	0.785	56%
J	1.287	64%
K	1.838	43%
L	2.853	55%
M	3.600	26%
N	4.340	21%
P	6.380	47%
Q	11.050	73%
R	16.000	45%
T	26.000	63%

Besides differences in designated areas, the size of the flange, as compared to the orifice area, affects inlet and outlet piping pressure drops [2]. API Standard 526 specifies flange sizes for each designated size (e.g., 4P6 has 4-inch inlet flange and 6-inch outlet flange). The smaller the ratio between flange area and orifice area, the higher the likelihood of high inlet and outlet pipe losses. Refer to Table 2.

- Inlet Line Losses based on inlet flange area to orifice area. Inlet line losses may be excessive when the ratio of inlet flange area to orifice area is below 3:1.
- Discharge Line Losses based on discharge flange area to orifice area. Discharge line losses may be excessive when the ratio of discharge flange area to orifice area is below 5.5:1 for conventional pressure relief valves.

In addition to pressure drops, oversizing may create choking in the relief valve body bowl (section of the body prior to discharge flange), choking at discharge expanders, and choking as the tailpipe or pipe lateral discharges into a larger header. Choking will release large amounts of energy via large pressure drops. The resulting high velocities increase erosion and intensify pipe vibration induced by flow or acoustic pressure waves.

Table 2 Flange to Orifice Areas

Designation [7]	Flange Diameter (nominal) (inch)		Orifice Area [7] (in^2)	Ratio: Nominal Flange Area / Orifice Area	
	Inlet	Outlet		Inlet	Outlet
D	1	2	0.110	7.1	28.6
E	1	2	0.196	4.0	16.0
F	1.5	2	0.307	5.8	10.2
G	1.5	3	0.503	3.5	14.1
H	2	3	0.785	4.0	9.0
J	3	4	1.287	5.6	9.8
K	3	4	1.838	3.8	6.8
L	4	6	2.853	4.4	9.9
M	4	6	3.600	3.5	7.9
N	4	6	4.340	2.9	6.5
P	4	6	6.380	2.0	4.4
Q	6	8	11.050	2.6	4.5
R	6	8	16.000	1.8	3.1
T	8	10	26.000	1.9	3.0

Other issues related to oversizing include personnel noise levels, increased piping sizes, relief valve opening stability and substantial engineering costs.

1.3 Alternatives to Restricting Lift

Many solutions are not appealing, including revising the process to reduce the required relief load or equipment to eliminate the relief case. Safety instrumented systems have been utilized to reduce the relief probability or required relief load. Purchasing a custom size pressure relief valve may be expensive. Studies to predict stability require time and expertise and may not predict definitive solutions. Basing the inlet piping on the required gas flow rather than rated gas flow to reduce pressure drops is not permitted per ASME Section VIII, Appendix M. Although the reason is not explained in code, the implication is that the inlet pipe and tailpipe is to be sized consistent with the relief valve rated flow for future debottlenecking. API Standard 520-I also provides such a warning but does describe several special instances where required flow may be utilized.

1.4 Disadvantages to Restricting Lift

There are several disadvantages to restricting the lift.

- A pressure relief valve with a required relief rate that is only slightly smaller than the full rated flow has essentially no extra capacity for expansion. If the process system is debottlenecked or alternate relief cases are identified, the required relief rate may need to be increased. Without extra capacity, the relief system, including pressure relief valve and inlet and outlet piping will require re-evaluation and most likely reconfiguration or replacement.
- Not all valves are certified per the ASME Section I or Section VIII requirements for restricted lift. For example, the major domestic vendors, with pressure relief valves manufactured according to API Standard 526, can provide direct spring-operated valves with restricted lift. Not all valves have restricted lift, including some specialty valves, and not all Section I valves and valves in non-compressible flow have this capability. Consult the manufacturer for the availability of restricted lift.
- Additional training is required for pressure relief valve repair personnel to assure proper inspection and reassembly of the restricted lift mechanism prior to final functional testing.

2. Code Requirements

This section discusses the requirements from the published Codes, ASME Code Cases that add to the Code but have not been yet written into the Code, and ASME interpretations of the Code.

2.1 ASME Boiler and Pressure Vessel Codes

2.1.1. ASME Section VIII-Division 1-2017 Edition, "*Rules for Construction of Pressure Vessels*", ASME BPVC.VIII.1-2017

This standard outlines the conditions for restricting lift for valves in service for Section VIII (pressure vessels). It has written the requirements from ASME Code Case 1945-3 (1994) into the pressure vessel code. Among the conditions are that a restraining device is only method to restrict the lift. The location of the restraining device must in an area of pressure relief valve that is sealed, similar to the set pressure adjustment. The minimum restricted lift of the pressure relief valve must be the greater of 30% of the full lift or 0.080 inch (2.0 mm). The restricted lift dimension is to be marked on the ASME nameplate in inches (millimeters)

2.1.2. ASME Section I -2017 Edition, "*Rules for Construction of Power Boilers*", ASME BPVC.1-2017

This standard outlines the conditions for restricting lift for valves in service for Section I (boilers). It has written the requirements from ASME Code Case 1923-3 (1995) into the boiler code. All conditions listed above for Section VIII Code are applicable for Section I.

2.1.3. National Board - List of Relief Valves with Restricted Lift

The National Board of Boiler and Pressure Vessel Inspectors <nationalboard.org> is composed of chief inspectors from enforcement jurisdictions of states, cities and provinces. It provides a list of relief valves with restrictions published in the document NB-18 *Pressure Relief Device Certification.*

2.1.4. National Board Interpretations

Interpretations of the *National Board Inspection Code* NBIC may be found on the website of the National Board of Boiler and Pressure Vessel Inspectors <nationalboard.org>. There are two interpretations on restricted lift.

• INTERPRETATION 95-56 (1995) Repairing of Relief Valves with Restricted Lift
Subject: RA-2231 (b)(1) Acceptance of Code Cases 1923 and 1945 when used in Original Construction. ASME Section VIII-Division 1-1995 Edition with the 1996 Addendum
Question: Under the provisions of paragraph RA-2231 (b)(1), is it permissible for a "VR" stamp holder to repair a restricted lift valve when ASME Code Case 1923-2 or 1945-3 was used in the original construction?
Reply: Yes, provided no change is made in valve lift.
Editorial Note: (1) The quoted clauses "RA" refer to an earlier edition of *National Board Inspection Code* NBIC, which has been reformatted and "RA" is no longer applicable to the lettering system. (2) Code Cases 1923 and 1945 have been written into ASME Section I and VIII respectively; see paragraph below on code cases.

• INTERPRETATION 98-24 (1998). Conversions
Subject: Restricted Lift per Code Case 1923 & 1945 1998 Edition
Question: May a "VR" Certificate holder perform a conversion, as defined in RA2242(c), of a pressure relief valve from one certified design type to another certified type which was certified by the manufacturer in accordance with ASME Code Cases 1923 or 1945?
Reply: Yes, provided all NBIC requirements pertaining to conversions are met and the certificate holder receives from the valve manufacturer specifications and instructions which include the additional marking requirements of the applicable code case.
Editorial Note: Refer to note in above interpretation.

2.1.5. ASME Code Cases

ASME Code Cases are additions to the Code but have not been yet written into the Code. The following three cases include on active Code Case and two which have been added to the Code.

• ASME CODE CASE 2787, Section VIII, Division 1. Approval Date: November 1, 2013. "*Multiple Markings of Certified Capacities for Pressure-Relief Valves.*"
This Code Case outlines the conditions for marking the relief valve with multiple certified capacities. The relief valve is to be capacity tested first on one of the certified medias (steam, air, gas, water) and then on the alternate media. No adjustments are to be made after each testing. The measured set pressures are to meet the tolerances in paragraph ASME Section VIII-UG-134. The full text of the code case is in Appendix 1.

- ASME CODE CASE 1945-3, Sections VIII, Division 1 and 2. Approval date August 8, 1994. "*Restricting Lift to achieve reduced relieving capacities of full lift, nozzle type, and flat seated safety and safety relief valves for compressible and incompressible fluid applications.*" This Code Case has been written into ASME Section VIII Code for pressure vessels.

- ASME CODE CASE 1923-2, Section I. Approval Date August 28, 1995. "*Restricting Lift to Achieve Reduced Relieving Capacities of Full Lift, Nozzle Type, Flat Seated Safety and Safety Relief Valves.*" This Code Case has been written into ASME Section I Code for boilers.

2.2 American Petroleum Institute Standards

2.2.1. API Standard 526 – 7th Edition, 2017 - Paragraph 7.7 "Restricted Lift" and Annex E "Valve Selection Example: Restricted Lift"

This standard states that the restricted lift relief device will follow all requirements of ASME Section VIII and this standard in Tables 3 through 30. The manufacturer is to be provided with a required capacity and the maximum rated capacity, and then the manufacturer shall determine the lift that will provide a rated capacity within these two values. An example is provided in Annex E of the standard.

2.2.2 API Standard 520-Part I - 9th Edition, 2014

There is no reference to restricted lift in the 9th Edition 2014. However the proposed 10th Edition will provide discussions on the requirements of ASME Section VIII and Tables 3 through 30 of API Standard 526. The manufacturer is to be provided with a required capacity and the maximum desired rated capacity, and then the manufacturer shall determine the lift that will provide a rated capacity within these two values. Instruction for the pressure relief valve specification sheets will be provided in Annex D. An informative Annex F will provide and example of valve selection for restricted lift.

2.2.3 API Standard 520-Part II - 6th Edition, 2015 - Paragraph 7.2.3.

This standard allows using restricted lift to reduce excessive built-up backpressure in Paragraph 7.2.3. The standard also allows using restricted lift to reduce excessive inlet pressure drop in Paragraph 7.3.7.5. The same information is in the proposed 7th Edition.

2.2.4 API Standard 521-7th Edition, 2019.

This standard provides documentation for restricted lift valves.

2.2.5. ISO 4126-1-2013

This standard states that the minimum restricted lift can be no less than 30% of full lift or 1 mm (0.039 inch) whichever is greater. The capacity certification is taken at three points between, at or above minimum lift, to below full lift. A curve fit is performed to establish the coefficient of discharge versus lift. The straight line capacity does not need to be checked. Otherwise, the standard mirrors ASME.

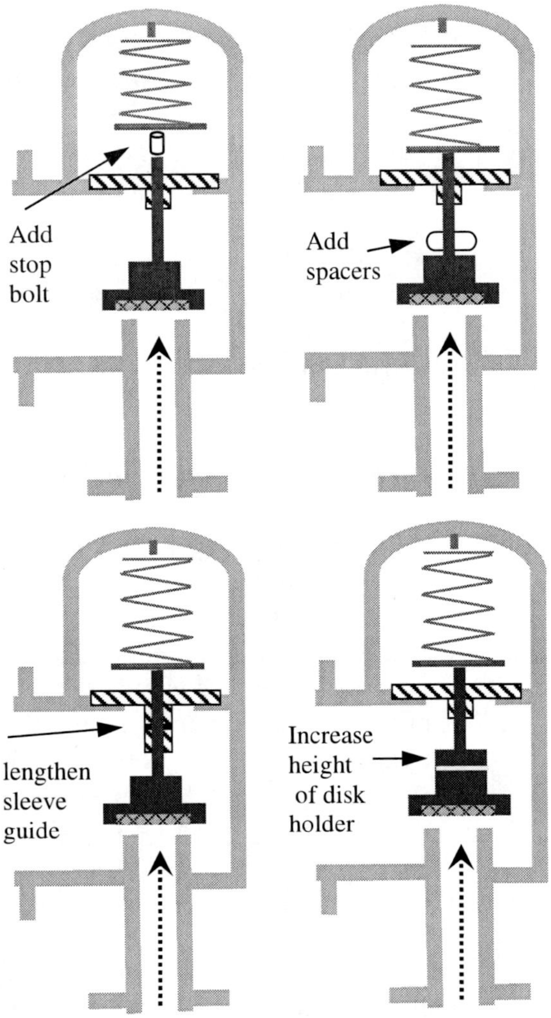

Figure 2 Methods to Restrict Lift

3. Mechanical Description

3.1 Methods to Restrict Lift for Spring Operated Relief Valves

There are several different *restraining devices*, which is the ASME term for the device that restricts lift. Flow is to be restricted only by the restraining device with no other changes made to the valve design [Section VIII-UG-136(a)(11) and Section I-clause PG-73.2.10].

Spacers or washers are typical since they are simple to add. Spacers may be provided in a series of thicknesses or they may be supplied in one thickness which must be machined to the correct length. Regardless of the type of restraining device, the device may physically wear over time reducing its size. Rechecking the lift restriction device for wear and measuring the restricted lift shown on the nameplate is part of the maintenance procedure.

3.2 Methods to Restrict Lift for Pilot Operated Relief Valves

Figure 3 illustrates how the lift is restricted in pilot operated valves.

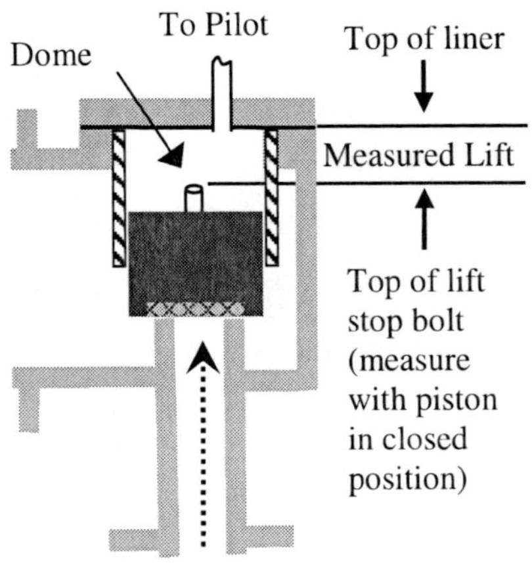

Figure 3 Pilot Operated Restricted Lift

3.3 Restricted Lift Capacity – Definitions and Criteria

The capacity is defined as a function of lift, where the length of the lift is physically measured. For example, for a spring-loaded relief valve with spacers as the restraining device, it is measured from the top of the spacer to the bottom of the sleeve guide. Figures 3 and 4 illustrate how the restriction is measured for pilot-operated and spring-operated valves respectively.

(1) The "restricted lift ASME nameplate capacity" is formally defined as a function of the ASME rated full rate lift and the measured restricted lift. The lift at "nameplate capacity" is sometimes informally referred to as certified lift. The ASME nameplate capacity is determined by the following equation [ASME Section VIII-UG-133(h) and Section I-PG-73.2.10(f).] ISO-4126 has similar rules, refer to paragraph 2.2.5. An example is provided in paragraph 4.1.

$$\text{Restricted Lift ASME Nameplate Capacity} = \text{ASME rated capacity at full rated lift} \times \frac{(\text{measured restricted lift})}{(\text{full rated lift})}$$

(2) The "restricted lift ASME nameplate capacity" is to be <u>less</u> than the "actual measured capacity at restricted lift". [Section VIII-UG-131(e)(1) and Section I-PG-69.2.3(a)]. Paragraph 5.1 of this paper describes the flow tests to measure the actual capacity and paragraph 4.1 with Figure 5 graphically illustrates the concept.

(3) The "restricted lift ASME nameplate capacity" value is between the "required capacity" and "maximum desired rated capacity", which are provided by the purchaser in their data sheet [API-526].

(4) The "actual measured lift" is that measured with a flow test as discussed in paragraph 4.1. This is to be greater than 30% of full rated lift and greater than 0.08 inches. These are the same for pressure vessels and power boilers [Section VIII-UG-136(a)(11) and Section I-PG-73.2.10(e)].

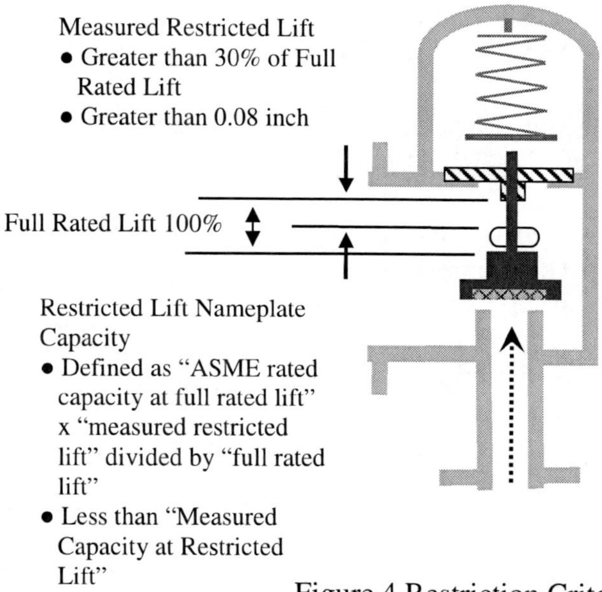

Figure 4 Restriction Criteria

4. Operational Factors

Operating factors include overpressure, blowdown, shape of the capacity curve, etc. These may be different for restricted PRVs than for full capacity PRVs. This section discusses differences, if any.

4.1 Flow Capacity and Lift Curves

Capacity curves are used to understand flow characteristics. Typically, they are shown as capacity as a function of overpressure. Capacity curves can also be developed as a function of lift. Figure 5 provides a typical example gas flow as a function of lift in a spring-operated PRV. This illustrates the actual capacity as a function of lift. Paragraph 3.3 provides the formal definitions of capacity.

EXAMPLE:
The following example will illustrate the procedure. The pressure relief valve in nitrogen service has the following specifications.

- Required capacity = 1,500 SCFM (to be provided by purchaser per API-526)
- Maximum desired rated capacity = 2,000 SCFM (to be provided per API-526)
- Set pressure = 100 psig

Using the test data in Figure 5, and the required and maximum desired rated capacity provided by the purchaser, the manufacturer recommends a 2J3 pressure relief valve with it lift restricted to 0.3 inches.

- Restricted Lift = 0.3 inches
- ASME Full Rated Lift = 0.5 inches (as measured – see Figure 5)
- ASME Full Rated Capacity = 3,000 SCFM (as measured – see Figure 5)

Thus, the restricted lift "nameplate capacity" = (3,000) * (0.3 / 0.5) = 1,800 SCFM [refer to equation 3.3 (1)]

To summarize the API and ASME Code criteria:

- The "nameplate capacity" per API-526 meets the "required capacity" to mitigate the over pressure and does not exceed the 2,000 SCFM "maximum capacity" to meet other system requirements.
- The "measured restricted lift" (0.3) is greater than 30% of "ASME Full Rated Lift" (0.15 = 30% of 0.5) and thus meets ASME Code

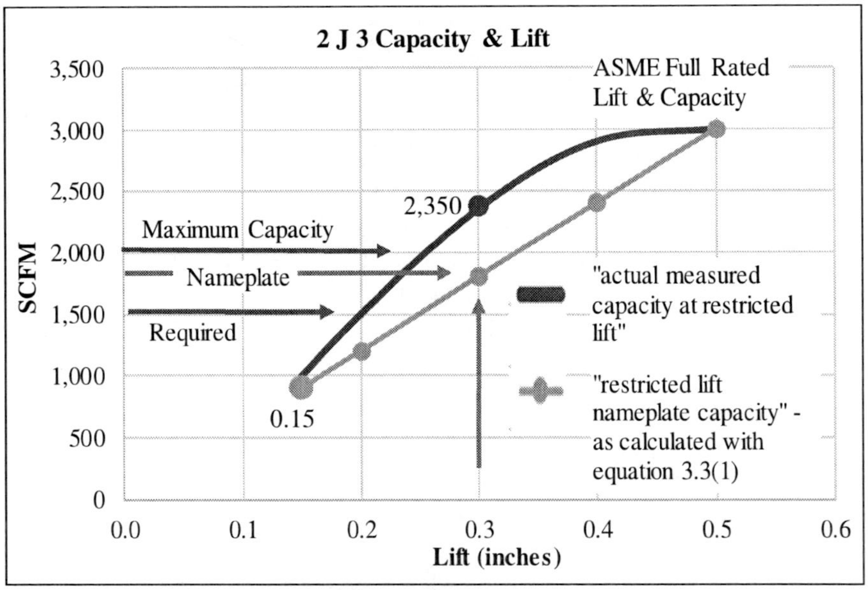

Figure 5 Capacity Test for Restricted Lift

- The "nameplate capacity" (1,800 SCFM) is less than the "actual measured capacity at restricted lift" (2,350 SCFM). Although physically the valve is oversized, the ASME "nameplate capacity" is the limiting ASME Code capacity.

4.2 Blowdown

The blowdown was not significantly affected by the restricted lift for direct spring-operated valves in both gas and liquid service. Test were run on three valves at four different lifts with different trims [1]. Graphs of the typical results are given in Figure 6.

Table 3 Tested Direct Spring Operated Valves for Blowdown

Orifice	Flange Size	Service & Trim & Pressure Range [psig]	Pressure Range [psig]	Lift Settings
"F"	1.5 x 2	• Gas with gas certified trim • Gas with dual certified trim • Liquid with dual-certified trim	246 - 282 psig	100% lift, 80% of full lift, 60% and 40% of full lift
"H"	2 x 3	• Gas with gas certified trim • Gas with dual certified trim • Liquid with dual-certified trim	201 - 223 psig	100% lift, 80% of full lift, 60% and 40% of full lift
"J"	3 x 4	• Gas with gas certified trim • Gas with dual certified trim • Liquid with dual-certified trim	147 - 165 psig	100% lift, 80% of full lift, 60% and 40% of full lift

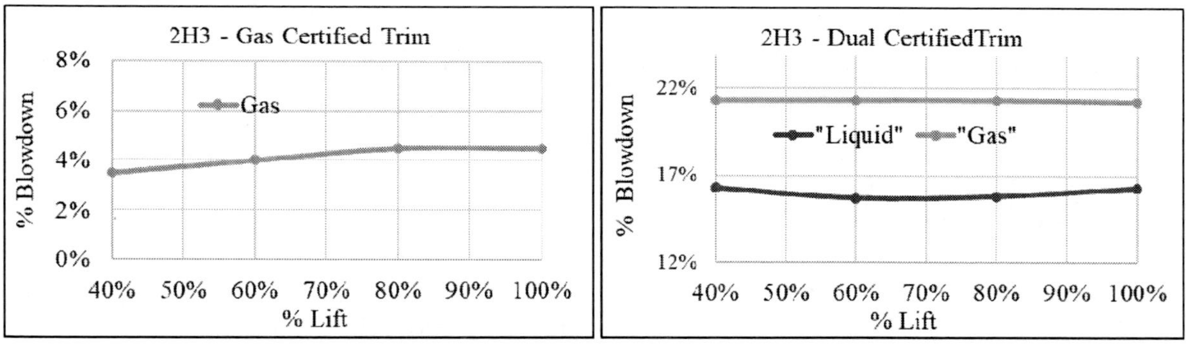

Figure 6 Blowdown Test Results - Set 202 psig

4.3 Overpressure

The capacity test for actual measured capacity at restricted lift is to be done per ASME standards.

- ASME Section VIII-Divisions 1 and 2. The capacity test is to be at overpressures not to exceed 10%.
- ASME Section I. The capacity test is to be at overpressures not to exceed 3%.

4.4 Opening Times

Relief valve opening times are not substantially affected by restricted lift. Estimates of opening times span a broad range and differ between the type, size and operation. For spring loaded valves, since design is not changing except with addition of a restraining device, the initial lift at set pressure may take the valve to its full travel stop when the restriction is below approximately 50%. The actual opening time however remains relatively unchanged.

4.5 Bellows Cycling

The bellows used for balancing a direct spring PRV for back pressure has a finite cycle life. The bellows may cycle more than 1,000 times or more during its service life. The reduced travel of the restricted lift design may help to reduce the early fatigue of the bellows. Avoiding flutter (opening and closing without the disk contacting the nozzle) can prolong the bellows service life.

4.6 Two-Phase Flow

Flow can be restricted for two-phase flow just as for single-phase liquid or single-phase vapor. Besides the engineering considerations, such as proper orifice sizing, selection of relief pressure, and considerations for blowdown and overpressure, the other important factors are certification and marking.

- single-phase liquid or single-phase vapor are certified and marked based on single phase fluids with either liquid trim, vapor trim or dual / multi-media trim.
- Fluids that are two-phase entering the relief valve or flashing liquids, are certified and marked with for dual / multi-media trim. Likewise, for restricted lift, the marking and certification are similar as outlined in the next paragraph 5.

4.7 Maintenance

During maintenance procedures, such as machining the nozzle or disk, the measure lift may widen [3]. The lift should be confirmed after such procedures

5. Certification Procedure

5.1 Discharge Coefficient Flow Test

Most manufacturers of API 526 pressure relief valves certify the capacity per ASME Section VIII-UG-131(e)(1) to establish a coefficient of discharge for a specific design family. A similar certification process is done for power boiler safety valves that are available with restricted lift; refer to ASME Section I paragraph UG-69.2.3. If this particular design family includes a restricted lift option, then the manufacturer is required to test three valves (each must be of a different size). Each of the three valves must be flow tested at the minimum restricted lift and two mid-points between this minimum and the full lift of the design. This measured and recorded capacity must meet or exceed the value calculated by the equation is paragraph 3.3(1) above.

5.2 Nameplate Capacity Calculation and Measurement

As mentioned previously in paragraph 2.2.1 (concerning API-526), and illustrated in paragraph 4.1 and Figure 5, the purchaser will provide the required capacity and the maximum capacity desired for the restricted lift valve. The manufacture will use one of the restriction lift techniques in paragraphs 3.1 / 3.2 to set the lift and provide a nameplate capacity (normally stated in standard conditions of air for gas valves or water for liquid relief valves). The nameplate capacity is defined by ASME Code in Section VIII-UG-136(h) and Section I-PG-73.2.10(f).

5.3 Schedule and Cost

Depending upon the technique used to restrict the lift the impact to schedule and cost is minimal. When using components such as bolts and washers to provide lift restriction, as illustrated in Figures 2 and 3, there is a minimal impact to delivery lead times and cost. These items are easily obtained and often are stocked parts for the manufacturer.

5.4 ASME and API Parameters

Using the capacity adjustment based upon the ratio of restricted lift to full lift, pressure relief valves may be sized using either ASME parameters, as published in NB-18 "Pressure Relief Device Certification" [11], or using API parameters, as published in vendor catalogues or API Standard 526 [7]. The API parameters (area and discharge coefficient) are often referred to as "effective" parameters. As with full lift sizing, caution should be taken not to mix the ASME and API parameters. That is, the ASME orifice area should not be used with the API effective discharge coefficient and similarly, the API effective orifice area should not be used with the ASME discharge coefficient. When the pressure relief valve is selected, the capacity using the API effective orifice area and discharge coefficient should be verified using the ASME certified orifice area and discharge coefficient [5].

5.5 Nameplate Markings

Marking instructions are provided for both pressure vessels and boilers.

Table 4 Markings for ASME Certification and API Orifice Letter Designation

Equipment	Code	Instructions
Power Boilers	ASME Section I - PG-110 "Stamping of Boiler Pressure Relief Valves	Does not require the length of the restricted lift to be provided, as is required for pressure vessels.
Pressure Vessels	ASME Section VIII-Division 1 - UG-129(a)(8) "Markings"	States the restricted lift in inches or mm is to be included on the ASME code nameplate
Pressure Vessels	Code Case 2787 (Section VIII vessels)	Provides information on multiple markings of certified capacities Section VIII relief valves.
Pressure Vessels & Boilers	API Standard 526	The effective orifice letter designations are provided in API-526. If a restricted lift valve design is chosen, the orifice letter designation shall include "-RL".

6. Conclusions

Mechanically restricting the opening - or the lift of the disk - is one method for a pressure relief valve to flow only the rate required to relieve overpressure, thus mitigating the disadvantages of oversizing. This paper explains the advantages and disadvantages of restricting the lift. Different methods to restrict the lift were illustrated. The effect on flow issues such as blowdown and overpressure were discussed.

The process for obtaining a restricted lift was described, including testing and certifying of the restricted lift relief valve design according to the regulatory Codes. National Board Certification, proper name plate marking, vendor calculations to determine the lift and capacity were presented.

7. References

7.1 Articles on Restricted Lift

[1] Emile Tezzo. "Effect of Restricted Lift on Blowdown - Direct Spring-Loaded Valves", API 2017 Fall Refining and Equipment Standards Meeting, November 2017, Seattle, Washington

[2] Dustin Smith and Shahar Yoram. "Address Inlet Pressure Loss Concerns with Restricted Lift Relief Devices", Hydrocarbon Processing Magazine, Volume 93(3), March 2014, pp 71-73

[3] Sean Croxford. "Restricted Lift", DIERS Spring 2018 Meeting, May 2018, Houston, Texas

7.2 Standards

[5] API Standard 520-I (9th Edition), "Sizing, Selection, and Installation of Pressure-Relieving Devices-Part I-Sizing & Selection". American Petroleum Institute, 2014

[6] API Standard 520 Part II (6th Edition), "Sizing, Selection, and Installation of Pressure-Relieving Devices-Part I-Installation". American Petroleum Institute, 2015.

[7] API Standard 526 (7th Edition), "Flanged Steel Pressure Relief Valves", American Petroleum Institute, 2017

[8] ASME Section VIII-Division 1-2017 Edition, "Rules for Construction of Pressure Vessels", ASME BPVC.VIII.1. American Society of Mechanical Engineers, 2017

[9] ASME Section I -2017 Edition, "Rules for Construction of Power Boilers", ASME BPVC.1. American Society of Mechanical Engineers, 2017

[10] ISO-4126-1:2013, "Safety devices for protection against excessive pressure -- Part 1: Safety valves", 2013

7.3 Books and Manuals

[11] NB-18 "Pressure Relief Device Certification", <www.nationalboard.org>. National Board of Boiler and Pressure Vessel Inspectors, Accessed January 2019. NB-18 includes assembler and manufacturer certifications, certified device types by manufacturer, rupture disk Combination Capacity Factors by manufacturer and other listings. The National Board is composed of chief inspectors from enforcement jurisdictions of states, cities and provinces.

[12] NBIC "National Board Inspection Code", <nationalboard.org>. National Board of Boiler and Pressure Vessel Inspectors. Accessed January 2019. This document includes interpretations and a Part 4, "Pressure Relief Devices", which provides information and guidance on installation, inspection, and repair of pressure relief devices.

Appendix 1 – ASME Code Case 2787 (2013)

Case 2787 Multiple Markings of Certified Capacities for Pressure-Relief Valves. Section VIII, Division 1. Approval Date: November 1, 2013

Inquiry: Under what conditions may a Manufacturer or Assembler of a pressure-relief valve place more than one certified capacity on the pressure-relief valve or the nameplate and comply with UG-129?

Reply: It is the opinion of the Committee that a Manufacturer or Assembler of a pressure-relief valve may place more than one certified capacity on the pressure-relief valve or the nameplate provided the following requirements are met:

(a) The pressure-relief valve has been certified by the Manufacturer per the requirements of UG-131 for each of the medias with the following additional requirements:

(1) During the certification of capacity testing required per UG-131, pressure-relief valve shall be tested first on one of the certified media (steam, air, gas or water) and then tested on all other media requested by the manufacturer.

(2) There shall be no adjustments to any of the pressure-relief valves after completion of the testing on the first media,

(3) The measured set pressure for the valve tested on the additional media shall meet the tolerance requirements of UG-134(d)(1) , based on the pressure at which the vale was set to operate in (a)(1), or marked set pressure for sample production valves.

(b) The pressure-relief valves shall meet all the requirements of UG-136 for certified capacities stamped on the valve or nameplate except as follows:

(1) For sample production pressure-relief valves selected for capacity certification per UG-136(c)(3)(a) the same requirements per (a) shall apply,

(2) Production testing per UG-136(d)(4) shall be performed using any one of the certified medias to be marked on the valve except steam shall be used when one of the certified media is steam,

(3) This Case number shall be on a plate permanently attached to the pressure-relief valve.

Optimal Placement of Relief Devices for Hydrogen Peroxide Pipeline Protection

Daniel Nguyen, P.E., PMP
ioMosaic Corporation
93 Stiles Road, Salem, New Hampshire 03079
nguyen@iomosaic.com

Prepared for Presentation at
American Institute of Chemical Engineers
2019 Spring Meeting and 15th Global Congress on Process Safety
New Orleans, LA
March 31 – April 3, 2019

Optimal Placement of Relief Devices for Hydrogen Peroxide Pipeline Protection

Daniel Nguyen, P.E., PMP
ioMosaic Corporation
93 Stiles Road, Salem, New Hampshire 03079
nguyen@iomosaic.com

Keywords: Hydrogen peroxide, pipeline, decomposition, contamination, overpressure, pressure relief protection

Abstract

The causes and effects of hydrogen peroxide decomposition have been widely researched and published. The two most common overpressure phenomena studied are contamination and thermal runaway decomposition. Depending on the cause and concentration, hydrogen peroxide decomposition can generate a large amount of oxygen, resulting in a volume expansion as many as 2,500 times the initial volume. Heat released from the reaction may exceed the rate of heat lost to the surroundings, resulting in temperature increase and further intensifying the rate of decomposition. Relief systems must thus be properly sized in order to protect the equipment from overpressure. While there are various guidelines circulated by leading industry organizations and manufacturers, most have focused on the handling and bulk storage of hydrogen peroxide.

This paper examines an overpressure relief protection for pipeline, specifically designed for hydrogen peroxide transport over an extended distance. Presented as a case study, it includes a series of sensitivity analyses, accounting for all credible overpressure scenarios, to obtain an optimal placement of relief devices along the pipeline. Both contamination and localized abnormal heating are examined for hydrogen peroxide concentrations up to 70 wt%. An assessment on an exposed and insulated pipeline will also be discussed.

1 Introduction

Hydrogen peroxide (HP) is a reactive oxidation chemical. It is known for its environmentally friendly decomposition products. HP is used in various industrial and commercial applications including pulp bleaching, chemical synthesis, and environmental remediation. In addition to the considerable potential energy, the industrial grades of hydrogen peroxide are classified as strong oxidizers and corrosive. Typical industrial products have concentrations with 35%, 50%, and 70 % HP by weight, as aqueous solutions[6]. Table 1 summarizes a general classification based on HP concentrations.

The causes and effects of HP decomposition have been extensively researched and published[1,2,5,6]. It is widely documented that the commercial grades of hydrogen peroxide are basically unstable[1]. HP continuously decomposes and generates heat. Most manufacturer guidelines indicate that HP typically loses less than 1% relative strength per year. At this rate, the heat of decomposition is dissipated readily to the surroundings. When the rate of heat loss to the surroundings is greater than the rate of heat generated, the hydrogen peroxide fluid temperature generally resembles the ambient conditions.

Table 1: General classification of hydrogen peroxide[3]

Concentration	Category	Typical Applications
< 8%	Non-hazardous	Toothpaste, general sterilizer, detergent bleach
8-28%	Class 1 Oxidizer	Swimming pool chemical treatment
28-52%	Class 2 Oxidizer; Corrosive and Class 1 Unstable (reactive)	Industrial strength grade
52-91%	Class 3 Oxidizer; Corrosive and Class 2 Unstable (reactive)	Specialty chemical processes
> 91%	Class 4 Oxidizer; Corrosive and Class 3 Unstable (reactive)	Rocket propellent

The two most common HP overpressure phenomena studied are contamination and thermal runaway decomposition. Depending on the cause and concentration, hydrogen peroxide decomposition can generate a large amount of oxygen, resulting in a volume expansion as many as 2,500 times the initial volume[5]. Heat released from the reaction may exceed the rate of heat lost to the surroundings, resulting in temperature increase and further intensifying the rate of decomposition. Relief systems must thus be properly sized in order to protect the equipment from overpressure. While there are various guidelines circulated by leading industry organizations and manufacturers, most have focused on the handling and bulk storage of hydrogen peroxide.

In this paper, we consider pipeline as a means for transporting hydrogen peroxide. We examine the overpressure relief protection for pipeline, specifically designed for hydrogen peroxide transport over an extended distance. Presented as a case study, it includes a series of sensitivity analyses, accounting for all credible overpressure scenarios, to obtain an optimal placement of relief devices along the pipeline. Both contamination and localized abnormal heating are examined for hydrogen peroxide concentration up to 70 wt%. An assessment on an exposed and insulated pipeline will also be discussed.

2 Hydrogen Peroxide Decomposition

Hydrogen peroxide undergoes the following decomposition reaction:

$$H_2O_{2\,(l)} \longrightarrow \frac{1}{2}O_{2\,(g)} + H_2O_{(l)} \qquad \Delta H = -98 \text{ kJ/mol or } -2{,}882 \text{ kJ/kg}$$

With stabilized, uncontaminated commercial grade HP, the decomposition is very slow[1]. The kinetics of the reaction can be described by the following equation:

$$\frac{d[H_2O_2]}{dt} = -k[H_2O_2]$$

Where:

$$k = 0.81 \cdot 10^7 k_F e^{\frac{-10357}{T}} \quad [\text{min}^{-1}]$$

T is temperature in Kevin and k_F is an acceleration factor [unitless]. For uncontaminated product, $k_F = 1$.

Besides being exposed to an abnormal heat source, contact of hydrogen peroxide with improper materials is one of the primary causes of decomposition. Some common sources of contaminants include transition metals, organic substances, strong acids or bases, groundwater, and dust. The rate of HP decomposition varies, depending on the strength and nature of the impurities exposed.

One way to classify the potential sources of contaminants is to assign each to a level of catalytic activity, represented by the acceleration factor, k_F. For a stabilized and uncontaminated hydrogen peroxide, the product is considered at a normal condition, with the k_F assigned equal to one (1). Table 2 provides some typical k_F values from experimental results for iron content[1].

Table 2: Iron content contamination impact for standard technical grade 70 wt% HP[1]

Iron Content [mg/kg]	$k \times 10^8$ [min^{-1}]$^{¥}$	Acceleration Factor k_F [unitless]
0.12	0.95	1.46
0.37	3	4.6
7.76	7.56	11.6
15.5	9.42	14.4
31	64.6	98.9
38.8	913	1,398
46.6	6,763	10,366

¥ k values calculated at 25°C

3 Case study

In this paper, we use a practical case study as the basis of our discussions. A major manufacturer has a need to determine how feasible it is to construct a pipeline, transporting hydrogen peroxide product to its customer within a couple of miles away. Since HP is self-reactive, both generating heat and decomposing to form oxygen and water, a major concern is to protect the pipeline from overpressure. The target HP concentrations range from 40 to 70 wt%. The proposed pipeline distance is approximately 10,000 feet, utilizing some existing pipe rack infrastructure. Part of the pipeline may be situated close to public roadway. In addition, certain portions of the pipeline can be in contact with or buried under the ground, a major concern for corrosion/erosion that may cause an increase in metal ions and further HP decomposition.

In this case study, our primary objectives are:

- Determine the critical (maximum) length of a pipe section that can be safely protected with one largest possible relief device
- Present a sound methodology for modeling pipeline relief protection
- Suggest some credible scenarios for pipeline protection considerations

As a result of the maximum pipe length established, we would be able to determine the proper number and placement of relief devices along the entire pipeline.

Table 3 provides a summary of the pipeline considered in this study. The pipeline is designed per ANSI B31. As a result, the pressure rating may be exceeded by 33%. It is assumed that the design scenario does not last more than 10 hours per upset and does not last more than 100 hours per year.

The pressure at the source of the pipeline is approximately 155 psig. All our analyses assume the pipeline to be blocked in at 155 psig and 110°F.

Table 3: Pipeline case study specifications

Parameter	Description
Pipeline size/length	3" schedule 40, 304L stainless steel; 10,000 feet
Pipeline pressure rating	1300 psig at 600°F
Pipe surface roughness	0.00046 m
Design basis	ANSI 31; maximum allowable accumulated pressure, 33%
HP concentrations considered	43 wt% and 70 wt%
Normal/max. operating pressure	150/155 psig
Normal/max. operating temperature	80/110°F
Relief protection device	3L4, set at 170 psig
Surrounding conditions	Stability class D, 5 m/s wind, 105°F ambient temperature

4 Study Approach

The case study was performed using SuperChems™, a detailed dynamic simulator and a component of ioMosaic's Process Safety Office®. The program consists of various models for simulating the fluid dynamics of vessels and pipeline containing multi-phase fluids.

To account for detailed equipment wall and fluid heat transfer dynamics, the simulated equipment is segmented into multiple zones, as shown in Figure 1[4,7]. Detailed heat transfer to/from the surroundings and between the zones are dynamically accounted for. There is no limit on the number of zones a user can specify. The ability to distribute an equipment into multiple segments allows users to closely examine the dynamics of the fluids and equipment wall thermal effects. Other valuable applications of the segmentation approach include the modeling of detailed insulation, external fire, localized heating, and flame jet impingement.

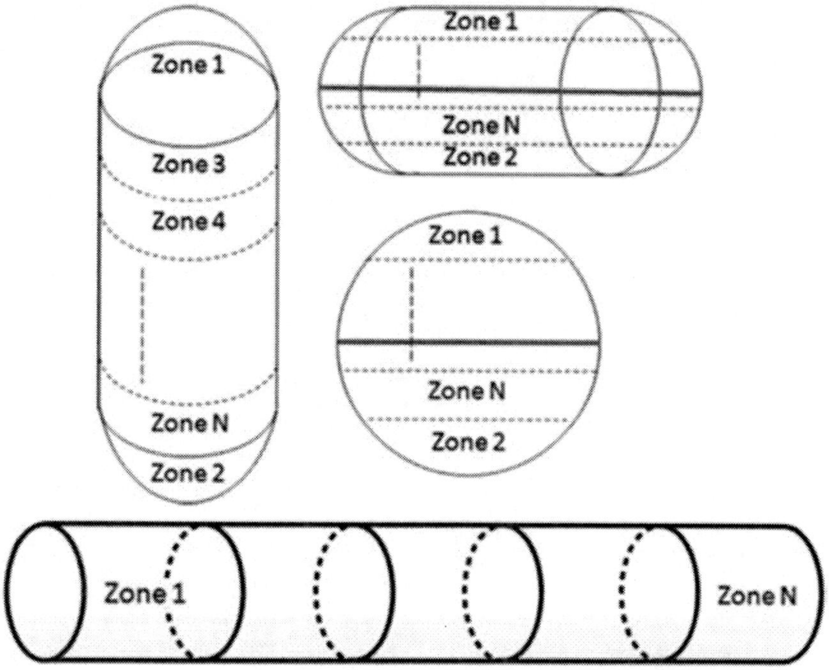

Figure 1. Samples of Equipment Segmentation Scheme

Figure 2 illustrates a simple representation of the pipeline. It illustrates the pipeline being divided into ten segments. As shown, various options are exposed allowing users to specify what to be considered. In our case study, we enable all piping to be visible to solar heating. For external fire, we want to examine what happens if a certain segment is engulfed and visible to fire. To consider insulation, its thickness and relevant properties must be specified (not shown in the figure).

Table 4 summarizes some calcium silicate insulation properties used in this study.

Heat Transfer Segment	Visible to Fire	Engulfed by Fire	Protected by Water Spray	Visible to Solar Heating
Bottom Segment 0 -> 1	[ ] Yes	[ ] Yes	[ ] Yes	[x] Yes
Segment 1 -> 2	[ ] Yes	[ ] Yes	[ ] Yes	[x] Yes
Segment 2 -> 3	[ ] Yes	[ ] Yes	[ ] Yes	[x] Yes
Segment 3 -> 4	[ ] Yes	[ ] Yes	[ ] Yes	[x] Yes
Segment 4 -> 5	[x] Yes	[x] Yes	[ ] Yes	[x] Yes
Segment 5 -> 6	[ ] Yes	[ ] Yes	[ ] Yes	[x] Yes
Segment 6 -> 7	[ ] Yes	[ ] Yes	[ ] Yes	[x] Yes
Segment 7 -> 8	[ ] Yes	[ ] Yes	[ ] Yes	[x] Yes
Segment 8 -> 9	[ ] Yes	[ ] Yes	[ ] Yes	[x] Yes
Top Segment 9 -> 10	[ ] Yes	[ ] Yes	[ ] Yes	[x] Yes

Figure 2. Sample of pipeline segmentation specifications

Table 3: Calcium Silicate Insulation Properties

Temperature [°C]	Density[γ] [kg·m^{-3}]	Specific heat capacity[γ] [J·kg^{-1}·C^{-1}]	Thermal Conductivity [W·m^{-1}·°C^{-1}]
93	288	710	0.078
149	288	710	0.084
204	288	710	0.088
260	288	710	0.092
315	288	710	0.097
370	288	710	0.101
427	288	710	0.105
482	288	710	0.108
538	288	710	0.111

[γ] *Both insulation density and specific heat capacity are considered constant*

Pipeline has a rather unique configuration. For emergency venting design, its geometry is not generally favorable for a complete vapor-liquid disengagement. On the other hand, its high surface area to volume ratio has a major advantage, providing ample medium for detailed heat transfer considerations. Given its uniqueness, we consider the following assumptions as part of the relief sizing modeling:

- A hydrogen peroxide runaway reaction in a designated pipe section will create a vapor bubble within that section
- The pressure accumulation in the designated section will push the liquid out on both sides of the pipe to accommodate the rapid volume increase caused by the gas generation
- Pressure loss due to friction and the presence of fittings influences the pressure accumulation in the pipe (volume created by driving the liquid out versus volume generated by the reaction)
- Given unfavorable geometry, two-phase flow is assumed, representing a more stringent design
- Since pipeline size is 3", the largest PSV can be installed is equivalent to an 3L4 API orifice

The primary cases considered in this study are summarized in Table 4. All cases assume that a designated pipe section is protected by a single relief device. Each designated pipe section has uniform decomposition and contamination. In the first case, we want to establish a baseline, worst-case scenario by assuming that the designated pipe section is under adiabatic conditions. It implies that there is no heat transfer considered between the pipeline fluid and the surroundings. Given the proposed relief device and its set point, we want to understand the overpressure impacts at different pipe section lengths and various contamination levels for HP concentrations of 43 wt% and 70 wt%. Ultimately, we want to find the maximum (critical) pipe section that can safely be protected by one 3L4 relief device.

Based on the insights obtained from the adiabatic analyses, we would pick a pipe length as the primary pipe section for the subsequent evaluations. Accounting for potential contact with soil and groundwater, we use 1,500 as the acceleration factor (k_F), forming the basis for product contamination. Though it is slightly conservative, k_F equaling 1,500 is reasonably consistent with the CEFIC's suggestion (1,000) for the design of a safety vent[1].

The subsequent cases consider detailed heat transfer to and from surroundings. Although we can consider partially buried or shaded segments, our primary focus in this case study is to examine a bared, fully exposed pipeline. We then want to compare with situations where a full insulation is considered.

While there are no flammable liquids neighboring the pipeline infrastructure, the pipeline exposed to an abnormal heat source is a possibility. Notice the case study states that the pipeline may be situated close to a public roadway. It is known that truck carrying flammable materials using the nearby public road is a common occurrence. A concern is that there might be incidents involved truck rollover or spills of flammable liquids along the pipeline. As a result, a localized fire or flame impingement is credible.

By evaluating all these sensitivity cases, our ultimate objective is to find an optimal replacement of relief devices along the pipeline proposed.

Table 4: Summary of Sensitivity Analysis Cases

Case	k_F	Description
Adiabatic Pipeline	Varied	Both 43 wt% and 70 wt% HP, assuming adiabatic conditions for various contamination levels and pipe sections
Bared Pipeline	1,500	70 wt% HP, exposed pipeline and accounted for immediate surroundings
Insulated Pipeline	1,500	70 wt% HP, assuming 1" calcium silicate insulation, accounted for immediate surroundings
Fire Exposed Pipeline	1	70 wt% HP, uncontaminated, with localized fire exposure, accounted for immediate surroundings

5 Results Analysis

The sensitivity analyses of different pipe section lengths and contamination levels are shown in Figures 3 through 5. All cases assume adiabatic conditions. Each pipe section is equipped with one relief device equivalent to an 3L4 API orifice. Figure 3 illustrates the results for a 4,000-feet pipe section with 43 wt% HP. At acceleration factor (k_F) equal to 1,500 or higher, the simulations show that accumulated pressures exceed the maximum allowable accumulated pressure (MAAP), which is 1729 psig. On the other hands, the pipe section can safely be protected with k_F equal to 150 or lower.

For a smaller pipe section with 2,000 feet, Figure 4 shows that it is still impractical to protect the pipeline with k_F at 15,000 or higher. At k_F equal to 1,500, the simulations show that the accumulated pressure is 1,300 psig, which is below the MAAP. Since we apply 1,500 as the k_F level for vent sizing, the results indicate that a 2,000-feet pipe section is acceptable for 43 wt% HP.

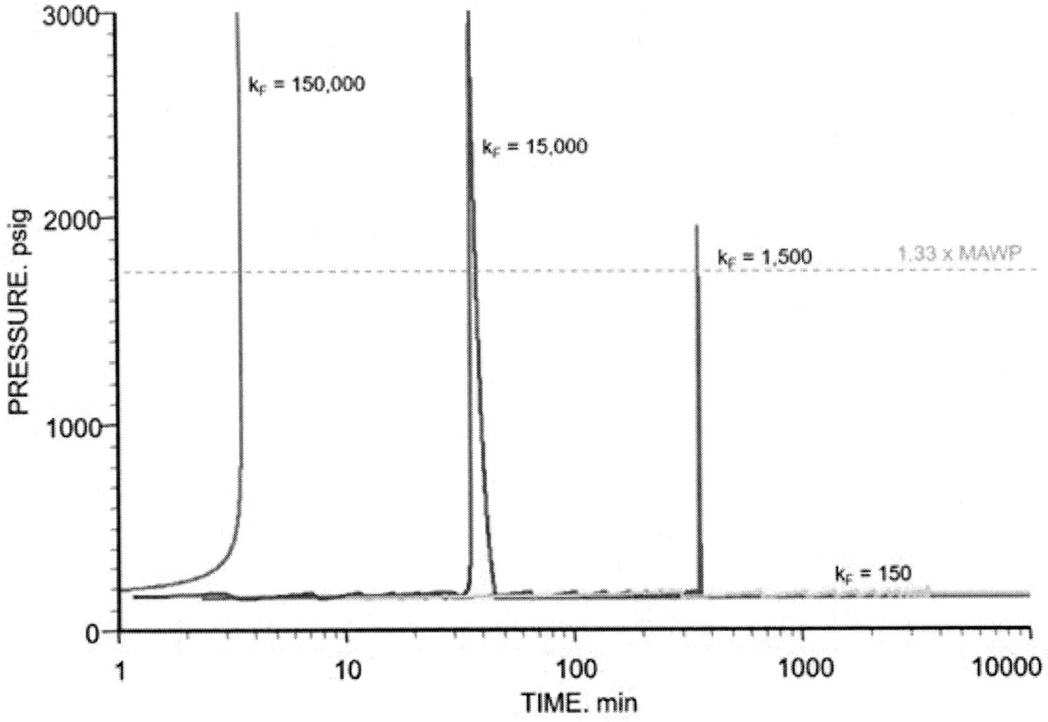

Figure 3. 4,000-feet, 43 wt% HP Pipeline at Different Levels of Contaminations (Adiabatic)

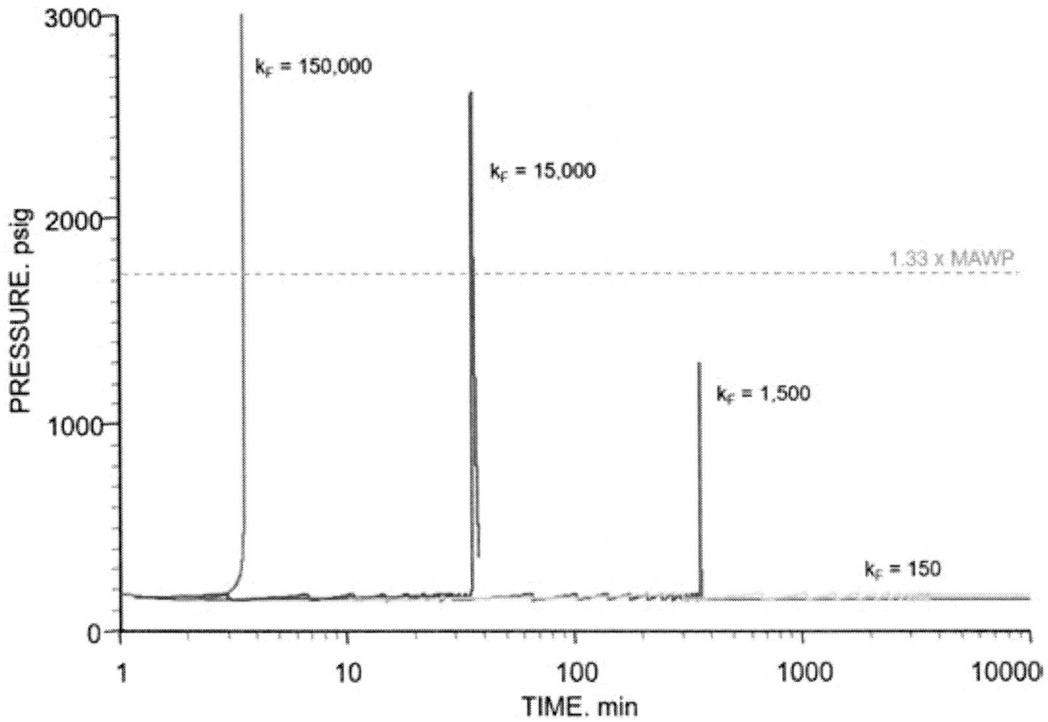

Figure 4. 2,000-feet, 43 wt% HP Pipeline at Different Levels of Contaminations (Adiabatic)

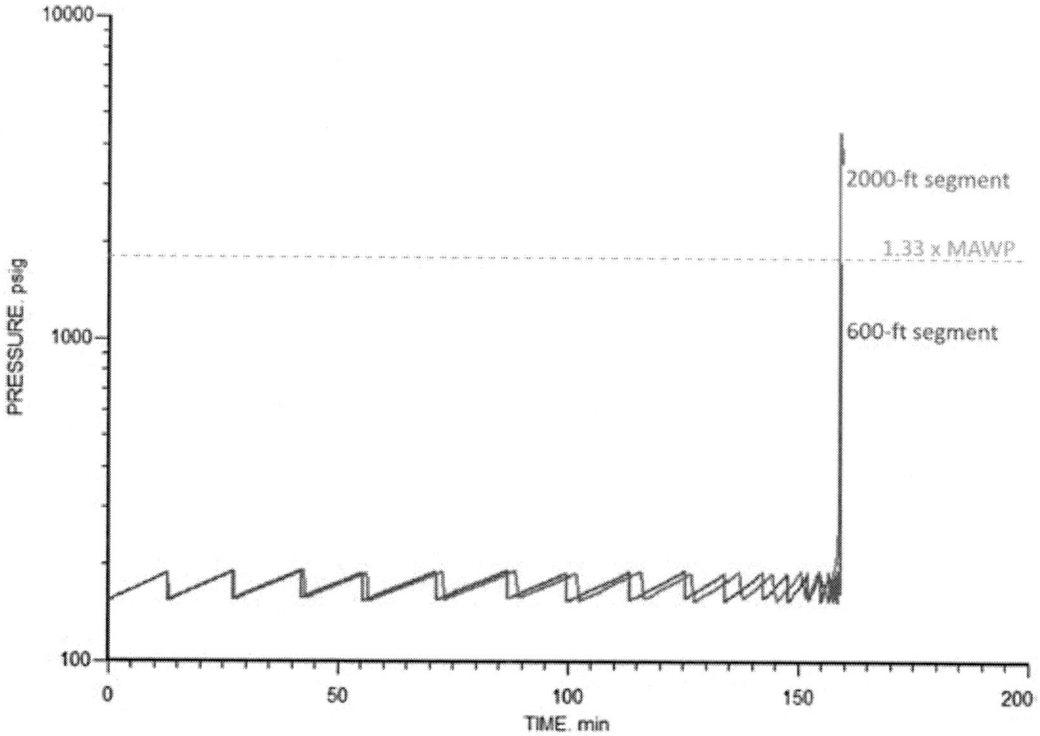

Figure 5. Comparison of different adiabatic pipeline segment lengths with 70 wt% HP (k_F = 1,500)

Figure 5 shows the simulations for 70 wt% HP concentration. As expected, higher HP concentrations would generate greater overpressure impacts. Based on the 2,000-feet pipe section, the simulation shows that the accumulated pressure reaches 4,500 psig, which far exceeds the MAAP. The overpressure is far worse than the longer 4,000-feet pipe section (43 %wt HP) at the highest considered acceleration factor (see Figure 3). In order to stay within the MAAP for 70 wt% HP, the pipe section must be shortened to 600 feet.

Although the adiabatic assumptions result in rather conservative accumulated pressure consequences, the simulations give us some insights. First, these cases confirm that it is almost impractical to protect the pipeline with high contamination level, where k_F is at or above 1,500. They illustrate the sensitivity and correlations among different concentrations and pipe sections considered. Second, these simulations help set the boundaries for the lower and upper worst-case scenarios. At 43 wt% HP, the critical (maximum) length that can be safely protected by a 3L4 relief device is 2,000 feet. As a lower worst-case limit, the 2,000-feet pipe section is a logical option selected as the basis for further evaluation, as described below.

Impact of detailed heat transfer considerations

Based on the 2,000-feet pipe section selected, Figures 6 through 8 show the pressure, temperature, and venting dynamics for the 70 %wt HP. The simulations consider the entire pipe section exposed to the given surrounding conditions. Detailed heat transfer from and to the surroundings are accounted for, including the impact of site location and full cycles of solar fluctuations. As shown in Figure 6, the pipeline fluid temperature rises and falls following the solar patterns. It should be noted that the pipeline fluid is heating up due to both self-heating and heat absorption from solar radiation. During the day, the heat generated and absorbed from solar radiation are greater than that dissipated to the surroundings. Consequently, both temperature and pressure rise.

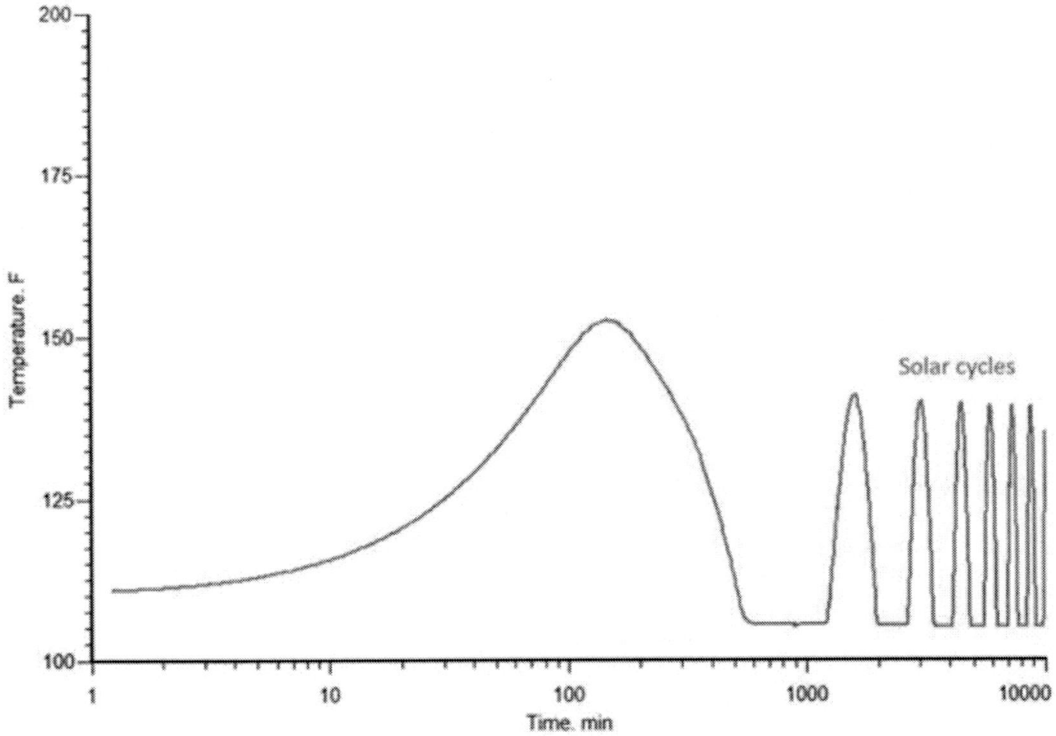

Figure 6. Exposed 2,000-feet pipeline with 70 wt% HP (k_F = 1,500) shows fluid temperature follow solar cycles

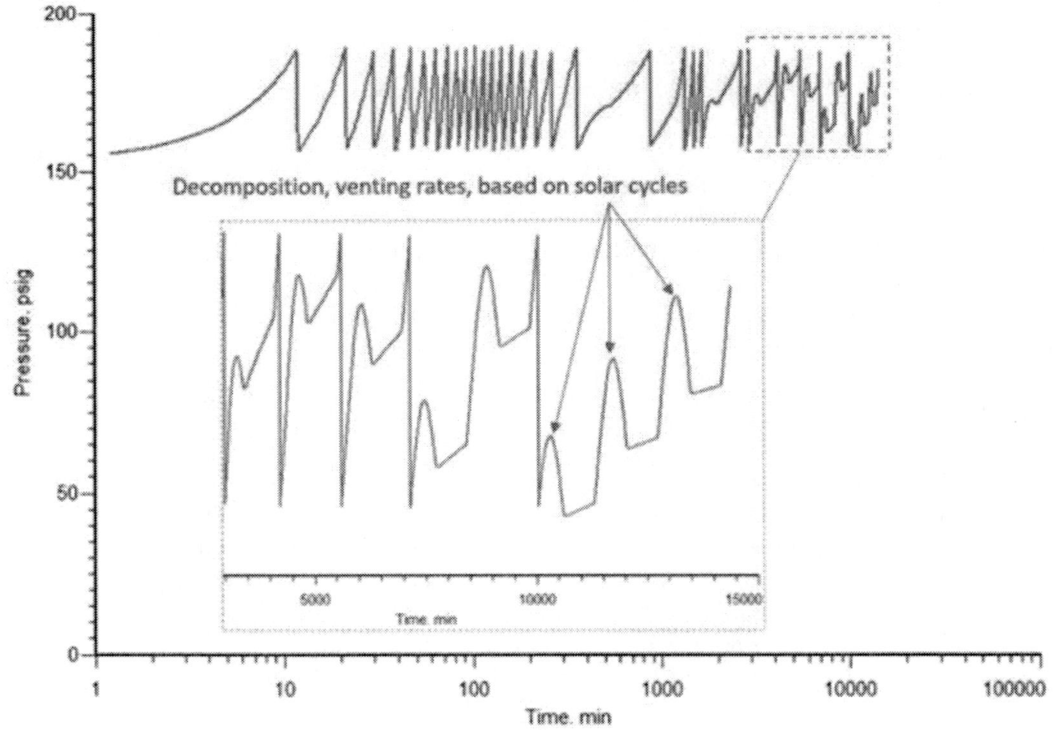

Figure 7. Exposed 2,000-feet pipeline with 70 wt% HP (k_F = 1,500) shows adequate protection

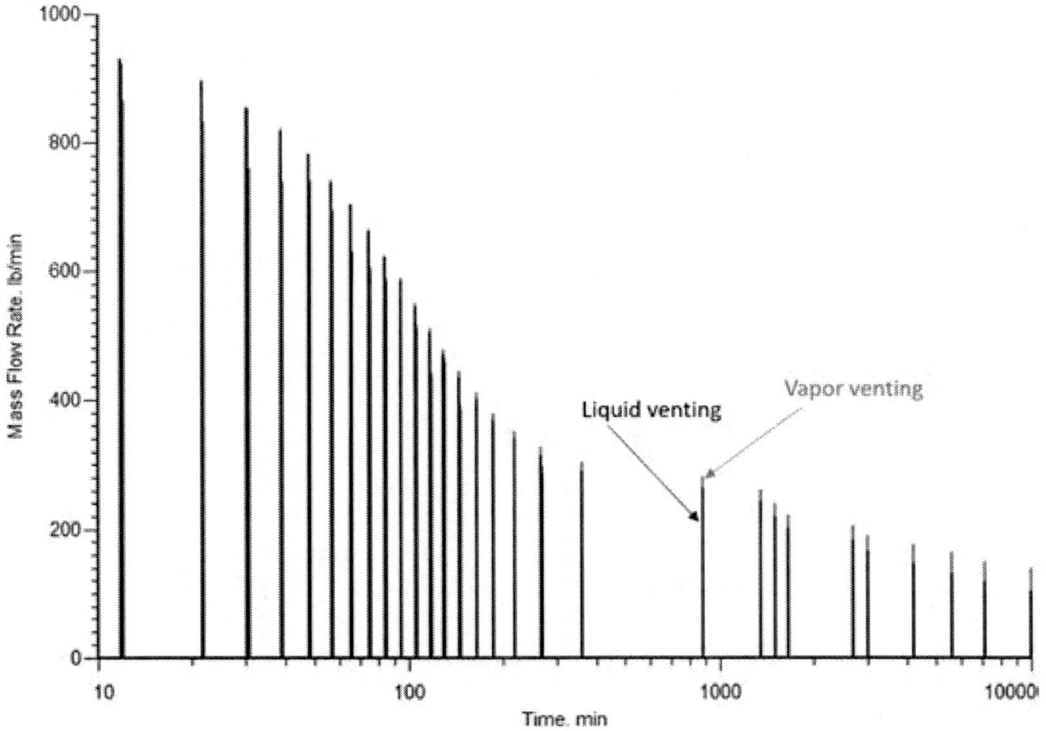

Figure 8. Venting profiles with the exposed 2,000-feet pipeline with 70 wt% HP (k_F = 1,500)

As the day turns into night, the solar impact is gradually diminished, resulting in reduced HP decomposition activities and self-heating. Ultimately, the pipeline fluid temperature equalizes with the surrounding temperature.

Figures 7 and 8 show that the relief valve opens as needed, dispelling the fluid to make room for the gas generated. Initially, there are mostly liquid being forced out. As the pipeline inventory depletes, there are more room for gas, resulting in higher vapor quality and lower mass venting. Note that while the fluid pressure may to follow the patterns of solar cycles, the valve lifting may not.

In this simulation, the results show that the 2,000-feet pipe section is adequately protected for up to at least 70 wt% HP. The analysis confirms the pipeline's unique advantage. Its high surface area to volume ratio indeed enhances heat transfer mechanisms and pipeline protection.

Impact of pipeline insulation

Although there are many practical applications and remedies for insulation, insulating pipeline with self-heating fluid such as hydrogen peroxide is generally not advisable. Figures 9 and 10 show what happen when insulation is considered. In this case, a 1-inch of calcium silicate insulation (see Table 3) is applied to the same 2,000-feet pipe section. As shown in both figures, the heat dissipated to the surroundings is considerably less than the heat self-generated, resulting in significant temperature rise. Note that comparing with the adiabatic assumptions, the insulated pipeline shows some heat shedding. First, the accumulated pressure does not reach the MAAP. Second, it is further evident by the delay in the temperature and pressure rise. One thing worth pointed out is that the bare (exposed) pipeline has a higher initial temperature rise. This is because the bare pipeline is initially heated by solar radiation, which accelerates HP self-heating decomposition resulting in temperature increase. On the other hands, the adiabatic and insulated cases, while not fully heated by solar, have no or limited means to dissipate the heat to the surroundings.

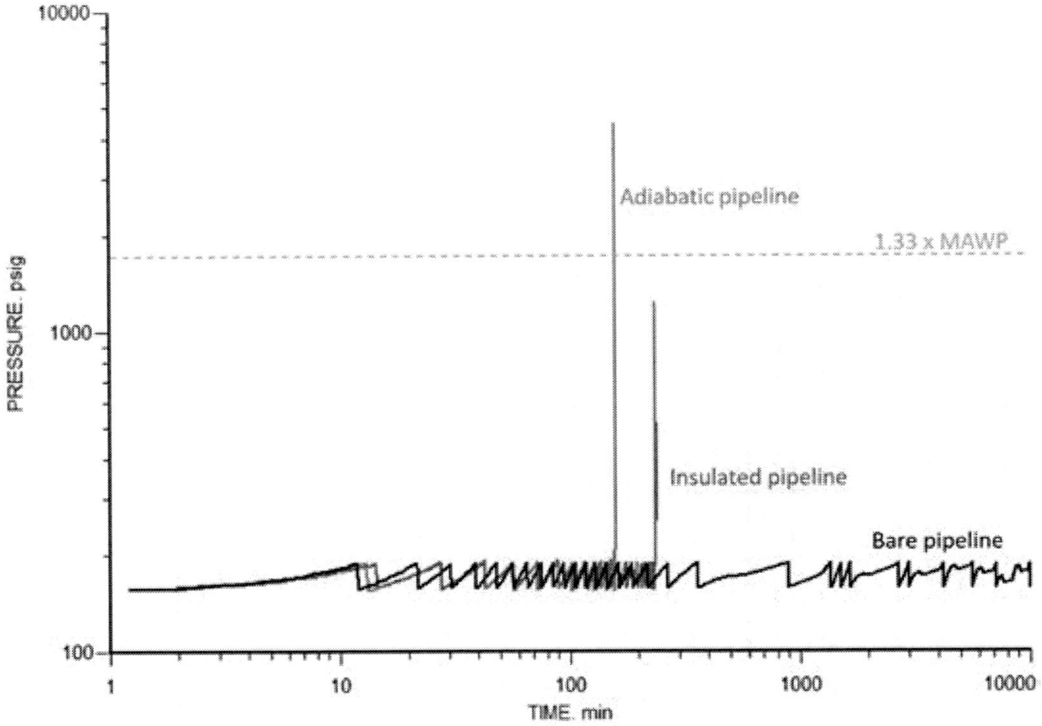

Figure 9. Pressure impact comparison for the 2,000-feet 70 wt% HP pipeline (k_F = 1,500)

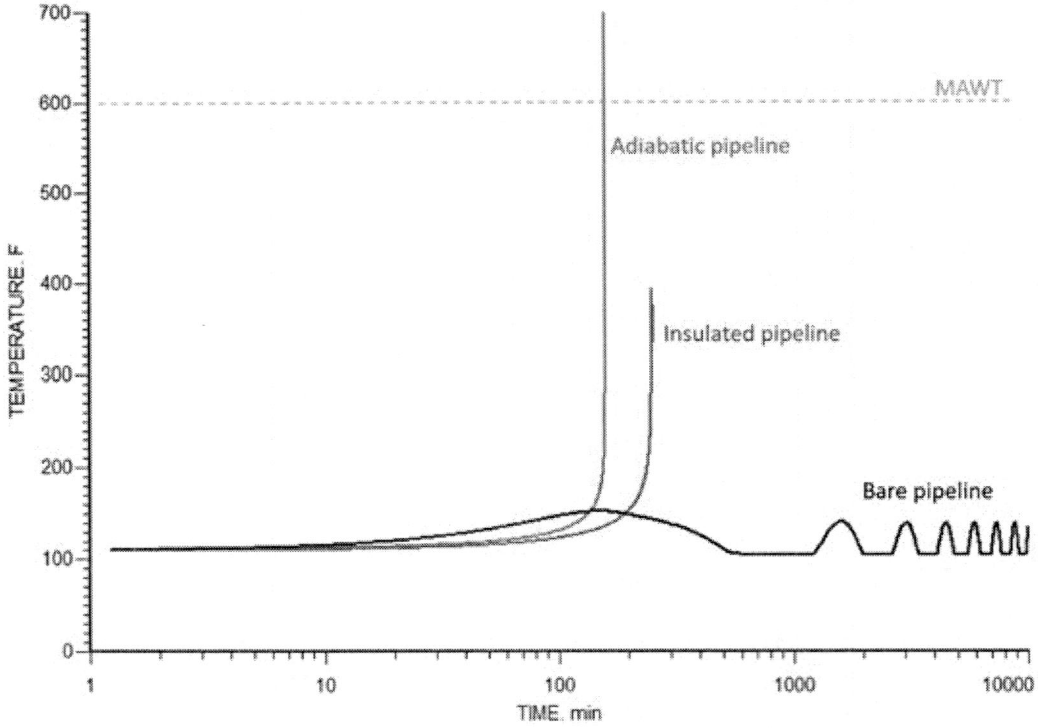

Figure 10. Temperature impact comparison for the 2,000-feet 70 wt% HP pipeline (k_F = 1,500)

Impact of localized fire exposure

Fire or abnormal heating input on pressure equipment is generally undesirable. It is particularly so when the system contains material with potential for exothermic runaway. In this case study, we want to examine what happens when a certain portion of the HP pipeline is exposed to fire. Note we consider hydrogen peroxide at its normal, uncontaminated condition (k_F=1) in all fire exposure cases. Using the same 2,000-feet pipe section, we want to further evaluate how much of the pipe section can be exposed to fire without overpressurizing the pipeline. As shown in Figure 2, a designated pipe section can be further divided into multiple segments. For example, dividing the 2,000-feet pipe section into ten (10) would result in ten 200-feet segments. Each segment has its own attributes. For instance, each segment can be designated as exposed to fire, visible to solar, being buried, or insulated. Likewise, if evaluating a 100-feet segment is required, the 2,000-feet pipe section should be divided into 20 segments.

Figure 11 shows a series of simulations, considering fire exposed on segments up to 10% (200-feet) of the 2,000-feet pipe section. All cases contain 70 wt% HP. The simulations indicate that 10% of the 2,000-feet pipe section exposed to fire would generate pressure up to 1,900 psig, which exceeds the MAAP. At 165-feet or shorter lengths exposed to fire, the accumulated pressures are well below the MAAP. It can thus be concluded that the given 2,000-feet pipe section is adequately protected for up to 8% (165-feet) of the pipe exposed to fire. Note that although detailed heat transfer is accounted in this situation, its impact is insignificant. This is because the heat rate dissipated the surroundings is diminutive compared to the intense heat from the fire and the self-heating generated from the accelerating reactions.

Figure 12 shows the pipeline wall temperatures as a function of time for the 200-feet fire exposure. Note that (a middle) segment 4 is picked as the portion of the pipeline exposed to fire. As shown, wall temperature for segment 4 is heated up rather quickly. It ultimately approaches the flame temperature while the other segments are hovering around the design temperature of the pipeline.

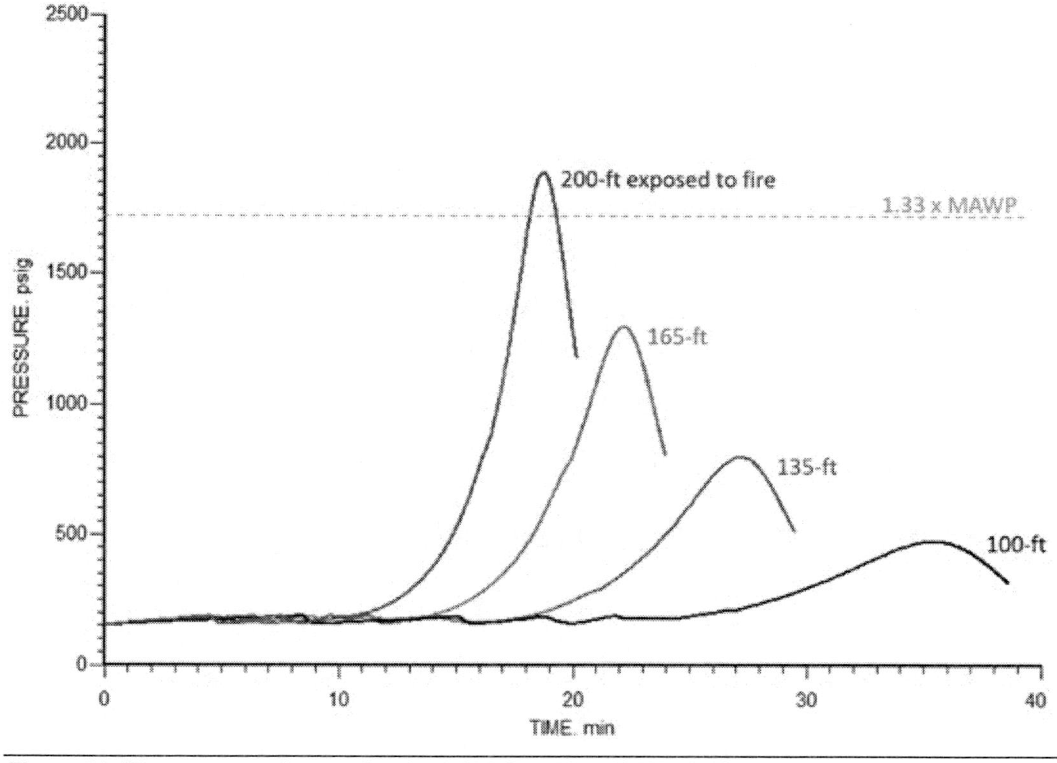

Figure 11. Fire exposure impact for the 2,000-feet 70 wt% HP pipeline (k_F = 1)

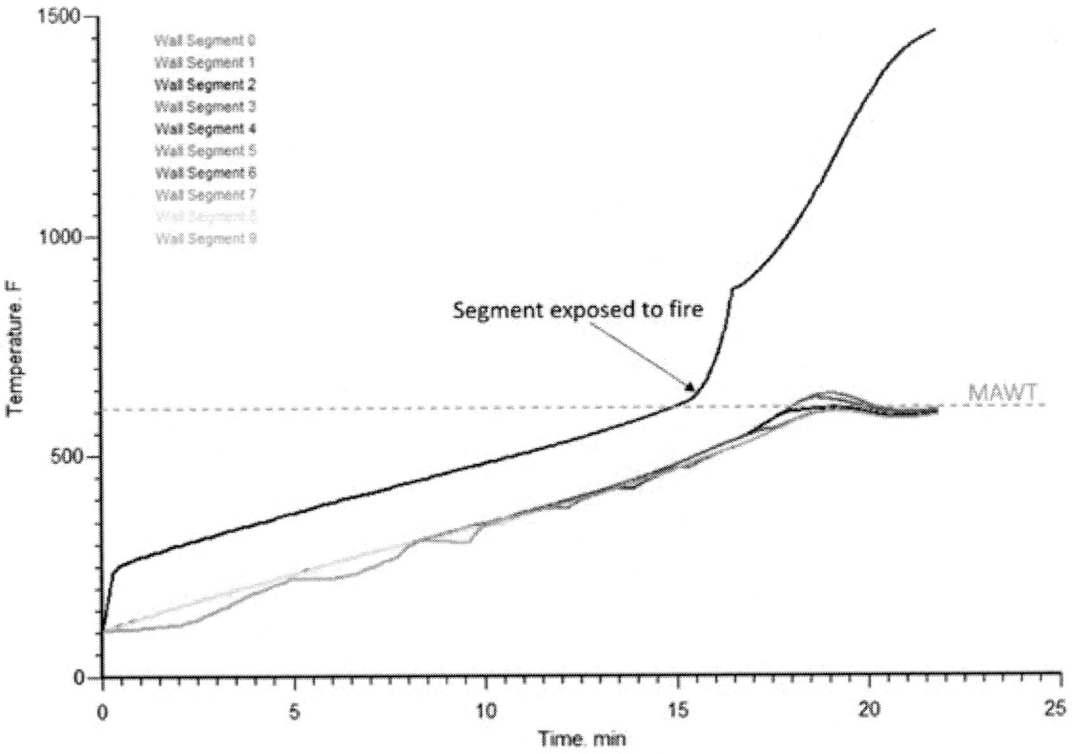

Figure 12. Segment temperature profiles of the 2,000-feet 70 wt% HP pipeline (k_F = 1)

In summary, all credible overpressure scenarios should be considered as part of pipeline relief protection. Based on the approach and assumptions stated, each relief device 3L4 is found adequate to protect a maximum length of 2,000 feet from its location. Accounting for both directions from its placement, one relief device is thus equipped to protect up to 4,000-feet section. The design basis in this case study is fire exposure for up to 165 feet per 2,000-feet pipe section.

Given the 10,000-feet pipeline, a total of three 3L4 relief devices are thus required. Figure 13 shows a proposed placement of relief devices along the pipeline. Starting from the left end (0 feet), the first device PSV-1 should be situated at 1,000 feet mark. It is designed to protect the pipeline from the left end up to mark point 3,000 feet. The second device (PSV-2 placed at 5,000 feet mark), is designed to protect the pipe sections from point 3,000 to 7,000 feet. Lastly, PSV-3 is placed at 9,000 feet mark, which is designed to protect the pipe sections from 7,000 feet to the end of the pipeline.

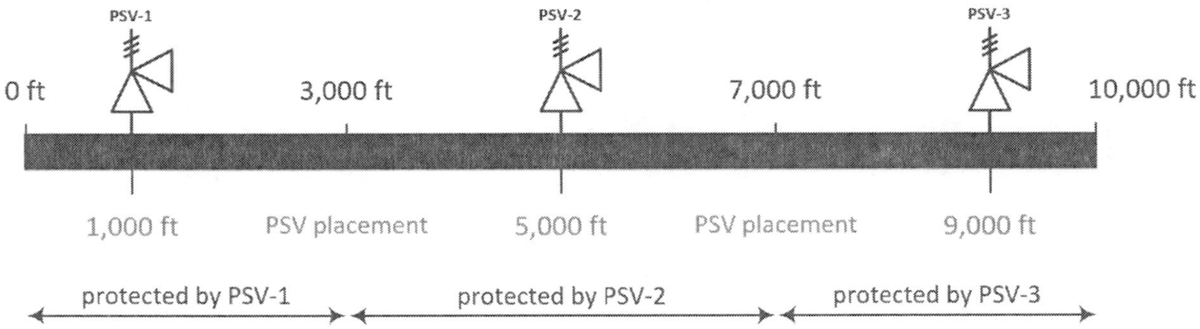

Figure 13. Proposed placement of relief devices along the 10,000-feet pipeline

6 Conclusions

Pipeline has a rather unique configuration. Its high surface area to volume ratio has a major advantage, providing ample medium for detailed heat transfer considerations. Given that hydrogen peroxide is self-reactive and self-heating, detailed heat transfer mechanisms from and to the surroundings should be accounted for, including the impact of site location and full cycles of solar thermal radiation.

The pipeline segmentation used in this paper demonstrates a reliable approach for modeling detailed pipe segments and wall dynamics. It provides engineers with sound tools to properly evaluate the causes and effectively address the consequences.

In order to protect the entire pipeline, all credible overpressure scenarios must be considered. In accordance with the basis and assumptions outlined, each relief device 3L4 is found adequate to protect a maximum length of 2,000 feet from within its location. Accounting for both directions from its placement, one relief device is thus equipped to protect up to 4,000-feet pipe section. Given the 10,000-feet pipeline, a total of three 3L4 relief devices are required.

7 References

[1] Peroxygens Hydrogen Peroxide Subgroup. "Bulk Storage Guideline", Cefic Sector Group, March 2012.

[2] W. Dejian, Q. Xinming, and H. Ping. "Safety Assessment on Hydrogen Peroxide for Storage and Transportation Based on Runaway Scenario," Applied Mechanics and Materials, pp. 215-220, July 2011.

[3] "Hydrogen Peroxide Material Compatibility Chart," https://www.industrialspec.com/images/files/hydrogen-peroxide-material-compatibility-chart-from-ism.pdf, Industrial Specialties Mfg. & IS MED Specialties, January 2019.

[4] D. Nguyen, J. Dunjo, H. Gilliam. "Mechanical Integrity Considerations in LNG Depressurization," paper presented at Global Congress on Process Safety, March 2015.

[5] Technical Data Sheet. "Hydrogen Peroxide Safety & Handling," Solvay Chemicals, Inc, Copyright 2006-2015.

[6] Construction, technical equipment and operations. "Tank installation for Hydrogen Peroxide," Evonik Industries, April 2008.

[7] G. Melhem, D. Gaydos. "Properly Calculate Vessel and Piping Wall Temperatures during Depressuring and Relief," paper presented at Global Congress on Process Safety, March 2014.

Low Pressure ≠ Low Risk

Strategies to Protect Tanks against Overpressure and Vacuum

Todd W. Drennen, P.E.
Baker Engineering and Risk Consultants, Inc. (BakerRisk)
8430 W Bryn Mawr Ave, Suite 720, Chicago, IL 60631
TDrennen@Bakerrisk.com

Prepared for Presentation at
American Institute of Chemical Engineers
2019 Spring Meeting and 15th Global Congress on Process Safety
New Orleans, LA
March 31 – April 3, 2019

Low Pressure ≠ Low Risk: Strategies to Protect Tanks against Overpressure and Vacuum

Todd W. Drennen, P.E.
Baker Engineering and Risk Consultants, Inc. (BakerRisk)
8430 W Bryn Mawr Ave, Suite 720, Chicago, IL 60631
TDrennen@Bakerrisk.com

ID# 546905

Keywords: Hazard Identification and Risk Analysis, Understanding Hazard & Risk, Venting & Emergency Relief

Abstract

Industry personnel often perceive low pressure tanks as being less dangerous than pressure vessels, particularly due to their less stringent standards for design and construction. However, this approach is mistaken as it overlooks the fact that a typical low pressure tank can often contain more material and be more easily opened to the atmosphere in response to a process upset when compared with a typical pressure vessel. This paper will discuss types of safeguards to prevent overpressure and vacuum of low pressure tanks, identify the strengths and weaknesses of these safeguards using consequence modeling techniques where applicable, and recommend overall safeguarding strategies in order to best utilize the strengths and mitigate the weaknesses of the available safeguards.

1. Introduction: Process Safety Hazards Associated with Low Pressure Tanks

Low pressure storage tanks abound in industry and are a necessity in maintaining a stable global supply chain for liquid raw materials, intermediates, and final products. These tanks allow large quantities of material to be stored at low pressure, often in remote locations relative to occupied buildings and public population centers. However, these advantages aside, it is irresponsible to assume that low pressure implies low risk.

While storage tanks operate at low pressure, their large volumes have the potential to pose greater inherent risk than pressure vessels of smaller volumes based solely on the maximum inventory of material that could be released in the event of loss of primary containment. In addition, tank pressure relief via a closed system is often impractical due to the backpressure that could develop in a closed relief header due to pressure losses to friction in the header piping, thus making relief of hazardous material to the atmosphere more common in the design of tanks than in pressure vessels. Additionally, inherently safer design with respect to equipment pressure and vacuum ratings is not generally feasible for storage tanks as compared to pressure vessels. Finally, a sense of complacency can develop around low pressure tanks at a facility, particularly if they are sited in relatively remote locations.

Table 1 below gives examples of Process Safety incidents involving low pressure tanks in the United States since 2000, based upon investigations completed by the United States Chemical Safety and Hazard Investigation Board (CSB) during that time. Note that many of the incidents listed involved ignition of flammable material inside tanks due to hot work being performed in areas adjacent to those tanks, thus reinforcing the additional risk of venting a tank to the atmosphere versus routing a relief device on a pressure vessel to a closed system.

Table 1: U.S. CSB Completed Investigations of Process Safety Incidents involving Low Pressure Tanks Since 2000 [1]

Date	Location	Incident Description	Consequences
February 8, 2017	DeRidder, LA	Flammable atmosphere in tank ignited by hot work	3 fatalities, 7 injuries
October 21, 2016	Atchison, KS	Incompatible material unloaded into incorrect tank causing chemical reaction	120 medical treatment cases onsite and offsite
January 9, 2014	Charleston, WV	Chemical storage tank leak	Contamination of municipal water supply
November 9, 2010	Buffalo, NY	Flammable atmosphere in tank ignited by hot work	1 fatality, 1 injury
October 23, 2009	Bayamón, Puerto Rico	Overfilling of gasoline storage tank causing fire and explosion	Pressure wave and damage to approximately 300 homes and businesses up to 1.25 miles from site
January 12, 2009	Woods Cross, UT	Release of flammable vapor cloud from storage tank causing flash fire	Blast wave and damage to homes beyond the plant fenceline
November 12, 2008	Chesapeak e, Virginia	Catastrophic fertilizer tank failure	2 injuries, community evacuation, contamination of public waterways
October 11, 2008	Petrolia, PA	Overfilling of acid storage tank	Evacuation of approximately 2,500 members of the public ordered by first responders
July 29, 2008	Tomahawk, WI	Flammable atmosphere in tank ignited by hot work	3 fatalities, 1 injury
June 5, 2006	Raleigh, MS	Flammable atmosphere in tank ignited by hot work	3 fatalities, 1 injury
January 11, 2006	Daytona Beach, FL	Flammable atmosphere in tank ignited by hot work	2 fatalities, 1 injury
July 17, 2001	Delaware City, DE	Flammable atmosphere in tank ignited by hot work	1 fatality, 8 injuries

Proper design of low pressure tanks requires consideration of the tank venting requirements for all credible causes of overpressure and vacuum, and the implementation of appropriate safeguards against these scenarios. Furthermore, it is entirely possible and sometimes likely that a low pressure tank may require different types of safeguards for each individual credible overpressure or vacuum scenario. When designing these pressure and vacuum relief systems, it is important to understand the strengths and weaknesses of each and to deploy these safeguards in a strategic manner that accentuates their strengths and mitigates their weaknesses.

1.1 Tank Overpressure and Vacuum Scenarios

API Standard 2000 defines the venting requirements for atmospheric and low pressure storage tanks for both overpressure and vacuum, including additional scenarios to be considered for refrigerated tanks [2]. Causes of overpressure as defined by API 2000 are listed in Table 2 below.

Table 2: Causes of Tank Overpressure per API 2000 [2]

API 2000 Overpressure Causes for All Tanks
Liquid Movement into a Tank
Weather Changes (Temperature Increase)
Fire Exposure
Pressure Transfer Vapor Breakthrough
Inert Pads and Purges (Supply Valve or Regulator Malfunction Open)
Abnormal Heat Transfer (Heating)
Internal Failure of Heat-transfer Devices (Heating/cooling Coil Mechanical Failure)
Vent Treatment Systems (Malfunction Causing Loss of Vent Flow)
Utility Failure
Change (Increase) in Temperature of the Input Stream to a Tank
Chemical Reactions (Exothermic)
Liquid Overfill
Control Valve Failure (Malfunction Open on Inlet or Closed on Vent)
Internal Explosion/Deflagration
Mixing of Products of Different Composition

Additional Causes for Refrigerated Tanks
Loss of Refrigeration
Heat Input Due to Pump Recirculation
Evaporation Due to Ambient Heat Input
Rollover (Unexpected Mixing of Two Liquid Phases due to Heat Input)
Overpressure of the Annular Space of a Double-wall Tank

Note: Brief Explanatory notes added in parentheses.

In reviewing the causes of overpressure listed in Table 2, these scenarios can be simplified to five general types: liquid inflow, vapor inflow, change in heat transfer, loss of venting capacity, and mixing of incompatible materials. Liquid inflow can displace the tank vapor space (often referred to as outbreathing) or eventually lead to overfilling of the tank. Overpressure due to vapor inflow can be caused by malfunction open of a control valve or regulator on a tank blanketing system, breakthrough of pressurized motive gas used for liquid transfer, or mechanical failure of an internal steam heating coil or steam jacket. Overpressure due to change in heat transfer can be caused by ambient conditions (i.e. increasing temperature or solar radiation), changes to process conditions upstream (i.e. hotter feed), external fire, malfunction of a heating coil or jacket temperature control system causing high temperature, or loss of cooling or refrigeration system functionality. Loss of venting capacity can cause overpressure if a continuous tank venting system malfunctions closed or plugs, or if an intermittent tank venting system fails to open on demand. Lastly, mixing of incompatible materials can cause overpressure in a tank due to exothermic mixing, exothermic reactions, deflagrations, or detonations.

When considering causes of vacuum, it may be noted that many of those cases listed in API 2000 are simply the inverse of analogous causes of overpressure. Table 3 lists these causes of vacuum.

Table 3: Causes of Tank Vacuum per API 2000 [2]

API 2000 Overpressure Causes for All Tanks
Liquid Movement Out of Tank
Weather Changes (Temperature Decrease and/or Precipitation)
Inert Pads and Purges (Supply Valve or Regulator Malfunction Closed)
Abnormal Heat Transfer (Cooling)
Internal Failure of Heat-transfer Devices (Cooling Coil Mechanical Failure)
Vent Treatment Systems (Malfunction Causing Vacuum in Vent System)
Utility Failure
Change (Decrease) in Temperature of the Input Stream to a Tank
Chemical Reactions (Endothermic)
Control Valve Failure (Malfunction Closed on Inlet)
Mixing of Products of Different Composition

Additional Causes for Refrigerated Tanks
Maximum Refrigeration

Note: Brief Explanatory notes added in parentheses.

As with the causes of overpressure, causes of vacuum can be simplified into more basic categories: liquid outflow, loss of vapor inflow, change in heat transfer, and mixing of incompatible materials. Liquid outflow can pull vacuum on the tank vapor space (often referred to as inbreathing). Vacuum due to loss of vapor inflow can be caused by malfunction closed of a control valve or regulator on a tank blanketing system.

Overpressure due to change in heat transfer can be caused by ambient conditions (i.e. decreasing temperature or solar radiation), changes to process conditions upstream (i.e. colder feed), malfunction of a heating coil or jacket temperature control system causing low temperature, an increase in cooling or refrigeration system heat duty, or the introduction of cooling media directly into the tank due to a mechanical failure of a cooling coil or jacket. Finally, mixing of incompatible materials can cause vacuum in a tank due to endothermic mixing or endothermic reactions.

While API 2000 presents fairly comprehensive lists of causes of overpressure and vacuum, it is worth noting that no code or standard can adequately address all conceivable processes. Therefore, it is incumbent upon those designing tank pressure and vacuum relief systems to seek out and apply the knowledge and experience of engineering and operations personnel familiar with the process associated with the particular tank under consideration.

One example of a scenario not explicitly covered in API 2000 is the case of vacuum due to a broken tank nozzle. In this case, it is possible that the capacity for gravity flow through a broken nozzle at the bottom of a tank (for example, caused by vehicle impact) could exceed the capacity of any pump used to move liquid out of that tank. Thus, it is possible that if the tank vacuum protection systems are only designed for the pump-out rate, the gravity flow from a broken nozzle could result in vacuum below the tank vacuum rating in spite of the vacuum protection systems in place. This example is illustrated in Figure 1 below; note that the gravity flow capacity for the broken 4" nozzle, 918 gpm, was calculated per the flow equations for compressible fluid through a nozzle as listed in Crane Technical Paper 410 [3] and the results of this calculation were independently verified using BakerRisk's SafeSite$_{3G}$© hazard assessment software. Refer to Appendix A for the detailed calculations used to determine the flow rate through the broken 4" nozzle.

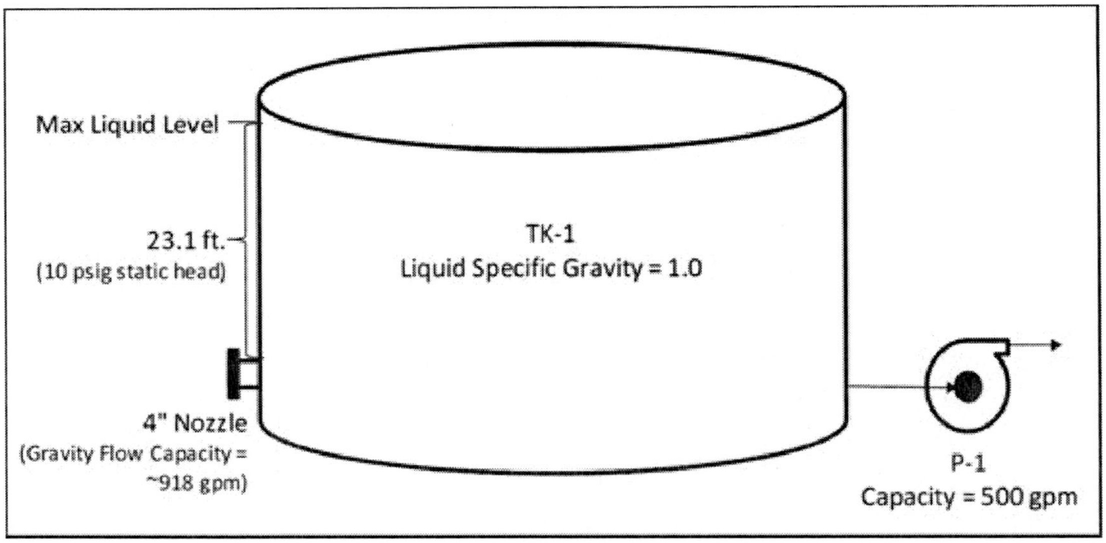

Figure 1: Broken Nozzle Gravity Flow Capacity vs. Pump Capacity [3]

However, for this case, the credibility (or lack thereof) of the broken nozzle case should be determined by personnel knowledgeable of the site, the location of the tank relative to vehicle traffic, and the credibility of any other potential causes of nozzle breakage, either external or internal to the tank. Also note that while the broken nozzle case is not explicitly listed in API 2000, it can be argued that is implicitly covered in Section 3.2.2 titled "Liquid Movement into or out of a Tank," which states "Liquid can enter or leave a tank by pumping, by gravity flow, or by process pressure." [1]

1.2 Tank Pressure and Vacuum Relief Systems – Strengths and Weaknesses

Once the credible causes of overpressure and vacuum have been established for a given tank, the next consideration is the best strategy to safeguard against each of these identified scenarios. A variety of safeguards are available for these purposes: some can be incorporated directly into the design of the tank or the surrounding process, some are elements of the Basic Process Control System (BPCS), and others are dedicated pressure and/or vacuum relief devices. Each of these safeguard types has its own specific strengths and weaknesses. Table 4 lists the strengths and weaknesses of several types of safeguards against tank overpressure. Please note that in this table, BPCS refers to a Basic Process Control System and HMI refers to a Human Machine Interface.

Table 4: Strengths and Weaknesses of Tank Pressure Safeguards [2]

Safeguard	Example	Strengths	Weaknesses
Free vent to atmosphere	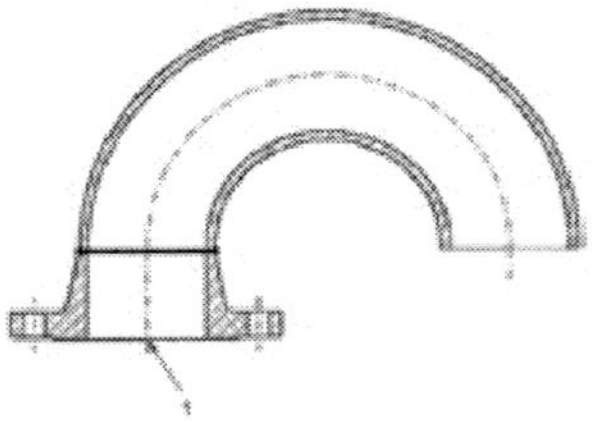	◆ Incorporated into tank design ◆ No Moving Parts	◆ Tank vapor space is continuously open to atmosphere. ◆ Vent elevation and/or discharge may prevent proper relief of liquid overfill scenario.
Flame/detonation arrestor vent		◆ Similar design simplicity to free vent ◆ Prevents external ignition sources from igniting flammable vapor in tank	◆ Similar drawbacks to free vent ◆ Potential for plugging if not properly maintained ◆ Generally designed for temporary and/or localized ignition sources, but not a sustained fire
Overflow Line	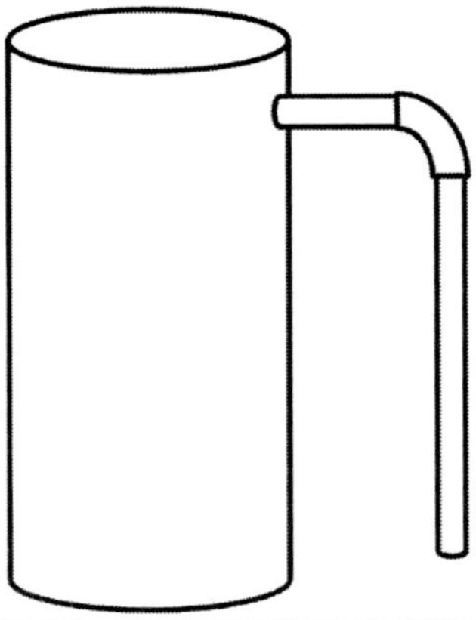	◆ Similar design simplicity to free vent ◆ Can be routed to a closed system or designed with a liquid seal if necessary ◆ Effective for overfilling cases	◆ Piping configuration may result in too much backpressure to provide effective vapor relief ◆ Potential static discharge issue if liquid is subjected to significant free-fall distance ◆ Tank liquid level limited by overflow line nozzle location
Overflow Vent		◆ Similar design simplicity to free vent ◆ Effective for overfilling cases	◆ Potential generation of significant flammable aerosols and vapors due to "waterfall" effect ◆ Tank liquid level limited by overflow vent location

Safeguard	Example	Strengths	Weaknesses
Continuous venting via pressure regulator	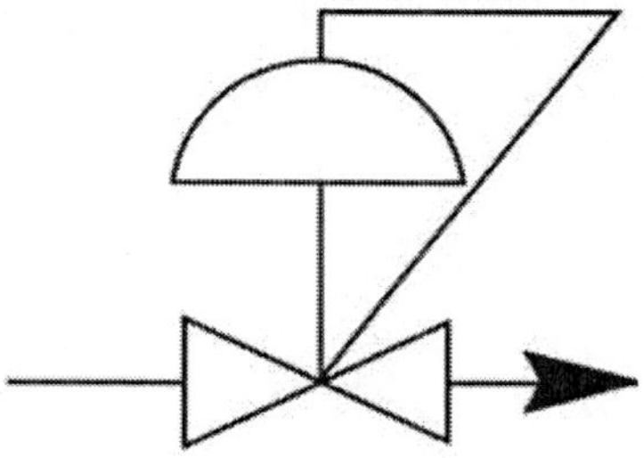	◆ Can be routed to a closed system if necessary ◆ Independent of BPCS	◆ Functionality can be compromised by high backpressure ◆ Regulator can malfunction (open or closed) if not properly maintained ◆ Regulator would typically only designed to vent vapor, and would not be effective for overfilling
Continuous venting via control valve	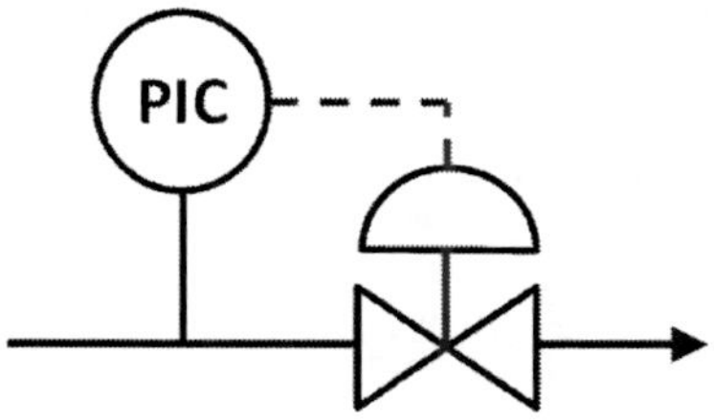	◆ Can be routed to a closed system if necessary ◆ Valve function could be observed remotely via HMI	◆ Functionality can be compromised by high backpressure ◆ Susceptible to common-cause failure with other BPCS functions ◆ Control valve can malfunction (open or closed) if any part of the loop is not properly maintained ◆ Unlikely to be effective for liquid overfilling (similar to regulator)
Reclosing pressure vent	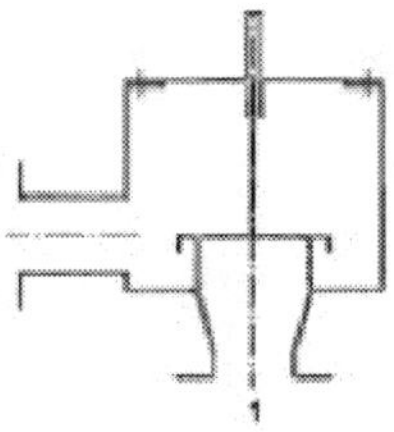	◆ Independent safeguard designed solely for pressure relief ◆ Keeps tank isolated from atmosphere when not venting pressure	◆ Can fail to function on demand if not properly maintained ◆ Not typically designed for liquid overfilling
Non-reclosing emergency vent	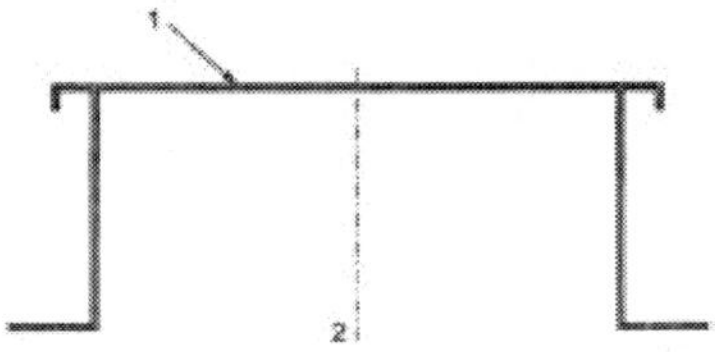	◆ Independent safeguard designed solely for pressure relief ◆ Can be fitted on a manway to provide significant relief capacity	◆ Can fail to function on demand if not properly maintained ◆ Not typically designed for liquid overfilling ◆ If the vent opens due to a relief event or malfunction, the tank remains open to atmosphere

Note that Table 4 includes a discussion of tank overflow vents, relief from which can cause a "waterfall" effect (i.e., high flow rate of liquid falling freely through the air outside the tank), generating large amounts of mist and aerosol. This mist and aerosol would have the potential to flow in all directions away from the tank due to calm low/no wind conditions, as occurred during the incident on October 23, 2009 in Bayamón, Puerto Rico described briefly in Table 1. A vapor cloud of this type, ignited by either an external ignition source or by static discharge from falling liquid, could pose significant flash fire and vapor cloud explosion (VCE) hazards.

A Tank Overflow Mitigation System (TOMS) [4] has recently been developed that can be installed on a tank with overflow vents in order to direct liquid flow to secondary containment via a closed gutter, thus preventing the waterfall effect and reducing the size of the potential vapor cloud. Figures 2 and 3 illustrate the overall schematic and gutter cross section, respectively, of a TOMS retrofitted onto an existing tank with an overflow vent.

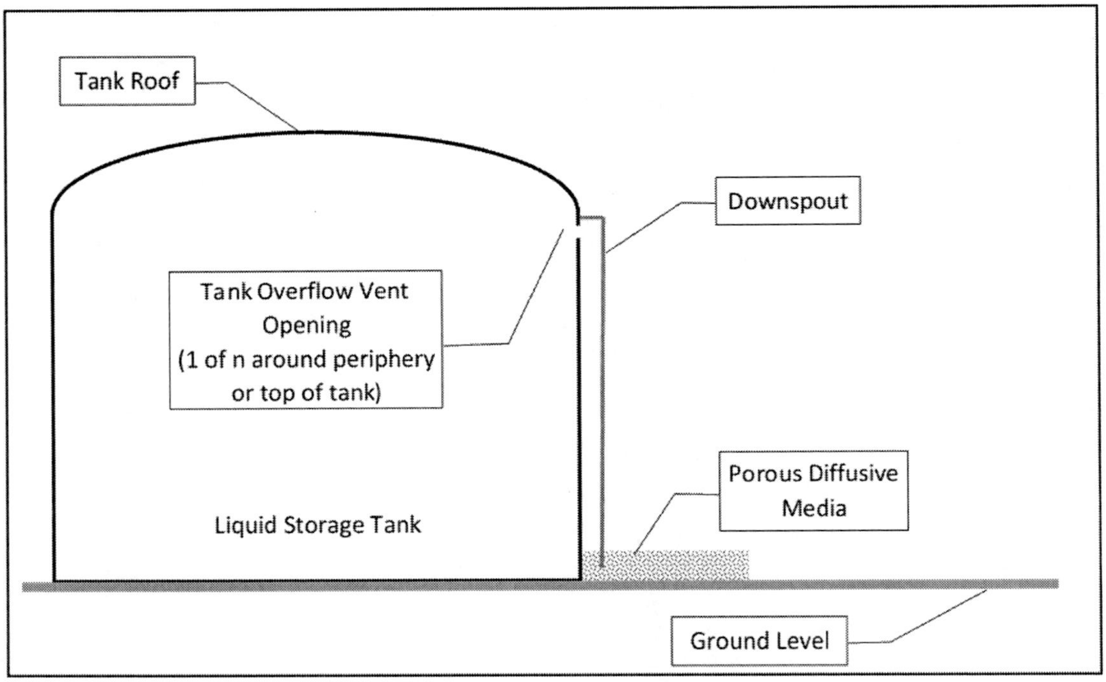

Figure 2: Schematic of Tank Overflow Mitigation System (TOMS) [4]

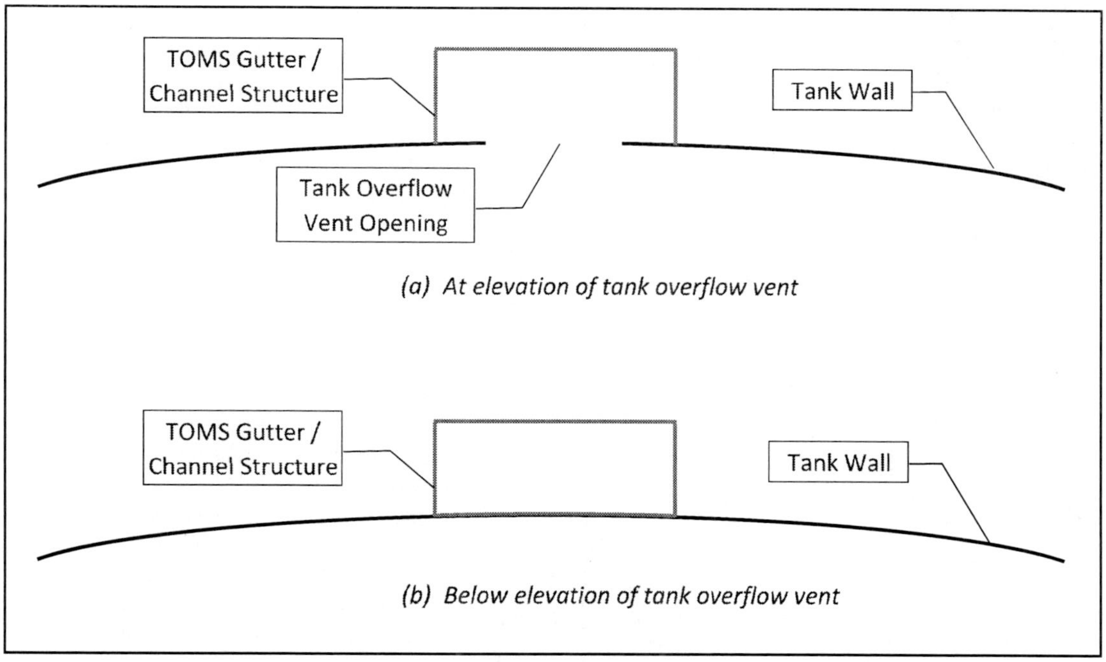

Figure 3: Gutter Cross Section of TOMS Added to Existing Tank (Looking Downward) [4]

As with safeguards against overpressure, safeguards against tank vacuum each have their own strengths and weaknesses; Table 5 lists these for several types of safeguards against tank vacuum. Note that in this table, as with Table 4, BPCS refers to a Basic Process Control System and HMI refers to a Human Machine Interface.

Table 5: Strengths and Weaknesses of Tank Vacuum Safeguards [2]

Safeguard	Example	Strengths	Weaknesses
Free vent to atmosphere	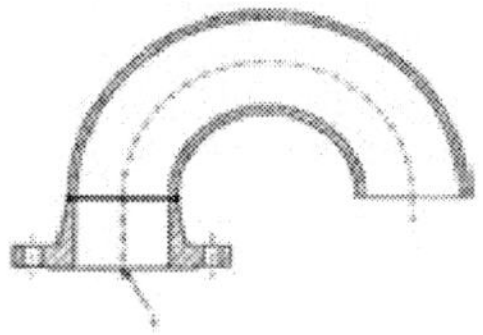	◆ Incorporated into tank design ◆ No Moving Parts	◆ Tank vapor space is continuously open to atmosphere.
Flame/detonation arrestor vent		◆ Similar design simplicity to free vent ◆ Prevents external ignition sources from igniting flammable vapor in tank	◆ Similar drawbacks to free vent ◆ Potential for plugging if not properly maintained

Safeguard	Example	Strengths	Weaknesses
Continuous inert blanketing via pressure regulator	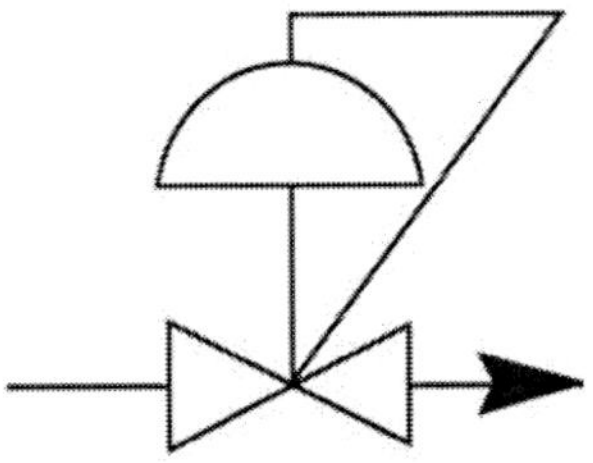	◆ Prevents air ingress from creating flammable mixture in tank vapor space ◆ Independent of BPCS	◆ Regulator can malfunction (open or closed) if not properly maintained ◆ Potential asphyxiation hazard if tank vapor space is vented to atmosphere
Continuous inert blanketing via control valve	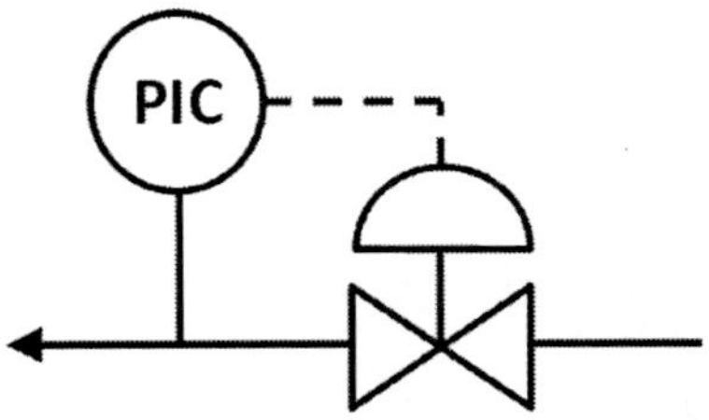	◆ Prevents air ingress from creating flammable mixture in tank vapor space ◆ Valve function could be observed remotely via HMI	◆ Susceptible to common-cause failure with other BPCS functions ◆ Control valve can malfunction (open or closed) if any part of the loop is not properly maintained ◆ Potential asphyxiation hazard if tank vapor space is vented to atmosphere
Reclosing vacuum breaker	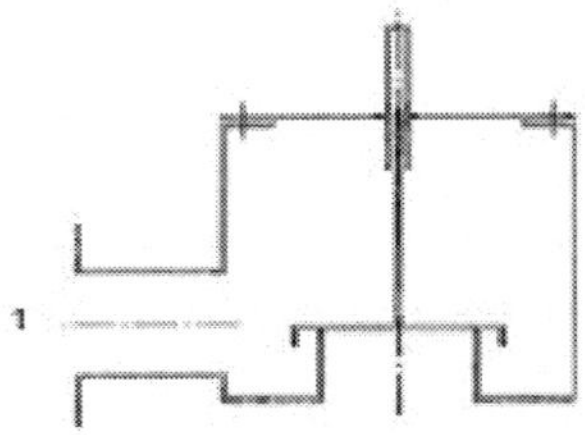	◆ Independent safeguard designed solely for vacuum relief ◆ Keeps tank isolated from atmosphere when not breaking vacuum	◆ Can fail to function on demand if not properly maintained ◆ Air ingress to tank when breaking vacuum could create a flammable internal atmosphere

In addition to the strengths and weaknesses of the safeguards listed in Table 4 and Table 5, it is important to consider the hierarchy of controls. Thus, protection of a tank against overpressure and vacuum by system design is the most desirable solution, followed by protection via the BPCS, then protection by alarms and human intervention, etc. Figure 4 provides a visual representation of this hierarchy.

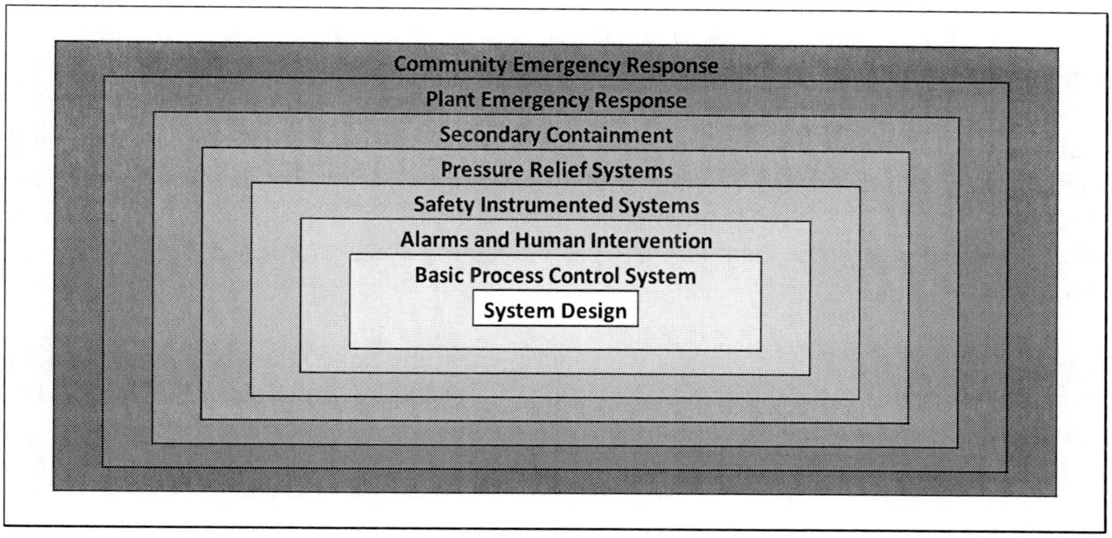

Figure 4: Hierarchy of Controls

Working through the hierarchy of controls with respect to tank overpressure and vacuum protection, it is then possible to discuss the applicable types of safeguards at each level. Beginning with system design, changing the pressure or vacuum rating of a tank to achieve inherently safer design is often impractical; inherently safer design with respect to pressure rating may require a pressure vessel rather than a tank. Open atmospheric vents, flame and detonation arrestor vents, and liquid overflow lines may be considered part of the system design; however, their mechanical integrity must be ensured by means of an appropriate Inspection, Testing, and Preventive Maintenance (ITPM) program. Additionally, even if these safeguards function as intended, the consequences of relief of any hazardous tank contents to the atmosphere cannot be overlooked, as discussed in a later section.

With respect to control systems, gas blanketing of a tank via the BPCS and/or mechanical pressure regulators can be an effective means of venting pressure and breaking vacuum caused by liquid movement into and out of the tank without the need to vent hazardous tank vapors to the atmosphere or to pull air into the tank and potentially create a flammable internal atmosphere. That said, such blanketing systems typically add another cause of overpressure and must be included in the ITPM program to ensure their continued functionality.

Regarding alarms and human intervention, a level alarm may provide a reliable safeguard against overfilling, so long as there is sufficient response time from the initial alarm activation before the tank overfills. Pressure alarms, however, generally would not allow sufficient response time before exceeding the tank pressure or vacuum rating.

A Safety Instrumented System (SIS) may be useful in preventing overpressure or vacuum of a tank, but this type of system would require a high-integrity safety interlock independent of the BPCS, necessitating significant investments in its design, installation, maintenance, and testing. Therefore, it is recommended that such a system not be installed without consulting engineering and operations personnel knowledgeable of the process associated with the tank and SIS subject matter experts in order to verify the proposed system's effectiveness for the application under consideration.

Tank pressure relief systems, such as atmospheric pressure/vacuum conservation vents and emergency vent hatches, can be employed as safeguards. However, if the tank contains hazardous material, these types of safeguards should only be considered as a last line of defense if other measures, such as inert gas blanketing for example, do not function as intended.

Secondary containment and emergency response, either onsite or in the larger surrounding community, would only be considered as means of reducing the impact of a hazard that was already present due to loss of primary containment. Thus, these types of controls could not be treated as preventive safeguards.

1.3 Dispersion Consequence Modeling for an Example Hexane Storage Tank.

With the credible overpressure and vacuum scenarios identified for a given tank and a safeguarding strategy developed for those scenarios in accordance with the hierarchy of controls, the next step in evaluation of the risk associated with a tank is the evaluation of the consequences of atmospheric dispersion of hazardous material releases from the relief devices.

In order to provide an example of this type of consequence modeling, BakerRisk's SafeSite$_{3G}$© hazard assessment software was utilized to analyze the consequences of relief scenario dispersion for a proposed hexane storage tank at the Wilfred E. Baker Test Facility. The specifications of the hypothetical tank are as follows:

Length: 16 feet	**Diameter: 4 feet**
Orientation: Horizontal	**Head Type: Flat**
Elevation above Grade: 1 foot	**Design Pressure: 10" H_2O (g)**

The following overpressure scenarios were identified for this tank:

- **Liquid overfilling by a transfer pump with capacity of 100 gpm**
- **Nitrogen blanket control valve malfunction open based on a 50 psig nitrogen supply via a generic 1" globe valve**
- **External fire based on a liquid level of 75% of the tank diameter have been identified as applicable overpressure scenarios for the tank.**

For safeguards, a 2" overflow line to grade has been selected for overfilling, a 3" gooseneck vent (which would require a flame arrestor) for nitrogen control valve malfunction open, and an 8" emergency vent hatch for external fire. Note that pressure and vacuum due to liquid movement into and out of the tank would also be vented via the 3" gooseneck vent, but those cases were not quantitatively evaluated for this example. Refer to Appendix A for further information regarding the calculation of the individual relief rates.

SafeSite$_{3G}$© was used to model the dispersion for each scenario listed in Table 6 under two different weather conditions for the Wilfred E. Baker Test Facility: F2.3 and D7.2. In the weather condition names, the letter represents Pasquill atmospheric stabilities and the number represents wind speeds in meters per second; for example F2.3 would refer to atmospheric stability F with a wind speed of 2.3 m/s. In the Pasquill stability classes, A is very unstable and F is very stable. These weather conditions were chosen based on statistical meteorological data for the San Antonio International Airport (SAT). For the purpose of this example, the figures shown represent the F2.3 weather conditions, as they generated more significant consequences than the D7.2 conditions for each individual analysis. Note that SafeSite$_{3G}$© uses a free-field dispersion model that does not account for specific obstructions such as a building; therefore, flammable clouds are reported up to 50% of the Lower Flammability Limit (LFL) as a conservative illustration of areas where obstructions could create localized flammable mixtures in air. The dispersion model used in SafeSite$_{3G}$© was developed based on one-dimensional turbulence theory [5].

Table 6: Release Sources for SafeSite$_{3G}$© Relief Dispersion Consequence Model

Name	Material	Hole Diameter (in)	Temperature (F)	Pressure (psig)	Height of Release (ft.)	Angle of Release (deg.)
Overfill	n-HEXANE	2	100	0.361	1	-90
Control Malfunction	NITROGEN	3	100	0.255	5	-90
Fire	n-HEXANE	8	157.2	0.2182	5	90

Figure 5 and Figure 6 show the flammable vapor cloud contours at grade and side view, respectively, for the relief of the overfilling scenario via a 2" overflow line terminating 1 foot above grade:

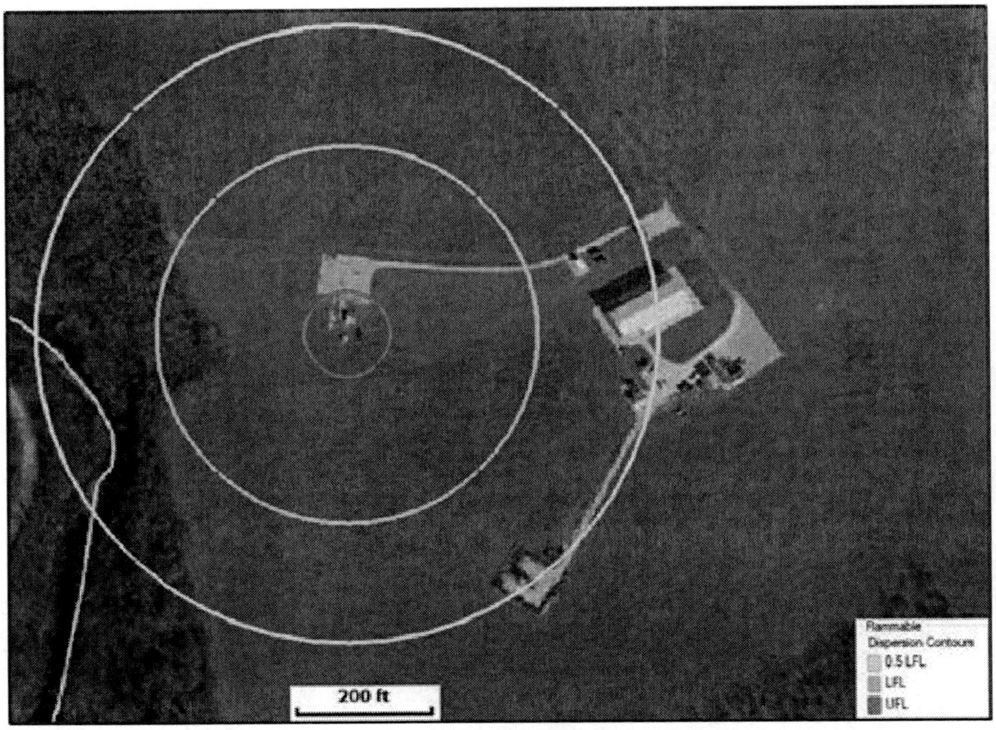

Figure 5: Flammable Contours for Overfilling Relief (F2.3 Weather)

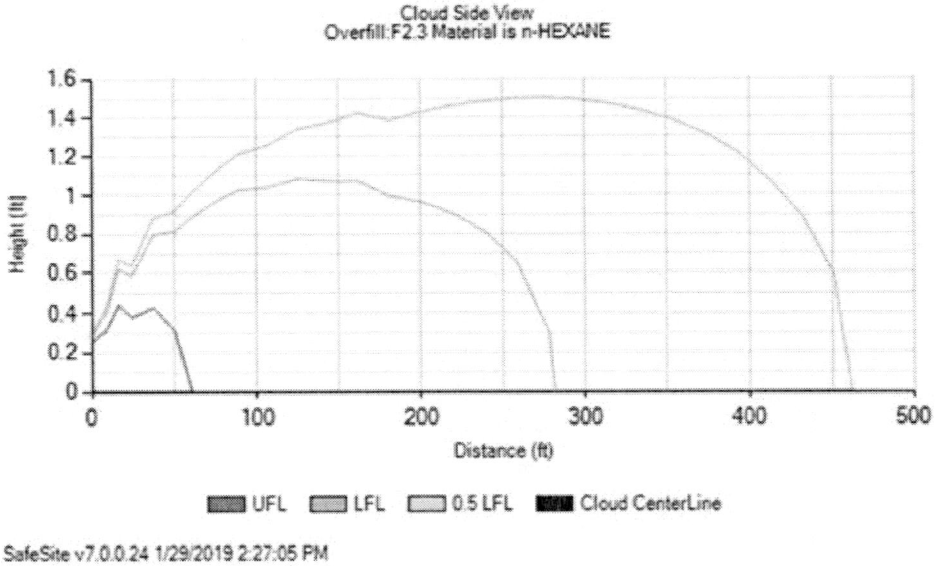

Figure 6: Flammable Vapor Cloud Side View for Overfilling Relief (F2.3 Weather)

The figures show that a flammable vapor cloud at grade could extend up to hundreds of feet downwind of the release point and potentially affect an occupied building, but that the cloud would remain within a 1.5-foot elevation above grade. Based on this consequence model, it may be appropriate to add a preventive measure to the system such as an interlock to shut down the transfer pump on high level in the tank and/or to add a means of mitigating the hazard such as a properly sized secondary containment dike with appropriate hazardous area classification. It is also worth noting that this model shows the steady-state dispersion and does not account for exhausting the supply of hexane, which may be relevant depending on the quantity of hexane available upstream of the transfer pump, which was not defined for the purpose of this example.

Looking to the malfunction open of the nitrogen blanket system control valve, the consequence model would be a bit different as nitrogen is neither flammable nor toxic but may pose an asphyxiation risk to personnel by creating a localized oxygen-deficient atmosphere. Figure 7 below shows the potential effects of oxygen-deficient atmospheres on the human body at several concentrations per Compressed Gas Association Safety Bulletin SB-2 – 2007 [6].

Atmospheric Oxygen Concentration (%)	Possible Results
20.9	Normal
19.0	Some unnoticeable adverse physiological effects
16.0	Increased pulse and breathing rate, impaired thinking and attention, reduced coordination
14.0	Abnormal fatigue upon exertion, emotional upset, faulty coordination, poor judgment
12.5	Very poor judgment and coordination, impaired respiration that may cause permanent heart damage, nausea, and vomiting
<10	Inability to move, loss of consciousness, convulsions, death

Figure 7: Effects of Oxygen-Deficient Atmospheres on the Human Body

Figure 8 shows the side view of the asphyxiation hazard cloud associated with the relief of the nitrogen control valve malfunction open scenario via a 3" gooseneck vent. Note that the reported concentrations of nitrogen in air in ppm correspond to 10.0%, 12.5%, and 14.0% oxygen from highest to lowest reported nitrogen concentration.

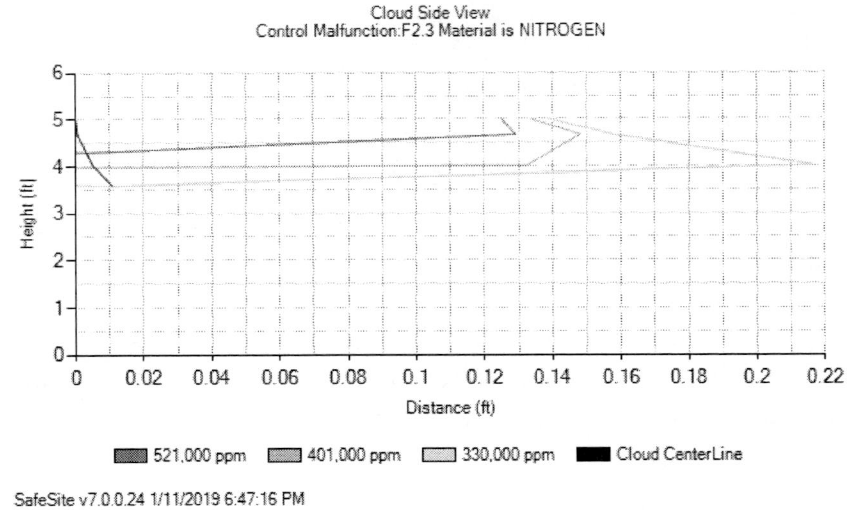

Figure 8: Asphyxiation Hazard Side View for Control Malfunction Relief (F2.3 Weather)

Note that the maximum distance of the oxygen deficient atmosphere is less than 1 foot; thus, the hazard posed by this release outdoors is minimal. However, if the release point were located indoors with air circulation limited by the building HVAC system, the oxygen concentration in the building could potentially drop to dangerous levels over time if the control valve malfunction went undetected.

Completing the modeled release results, Figure 9 shows the flammable vapor cloud side view for the relief of the external fire scenario via an 8" emergency vent hatch on top of the tank:

Figure 9: Flammable Vapor Cloud Side View for External Fire Relief (F2.3 Weather)

Based on this result, the flammable cloud of vaporized hexane due to the external fire rises rapidly within the first 5 feet of horizontal distance from the release point and then remains at least 20 feet above grade before dissipating below the flammable range. For the location of the tank, this would not pose a hazard beyond the potential escalation of the fire due to the hexane vapor cloud; while this is not desired, it would still be preferable to a full rupture of the tank due to overpressure while exposed to the fire.

Summarizing the consequence modeling, for the tank in hexane service, the liquid overfilling relief case resulted in potential escalation of consequences beyond the immediate area of the tank, whereas the relief cases due to the nitrogen blanket control valve malfunction open and the external fire scenario did not. Had this been an actual chemical plant pressure relief dispersion study and not a hypothetical example, plant personnel would also have the option to pursue further risk analysis using more comprehensive methods, such as a site-wide facility siting study and/or a Quantitative Risk Assessment (QRA).

2. Conclusion

When evaluating the potential risks associated with overpressure and vacuum of low pressure storage tanks, scenario applicability should be evaluated in accordance with API 2000. However, consideration should also be given to scenarios not explicitly covered in that standard based on consultation with engineering and operations personnel knowledgeable of tank locations relative to other equipment and activities at a site and of the process systems associated with the tanks under consideration. Furthermore, the strengths and weaknesses of potential safeguards against each scenario should be evaluated, and an overall safeguarding strategy should be developed for each tank with consideration given to the hierarchy of controls. Finally, even if appropriate safeguards are in place, the potential consequences of these safeguards functioning as intended in an emergency should be evaluated, and additional preventive safeguards and/or consequence mitigation measures should be implemented if additional risks are identified as a result of these evaluations. In these cases, plant personnel would also have the option of pursuing more comprehensive site-wide risk evaluations such as facility siting and QRA studies in order to better understand the process safety risks associated with their low pressure tanks.

Appendix A: Broken Nozzle Flow Calculation and References

Liquid Flow Through Broken Nozzle - No Phase Change

Dimensions of broken nozzle:

Nozzle Wall Thickness Input: Measured Wall Thickness (in)

OD =	4.5	in	(Dimensions based on 4" Schedule 40 Steel Pipe)
z =	0.237	in	Wall Thickness (Manual Input)
ID =	4.026	in	

Liquid rate across the broken nozzle:

$W = 1{,}891\ d_1^2 * C * \sqrt{(\Delta P * \rho_1)}$ (Crane Technical Paper No. 410, pg. 3-24; Note that the flow coefficient C is obtained from pg. A-20)

P_1' {	P_1 =	10.0	psig	High pressure side operating pressure
	P_2 =	0.0	psig	Atmospheric pressure
	ΔP =	10	psi	
	$\Delta P/P_1'$ =	0.4049		
	Flow Type:	Nozzle		
	C =	0.60		Flow Coefficient C for Nozzles (Crane A-20)
	ρ_1 =	62.4	lb/ft^3	Tank liquid density
	W =	459,391	lb/hr	Liquid Rate through broken nozzle

Volumetric Liquid Rate = 917.9 **GPM**

Notes:

#1 For reading the "C" chart, turbulent flow was assumed.
#2 See Subsequent Pages for Crane Technical Paper 410 Charts
#3 Used $Y = 1 - 0.317dP/P_1$ per "PRV sizing for exchanger tube rupture" by Wing Y Wong in Hyd. Proc. Feb 1992. Note that flow through a nozzle is one of multiple typical methods of calculating exchanger tube rupture relief rates.

Figure A-1: Broken Nozzle Flow Calculation [3]

Note that the above calculation results were independently verified using BakerRisk's SafeSite$_{3G}$© hazard assessment software.

Appendix B: Pressure Relief Calculations

Gas Control Valve Malfunction Open - Critical Flow

Critical Flow Check Calculation

Inlet Pressure of Device	$P =$	50	psig
Absolute Inlet Pressure of Device	$P_1 =$	64.7	psia
Ratio of Specific Heats (Cp/Cv)	$k =$	1.40	
Critical Flow Nozzle Pressure	$P_{cf} =$	19.5	psig
Pressure at Outlet of Device (Backpressure)	$P_b =$	0.361	psig

Critical Flow Will Occur: Proceed to Critical Flow Spreadsheet

FOR GASES WITH 50% PRESSURE DROP OR MORE:

Valve Type:	Generic 1" Globe Valve	
Flow Coefficient C_v	12	
Flow Coefficient C_1	38	
Inlet Pressure (P_{in})	50	psig
Molecular Weight (MW)	28.0	lbm/lb-mole
Gas Constant	1545	ft lbf/lbmole °R
Density at Relief Conditions (ρ_r)	0.074	lb/ft3
Density (ρ)	0.0738	lbm/ft3 at 14.7 psia and 60°F
Temperature (T)	100	°F
Temperature (T)	560	°R
Specific Gravity (SG)	0.97	
Volumentric Flow (W)	28,919	SCFH
Mass Flow rate (W)	2,133	**lb/hr relieving fluid**
Volumetric Flow rate (Q)	481	**ACFM relieving fluid**
Air equivalent	28,416	**SCFH air**

Equations Used:

$$W\,(\text{scfh}) = C_1 * Cv * (P+14.7) * \text{SQRT}(520/(SG * T°R))$$

$$W\,(\text{lb/hr}) = \frac{W(\text{scfh}) * 14.7 * 144 * MW}{1545 * 520}$$

$$\text{Air equivalent} = W(\text{scfh}) * (MW/29)^{0.5}$$

Figure B-2: Nitrogen Blanket Control Valve Malfunction Open [3]

Detailed Wetted Area Calculation - Horizontal Tank

Horizontal Tank Dimensions:

Diameter (D)	4.00	ft	
Tangent to tangent (L)	16.00	ft	
Grade to Bottom of vessel (F)	1.00	ft	
Liquid Level (d)	3.00	ft	
Vapor Space Height (h)	1.00	ft	
Tank Fill Level	75.0	%	
Max Liquid Level	3.00	ft	
Method to be Used to Calculate Relief Rate	**API 2000**		
Max. Fire Exposure Limit	30	ft > F + D	
Level for Wetted Area Calculation	3.00	ft	
Partial Surface Area Factor for Horizontal Cylinder ϕ	4.19		$\phi = 2\cos^{-1}\left(\frac{D-2h}{D}\right)$
Straight Wetted Surface Area	134.0	ft^2	$A = (^D/_2) * \phi * L$
Type of Heads	**Flat**		
Heads Wetted Surface Area (If 100% full) *(If one head is flat, conservatively assume the greater-area head type.)*	25.1	ft^2	*Flat Heads;* $A = 2\pi * (^D/_2)^2$ *2:1 Elliptical Heads;* $A = 2.61 * D^2$ *Hemispherical Heads;* $A = 4\pi * (^D/_2)^2$
Head Surface Area Factor Based on Level	1.0		(0.5 if vessel is 50% full or less, otherwise 1.0)
Piping Allowance	15.9	ft^2	(Lower of 10% initial calc. area and 100 ft^2)
Total Wetted Surface Area	175.1	ft^2	

Design Conditions (For Reference Only)

MAWP = 0.4 psig

Set Pressure = 0.4 psig

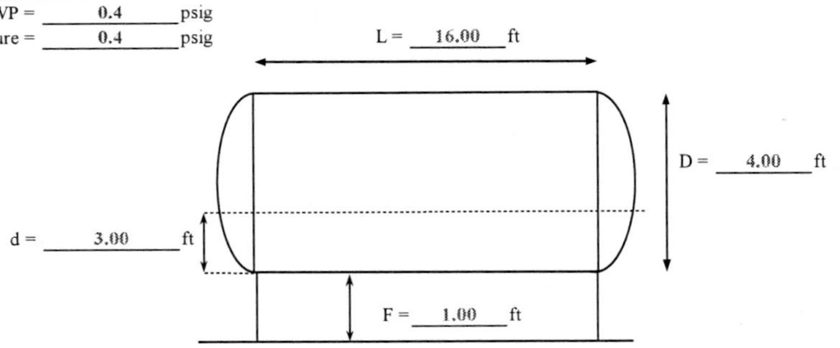

Figure B-3: External Fire Wetted Area Calculation

Liquid Vaporization Per API 2000 - Wetted Area Manual Input

Total Area Exposed:

Area Exposed, A 175.1 ft^2

Calculation Comments: -

Heat Input During Fire Exposure (See Table 4, API 2000):

A, ft^2	Q, BTU/hr	(Equations from API 2000, Table 4)
175.1	3,502,000	Q = 20,000(A) for Area > 0 and < 200
-	-	$Q = 199{,}300(A)^{0.566}$ for Area > 200 and < 1000
-	-	$Q = 963{,}400(A)^{0.388}$ for Area > 1000 < 2800
-	-	$Q = 21{,}000(A)^{0.82}$ for Area > 2800 (1 psig < Design Pressure < 15 psig)
-	-	Q = 14,090,000 for Area > 2800 (Design Pressure < 1 psig)

Q_H = 3,502,000 **BTU/hr**

Protection Factor, K_p:

			K_p Comments
X	1.00	Bare Vessel	-
		Approved Drainage	-
		Approved Insulation	-
		Approved Water Spray	-
		Approved Water Spray and Insulation	-

Kp = 1.00

Physical Properties:

λ, Latent Heat of Vaporization =	146.6	BTU/lb
k, Ratio of Specific Heats at STP =	1.06	
C, Dimensionless Flow Factor =	322.3	
Z, Compressibility Factor =	0.96	
T_F, Temperature of Vapor =	157.2	°F
T, Absolute Temperature of Vapor =	617.2	°R
MW, Molecular Weight of Vapor =	86.2	

Required Relieving Flow:

W, required Relief Capacity in lb/hr = Q * (K_p/λ)

W = 23,888 **lb_m/hr**

V_a, required Relief Capacity SCFH Air Equivalent = (1,100 * Q * (K_p/(C * λ)) * ((Z * T)/MW)$^{0.5}$

V_a = 213,629 **SCFH Air**

Figure B-4: External Fire Relief Rate Calculation

References

1. United States Chemical Safety and Hazard Investigation Board Completed Investigations; http://www.csb.gov/investigations/completed-investigations/?Type=2

2. American Petroleum Institute; *API Standard 2000: Venting Atmospheric and Low-pressure Storage Tanks (Seventh Edition)*; March 2014; API Publishing Services, 1220 L Street, NW, Washington, DC 20005.

3. Crane Company, Engineering Department; *Technical Paper No. 410: Flow of Fluids Through Valves, Fittings, and Pipe (25th Printing)*; 1991; Crane Company, 104 North Chicago Street, Joliet, IL 60434.

4. Thomas, J.K., Baker Engineering and Risk Consultants; *U.S. Patent Application No. 16/173,445; Tank Overflow Mitigation System*; filed 29-October-2018.

5. Rowley, J.; *A New Integral Dispersion Model based on One-Dimensional Turbulence Theory*; 7-9 May 2014; Institution of Chemical Engineers (Great Britain) Symposium; 24th Institution of Chemical Engineers Symposium on Hazards 2014: (HAZARDS XXIV); Edinburgh, United Kingdom.

6. Compressed Gas Association; *Safety Bulletin SB2-2007: Oxygen Deficient Atmospheres;* 2007; Compressed Gas Association, Inc., 4221 Walney Road Chantilly, VA 20151.

Comparison between Steady State and Dynamic Pressure Relief Calculations

Neil Prophet
ioMosaic Corporation
1900 St. James Place, Suite 700, Houston, Texas 77056
nprophet@iomosaic.com

Carlos Olivares
ioMosaic Corporation
1900 St. James Place, Suite 700, Houston, Texas 77056
olivares.c.tx@iomosaic.com

Micah Wessels
ioMosaic Corporation
1900 St. James Place, Suite 700, Houston, Texas 77056
wessels.m.tx@iomosaic.com

Prepared for Presentation at
American Institute of Chemical Engineers
2019 Spring Meeting and 15th Global Congress on Process Safety
New Orleans, LA
March 31 – April 3, 2019

Comparison between Steady State and Dynamic Pressure Relief Calculations

Neil Prophet
ioMosaic Corporation
1900 St. James Place, Suite 700 | Houston, Texas 77056
nprophet@iomosaic.com

Other Authors: Carlos Olivares, Micah Wessels
ioMosaic Corporation

Keywords: pressure relief systems design, overpressure protection, dynamic calculations

Abstract

API Standard 521 is open to the application of dynamic simulations when considering pressure relief system calculations. Some overpressure scenarios are very dynamic in their nature, such as external fire exposure, abnormal heat input, vapor breakthrough and runaway reaction, to name a few.

Modeling overpressure scenarios dynamically can allow for a more accurate, less conservative, representation of the affected system; often resulting in a lower relief requirement, as well as an understanding of how the system pressure and temperature changes over time.

This paper compares an external fire overpressure scenario which is modeled using steady state calculations versus the same system modeled dynamically. It is intended to give the user a better understanding of the differences in the two approaches, as well as the benefits of applying dynamic simulation for specific scenarios.

1 Introduction

Traditional pressure relief sizing calculations have typically used steady state equations to determine both the required relief rate and the pressure relief system capacity, for every applicable overpressure scenario. These equations are well accepted and understood and are available in API Standards 520 and 521.

API Standard 521 is open to the application of dynamic simulations when considering pressure relief system calculations. Some overpressure scenarios are very dynamic in their nature, such as external fire exposure, abnormal heat input, vapor breakthrough and runaway reaction, to name a few.

Modeling overpressure scenarios dynamically can allow for a more accurate, less conservative, representation of the affected system; often resulting in a lower relief requirement as well as an understanding how the system pressure and temperature changes over time.

This paper compares an external fire overpressure scenario which is modeled using steady state calculations versus the same system modeled dynamically. It is intended to give the user a better understanding of the differences in the two approaches, as well as the benefits of applying dynamic simulation for specific scenarios.

2 Overpressure Scenarios

Some overpressure scenarios are very dynamic in their nature, such as external fire exposure, abnormal heat input, vapor breakthrough, transient analysis of a tube rupture, and runaway reaction. In cases such as these, it can be advantageous to utilize dynamic modeling.

External Fire

Pressure vessels, other pressure-containing equipment, or storage tanks may be exposed to fire at some time in the life of an operating plant that handles or processes flammable liquids or gases. Unless external fire can be specifically excluded, the ASME Boiler and Pressure Vessel Code and NFPA 30 require consideration of a fire scenario for such equipment.

For pressure vessels in process areas, the maximum fire height is usually taken as 25 feet above a height where liquids can pool. API Standard 521 heat flux calculations for process vessels typically specify a wetted area based on liquid level, plus an extra allowance for process piping and instrumentation exposed to fire.

As the fire burns, heat is transferred to the vessel wall. Where the vessel wall is in contact with the liquid contents, the heat is transferred directly to the liquid inside the vessel. In the case where the vessel wall is in contact with the vapor contents, the heat is transferred to the metal wall, resulting in a significant temperature rise of the metal wall.

Dynamic modeling of such an event allows the following phenomena to be analyzed:

- Temperature and pressure rise of vessel contents over time
- Vessel contents composition changes over time
- Liquid level changes over time, resulting in a change to the wetted area
- Pressure relief valve venting time, and duration
- Vessel wall temperature effects

Abnormal Heat Input

Abnormal heat input is an overpressure scenario where excess vapor volume is generated as a result of unexpected high process heat input. This scenario is typically considered as a case for reboilers in the event of temperature control valve failure. Vapor generation can exceed the system's ability to condense or otherwise absorb the build-up of pressure, which may include non-condensables caused by overheating.

Abnormal heat input may occur as a result of numerous initiating events such as: failure of automatic controls, power failure, inadvertent opening of normally closed valves, or general operator error.

The relief requirement for abnormal heat input is typically based on the amount of material that can be vaporized due to the additional heat load. For fired heaters, when evaluating the fuel control valve fails wide-open case, the duty can be expected to increase to 150% of normal value unless manufacturer data dictates otherwise. For heat exchangers, the duty can be expected to increase to no more than 200% of design duty.

Similar to an external fire scenario, dynamic modeling of such an event allows the following phenomena to be analyzed:

- Temperature and pressure rise of vessel contents over time
- Vessel contents composition changes over time
- Liquid level changes over time
- Pressure relief valve venting time, and duration

Vapor Breakthrough

Vapor breakthrough, also known as gas blow-by, can result from the failure open of a liquid outlet level control valve that flows to a downstream system with a design pressure less than the upstream maximum normal operating pressure.

The vapor breakthrough case relief load could be based on the maximum flow through a control valve and bypass valve simultaneously wide open, or either separately. When modeling such a scenario dynamically, consideration can be given to the relative liquid volumes of the two vessels to determine if a liquid release may occur in the downstream vessel. The dynamic simulation will determine if the downstream vessel can be hydraulically filled, and the relief device sizing calculation then evaluated for liquid displacement for gas blow-by.

The dynamic mode can also determine if the downstream vessel can be filled to a level that is higher than the height of the inlet nozzle at the start of the gas blow-by scenario. If so, two phase flow would be evaluated.

Dynamic modeling of such an event allows the following phenomena to be analyzed:

- Temperature and pressure rise of vessel contents over time
- Phase, liquid level and volume changes over time
- Pressure relief valve venting time, and duration

Runaway Reaction

Runaway chemical reactions may result in overpressure due to the rapid increase in pressure developed during an exothermic reaction.

Such an overpressure scenario really lends itself to dynamic modeling.

The reaction chemistry should first be qualified using small scale experiments. This data can then be coupled with fluid dynamics to design the emergency relief system. In addition to vent sizing information, the dynamic modeling approach can provide insights into process modification and refinement as well as the establishment of a safe operating envelope.

Modeling the runaway reaction dynamically can account for vessel insulation, a glass lining, jacketing or protection by water spray.

For dynamically modeled reactive systems, essential parameters that can be evaluated include:

- System Thermal Stability, i.e. nature and relative magnitude of any exothermic (or endothermic) effects. These include primary thermal runaway and any subsequent secondary decomposition reactions.
- Total Heat Released, characterized by the specific reaction heat release per (mass or molar) unit of reacting material.
- Rate of Heat Release, characterized by global kinetic parameters describing heat generation as a function of temperature and residual reactant concentration.
- Rate of Gas Generation, due to volatilization of liquid components and vapor expansion effects with increasing reaction temperature or generation of additional gaseous reaction or decomposition products.
- Scale-Up or Loading Factors, including credible heat influx from external fire events and physical characteristics of a full-scale containment vessel.
- Vent Flow Characteristics, including multiphase/multicomponent system properties and effects of continuing chemical reaction.

Tube Rupture Transient Analysis

A heat exchanger single tube failure may be considered when the high-pressure side maximum operating pressure exceeds the low-pressure side corrected hydrostatic test pressure.

The required relieving rate should be calculated to consider displacement of the pre-existing fluid, any mixing of high-pressure side and low-pressure side fluids, including flashing; and high-pressure side fluids after displacement has been completed.

API Standard 521 discusses the benefits of a dynamic analysis for a tube rupture scenario; and also acknowledges that in some cases where there is a wide difference in design pressure between the two exchanger sides, then transient conditions can produce overpressure above the test pressure. A dynamic pressure relief analysis gives the user an understanding of the pressure rise rate, and whether a pressure relief device can act fast enough to provide sufficient overpressure protection for this scenario.

3 Example Case – External Fire Comparison Between Steady State and Dynamic Models

In order to examine the differences between a steady state and a dynamic model for an external fire exposure scenario, the following system was studied:

A vertical vessel with a straight side length of 30 feet (tangent to tangent) and inner diameter of 10 feet, at 8 feet above grade level, was modeled being exposed to a three-hour external fire. The vessel was initially 45% liquid full with a hydrocarbon mixture, with an additional 10% wetted area added to account for interconnected piping. Adequate drainage and firefighting consideration was applied to both the dynamic and steady state models. The design pressure of the vessel is 250 psig, initially operating at 100 psig and 97 °F.

Calculations were based on an allowable accumulation of 21% for an external fire exposure scenario. API discharge coefficients were used for sizing the required relief area. The relief fluid for the steady state model is based on a 5 mol% vaporization, while the relief fluid for the dynamic model was determined with respect to time, heat input, and vaporization of the residing fluid. Table 3 provides a summary of the inputs used in this calculation.

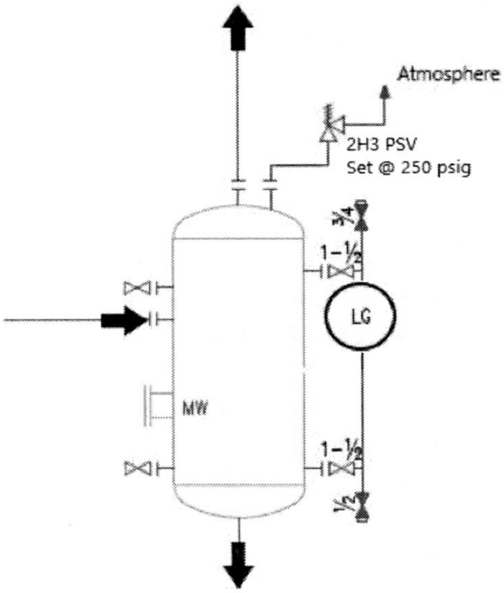

Figure 1: Vessel layout

The dynamic model also takes the currently installed relief system into account during an overpressure event. For this example, a simple relief system installation was used. The relief device modeled in the dynamic calculation was a 2" H 3" bellows relief device certified for vapor service, with 1.25' of 2" Nominal Pipe Size (NPS) Schedule 40 vertical inlet piping and 1' of 3" NPS Schedule 40 horizontal outlet piping discharging to the atmosphere. The lift versus overpressure and lift versus backpressure curves used, are the API curves for a standard API 526 valve with vapor trim (a relief device certified for vapor service).

The use of a simple layout means that there are no fitting losses due to any bends, elbows, or manual valves. The relief device has a set pressure of 250 psig, and a reseat pressure of 241.8 psig which corresponds to a 7% blowdown. API discharge coefficients were used for the relief device, to match the API discharge coefficients used in the relief requirement calculations (both steady state and dynamic models).

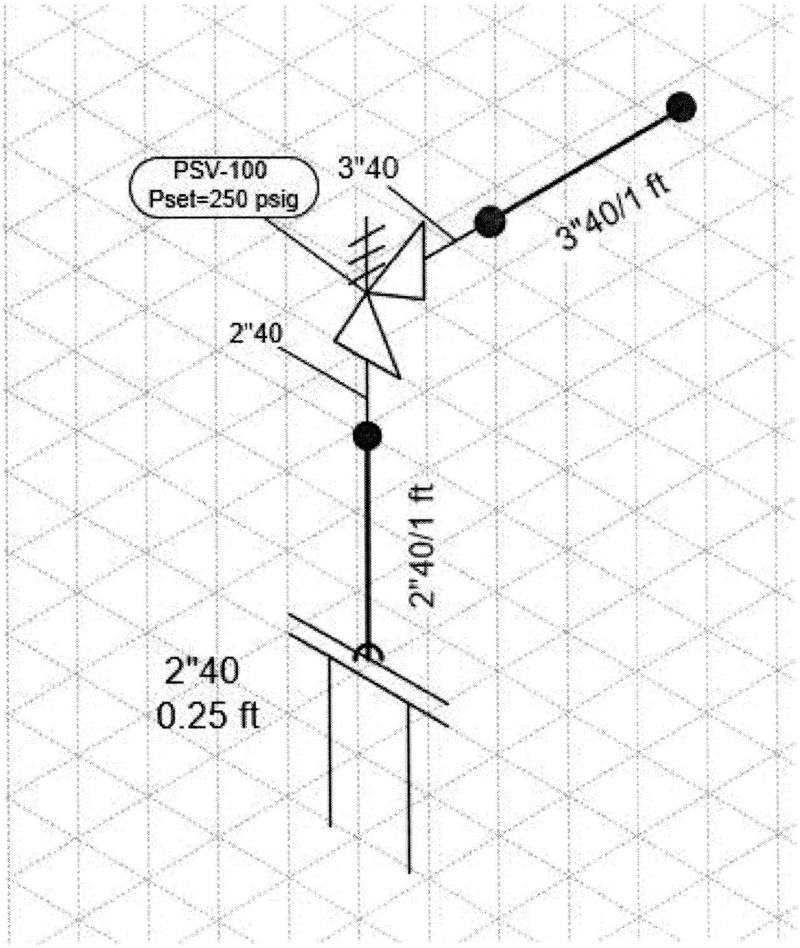

Figure 2: Pressure relief system layout

Table 1: Initial Vessel Liquid Inventory

Component	Chemical Formula	Mole Fraction	Molecular Weight
Propylene	C3H6	0.2358	42.08
Propane	C3H8	0.2147	44.10
1-Butene	C4H8	0.1768	56.11
1,3-Butadiene	C4H6	0.1902	54.09
n-Butane	C4H10	0.1236	58.12
1-Hexene	C6H12	0.0589	84.16

External Fire Heat Input Equations

Fire exposure on process equipment can result in overpressure due to the vapor generation from boiling of the liquid mixture. In this example case, the external fire is based on an open pool fire - both API 521 and API 2000 are based on open pool fires, with a heat flux comparable to a heavy hydrocarbon, as the basis for sizing pressure relief devices for fire cases.

In the case of an external fire, it is necessary to determine the portion of the vessel that is wetted by its internal liquid level up to 25 ft above the source of the flame, meaning any wetted area above the 25 ft fire limit is not considered with respect to heat input from external fire exposure.

The wetted area for the vertical vessel can be calculated with the following equations:

$$A_{cylinder} = \pi * z * D$$

$$A_{Elliptical\ head} = \pi * b^2 + \pi * \frac{a^2}{2 * \epsilon} * \ln(\frac{1 + \epsilon}{1 - \epsilon})$$

$$\epsilon = \sqrt{1 - \frac{a^2}{b^2}}$$

Z is the liquid height from the tangent line of the vessel

D is the inside diameter of the vessel

a is the minor semi-axis

b is the major semi-axis

The wetted area would include the areas exposed to the liquid contained within the vessel, up to 25 ft. This would include the cylinder area up to the liquid level (z) plus the bottom head, as they are within the height limit.

It is common practice to assume the vessel to be isolated to simplify the analysis. This assumption is based on rotating equipment being stopped; and emergency procedures being implemented, which would include closing appropriate manual valves.

The heat input rate caused by an external fire is calculated with the following equation if there are prompt firefighting efforts and drainage of flammable materials away from the vessels. API Standard 521, 6th edition, Section 4.4.13.2.4.2.

$$Q = C * F * A^{0.82}$$

Q is the total heat absorption to the wetted surface (Btu/h)

C is a constant 21,000 in USC units (or C can be 34,500 if adequate drainage and firefighting are not present)

F is an environmental factor; F=1 for vessels without insulation credit (Table 5 of API Std 521 provides examples based on different insulation thermal conductivity)

A_s is the total wetted surface (ft2)

The fire relief rate is equal to the fire heat input rate divided by the latent heat of vaporization of the relief vapors.

$$W = Q/\lambda$$

W is the rate of vapor generation (lb/h)

Q is the total heat absorption to the wetted surface (Btu/h)

λ is the latent heat of vaporization at relieving conditions (BTU/lb)

In order to establish the vapor relief requirement for a mixture under external fire exposure, the latent heat of vaporization is required. For pure liquid mixtures it is relatively simple to establish the latent heat, as data for single components can be obtained from literature (see Figure A.6 in API 521). However, in the case of our example, a vessel containing multicomponent liquid mixture, calculating the latent heat contains many pitfalls. In the absence of advanced tools such as SuperChems Expert, calculating the heat of vaporization of the mixture presents various challenges. The latent heat for mixtures of components with a wide range of boiling points will require a more complex calculation due to the constantly changing composition, temperature, and heat of vaporization as the mixture changes over time. This calculation is best modeled using a dynamic analysis as noted by API 521 section 4.4.13.2.5.2.

> "For these and other multicomponent mixtures that have a wide boiling range, it might be necessary to develop a time dependent model where the total heat input to the vessel not only causes vaporization but also raises the temperature of the remaining liquid, keeping it at its boiling point."

In this paper, results using an integral isobaric latent heat established at the first 5 mol% of the liquid vaporized at maximum allowable venting pressure is compared against a dynamic solution using the SuperChems software. Note that the integral isobaric latent heat calculation will result in a relief fluid biased toward the light ends of the mixture. One potential flaw is that the maximum relief requirement may not be established until after the light ends are vaporized.

Note that in our steady state calculation the relief requirement is adjusted by accounting for the density correction factor. Calculating the required rate by dividing the rate of heat absorption by the latent heat of vaporization becomes overly conservative near the critical point where the latent heat goes to zero. In order to calculate a more accurate requirement, the relief requirement can be adjusted by the specific volume.

$$W_r = W * (\frac{\rho_l - \rho_v}{\rho_l})$$

Wr is mass relief rate (lb/h)

ρl is liquid density (lb/ft3)

ρv is vapor density (lb/ft3)

The steady state assumption previously discussed may be deemed inappropriate, and hence more rigorous methods of calculations may be used, such as dynamic analysis.

4 Results Comparison

Running the same external fire calculation both at steady state and dynamically yields results, as follows:

Table 2: External Fire Results Comparison

	Relief Requirement	Pressure Relief Capacity
Steady State External Fire	27,583 lb/hr	24,298 lb/hr
Dynamic External Fire	24,007 lb/hr	24,298 lb/hr

As can be observed, the dynamic calculation results in a lower relief requirement (by approximately 13 % less) than the steady state calculation. In both cases, the pressure relief capacities are the same, as this is based on the steady state system rated capacity at 21% overpressure.

The following tables summarize calculation inputs and outputs for an external fire calculation, modeled both steady state and dynamically.

Table 3 shows steady state detailed calculation results for the external fire scenario.

Table 4 shows dynamic simulation calculation results for the external fire scenario.

Table 5 shows the steady state pressure relief system capacity results for the external fire scenario.

Table 3: External Fire Steady State Detailed Calculation Results

GAS/VAPOR RELIEF UNDER FIRE/USER HEATING USING CCPS/API 520 (SUPPORT) INFORMATION		
Inputs		
Vessel wetted surface area	75,927.97	in²
Additional wetted surface area	7,592.80	in²
Heating rate	3,877,954.29	BTU/hr
Set pressure	250.00	psig
Back pressure	0.00	psig
Percent overpressure	21.00	%
Discharge coefficient	0.975	
Outputs		
Temperature at maximum pressure	187.37	°F
Compressibility factor	0.70	
Density	3.16	lb/ft³
Heat of vaporization	124.28	BTU/lb
Average Molecular Weight	48.14	
Specific Volume Correction Factor	0.88	
Correction factor vapor density	3.16	lb/ft³
Correction factor liquid density	27.20	lb/ft³
Back pressure correction factor (Kb)	1.00	
Flow is critical (choked)	Yes	
Evaporation/relief rate	**27,582.44**	**lb/h**
Volumetric flow rate	**212,817.51**	**SCFH**
Required relief device orifice area	**0.834**	**in²**

Table 4: External Fire Dynamic Simulation Calculation Results

VESSELS CONTAINING TWO PHASES (DYNAMIC) (SUPPORT) INFORMATION	
Inputs	
Initial total mass	41,474.24 lb
Volume full of liquid	45.00 %
Temperature	97.27 °F
Pressure	100.00 psig
User defined top flow type	Vapor
Top flow user specified backpressure	0.00 psig
User defined bottom flow type	No flow
Bottom flow user specified backpressure	0.00 psig
Consider external fire exposure	YES
Set pressure	250.00 psig
Discharge coefficient	0.975
Outputs	
Top: First Flow	
Time	0.608 h
Temperature	172.77 °F
Pressure	257.85 psig
Liquid flow rate	0.00 lb/h
Vapor flow rate	13,502.89 lb/h
Total flow rate	**13,502.89 lb/h**
Top: Flow at Maximum Pressure Data	
Time	1.874 h
Temperature	234.18 °F
Pressure	301.41 psig
Liquid flow rate	0.00 lb/h
Vapor flow rate	24,007.01 lb/h
Total flow rate	**24,007.01 lb/h**
Required relief device orifice area	**0.7850 in^2**

Table 5: External Fire Pressure Relief System Capacity Results

RELIEF CAPACITY - PSV-100 LAYOUT - 250 PSIG	
Relief Conditions:	
Temperature	187.37 °F
Pressure	302.50 psig
Superimposed backpressure	0.00 psig
Exit Conditions:	
Exit Temperature	104.82 °F
Exit Pressure	6.50 psig
Exit Mach Number	0.96
Kb correction	1.00
Kp correction	1.00
Pressure at device inlet	302.07 psig
Irreversible inlet pressure drop	0.40 psi
Backpressure	10.14 psig
% Irrev. inlet pressure drop relative to actual set point	**0.16 %**
% back pressure relative to actual set point	**4.05 %**
Vapor Mass Flow Rate	**24,312.05 lb/h**
Volumetric Flow Rate	**187,582.21 SCFH**
Relief Device Flow Area	**0.7850 in^2**

Significantly, in the case of the steady state calculation, the pressure relief valve would be considered undersized and to provide inadequate pressure relief flow capacity; whereas with the dynamic calculation would result in the pressure relief system being considered adequately sized.

This of course could result in unnecessary discussion list items which could in turn result in unnecessary mitigation actions and unnecessary capital expenditure.

The following figures provide additional understanding of conditions inside the affected process equipment during an external fire scenario modeled dynamically.

Figure 3 shows a dynamic graph of vessel pressure versus time for the external fire scenario; and demonstrates pressure rise rate until the pressure relief device opens after approximately 0.6 hours. The pressure continues to build even as the pressure relief device opens and continues to vent vapor until the vessel empties.

Figure 4 shows a dynamic graph of vessel contents temperature versus time for the external fire scenario. After two hours of fire exposure, the vessel is vapor-full and the vessel wall temperature rises more rapidly.

Figure 5 shows the dynamic graph of liquid height versus time for the external fire scenario, showing that after approximately two hours there is no liquid left in the vessel.

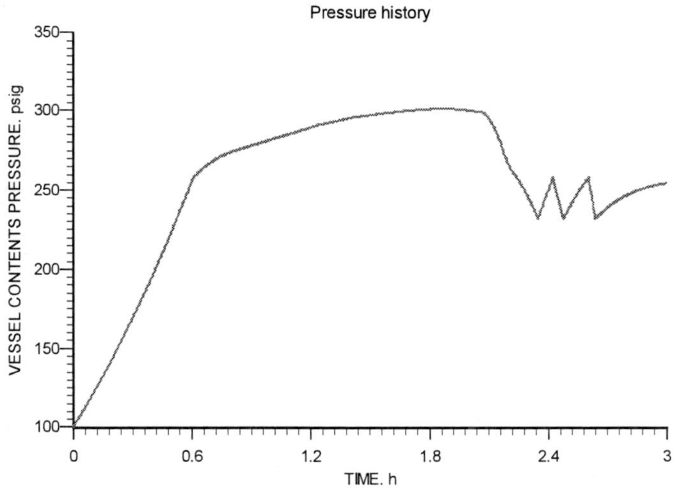

Figure 3: Dynamic graph of vessel pressure versus time

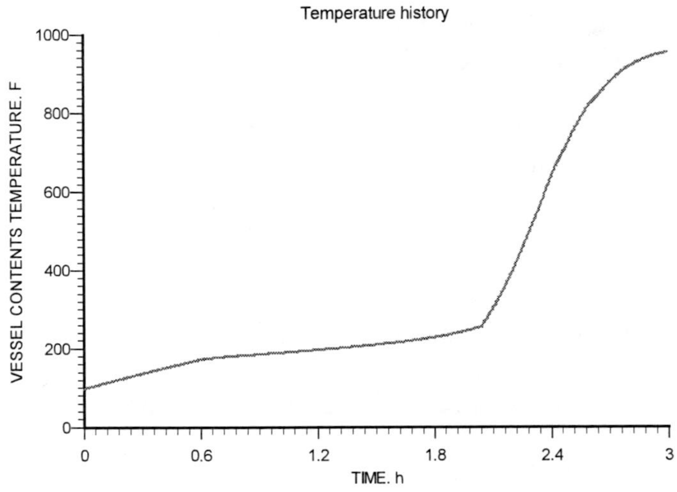

Figure 4: Dynamic graph of vessel contents temperature versus time

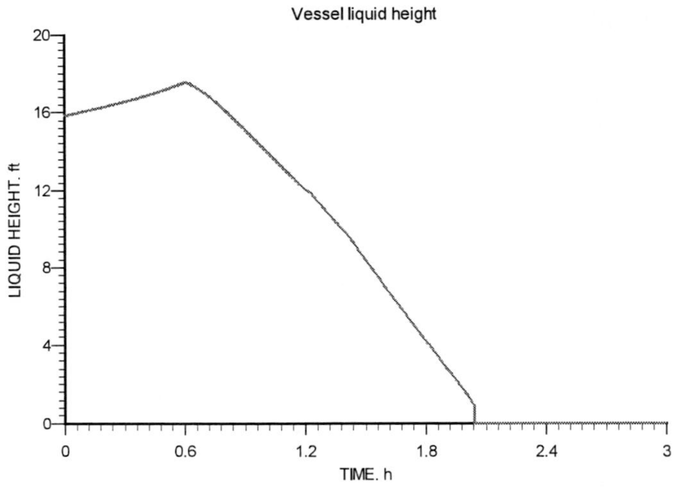

Figure 5: Dynamic graph of liquid height versus time

Additionally, external fire calculations can also be used to consider the effects of fire impinging on the vessel metal wall. From this, it can be determined if, or when, the vessel may be likely to fail due to increased temperature rather than pressure. Figure 6 illustrates an example of dynamic simulation to predict vessel wall failure due to fire exposure.

Various parts of the vessel can be represented by "segments" which have differing levels of heat flux exposure due to changing liquid levels in the vessel. The vessel stress due to internal pressure and temperature can be predicted for each segment.

At the point where the wall failure stresses are less than the vessel stress caused by internal pressure (approximately 34 minutes), failure of the vessel could be expected.

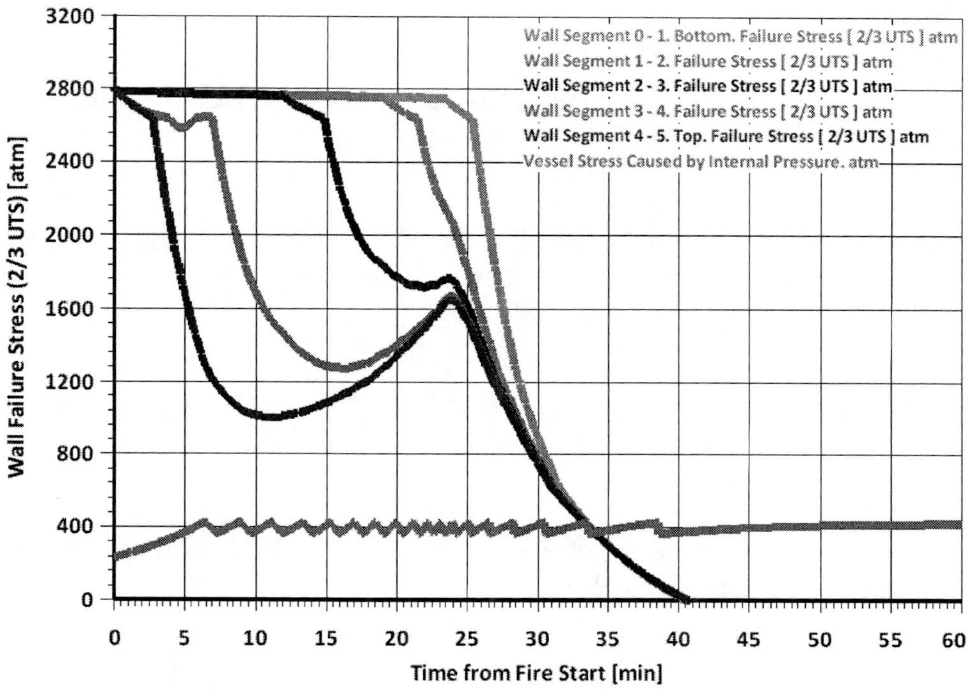

Figure 6: Dynamic graph of metal wall temperature and vessel stress versus time

5 Conclusions

Modeling overpressure scenarios dynamically can allow for a more accurate, less conservative, representation of the affected system; often resulting in a lower relief requirement as well as an understanding how the system pressure and temperature changes over time.

The availability of dynamic simulation software tools allows the relief systems analyst a broad range of options for performing pressure relief sizing calculations.

The option of dynamic pressure relief calculations can ensure that excessive design conservatism is removed, and that any unnecessary capital expenditure changes can be avoided.

6 References

[1] Pressure-Relieving and Depressuring Systems, API Standard 521, 6th Edition, January 2014

[2] Sizing, Selection and Installation of Pressure Relieving Devices - PART 1 - Sizing and Selection, API Standard 520 Part 1, 9th Edition, July 2014

[3] G. A. Melhem. Advanced Consequence Analysis: Fluid Flow, Emergency Relief System Design, Thermal Hazards Assessment, Emission, Dispersion, Fire, and Explosion Dynamics. ioMosaic Corporation, November 2015

[4] CCPS. Guidelines for Pressure Relief and Effluent Handling, 2nd edition, Center for Chemical Process Safety, American Institute of Chemical Engineers, New York, NY, 2017.

[5] Sizing, Selection, and Installation of Pressure-Relieving Devices – PART 2 – Installation, API Standard 520 Part 2, 6th Edition, March 2015

[6] ASME - Boiler and Pressure Vessel Code VIII, Division 1, - Rules for Construction of Pressure Vessels, Edition 2017

Common Misunderstanding of Overpressure Protection Scenarios for Pressure Relief Systems

Anupol Paiboonruamsilp

SCG Chemicals Co. Ltd,
271 Sukhumvit Road, Map Ta Phut, Rayong,
Thailand, 21150
anupolpa@scg.com,
anupol_che@hotmail.com

Prepared for Presentation at
American Institute of Chemical Engineers
2019 Spring Meeting and 15th Global Congress on Process Safety
New Orleans, LA
March 31 – April 3, 2019

Common Misunderstanding of Overpressure Protection Scenarios for Pressure Relief Systems

Anupol Paiboonruamsilp

**SCG Chemicals Co. Ltd,
271 Sukhumvit Road, Map Ta Phut, Rayong,
Thailand 21150**

**anupolpa@scg.com,
anupol_che@hotmail.com**

Keywords: Overpressure, Pressure Relief Systems, LOPA, Control Valve Failure, Vapor Breakthrough, Liquid Overfilling.

Abstract

Many process engineers who develop design scenarios for relief systems often have certain common misunderstandings, especially during HAZOP meetings where some credible overpressure scenarios are often considered either over-conservative or under-conservative. For instance, some engineers tend to over-credit the safety interlock function, thereby ignoring certain overpressure scenarios without considering Layer of Protection Analysis (LOPA) for the protected system,or neglecting a control valve failure scenario in either a fully closed or fully opened position. The latter scenario is also often confused with the instrument air failure scenario. Therefore, this presentation will highlight the common misconceptions of overpressure relief systems gathered from field engineers, which are related to the following issues;

- What is the relationship between Layer of Protection Analysis (LOPA) and pressure relief scenarios?
- How to identify whether an overpressure scenario is applicable or not for the protected system?
- What is the fundamental difference between control valve failure and instrument air failure?
- Does a vapor breakthrough case always generate vapor relief?
- Should a liquid overfilling scenario be considered for a vessel or column?

1 What is the relationship between Layer of Protection Analysis (LOPA) and pressure relief scenarios?

Layer of Protection Analysis (LOPA) is a semi-quantitative risk assessment measure that usually follows a qualitative risk assessment tool such as HAZOP. LOPA consists of: 1) an initiating event (IE) responsible for a chain of inadvertent events that leads to the unwanted impact, 2) Independent Protection Layers (IPL's) that help reduce the likelihood of the undesirable events, and 3) a comparison with the tolerable frequency of As Low As Reasonably Practicable (ALARP).

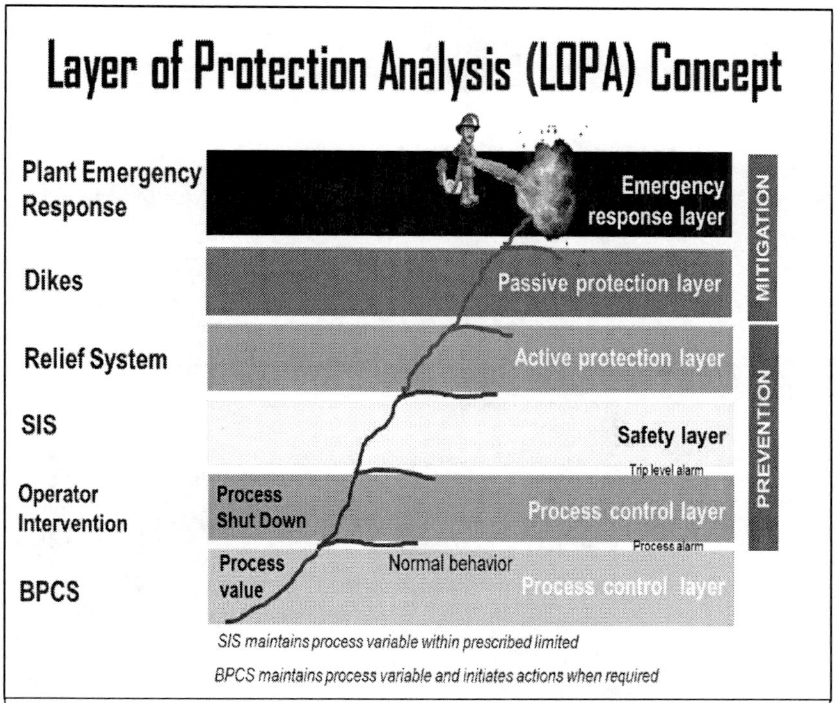

Figure 1a- Concept of Layer of Protection Analysis (LOPA)

Overpressure protection by relief system is just one of the independent protection layers (IPLs) in LOPA (Figure 1a). The purpose of an independent protection layer is to effectively prevent the consequences of the scenarios. IPLs have an associated probability of failure on demand (PFD). IPLs must be auditable and independent of all other IPLs and the initiating event being considered. The independence requirement of IPLs demands that safeguards credited as IPLs do not depend on each other. For example, the credit of a basic process control system (BPCS) response or interlock trip response will normally not be taken into account to decrease the relief load for an overpressure scenario.

If the safety instrumented system (SIS) is required to limit the relief load in the specific overpressure scenario, the probability of failure on demand of the safety interlock shall cover that of the relief device.

The example in figure 1b below considers a failure of level control loop with a wide open level control valve. This enables a higher flashing liquid to enter to the separator.

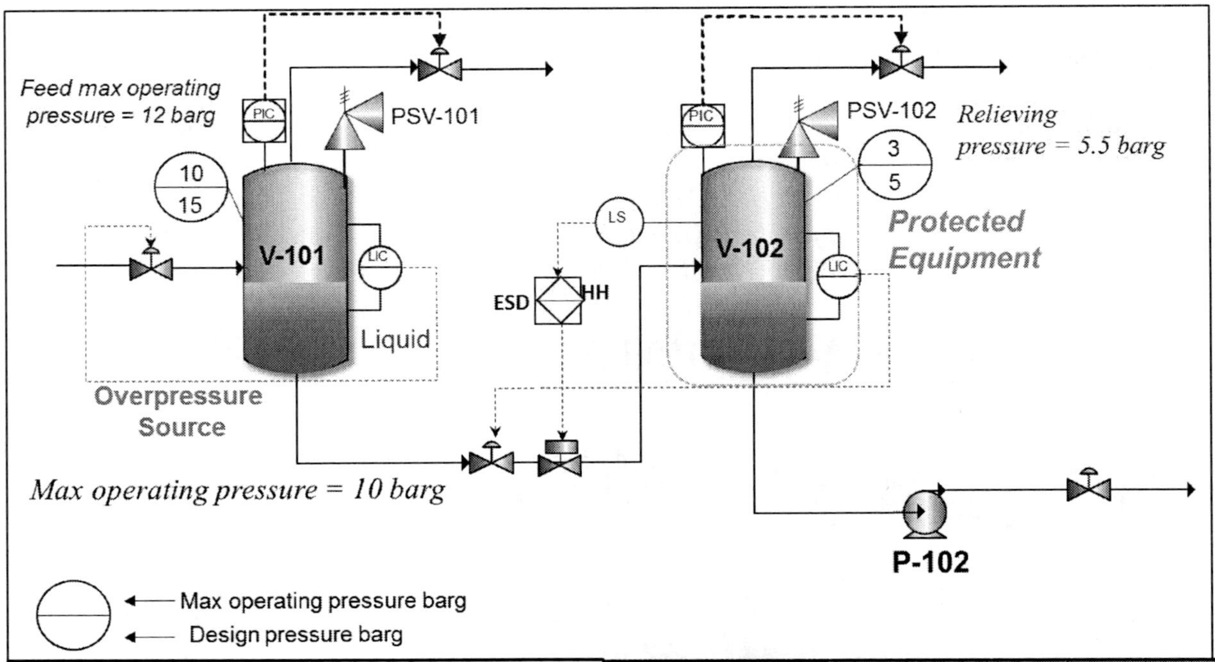

Figure 1b – Example of LOPA and the Risk Matrix

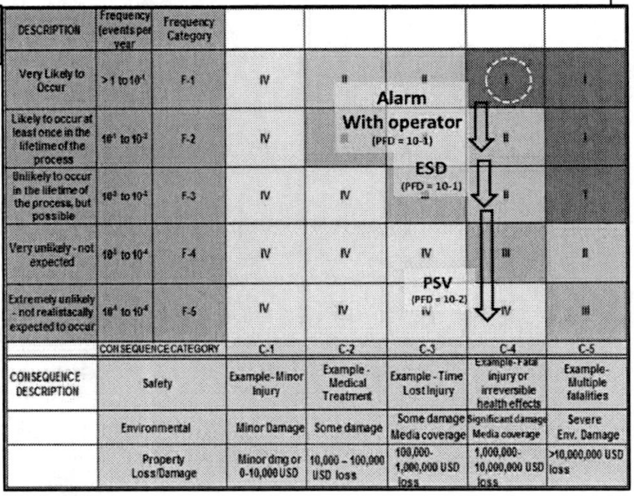

DESCRIPTION	Frequency (events per year	Frequency Category					
Very Likely to Occur	>1 to 10⁻¹	F-1	IV	II	II	I	I
Likely to occur at least once in the lifetime of the process	10⁻¹ to 10⁻²	F-2	IV	III		II	I
Unlikely to occur in the life time of the process, but possible	10⁻² to 10⁻³	F-3	IV	IV		II	I
Very unlikely - not expected	10⁻³ to 10⁻⁴	F-4	IV	IV	IV	III	II
Extremely unlikely - not realistacally expected to occur	10⁻⁴ to 10⁻⁶	F-5	IV	IV	IV	IV	III
	CONSEQUENCE CATEGORY		C-1	C-2	C-3	C-4	C-5
CONSEQUENCE DESCRIPTION	Safety		Example-Minor Injury	Example - Medical Treatment	Example - Time Lost Injury	Example-Fatal injury or irreversible health effects	Example-Multiple fatalities
	Environmental		Minor Damage	Some damage	Some damage Media coverage	Significant damage Media coverage	Severe Env. Damage
	Property Loss/Damage		Minor dmg or 0-10,000 USD	10,000 – 100,000 USD loss	100,000-1,000,000 USD loss	1,000,000-10,000,000 USD loss	>10,000,000 USD loss

Typically, a frequency of failure for BPCS is approximately 10^{-1} event per year which is very likely to occur. The consequence of overpressure in the separator possibly generates loss of containment and ends up with fire and explosion. Therefore, the unmitigated risk of the initial event is assessed as Risk Level 1, which is strongly unacceptable as per the Risk Matrix. In order to reduce the risk to at least the Risk Level 3 (ALARP), additional IPLs which consist of the following shall be provided.

- High-high level alarm and operator action, PFD = 10^{-1},
- A typical emergency shutdown (ESD) from the level interlock, PFD = 10^{-1},
- A relief valve, PFD =10^{-2}.

The reduced Risk Level will be decreased to a risk Level 4 which is acceptable.

However, there may be some design cases, in which the relief device will be avoided because of the huge relief load generated and also affecting flare capacity. Then, the PFD of the emergency shutdown (originally 10^{-1}) shall be replaced with the PFD of the relief valve (10^{-2}) to satisfy the safety integrity level (SIL) with appropriate risk analysis approach such as LOPA. Therefore, the SIL of this ESD loop shall have a final PFD of 10^{-3}.

Alternatively, as per ASME VIII paragraph UG-140, a pressure vessel does not require a pressure relief device if the pressure is self-limiting (e.g., the maximum discharge pressure of a pump or compressor), and this pressure is less than or equal to the MAWP of the vessel at the coincident temperature. Increasing the design pressure of the protected equipment to the maximum design pressure of all inlet sources will eliminate overpressure scenarios due to inlet flows. This is an inherently safer design approach to eliminate overpressure scenarios caused by a higher source pressure

2 How to identify whether an overpressure scenario is applicable or not for the protected system?

Prior to determining whether certain relief scenarios will be applicable or not, the following logical questions have to be answered;

- ✓ What is the specific protected equipment? A system boundary of the protected equipment should be properly identified.
- ✓ What is the overpressure source? All potential overpressure sources affecting the protected system should be listed.

As illustrated in Figure 2, potential overpressure sources attached to the protected equipment may come from;

1) Direct, higher design pressure sources than that of the protected equipment,

2) Inadvertent external heat input such as failure of a control valve to supply inadvertent heat input to heat exchangers, etc.

3) External fire case or thermal solar radiation on long transfer pipes.

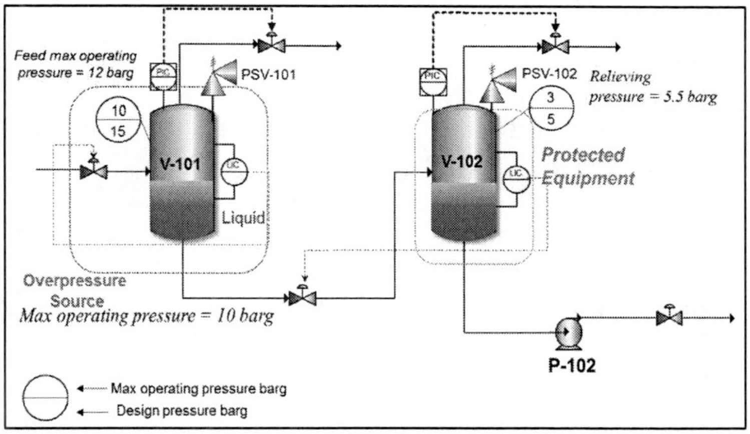

Figure 2 – Identification of the protected equipment and overpressure sources

Even through, UG-140 (1) allows an increase the design pressure of the equipment to withstand the higher inlet pressure source as an inherent safe design, an external fire case still cannot be avoided if the location of the equipment is potentially exposed to a pool fire source at or less than 7.6 m from the base of a pool fire. This is an average height of pool fire flame. An external fire case is usually considered as a basic overpressure case for equipment designed as per ASME code.

In case the location of the equipment is elevated beyond the average height of the pool fire flame and the design pressure of the equipment equal to all the higher pressure sources, a pressure relief system is no longer required.

Obviously, Figure 3 shows that not all multiple inlet sources are able to overpressure the protected equipment (V-102). Only the inlet stream with the design pressure of 25 barg is considered capable of pressurizing the protected equipment up to its relief pressure of 16.5 barg.

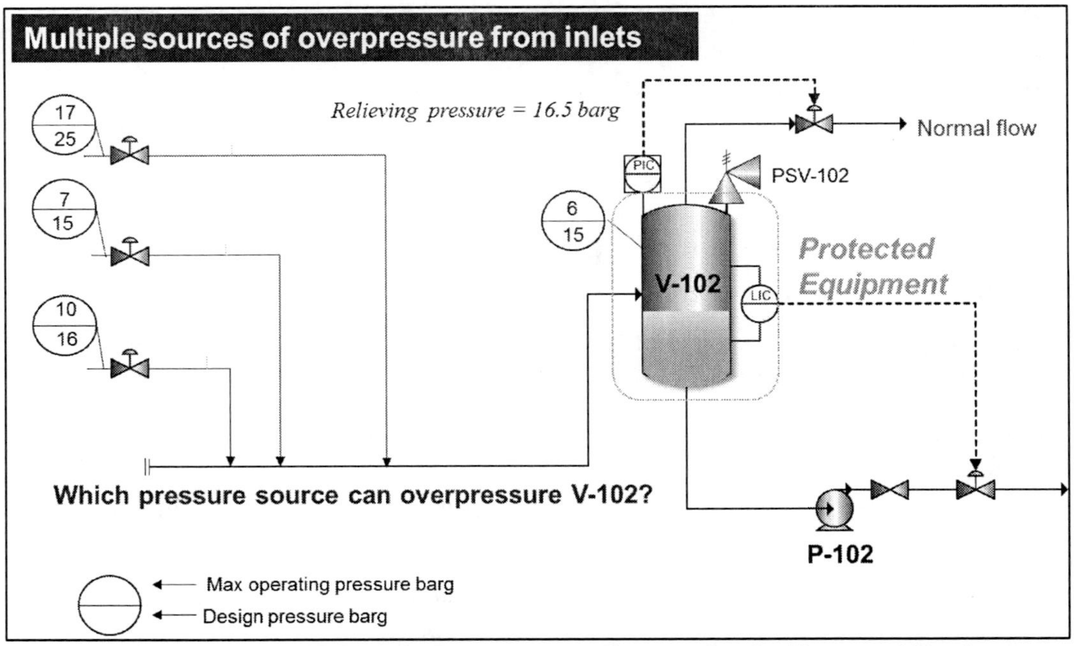

Figure 3 – Example of Multiple Overpressure Sources for the Protected Equipment

3 What is the fundamental difference between control valve failure and instrument air failure?

Any component of a basic process control loop such as instrument element, signal, DCS system, or even control valve body can fail. When speaking of a control valve failure scenario, this usually covers any failure in the whole basic process control loop. Even the control valve itself can mechanically fail by getting stuck-closed, or stuck-open regardless of its fail position during instrument air failure. What follows is a basic concept for relief load calculation in case of control valve failure.

When the transmitted signal or operating medium to a control valve fails, the conservative approach is to assume that the valve is driven to the fully open or fully closed position, whichever is more serious. Credit of the affected BPCS to decrease relief load shall not be considered. Other BPCS loop shall be considered to be in normal operation.

Failure of each control valve at the inlets and outlets around the protected equipment should be considered when determining all possible overpressure cases. When considering overpressure scenarios caused by any control valve, failure of the valve in the fully opened position or fully closed position should be assumed based on the position that causes the worst case. For example, in figure 4 the outlet level control valve of the vessel (V-101) is designed as fail-close, but the control valve should be considered wide open at the rated Cv when overpressure at the downstream equipment (V-102) is considered.

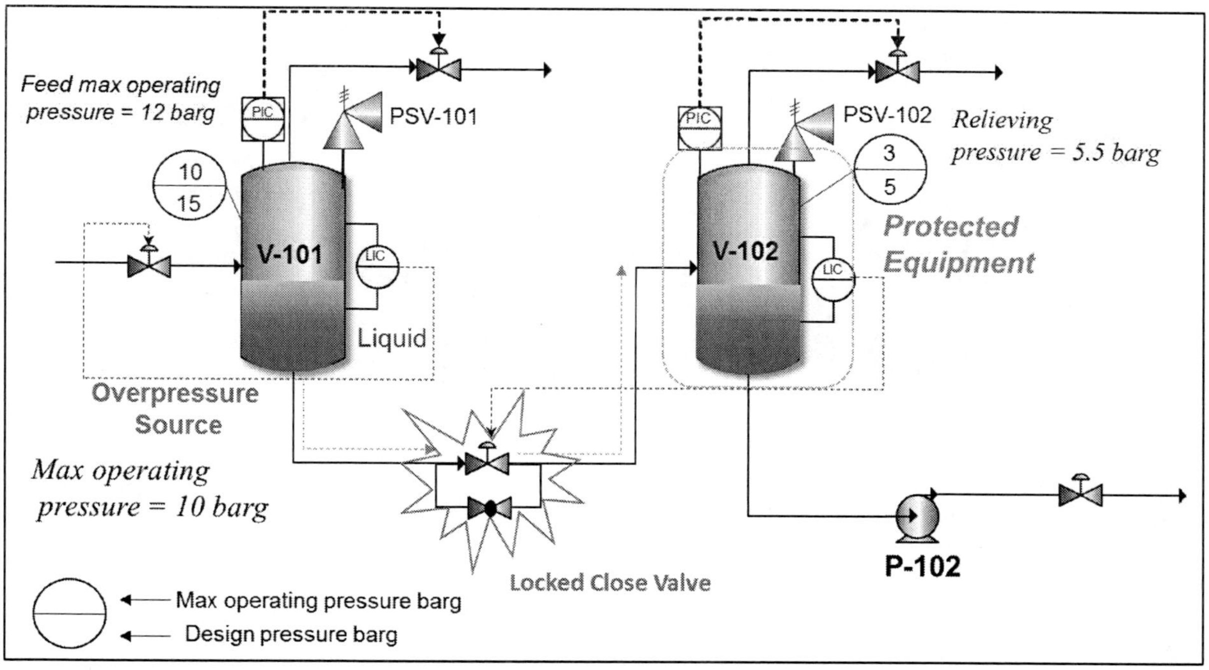

Figure 4 – Example of Control Valve Failure Case

An inadvertent opening of the bypass valve and the control valve widely opened is a credible scenario and usually governs the relief load for the low-pressure system. As per API 521, the administrative procedure such as a locked closed bypass valve can be in place if the upstream pressure sources cannot exceed the corrected hydrostatic test pressure of the downstream equipment. Since failure of the administrative control leads to unacceptable risk from a severe consequence due to loss of containment. Otherwise, both the control valve and bypass valve being wide open should be considered. Some design practices allow to consider this case of both valves inadvertently opened as a remote contingency with an allowable overpressure up to its hydrostatic test pressure. If the calculated relief load from both the control valve and the bypass valve results in an impractical relief valve size, a safety instrumented system should be considered instead.

Instrument air failure is a common utility failure case which results in simultaneous loss of instrument air to all instruments. This means that all control valves and any air-operated valves will go to their failure positions. The following concepts of total instrument failure are to be considered;

- ✓ Total relief load results from simultaneous loss of instrument power to all instruments.
- ✓ This implies the loss of instrument power + failure of all UPS supply.
- ✓ All control valves would then adopt their IA failure positions (FO or FC).

Basically, if all control valves fail positions have been designed properly, there will be no relief load through relief valves generated during instrument air failure, apart from pressure control valves intentionally deigned to vent to the flare system.

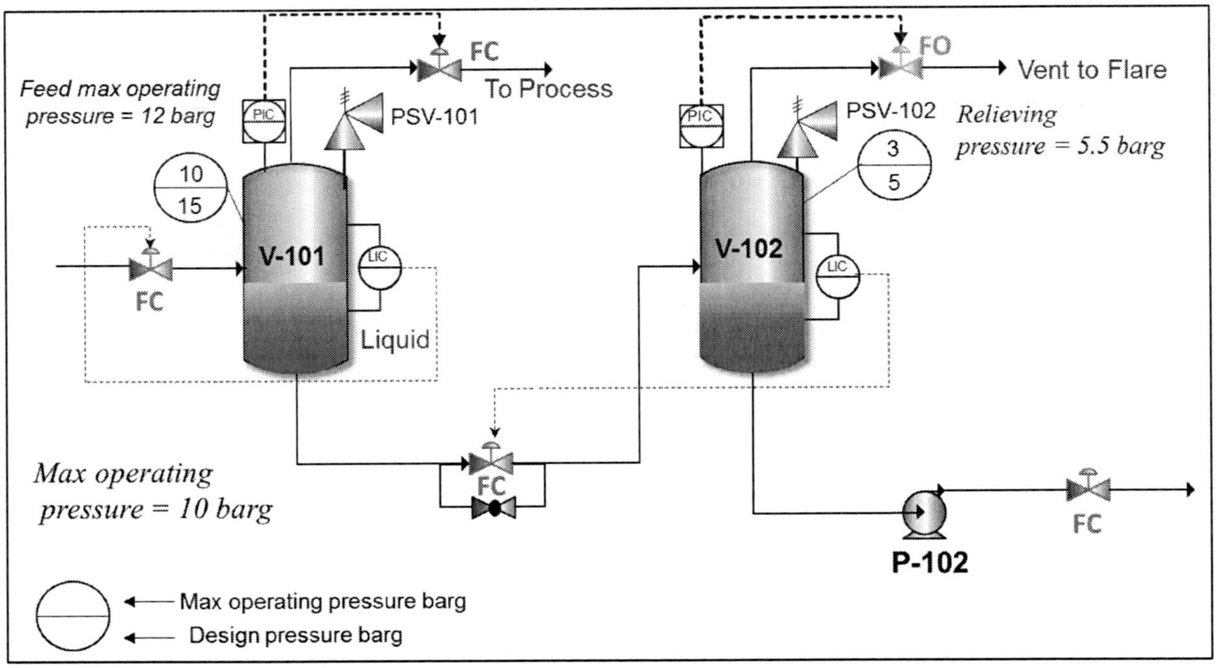

Figure 5– Example of Instrument Air Failure Case

Generally, the failure positions of all control valves will not allow any heat input into the system, while isolating the protected system by cutting off all feeds and outputs. Exceptions are vent streams to the flare or safe location (case by case). Hence, it is usually found that, after total instrument air failure, there will be no relief load through the relief valves except for pressure control loops normally venting to the flare system, whereby a high vent gas rate will pass through the pressure control valve (designed as Fail-Open) calculated with the rated Cv of the valve. In case the pressure control valve size is large, this sometimes governs the flare load capacity.

4 Does a vapor breakthrough case always generate vapor relief?

A wide open liquid control valve of the vessel (V-102) as shown in Figure 6, also causes a vapor break through case, which usually governs a relief valve sizing for V-102, especially when its manual bypass valve is also taken into account with the same Cv as the control valve.

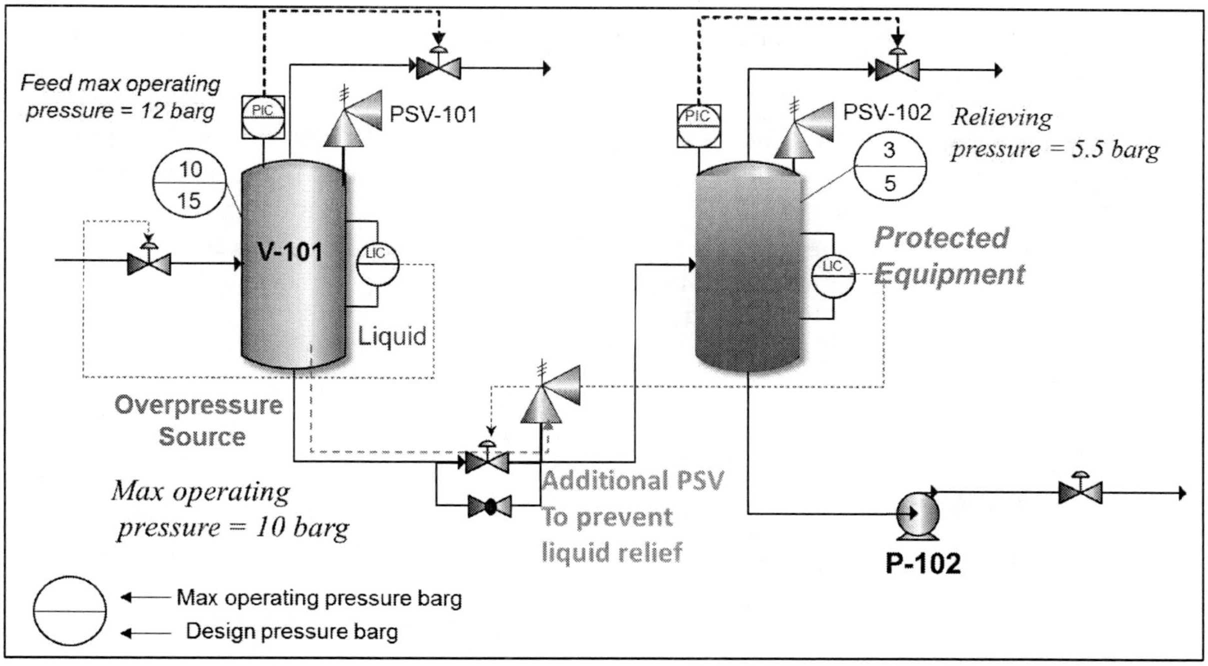

Figure 6– Example of Vapor Breakthrough Case

The general assumption for vapor breakthrough case is that only a vapor flow will break through the liquid control valve at the rated Cv and relieves at the low-pressure system. In fact, the vapor break through will happen after loss of liquid level from the high-pressure system. The liquid inventory from the high-pressure system will be flashed via the control valve and remains in the low-pressure system. It is important to also consider the liquid volume gained in the low-pressure system whether flooding liquid exists in the low-pressure system. Also, flooding liquid above the inlet nozzle results in inappropriate vapor-liquid separation and generates a two-phase relief at the low-pressure system.

The worst case scenario is total flooding in the downstream equipment prior to vapor breakthrough, which requires a liquid relief instead. The latter will result in a large relief valve size since the required relief is based on a volume of liquid equal to the volume gained by vapor entering the low-pressure system. The additional measure to avoid impractical size of the relief device on the low-pressure system is to install a relief deice immediately downstream of the control station to allow only vapor relief.

5 Should a liquid overfilling scenario be considered for a vessel or column?

Liquid overfilling especially in vessels or columns might happen as a result of stopped liquid outflows while liquid inflow continues at the design flow rates. The criteria for ignoring liquid overfilling are as follows;

- Check overfilling time in the vessel against an operator's intervention (10-30 mins depending on complexity).
- The alarm must be independent of the possible causes of failure.
  Or

- A provision of safety interlock with high integrity level to eliminate the liquid overfilling case.

If the above criteria are satisfied, liquid overfilling case can be considered mitigated. It is important to ensure that the instrument for level alarm shall be independent of other liquid level controls that cause the liquid overfilling scenario. For example, the level alarm instrument in figure 7 which is not installed on a separate bridle from the level control system cannot be taken as an independent level alarm to warn an operator in case of liquid overfilling case.

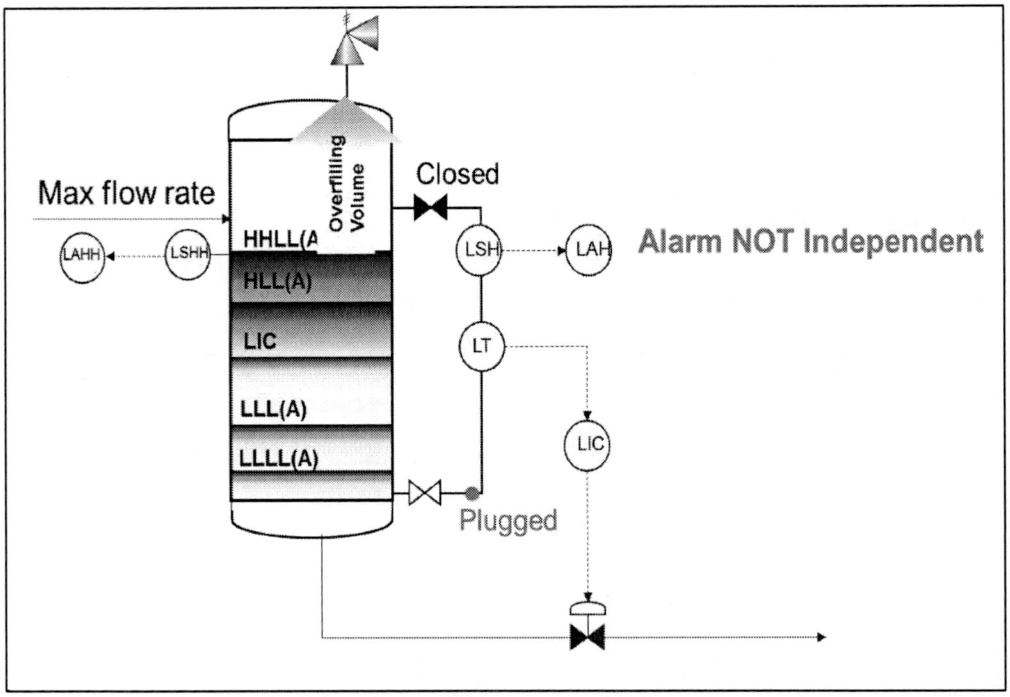

Figure 7 – Liquid overfilling case with instrumentation

6 Conclusion

All things considered, most common misconceptions of overpressure protection scenarios for relief systems are induced by misunderstanding of the concept of the independent protection layers (IPL) in the layer of protection analysis (LOPA). Process engineers should determine that each layer of protection is independently designed and do not depend on any other IPL, especially in the case of a basic process control system (BCPS) loop, a typical safety interlock and a relief valve. The safety interlock can be considered to replace some overpressure scenarios of the relief valve if the reliability of the safety instrumented system (SIS) meets the safety integrity level specified by an appropriate risk analysis such as a LOPA, per ASME Code Section VIII, UG-140.

Prior to considering any possible overpressure scenario, all inlet sources affecting a protected equipment should be identified. Basically the protected equipment can be overpressured by any inlet streams having a higher design pressure than that of the equipment. Increasing the design pressure of the low-pressure equipment to withstand the higher pressure source is also acceptable as per ASME Code Section VIII, UG-140. However, an external fire case still exists unless the equipment will be elevated above the average flame height from a pool fire

In addition, the fundamental difference between control valve failure and instrument air failure is usually confusing in their definitions. API -521 defines the concept of control valve failure case as any failure from some instrument element, signal, DCS system, or even control valve body. This definition is unlike instrument air failure which is defined as a common utility failure. This generates multiple relief loads from relief valves in case control valves are not designed properly in a fail-safe position.

Both inadvertent opening of a bypass valve and a wide open control valve are also credible scenarios. Because full opening of both control valve and bypass valve usually provides a huge relief load. An administrative procedure to prevent an inadvertent opening of the bypass is allowed in case the upstream pressure inlets cannot exceed the hydrostatic test pressure of the low-pressure system.

The failure of liquid control loop system results in a vapor breakthrough case at the low-pressure system. The vapor from the high-pressure system will pass through the liquid control valve at the rated Cv or simultaneously through the bypass valve when loss of the liquid level at the high pressure system. It is important to determine whether the low-pressure system will be flooded with remaining liquid level in the low-pressure pressure including the liquid inventory from the high-pressure source. The remaining liquid level can cause inappropriate vapor-liquid separation and results in two-phase relief. In case of total flooding at the low-pressure system, vapor breakthrough will push the flooding liquid through the relief device as a liquid relief. The latter will provide a large size of relief device. The alternative solution is to install the relief device immediately after the control valve station to allow vapor breakthrough relief only.

A liquid overfilling scenario for a vessel or column is a credible scenario unless a response time for operator's intervention should be sufficient and over about 10-30 minutes depending on system complexity and operating skills. Moreover, it is important to ensure that the level alarm instrument to warn the operator shall be completely independent of other level control systems, including installation on a separated bridle. Otherwise, a safety instrumentation with a high integrity level shall be designed to eliminate the liquid overfilling scenario or protected by the relief valve instead.

7 References

[1] ASME Section VIII, Division 1, UG-140, Overpressure Protection by System Design.

[2] Pressure-Relieving and Depressurizing Systems, API STD 521, Sixth Edition, January 2014.

[3] CCPS. *Layer of Protection Analysis, Simplified Process Risk Assessment.* Center for Chemical Process Safety, American Institute of Chemical Engineers, New York, NY. (2001)

Evaluating Reactive Relief Systems with Unstable Intermediates

Stacy Fields
Eastman
200 S Wilcox Dr, Kingsport, TN 37660
sfields@eastman.com

Prepared for Presentation at
American Institute of Chemical Engineers
2019 Spring Meeting and 15th Global Congress on Process Safety
New Orleans, LA
March 31 – April 3, 2019

Approved For External Use

Evaluating Reactive Relief Systems with Unstable Intermediates

Stacy Fields
Eastman
200 S Wilcox Dr, Kingsport, TN 37660
sfields@eastman.com

Keywords: Reactive relief, PSI

Abstract

Evaluating reactive systems for relief presents many complexities when compared with more common single-phase flow, non-reactive systems. In such cases it is important that the designer understand the physical and reactive characteristics (including kinetics) of the materials as well as all reactions that could occur. Obtaining an accurate model of the kinetics of a reaction presents its own unique challenges. However, even when kinetics can be measured, the reaction mechanism may include unstable intermediates for which little or no physical property data exists. Due to the instability of some of these species, measuring these parameters may be costly, difficult, or impossible. Consequently, while the reactive dynamics of the material can be predicted using known kinetics, the physical behavior cannot. Even using the most sophisticated dynamic simulation tools available, there is still a critical knowledge gap preventing the proper evaluation of the system. This paper details one method of developing useable physical property predictions for unknown components and applying them, in conjunction with measured kinetic data, to more fully evaluate a system for reactive relief cases.

1 Introduction

Evaluating reactive systems for relief is known to present many complexities compared with more common single-phase flow and non-reactive systems. In such cases it is important that the designer understand the physical and reactive characteristics of the materials, as well as all reactions that could occur and the kinetics of those reactions. Obtaining an accurate model of the kinetics of a reaction presents its own unique challenges. However, even when kinetics can be measured, the reaction mechanism may include unstable intermediates for which little or no physical property data exists. Due to the instability of some of these species, measuring these parameters may be costly, difficult, or impossible. Consequently, while the reactive dynamics of the material can be predicted using known kinetics, the physical behavior cannot. It is critical that this knowledge gap be filled thoughtfully and intentionally. Any evaluation, no matter how sophisticated, that is based on inaccurate physical properties is meaningless or as the computer science adage goes, "garbage in, garbage out".

2 Review of Physical Properties and Thermodynamics for Relief Applications

To model the physical behavior of any chemical in a simulation, both pure component physical properties and vapor-liquid equilibrium (VLE) must be accurately predicted. To understand how components will behave during a runaway reaction scenario, it is critical to understand the volatility and heat transfer. Consequently, the key pure component temperature dependent parameters that must be estimated for relief applications are heat of vaporization, vapor pressure, heat capacity, liquid viscosity, vapor and liquid density, and thermal conductivity. Pure component properties are widely available in published databanks (DIPPR, NIST, Dortmund, etc.) and can also be measured experimentally.

There are three widely accepted options for phase equilibrium estimations: ideal, equation of state, and activity coefficient. The ideal method, as the name implies, assumes that both the liquid and gas phases behave ideally. The ideal gas equation works well for gases at vacuum/ low pressure or very high temperatures. Ideal liquid behavior can be assumed when interactions between molecules are known to be small, such as when molecules of similar size and character are mixed together.

Equation of state models characterize the state of matter of a material under a given set of physical conditions (usually pressure, volume, and temperature). They assume that the non-ideality in the vapor phase and the liquid phase is the same. There are many different equation of state models including Van der Waals, Redlich-Kwong, and Peng-Robinson to name just a few because there is no one model that accurately predicts properties for all substances at all conditions. These models work well for many applications, particularly in the oil and gas industry, but do not work well for systems with two liquid phases.

Finally, phase equilibrium can be estimated using activity coefficient models such as the extended Raoult's law, which characterize a substance's deviation from ideal behavior:

$$y_i \varphi_{P,i} P = x_i \gamma_i P_i^{sat}$$

where y_i is the mole fraction of component *i* in the gas phase, $\varphi_{P,i}$ is the fugacity coefficient, P is the pressure, x_i is the mole fraction of component *i* in the liquid phase, γ_i is the activity coefficient, and P_i^{sat} is the vapor pressure of component *i*.

Activity coefficient models differ from equations of state in that they allow for the predicted non-ideality in the liquid phase to be different from the predicted non-ideality in the vapor phase. Other examples of activity coefficient models are UNIQUAC, NRTL, and UNIFAC. These types of models tend to be a better choice for systems with a high degree of non-ideality, such as those containing polar compounds.

Each chemical in the system has unique physical properties that are related to the structure of the molecule. Examination of homologous series reveals that these properties vary predictably. As such, a method for approximating physical properties of unstable intermediates can be developed using a group contribution approach. Probably the most well-known example of the application of a group contribution approach is UNIFAC. This model uses functional groups attached to the model to predict molecular interactions.

3 Discussion of Method

3.1 Tools

SuperChems™ Expert is one commercially available software package available for pressure relief and flare systems evaluation and design [1] and is the primary tool used for relief evaluation at Eastman. It is particularly useful in evaluating reactive relief scenarios because of its dynamic capabilities. One major drawback is that there is only one option for estimating phase equilibria available in SuperChems™, a modified Peng-Robinson equation of state. While this eliminates the need to decide which method of estimating phase equilibria best fits the system being evaluated, it is somewhat limiting, particularly for specialty chemical applications. SuperChems™ does have the ability to regress parameters for its equation of state from imported pressure, temperature, and equilibrium composition (PTxy) data, but the process for importing this data is time consuming and prone to error.

The other software package used in relief evaluations is Aspen Plus® [2]. Aspen has the advantage of offering multiple phase equilibrium models, both equation of state and activity coefficient. Aspen also facilitates adjusting and comparing properties of interest, regressing parameters from easily imported data sets, and exporting PTxy data sets which can then be imported into SuperChems™. However, Aspen Plus® was primarily developed as a steady-state design tool and is not well suited to dynamic simulations such as reactive relief evaluation.

3.2 Methodology

To develop physical property estimates for unknown components, a pseudo component is created to represent each unknown component by starting with a reference component and adjusting the physical properties to approximate the unknown component, as illustrated in Figure 1. The reference component should be the most structurally similar chemical with data available, either in existing databanks or measured experimentally. Estimates must be developed for both key pure component properties and vapor-liquid equilibrium (VLE). This analysis is usually done in Aspen Plus® first. Once the properties are developed, PTxy data can be exported from Aspen. This data must then be carefully manipulated into the required format for the input file SuperChems uses to read in data and then regress binary interaction parameters for use in its equation of state. From there, the foundation is set for the rest of the relief evaluation.

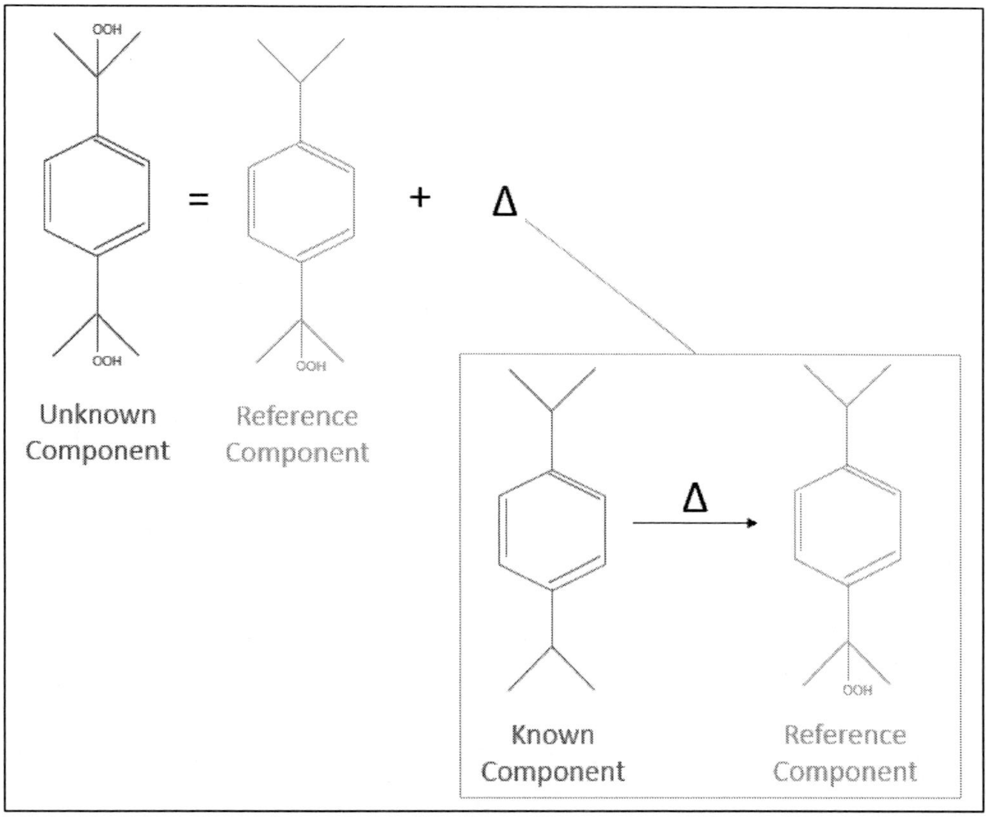

Figure 1. Illustration of Method

4 Pure Component Properties

It is possible to use functional groups and a group contribution method to estimate pure component physical properties, much the same way that UNIFAC estimates liquid phase activity coefficients. This task may seem daunting when one considers the often-complex reaction networks with multiple side and degradation reactions common in reactive relief evaluation. However, the products of these reactions are often isomers or oligomers with a small number of functional groups arranged in different configurations, consequently the number of analyses that need to be completed collapses quickly.

To understand how components will behave during a runaway reaction scenario, it is critical to understand the volatility and the enthalpy balance. Consequently, the most important pure component temperature dependent parameters that must be estimated are heat of vaporization, vapor pressure, and heat capacity.

The study of homologous series of chemicals of similar size with different types of functional groups reveals that the change in heat of vaporization with the addition of each type of functional group is fairly constant, as shown in Figure 2.

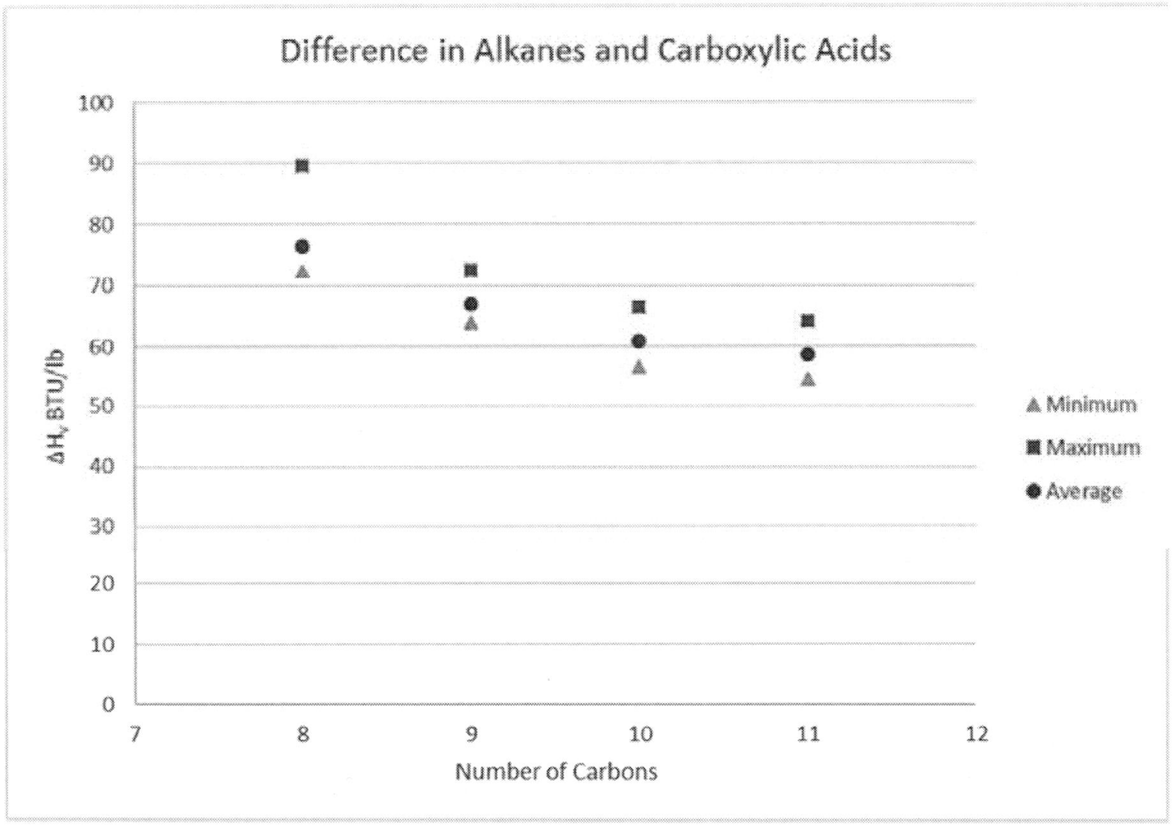

Figure 2. Change in ΔHv with the Addition of a Carboxylic Acid group to an Alkane over 50-250 °C Range

Therefore, a simple additive group contribution method can be used for this property. The estimated heat of vaporization for each pseudo component is the value for the reference component modified to account for the additional functional groups the unknown component has relative to the reference component.

The additive group contribution method is inappropriate for estimating the vapor pressure. The simplified Antoine equation shows the relationship between heat of vaporization and vapor pressure:

$$\ln P_{sat} = A + \frac{B}{T_{absolute}} \text{ where } B = \frac{-\Delta H_{vap}}{R}$$

With this in mind, the B parameter can be calculated using the molar heat of vaporization estimated previously. The A parameter can then be varied to match the predicted vapor pressure curve.

The heat capacity correction factor can be determined, as with heat of vaporization, by studying homologous series of chemicals similar in size to the unknown chemicals of interest. For large molecules, it has been observed that the mass basis heat capacity does not vary significantly with increasing molecule size. Table 1 shows the absolute value and percent difference in heat capacity on a mass basis for two large alcohols, 1-nonanol and 1-tridecanol, over a 130°C temperature range. For most of the range, the percent difference between the heat capacity of the C9 and the C13 is less than 2% and at no point is it greater than 10%.

Table 1. Comparison of Liquid Heat Capacity Across a Homologous Series

Temperature °C	1-nonanol $C_9H_{20}O$ BTU/lb·F	1-tridecanol $C_{13}H_{28}O$ BTU/lb·F	% difference
50	0.610917	0.561649	8.06
56.84	0.627101	0.591394	5.69
63.68	0.643152	0.617504	3.99
70.53	0.658738	0.640234	2.81
77.37	0.67358	0.659837	2.04
84.21	0.687446	0.676567	1.58
91.05	0.700151	0.690676	1.35
97.89	0.711561	0.702419	1.28
104.74	0.721592	0.71205	1.32
111.58	0.730207	0.719823	1.42
118.42	0.737421	0.725989	1.55
125.26	0.743295	0.730805	1.68
132.11	0.747939	0.734523	1.79
138.95	0.751512	0.737397	1.88
145.79	0.754226	0.73968	1.93
152.63	0.756337	0.741627	1.94
159.47	0.758155	0.74349	1.93
166.32	0.760032	0.745525	1.91
173.16	0.762375	0.747983	1.89
180	0.765639	0.751119	1.90

If this is found to be the case, it may be acceptable to use the uncorrected values from the reference component. However, it is important to always verify that this assumption is valid for the system being evaluated before applying it. For example, in relief scenarios with a large temperature rise, heat capacity varies more over the range and therefore becomes more important. Other parameters which impact the relief design to a lesser extent than the key parameters previously discussed include liquid viscosity, vapor and liquid density, and thermal conductivity. Most of these parameters should not vary to an extent which would invalidate the relief evaluation. Obviously, this assumption should be validated before applying it to any calculation, but should it prove to be true, using the uncorrected reference component values should be adequate.

In some systems, there are chemical species with very low vapor pressure that will not vaporize, even under runaway reaction conditions, and therefore do not participate in VLE. For these species, the heat capacity is not as critical. Time can be saved on the analysis by assuming the reference component's heat capacity, particularly if the reference component is smaller than the unknown component. It is not acceptable to leave these components out of the analysis entirely as they will act as diluent in the reactive relief scenario in both the phase equilibria and the enthalpy balance.

5 Vapor Liquid Equilibrium

When thinking about vapor liquid equilibrium in the context of evaluating reactive relief scenarios, the first thing to consider is what non-idealities the system is likely to exhibit. It is often the case in these scenarios that assuming ideal gas behavior for the vapor phase is acceptable and only the liquid phase non-idealities need to be quantified. In such cases, Raoult's Law can be modified to:

$$y_i P = x_i \gamma_i P_i^{sat}$$

The liquid and vapor mole fractions and the partial pressures can be calculated if the activity coefficient is known. Since activity coefficient models simplify at infinite dilution, infinite dilution activity coefficients are the most commonly used in industry.

There are several different activity coefficient models, including UNIQUAC and NRTL, which can be employed for known components with available model parameters. While Aspen Plus® has built-in parameters for a large library of components, it has been found that they are not reliable at the extremes of composition nor for all operating conditions. As such, the most accurate method of obtaining parameters for these models is to regress them from measured PTxy data across the expected range of conditions.

If no data is available, UNIFAC can potentially be used to estimate parameters. However, UNIFAC does have limitations because it requires parameters for the interactions between functional groups for its activity coefficient predictions. While researchers have generated hundreds of interaction parameters for pairs of functional groups, there are millions of different chemicals. Developing interaction parameters for every possible functional group pair is infeasible if not impossible. If no parameters have been developed in the literature for two functional groups, the results from UNIFAC can be inaccurate to the point of being unusable. It is important to check the UNIFAC interaction mapping for each molecule to be sure that it is representative of the component's structure.

When determining the plausibility of activity coefficients, it may be useful to keep some guideline values in mind. Very dissimilar, strongly non-ideal molecules have coefficients greater than 5. Moderately non-ideal pairs have coefficients in the 2 to 5 range. Similar components will have activity coefficients in the 1.1 to 1.5 range. An activity coefficient equal to 1 means the components interact ideally. Coefficients less than 1 exist, as is the case with formic acid and water, but are very rare in industry.

For example, these molecules differ only in the addition of one hydroperoxide group:

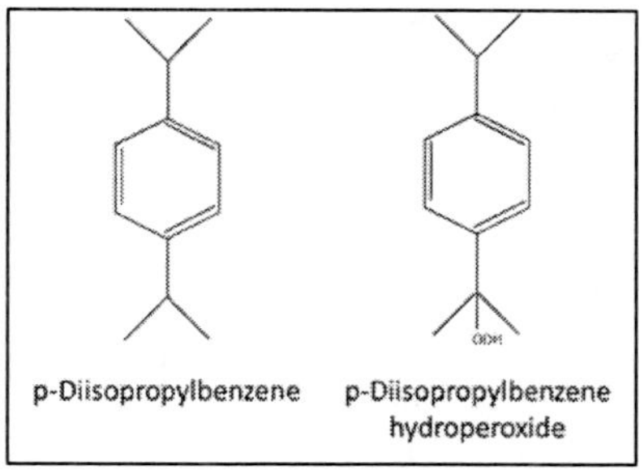

Figure 3. Example Molecules

By inspection, it would be reasonable to expect the similar molecules in Figure 3 to interact in a moderately non-ideal way. However, when Modified UNIFAC 2018 [3] is used to estimate activity coefficients, the result is surprising, as seen in Figure 4.

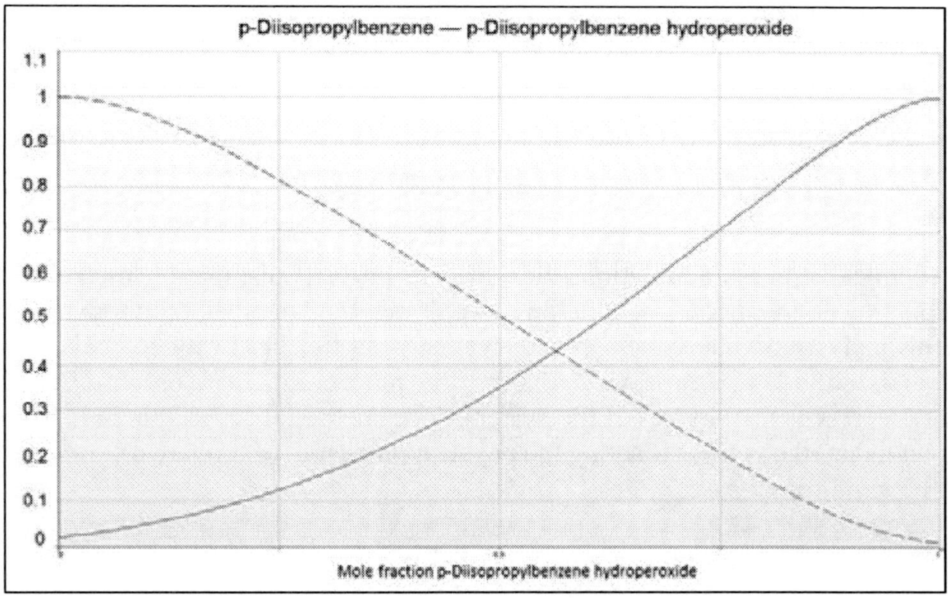

Figure 4. Initial Modified UNIFAC 2018 Estimate of Activity Coefficients

Based on what is known about activity coefficients, this result is highly suspect. The most likely cause is that the molecule has not been correctly mapped. When the functional groups that the UNIFAC utility mapped were examined, it was found that the utility had mapped the wrong peroxide functional group to represent this molecule. After this was corrected, the result was in line with expectations, as per Figure 5.

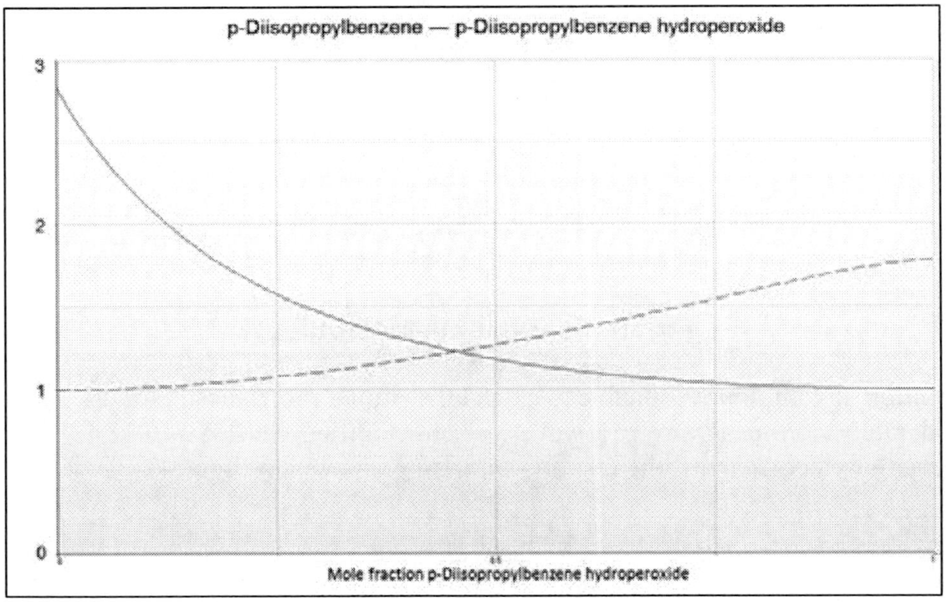

Figure 5. Corrected Modified UNIFAC 2018 Estimate of Activity Coefficients

6 Conclusion

In any chemical process simulation, the physical property estimation methods are the foundation the entire model is built on. Applying fundamental thermodynamics and thoughtful analysis of the system being evaluated is the best way to make sure that foundation is solid. It is critical that both the pure component properties and VLE accurately represent the physics of the system. The method discussed here is just one of many potential methods for estimating the physical properties of unknown components and may not be valid for all cases.

7 References

[1] ioMosaic Corporation, Houston, TX, USA.

[2] Aspen Technology, Inc., Cambridge, MA, USA.

[3] Dortmund Data Bank Software & Separation Technology GmbH, Oldenburg, Germany.

The Effects of Oxygen Enrichment on the Explosibility Properties of Flammable Gases and Vapors

Richard W. Prugh
DEKRA Process Safety
113 Campus Drive
Princeton, NJ 08540
www.dekra-process-safety.com
prughrw@aol.com

Prepared for Presentation at
American Institute of Chemical Engineers
2019 Spring Meeting and 15th Global Congress on Process Safety
New Orleans, LA
March 31 – April 3, 2019

The Effects of Oxygen Enrichment on the Explosibility Properties of Flammable Gases and Vapors

Richard W. Prugh
DEKRA Process Safety
113 Campus Drive
Princeton, NJ 08540

Keywords: Flammables, Oxygen, Explosion

Abstract

In the chemical industry, oxygen-containing chemicals are used in the production of partial-oxidation products, including many in the pharmaceutical industry. The vapor spaces in oxidation reactors can contain high concentrations of oxygen gas, in combination with flammable vapors. Thus, it is important to determine the explosibility properties of organic-compound vapors in atmospheres containing high oxygen concentrations.

1. Introduction

In the chemical industry, "nascent" oxygen is involved in the production of partial-oxidation products, including many in pharmaceutical operations. The oxygen is initially in the form of peroxides, such as hydrogen and sodium peroxides, and also as chlorates and perchlorates, periodates, nitrates, and dichromates [1]. The vapor space in oxidation reactors can contain high concentrations of oxygen gas, in combination with the flammable vapors of organic compounds. To evaluate the possible explosion hazards of such mixtures of fuel and oxidizer, it is important to determine the explosibility properties of organic compounds in atmospheres containing high oxygen concentrations.

The basic flammability/explosibility data for this study are from the following references:

Table 1. References for Flammability Data

REFERENCE	FLAMMABILITY/EXPLOSIBILITY PROPERTY					
	Flammable Limits	Auto-Ignition Temperature	Minimum Ignition Energy	Maximum Explosion Pressure	Deflagration Index Kg	Oxygen Conc. Limit
NFPA 49; NFPA 325 [2]	√	√				
NFPA 68 [3]				√	√	
NFPA 69 [4]						√
NFPA 77 [5]	√ *		√			
NFPA 497 [6]	√	√	√			
USBM 503 [7]	√ *					√
USBM 627 [8]	√ *	√ *				
Babrauskas [9]	√	√ *	√ *			
Britton [10]	√ *		√ *			√
Crowl and Louvar [11]	√ *	√	√	√	√	√
Yaws [12]	√	√				

* Limited data for oxygen

Several methods are presented in this paper for estimating explosibility properties for preliminary process design. However, it is prudent to confirm the fuel/oxygen relationships with tests where process safety depends on the reliable control of oxygen concentration.

2. Flammability/Explosibility Properties

The explosibility properties of combustible materials can be divided into Ignition Sensitivity and Explosion Severity [13, 14]. The Lower Flammable Limits, the Autoignition Temperatures, and the Minimum Ignition Energies would be important in evaluating the Ignition Sensitivity. The Maximum Explosion Pressure and the Deflagration Index – a measure of burning rate – would be important in evaluating the Explosion Severity. The following sections describe the relationships of these properties to oxygen concentration.

2.1 Lower Flammable Limit

Data from many sources [15, 16, 17, 18] indicate that there is no significant effect of oxygen concentration on the Lower Flammable Limit [LFL] from the 21 volume-percent oxygen in air to 100% oxygen. At oxygen concentrations greater than the LFL in air, the specific heat of oxygen is about the same as the specific heat of nitrogen; that is, excess oxygen behaves as a diluent at the LFL.

For comparison with the Minimum Explosible Concentrations of combustible dusts and for evaluating the hazards of mixtures of flammable gases/vapors of combustible dusts, the values of LFL can be converted to mass densities by:

$$C = [\,LFL / (100 - LFL)\,]\,[\,MW / 0.024\,] \quad \text{grams per cubic meter}$$

where 0.024 is the number of cubic meters per gram-molecular weight [at 25°C and one atmosphere pressure (14.7 psia)].

For example, the LFL for propane in air is 2.1 volume-percent and, for the molecular weight of 44.1, the Minimum Explosible Concentration would be given by:

$$C = [\,2.1 / (100 - 2.1)\,]\,[\,44.1 / 0.024\,] = 39 \text{ grams per cubic meter}$$

Since the LFL is not significantly affected by oxygen concentration, oxygen enrichment would not affect the ventilation and dilution or inerting precautions that should normally be taken for a flammable vapor or gas.

2.2 Autoignition Temperature

The Autoignition Temperature [AIT] of a fuel is a kinetic phenomenon; that is, the AIT is low for very-long exposures to a given temperature, and it is quite high for short exposures to a given temperature. This relationship can be shown to follow the Arrhenius "law" [19], as:

$$r = A\,e^{-Ea/RT} \quad \text{fraction per second}$$

where A is the pre-exponential factor (in "per second");
Ea is the activation energy (in calories per gram-mole);
R is a constant (1.987 calories per gram-mole per °K); and
T is the temperature (in °K).

If equal reaction rates are assumed for combustion in air and oxygen, at two different temperatures, a value for the activation energy can be obtained, from:

$$Ea = 1.987\,[\,\ln(100/21)\,] / [\,(1/T_{AIR}) - (1/T_{O2})\,] = 3.1 / [\,(1/T_{AIR}) - (1/T_{O2})\,] \quad \text{cal./gmol}$$

The Autoignition Temperatures for 90 chemical compounds in air and in 100% oxygen are available [20]. The average AIT in air was calculated to be about 394°C [about 667°K], and the average AIT in oxygen was about 361°C [about 634°K]. Thus, the differences in AIT for air and oxygen is quite small, at about 5% lower.

The calculated activation energy Ea, as based on the referenced data, would be about 40,000 calories per gram mole, which is a "reasonable" value for an oxidation/ combustion reaction [21].

Thus, the AIT for a given oxygen concentration can be estimated from:

$$AIT_{O2} = AIT_{AIR} / [\ 1 + (\ 1.987\ AIT_{AIR} / 40,000\ )\ (\ln (C_{O2}/21)\ ] - 273 \qquad {}^{o}C$$

Oxygen enrichment would not affect the precautions that should normally be taken for a flammable or combustible vapor, such as avoiding exposure to hot surfaces in ovens, in dryers, and on electrical equipment such as motors.

2.3 Minimum Ignition Energy

Previous investigators had found that the Minimum Ignition Energy [MIE] was inversely related to oxygen concentration [22]. The Minimum Ignition Energies for 13 chemical compounds in air and in 100% oxygen are available [23]. The average MIE in air [excluding hydrogen (very low) and trichloroethylene (very high)] was calculated to be about 0.30 millijoule, and the average MIE in oxygen was about 0.0027 millijoule. Thus, oxygen concentration has a very strong effect on MIE, with about a factor of 100 reduction for 100% oxygen. The exponent for oxygen concentration for the 13 chemical compounds [including hydrogen and trichloroethylene] is about 2.9.

Thus, the MIE for a given oxygen concentration can be estimated [+/- 50%] from:

$$MIE_{O2} = MIE_{AIR} / [\ (\ C_{O2} / 21\ )^{2.9} = 6,200\ MIE_{AIR} / C_{O2}^{\ 2.9} \qquad \text{millijoule}$$

As indicated above, oxygen enrichment would have a significant effect on the precautions [such as inerting] that should be taken for flammable gases and liquid against very weak electrostatic [and other] ignition sources, including corona discharges [24].

2.4 Maximum Explosion Pressure

The Maximum Explosion Pressure [Pmax] – for combustion in air or in an oxygen-enriched atmosphere – can be calculated rather accurately if the heat of formation of the fuel is known and the degree of dissociation of the carbon dioxide and steam combustion products can be determined. For combustion in an oxygen-enriched atmosphere, the degrees of dissociation are very high; thus, the Pmax in 100% oxygen could be expected to show a rather small increase in pressure as compared to combustion in air.

There is a direct relationship between flame temperature and explosion pressure, as governed by the Ideal Gas Law, as:

$$P_{Max} = 14.7 \, (T_{FLAME} + 273) \, / \, (25 + 273) - 14.7 \qquad \text{pounds/sq.in. [psig]}$$

The available flame-temperature explosion-pressure data are as follows [25]:

Table 2. Flame Temperatures and Explosion Pressures for Fuel/Oxygen Mixtures

FUEL	FLAME TEMPERATURES (°C)			MAX. PRESSURES (barg)		RATIO (bara/ bara)	IDEAL-GAS-LAW RATIOS (°K/bara)	
	In Air	In Oxygen	Ratio [°K/°K]	In Air	In Oxygen		In Air	In Oxygen
Methane	1,950	2,705	1.37	7.1	10.4	1.41	274	261
Propane	1,988	2,822	1.37	7.9	10.2	1.26	254	276
Ethylene	2,088	2,902	1.34	8.0	10.7	1.30	262	271
Acetylene	2,262	3,069	1.32	10.6	11.2	1.05	219	274

Thus, approximate values for Pmax in mixtures of oxygen and in air would be given by:

$$P_{Max,O2} = [\, (T_{FLAME} + 273) \, / \, 261 \,] - 1 \qquad \text{barg}$$

Thus, if the flame temperature can be calculated, from heats of formation and specific heats – with consideration of dissociations – the maximum explosion pressure for a hydrocarbon can then be calculated. It is likely that this relationship would also apply to other hydrocarbons and carbohydrates. Also, as indicated in the preceding Table 2, the values of P_{MAX} in 100% oxygen are about 30% greater than the value in air [excluding acetylene].

The data in Table 2 above can be used to estimate the Maximum Explosion Pressure [P_{MAX}] for a given concentration of oxygen as:

$$P_{MAX,C} = 0.035 \, C_{O2} + 0.92 \, P_{MAX,AIR} \qquad \text{bara}$$

The following Table presents the results of calculations, based on the above relationship, for two typical values for P_{MAX} in air [26]:

Table 3. Maximum Explosion Pressures for Fuel/Oxygen Mixtures

$P_{MAX,AIR}$	MAXIMUM EXPLOSION PRESSURES IN FUEL/OXYGEN MIXTURES (barg)							
	30 %O_2	40 %O_2	50 %O_2	60 %O_2	70 %O_2	80 %O_2	90 %O_2	100 %O_2
7 barg	7.4	7.7	8.1	8.4	8.8	9.1	9.5	9.8
8 barg	8.3	8.6	9.0	9.3	9.7	10.0	10.4	10.7
9 barg	9.2	9.6	9.9	10.3	10.6	11.0	11.3	11.7
10 barg	10.1	10.5	10.8	11.2	11.5	11.9	12.2	12.6

Thus, oxygen enrichment would have a significant effect on the extent of damage that could result from ignition of confined flammable gases and vapors, and could affect the design of explosion-protection systems and devices, such as explosion vents.

2.5 Deflagration Index, Kg

The Deflagration Index [Kg for flammable gases and vapors] is a measure of the speed of combustion and, thus, is directly related to the speed of flame through a mixture of fuel and oxidant. In the absence of strong turbulence, the flame speed is related to the fundamental burning velocity [FBV] by the ratio of the densities of the unburned mixture and the density of the combustion gases, by [27]:

$$S_{FLAME} = FBV\ (d_{UNBURNED}/d_{BURNED}) \quad \text{centimeters per second}$$

Typically, the ratio of the flame speed to FBV is about 6.

Previous investigators had observed that there is a relationship between the FBV and the Minimum Ignition Energy [MIE], as given by [28]:

$$FBV = K/MIE^{0.8} \text{ centimeters per second}$$

Some data are available for both the FBV and the MIE, as indicated in the following Table [29] :

Table 4. The Relationship Between FBV And Oxygen Concentration

GAS/VAPOR	FBV, Air (cm./sec.)	Kg, Air (bar-m/s)	MIE, Air (millijoule)	FBV, O_2 (cm./sec.)	MIE, O_2 (millijoule)	Kg, O_2 (bar-m/s)	Kg Ratio O_2/Air
Hydrogen	312	638	0.016	1400	0.0012	3,000	4.5
Methane	45	55	0.21	450	0.0027	450	4.6
Ethane	47.5	103	0.24	375	0.0019	670	6.5
Propane	46	100	0.25	370	0.0021	720	7.2
Butane	-	92	0.25	-	0.009	-	-
Hexane	-	-	0.24	-	0.006	-	-
Heptane	42.5	-	0.24	-	-	-	-
Averages	-	-	-	-	-	-	5.7

As indicated in the above Table, the Deflagration Index – and the speed of combustion – in oxygen is almost six times greater in 100% oxygen than it is in air [at 21% oxygen]. The data in the preceding Table can be combined to yield two approximate expressions for the Deflagration Index in given concentrations in oxygen, as:

$$Kg_{,O2} = 0.04 \; (K_{G,AIR}) \; (C^{1.2}) \qquad \text{bar-meters/second}$$

$$Kg_{,O2} = 1.85 \; C^{1.1} / MIE_{AIR}^{0.46} \qquad \text{bar-meters/second [excluding } H_2 \text{]}$$

Thus, if the value of the Deflagration Index in air is known, or if the Minimum Ignition Energy in air is known, an approximate value for the Deflagration Index at a given oxygen concentration can be determined.

3. Combined Relationships

The explosibility/flammability properties of a given flammable gas or vapor can be compared with a standard or typical flammable gas, through the use of Ignition Sensitivity and Explosion Severity. These property combinations were first developed and utilized for the comparisons of the hazards of coal mines, starting in 1940, where combustible dusts presented an explosion hazard [30].

3.1 Ignition Sensitivity

The product of LFL [in terms of grams per cubic meter], AIT, and MIE form the basis for a calculation of Ignition Sensitivity. A high value of this product indicates a low likelihood of ignition by relatively-weak ignition sources, such as electrostatic discharges, friction, impact, and hot surfaces. For propane in a mixture with 40% oxygen, the product is only 39 x 445 x 0.035 = 600. For comparison, the similar product for typical combustible dusts is about 45,000.

To determine the Ignition Sensitivity of a combustible dust, the LFLxAITxMIE product would be divided into the corresponding product for Pittsburgh [bituminous] Coal Dust, at 2,000,000. Thus, the Ignition Sensitivity for typical combustible dusts is about 2,000,000/45,000 = 45. By comparison, the Ignition Sensitivity for propane in a 40% oxygen mixture would be about 2,000,000/600 = 3,300, indicating the much greater sensitivity of this propane mixture to ignition sources.

As another indication of the effects of oxygen enrichment, the Ignition Sensitivity for a stoichiometric [worst-case] mixture of propane and air would be about 450. Thus, the 40% oxygen mixture would be about 7 times more sensitive to ignition [3,300/450].

3.1 Explosion Severity

The product of Deflagration Index Kg and the Maximum Explosion Pressure Pmax [in terms of barg] forms the basis for a calculation of Explosion Severity. A low value of this product indicates a low likelihood of severe damage from explosion. For propane in a mixture with 40% oxygen, the product is about 275 x 9 = 2,500. For comparison, the similar product for typical combustible dusts is about 1,500.

For combustible dusts, the product is divided by the corresponding product for Pittsburgh [bituminous] Coal Dust, at 840. Thus, the Explosion Severity for typical combustible dusts is about 1,500/840 = 1.8. By comparison, the Ignition Sensitivity for propane in a 40% oxygen mixture would be about 2,500/840 = 3.0, indicating the greater likelihood of damage if a confined mixture of propane and 40% oxygen were to be ignited.

The effect of oxygen enrichment can be shown by the Explosion Severity for a stoichiometric [worst-case] mixture of propane and air, at about 0.9. Thus, the 40% oxygen mixture could be about 3 times more damaging than an air mixture [3/0.9].

3.1 Combination of Ignition Sensitivity and Explosion Severity

If the Ignition Sensitivities of fuel/oxidant mixtures are less than 0.2 AND the Explosion Severities are less than 0.5, then such mixtures "are generally not considered to be significant explosion hazards" [31].
For the mixture of propane and 40% oxygen – for which the properties are shown above – the combination of Ignition Sensitivity and Explosion Severity is about 5 orders of Magnitude [a factor of almost 100,000] above the "no significant explosion hazard" criterion. Thus, it is unlikely that a mixture of air with any flammable or combustible vapor could be considered to present "no significant explosion hazard".

3.1 Combination of Ignition Sensitivity and Explosion Severity as Explosibility Index

The product of Ignition Sensitivity and Explosion Severity can be considered to be an Explosibility Index [32 to 37]. This Index has been found to be useful in comparing the hazards of combustible dusts, but had not yet been applied to flammable gases and vapors.

4. Graphical Presentation of Flammability/Explosibility Properties

Figure 1 presents the flammability/explosibility properties of a mixture of flammable fuel, oxygen, and nitrogen, in terms of Ignition Sensitivity [IS], Explosion Severity [ES], and Explosibility Index [EI] at atmospheric pressure and at an initial temperature of 25°C, as compared to the similar explosibility properties of propane at room temperature and atmospheric pressure. The ratios are based on the available data and the results of the calculations that are described above. As could be expected, the explosion hazards increase as the stoichiometric line – where all carbon is oxidized to carbon dioxide, and all hydrogen is oxidized to steam – is approached.

5. Conclusions.

To evaluate the explosion hazards of mixtures of fuels with oxygen-enriched atmospheres, the available data were utilized to obtain methods for estimating the values of pertinent explosibility properties. The relationships are as follows:

Lower Flammable Limit	Equal to that of fuel/air mixtures	
Autoignition Temperature	$AIT_{O2}=AIT_{AIR}/[1+(2.0AIT_{AIR}/40{,}000)(\ln(C_{O2}/21)]-273$	°C
Minimum Ignition Energy	$MIE_{O2} = 6{,}200\ MIE_{AIR} / C_{O2}^{\ 2.9}$	millijoule
Max. Explosion Pressure	$P_{Max,O2} = [[0.035\ C_{O2}) + (0.92\ (P_{Max,AIR} + 1))] - 1$	barg
Deflagration Index	$Kg_{,O2} = 0.04\ (K_{G,AIR})\ (C^{\ 1.2})$	bar-meters/second
	$Kg_{,O2} = 1.85\ C^{\ 1.1} / MIE_{AIR}^{\ 0.46}$	bar-meters/second

In addition, a method is presented for evaluating the Ignition Sensitivity and the Explosion Severity of fuel/oxygen mixtures, using the above relationships, together with a method for graphically showing the relative hazards of such mixtures.

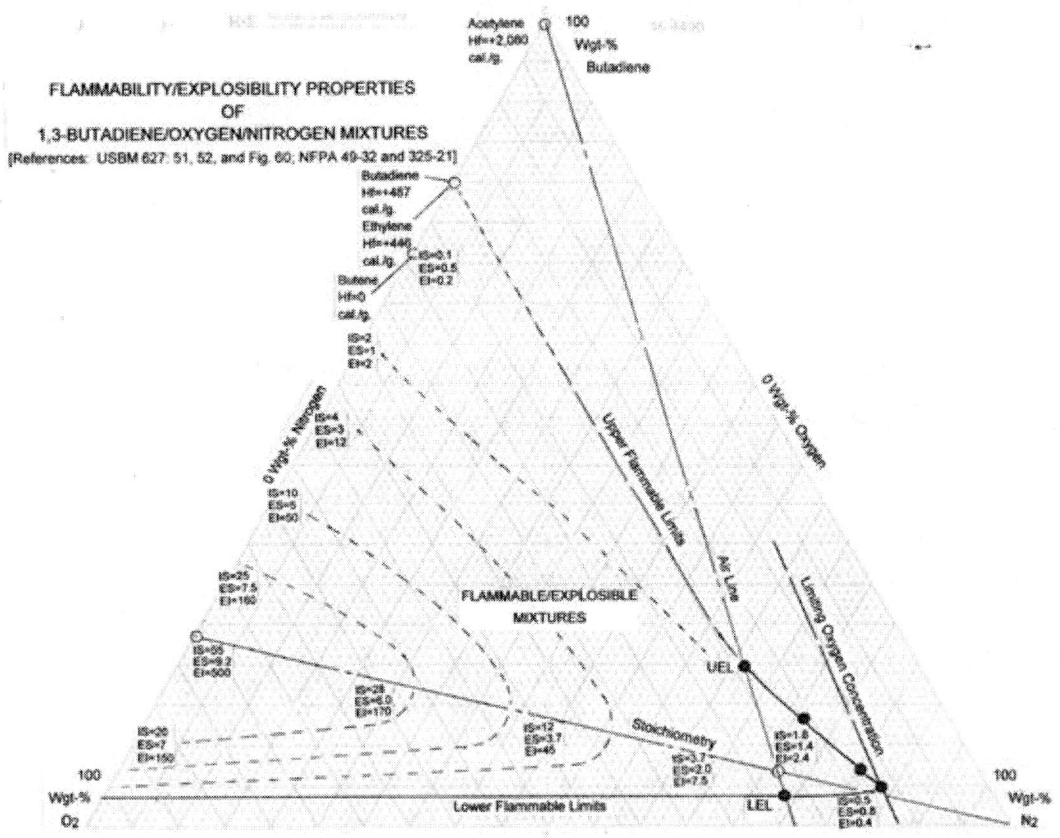

Figure 1 – Weight Concentrations of Gas Mixtures
with
Ignition Sensitivities, Explosion Severities, and Explosion Index
as compared to the properties of Propane

6. References.

1. Urben, P. G. [editor], "Bretherick's Handbook of Reactive Chemical Hazards", 6th Ed., Vol. 2, Page 287 (1999).

2. National Fire Protection Association, "Fire Protection Guide to Hazardous Materials", 13th Ed., including NFPA 49 and NFPA 325 (2002).

3. National Fire Protection Association, "Guide for Venting of Deflagrations", NFPA 68, Table D.1 (2002).

4. National Fire Protection Association, "Standard on Explosion Prevention Systems", NFPA 69, Table C.1(a) (2014).

5. National Fire Protection Association, "Recommended Practice on Static Electricity", NFPA 77, Table B.1 (2014).

6. National Fire Protection Association, "Recommended Practice for the Classification of Flammable Liquids, Gases, or Vapors and of Hazardous (Classified) Locations for Electrical Installations in Chemical Process Areas", NFPA 497, Table 4.4.2 (2017).

7. Coward, H. F., and G. W. Jones [U. S. Bureau of Mines], "Limits of Flammability of Gases and Vapors", Bulletin 503, Table 44 [and many "nose" Figures] (1952).

8. Zabetakis, M. G. [U. S. Bureau of Mines], "Flammability Characteristics of Combustible Gases and Vapors", Bulletin 627, Table A-1 [and many "nose" Figures] (1965).

9. Babrauskas, V., "Ignition Handbook", Table 1A (1992).

10. Britton, L. G., "Avoiding Static Ignition Hazards in Chemical Operations", Page 223 (1999).

11. Crowl, D. A., and J. F. Louvar, "Chemical Process Safety", 3rd Ed., Pages 251, 259, 261, 287, and 654 (2011).

12. Yaws, C. L., "Chemical Properties Handbook", Table 25-1 (1999).

13. National Fire Protection Association, "Recommended Practice for the Classification of Combustible Dusts and of Hazardous (Classified) Locations for Electrical Installations in Chemical Process Areas", NFPA 499, Paragraph A.5.2.3 (2017).

14. Frank, W. L., and S. A. Rodgers, "NFPA Guide to Combustible Dusts", Page 92 (2012).

15. Britton, L. G., "Avoiding Static Ignition Hazards in Chemical Operations", Page 223 (1999).

16. Babrauskas, V., "Ignition Handbook", Page 1023 (1992).

17. Zabetakis, M. G. [U. S. Bureau of Mines], "Flammability Characteristics of Combustible Gases and Vapors", Bulletin 627, Page 9 (1965).

18. Crowl, D. A., and J. F. Louvar, "Chemical Process Safety", 3rd Ed., Pages 263, 265 to 267, and 269 (2011).

19. Crowl, D. A., and J. F. Louvar, "Chemical Process Safety", 3rd Ed., Pages 398 and 400 (2011).

20. Babrauskas, V., "Ignition Handbook", Pages 1023 and 1024 to 1058 (1992).

21. Green, D. W., and M. Z. Southard, "Perry's Chemical Engineers' Handbook", 9th Ed., Page 19-24 (2019).

22. Babrauskas, V., "Ignition Handbook", Pages 75, 76, 203, and 625 (1992).

23. Babrauskas, V., "Ignition Handbook", Pages 1024 to 1055 (1992).

24. Babrauskas, V., "Ignition Handbook", Page 625 (1992).

25. Lewis, B., and G. von Elbe, "Combustion, Flames, and Explosions of Gases", 3rd Ed., Pages 717 and 719 (1987).

26. National Fire Protection Association, "Guide for Venting of Deflagrations", NFPA 68, Table D.1 (2002).

27. Babrauskas, V., "Ignition Handbook", Page 35 (1992).

28. Babrauskas, V., "Ignition Handbook", Page 74 (1992).

29. Zabetakis, M. G. [U. S. Bureau of Mines], "Flammability Characteristics of Combustible Gases and Vapors", Bulletin 627, Page 43 and Figures 44 and 45 (1965).

30. Nagy, J., et al. [U. S. Bureau of Mines], "Explosibility of Carbonaceous Dusts", Report of Investigations 6597, Page 2 (1965).

31. National Fire Protection Association, "Recommended Practice for the Classification of Combustible Dusts and of Hazardous (Classified) Locations for Electrical Installations in Chemical Process Areas", NFPA 499, Paragraph A.5.2.3 (2017).

32. Dorsett, H. G., et al. [U. S. Bureau of Mines], Laboratory Equipment and Test Procedures for Evaluating Explosibility of Dusts", Report of Investigations 5624, Page 2 (1960).

33. Jacobson, M., et al. [U. S. Bureau of Mines], "Explosibility of Dusts used in the Plastics Industry", Report of Investigations 5971, Tables A-1, A-2, and A-3 (1962).

34. Jacobson, M., et al. [U. S. Bureau of Mines], "Explosibility of Metal Powders", Report of Investigations 6516, Tables A-1, A-2, and A-3 (1964).

35. Nagy, J., et al. [U. S. Bureau of Mines], "Explosibility of Carbonaceous Dusts", Report of Investigations 6597, Page 2 (1965).

36. Dorsett, H. G., and J. Nagy [U. S. Bureau of Mines], "Dust Explosibility of Chemicals, Drugs, Dyes, and Pesticides", Report of Investigations 7132, Tables A-1, A-2, and A-3 (1968).

37. Grossel, S. S., "Safety Considerations in Conveying of Bulk Solids and Powders", Page 62 in the "Journal of Loss Prevention in the Process Industries", Butterworths, Vol. 1, No. 2 (April 1988).

7. Legal Disclaimer and Liability

(a) Limitation of Liability. The consulting services conducted by DEKRA Process Safety (the "Company") were performed using generally accepted guidelines, standards, and/or practices, which the Company considers reliable. Although the Company performed its consulting services pursuant to reliable and generally accepted practices in the industry, the Company does not guarantee or provide any representations or warranties with respect to Client's use, interpretation or application of the findings, conclusions, and/or suggestions of the consulting services provided by the Company. Moreover, the findings, conclusions, and the suggestions resulting from the consulting service are based upon certain assumptions, information, documents, and procedures provided by the Customer. AS SUCH, IN NO EVENT AND UNDER NO CIRCUMSTANCE SHALL THE COMPANY BE LIABLE FOR SPECIAL, INDIRECT, PUNITIVE OR CONSEQUENTIAL DAMAGES OF ANY NATURE WHATSOEVER, INCLUDING WITHOUT LIMITATION, ANY LOST REVENUE OR PROFITS OF THE CUSTOMER OR ITS CUSTOMERS, AGENTS AND DISTRIBUTORS, RESULTING FROM, ARISING OUT OF OR IN CONNECTION WITH, THE SERVICES PROVIDED BY THE COMPANY. The Customer agrees that the Company shall have no liability for damages, which may result from Client's use, interpretation or application of the consulting services provided by the Company.

(b) The Company's pricing of the consulting services provided does not contemplate that the Company shall have any liability resulting from its performance of the consulting services, except as otherwise set forth in the Quotation from the Company. Accordingly, the Customer shall indemnify and hold harmless the Company, its shareholders, directors, officers, employees and agents (the "Indemnified Parties") from and against any and all loss, cost, liability and expense, including reasonable attorney's fees and costs, which any of the Indemnified Parties may incur, sustain or be subject to, as a result of any claim, demand, action, investigation or proceeding arising out of or relating to either: (a) the consulting services provided by the Company; or (b) any material, equipment, specifications or safety information (or lack thereof) supplied to the Company (or which should have been supplied to the Company) by Customer and/or any failure of such materials, equipment, specifications and safety information to comply with any federal, state or local law or safety standard.

(c) For additional terms and conditions, which apply with respect to the provision of this report, see the Quotation provided by the Company and executed by Customer. If any terms set forth in the Quotation conflict with the terms set forth herein, the terms set forth herein shall apply.

Bridging Chemical Engineering (Risk Based Process Safety) to the Humanities – 7 Mega Elements' Approach (Protocol) of Accident Investigation in Light of Ethical Dimension, Architecture & Epistemology

Long(Steven) Zhang
No affiliation since Jul 2014 off-hiring from COP
11455 Bissonnet St Apt #514, Houston TX 77099
longz.houston@gmail.com

Prepared for Presentation at
American Institute of Chemical Engineers
2019 Spring Meeting and 15th Global Congress on Process Safety
New Orleans, LA
March 31 – April 3, 2019

Bridging Chemical Engineering (Risk Based Process safety) to the Humanities – 7 Mega Elements' Approach (Protocol) of Accident Investigation in Light of Ethical Dimension, Architecture & Epistemology

Long(Steven) Zhang
No affiliation since Jul 2014 off-hiring from COP
11455 Bissonnet St Apt #514, Houston TX 77099
longz.houston@gmail.com

Keywords: Investigation, Diagnosis, Humanities, *the Art of War, Narrative, Epistemology*

Abstract

In this paper, facing the challenges of Incident Investigation, actually whole CCPS 4th Pillar "Learn from Experience", chemical engineers' (process safety practitioners') brand new vocabulary (7 Mega Elements' narrative on failure diagnosis) would be presented in context of bridging chemical engineering (RBPS) to the Humanities, of which 3 core topics i.e. ethical dimension, architecture and epistemology would be explored and reflected upon so that a new era of RBPS diagnosis (Learn from Experience) predicted.

1 Introduction

In 2007 Center for Chemical Process Safety (CCPS) unprecedentedly presented to process industry Risk Based Process Safety (RBPS) 4 Pillars (and relevant 20 Elements), among which 4th Pillar "Learn from Experience" consists of 4 Elements: Incident Investigation, Measurement & Metrics, Auditing, and Management Review & Continuous Improvement[1].

In latest decades chemical engineers (process safety practitioners) have been facing some big challenges from accident/incident investigations, actually from whole CCPS 4th Pillar (Learn from Experience), for which there might have been many efforts, discussions, and even some 'solutions'. The author argues, however, the challenges comes not only from the needs of high quality and consistency of accident investigation, but, more fundamentally, from building up a brand new vocabulary (protocol) for chemical engineers to competently communicate of, and negotiate on, accident investigations or "why they occur and how ultimately to prevent".

Toward building a new vocabulary/protocol/narrative on accident investigation or, more broadly, on how Experience could be truly Learned (CCPS 4th Pillar), since 9th Global Congress, the author has had a series of efforts for exploration and reflections on the possible way forward, e.g. RBPS Core value (landlord for temple), Process Safety Cultural Communication (landlord's livelihood) & Egg model[2]; 4th Pillar as "espionage" not to enemies but ourselves[3][4]; 7 Mega Elements & 28 infra cultural activities (practices); RBPS Socio-Technical systems[5]; an ethical dimension of Vision 20/20[6] etc.. And in broader context of bridging chemical engineering (Risk Based Process Safety) to the Humanities, the author finally arrived at 7 Mega Elements' vocabulary/ protocol/ narrative/ "cosmology" on RBPS diagnosis for chemical engineers (process safety practitioners)[7].

2 7 Mega Elements' Protocol – Chemical Engineers' Vocabulary

Louis L. Bucciarelli claimed that he had taken seriously the "two cultures" challenge set by C. P. Snow and been working at bridging engineering to the Humanities for 35 years[8]; instrumental rationality does not suffice itself in engineering design, diagnosis and education[9]. And the author's experiences during the humble quest (through relevance of anthropology, intercultural communication, *the Art of War* etc.) for an ultimate solution to catastrophic (process safety) accidents (chemical engineering failure) seems not only resonating with Bucciarelli's bridging journey but a little bit unique, not only toward chemical engineering (Risk Based Process Safety) but also to engineering discipline at large, i.e., through benchmarking RBPS approach against Sun Tzu's *the Art of War,* 7 Mega Elements (vocabulary/ protocol/ narrative) not only were discovered for (chemical) engineers but function as 7 fundamental (chemical) engineering thoughts, which eventually sheds light on those entailed 28 practices of (chemical) engineering discipline.

Unlike Bucciarelli who tried to bridge philosophy and engineering, the author proceeded another way, i.e. first making relevance of anthropology & intercultural communication to chemical engineering (Risk Based Process Safety), then benchmarking Egg model against *the Art of War*, what the author calls 1st main ancient source for bridging (chemical) engineering to the Humanities, compared to 1st main modern one - Bucciarelli's *Engineering Philosophy*.

When guiding all visitors to some biggest lessons ("more original" two cultures) of ancient cultural traditions or civilizations of east and west, the author claimed that there should be an landlord (core value) to live within the temple building (CCPS 4 Pillars) and, furthermore, the landlord should have livelihood (Process Safety Cultural Communication (PSCC)) as well, and that should the landlord died the building seems still beautiful outwardly, however, with a purpose of tomb. In another words, when (chemical) engineers' vocabulary/ protocol/ narrative becomes deficient, or landlord and his livelihood are absent in the vocabulary, catastrophic (process safety) accidents are bound to happen, no matter how technically over competent relevant chemical engineers (process safety practitioners) might be, e.g., the lesson from the case of Bian Que–Duke Cai.

Through anthropology and intercultural communication, **RBPS Core Value** was defined in two ways, an unfolded one: *RBPS Core Value functions as the root of the tree of Process Safety*

Culture. Deficiencies are symptoms or signals for a dying or dead state of a core value, which means that the organization has strategically failed at the highest levels in managing process safety through a risk based approach – even if there has been no occurrence of process safety accidents till that moment. In another words, the RBPS temple is bound to collapse, sooner or later, due to the change or drift from the original design intent of the temple. And RBPS Core Value's simpler definition, through the concept of Power Distance in value studies: *there is no power, authority or governance on process safety except on behalf of process safety expertise (PSE)*. (PSE means the divine essence of the highest strategic insights and vision of process safety management; the Tao for process safety culture; and the soul of *The Art of War* against catastrophic process safety accidents)

And process safety culture does not exist except through ***PSCC****, whose ultimate purpose is to convey the message of RBPS Core Value. Without nourishment from RBPS Core Value, the tree of Process Safety Culture or PSCC would get dry and dead.* So the temple building was eventually upgraded to Egg model, as showed within the scheme 1st. And these 2 new universes (or 2 Mega Elements) might be deemed what Bucciarelli called the rhetoric of engineering as the parcel of instrumental explanation[10], and probably 2 most fundamental (chemical) engineering thoughts as well.

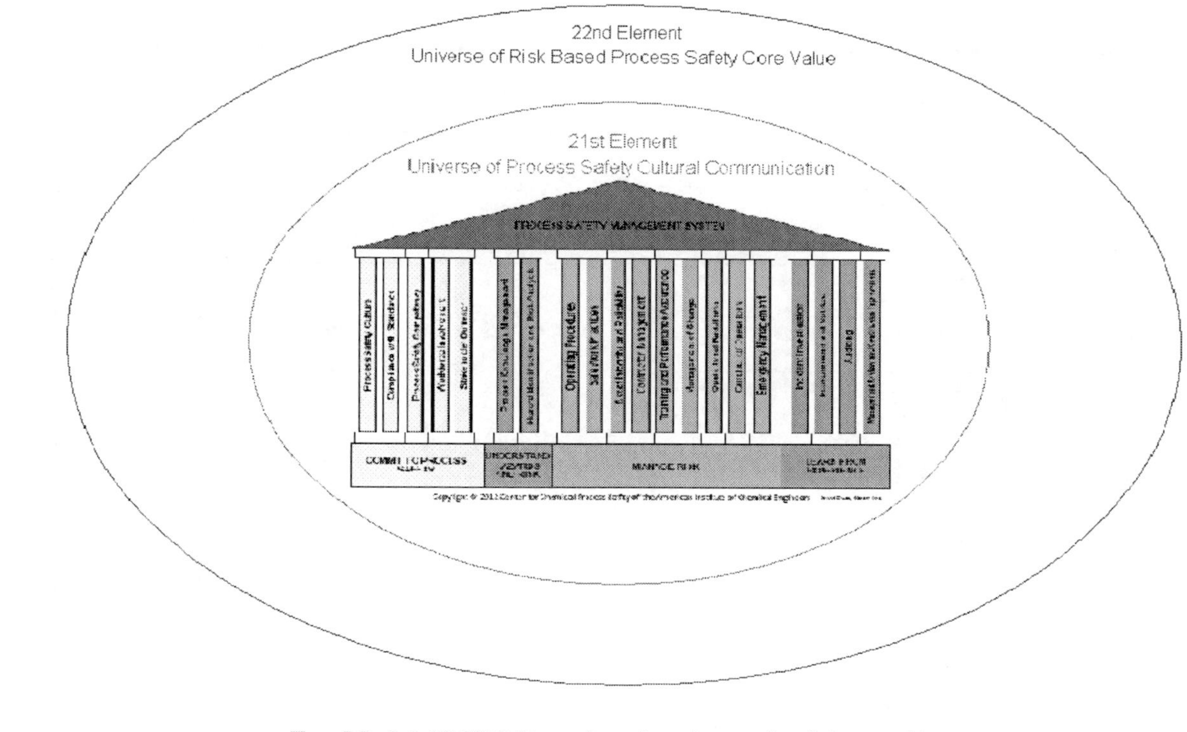

Egg Model (CCPS Temple + 2 universes) - Scheme 1st

The Temple (4 pillars) Figure Source: CCPS, Process Safety Boot Camp, contributed by David Guss, Nexen Inc. Copyright 2010 AIChE and used with permission

Then, in spite of the formidable strengths of most, actually 17, elements of CCPS 4 Pillars, except Process Safety Culture, Process Safety competency & Stakeholder Outreach, the outcome of benchmarking Egg model (4 Pillars + 2 Universes) against *The Art of War,* what the author calls a matured ancient vocabulary on engineering (military) diagnosis, not only shows that those 4 Pillars' (20 elements') narrative, even together with the new 2 Universes, would probably be not so architecturally organized at negotiation and communication on why process safety (chemical engineering) accidents (failure) occur & how to prevent, but makes the author further believe that Sun Tzu's vocabulary (6 Mega Elements, refer to Table 1st) might potentially become a narrative/ protocol paradigm for future (chemical) engineering diagnosis and *The Art of War* an main ancient source of bridging engineering to the Humanities.

For the fact that, however, we could not neglect the historical context of narrative/ protocol of *The Art of War*, the author, in light of modern globalization and global operations, fundamentally redefined Process safety Culture into Process Safety Cultural Communication (PSCC) as the 7th Mega Element, and upgraded another 3 ones i.e. Leadership, Weather/Climate & Espionage as below:

Leadership: a new chain of command, inherently from RBPS Core Value, rather than blindly from superiors;

Weather/Climate: RBPS Core Value enlightened outreach to the stakeholders & community, rather than only tactically maintaining seemingly good relationship with community; &

<table>
<tr><th colspan="2">RBPS Egg</th><th>The Art of War</th><th>Philosophy of Process Safety</th><th></th></tr>
<tr><td colspan="4">RBPS Core Value - Tao - RBPS Worldview</td><td>Worldview</td></tr>
<tr><td rowspan="4">Commit to Process Safety</td><td>Process Safety Competency</td><td rowspan="2">Leadership</td><td rowspan="2">People</td><td rowspan="7">Process Safety Cultural Communication (Process Safety Culture) RBPS Methodology</td></tr>
<tr><td>Workforce Involvement</td></tr>
<tr><td>Compliance with Standards</td><td rowspan="2">Weather/Climate</td><td rowspan="3">Plant</td></tr>
<tr><td>Stakeholder Outreach</td></tr>
<tr><td>Understand Hazards & Risk</td><td></td><td>Terrain</td></tr>
<tr><td>Manage Risk</td><td></td><td>Discipline</td><td>Procedure</td></tr>
<tr><td>Learn from Experience</td><td></td><td>Espionage</td><td>Engineering & Humanities</td></tr>
</table>

Table 1st

Espionage: 4th Pillar *Learn from Experience* (4 elements: *investigation, metrics, audit & review*) plays role "espionage", not to the enemies but to ourselves our organizational "national security"

And 7 Mega Elements might also be deemed as 7 (chemical) engineering thoughts.

3 Ethical Dimension (2 Universes) - What Chemical Engineers Count & How

While declared *there is no power, authority or governance on process safety except on behalf of Process Safety Expertise (PSE)* the author claimed that PSE shall eternally and exclusively chair at the highest strategic levels throughout all the processes of decision making and behavior determining, i.e. chemical engineers (process safety practitioners) need getting inspired from RBPS Core Value and persistently conveying the message of RBPS Core Value, both of which make up the ethical dimension (of 7 Mega elements' vocabulary/ protocol/ narrative), probably what Bucciarelli calls "best left sacred"[11], which ensures, or anthropologically designs, all relevant decision making and behavior determining correct. In this sense, at highest strategical levels, chemical engineers do not design at all but RBPS Core Value does.

From philosophical perspective, Bucciarelli claimed that the tensions at the interfaces between different design participants could not be solved by instrumental means alone [12]. And imitating American anthropologist Edward T. Hall's saying the author ever described process safety culture as a prison, unless we have RBPS Core Value as the key to unlock it and PSCC the brightness of the key. And Bucciarelli pointed out that *designing is a social process of negotiation, of iteration, of rectifying mis-steps, even misconceptions – a process rich in ambiguity and uncertainty* [13]. Similarly, the author in metaphor described PSCC as the photosynthesis of tree of process safety culture, the impulses from the heart (RBPS Core Value) and nourishment (highest designing principles) transportation from the root (RBPS Core Value). What does Bucciarelli mean by that 'social process'? What does he mean by "contribute to a sense of community among the citizenry" [14]? And what does the author mean by "photosynthesis" and "nourishment"? With hesitation & obligations, the author claims that the said social process, what the author puts PSCC, traditionally should be called Constitutional Democracy; and RBPS Core Value a nickname for Universal Value in chemical engineering.

As Universal Value & Constitutional Democracy in (chemical) engineering seem more subtle than those traditional ones the author here might call them Universal Value & Constitutional Democracy 2.0, what Bucciarelli described "old themes but in a new light" [15]. And this 2.0 might be acknowledged as two fundamental (chemical) engineering thoughts and the ethical dimension of (chemical) engineers' vocabulary/ protocol/ narrative on accident investigation and whole (chemical) engineering diagnosis (4th Pillars) as well. And probably only in this way could (chemical) engineers truly acquire what Bucciarelli called "new ways of perceiving and reading the world" [16], eventually leading the industries and communities into a new enlightened era of accident investigation and (chemical) engineering diagnosis. And the author might further states that, as the ethical dimension of (chemical) engineers' vocabulary (protocol), Universal Value & Constitutional Democracy 2.0 should function as the highest communication (designing) principles on accident investigation and (chemical) engineering diagnosis, and on accident prevention (foreknowledge) as well.

4 7 Mega Elements' Architecture – 28 Practices' Diagnosis Map

Inspired from Edward T. Hall's map of culture, the author developed a map of 28 infra cultural activities of Process Safety Culture (refer to 7 ones in the diagonal from upper left to lower right and 21 ones above), what Bucciarelli might like to call another, slightly different, "more complex vision of engineering thoughts and practices" [17]. And each infra cultural activity was defined as the critical statements & behaviors related to the interaction between Mega Element A and Mega Element B, e.g. infra activity 04 means: verbally & behaviorally communicate that the actual death of RBPS Core Value might be one of the root causes of accidents. And those 7 ones, extremely, are in fact of the critical verbal statements & behaviors related to the definition of the Mega Element (or engineering thought) concerned, e.g. infra activity 22: verbally and behaviorally embrace and reflect on Leadership - the engineering thought of new chain of command inherently from RBPS Core Value rather than blindly from superiors.

This matching between verbal statements & behaviors, or the inherent relationship between the message of RBPS Core Value and all relevant decision making & behavior determining, seems some resonance with Wittgenstein's recognition that rules and meanings (what the author calls ***tone***) in talk and texts (what the author calls vocabulary/ protocol/ narrative) were mixed with practice [18], or with what Bucciarelli put another way "the meaning of statements is embedded in practice" [19]. So these 28 infra cultural activities' map, an unfolded version of what & how chemical engineers (process safety practitioners) count, might accommodate as highest strategic encompassing diagnosis protocol (map).

	RBPS C.V. 0	P.S.C.C. 1	Leadership 2	Weather/C. 3	Espionage (Diagnosis) 4	Terrain 5	Discipline 6
RBPS Core Value 0	00	01	02	03	04	05	06
Process Safety Cultural Communication (PSCC) 1	10	11	12	13	14	15	16
Leadership 2	20	21	22 4 dimensions	23	24	25	26
Weather/ Climate 3	30	31	32	33	34	35	36
Espionage (Diagnosis) 4	40	41	42	43	44	45	46
Terrain 5	50	51	52	53	54	55	56
Discipline 6	60	61	62	63	64	65	66

28 Infra Cultural Activities' Diagnosis Map (Table 2nd)

As the Humanities e.g. philosophy, anthropology & intercultural communication, Sun Tzu's *The Art of War* etc. sheds more lights on the old themes (i.e. RBPS Core value, PSCC, Leadership and Weather/Climate), upon this Diagnosis Map, "another" complex vision of engineering thoughts and practices, the author claims that *the yellow grounded, or more exactly red number lettered area, are all about organizational "national security", and that the infra activity 44 (Espionage)'s 2nd rationality, next to the instrumental one, whose protocol area is painted green, SHALL be reactivated so that "espionage" to our organization would shift to a more proper name - "counter-intelligence", on accident investigation & prevention and on engineering diagnosis at large.*

Through maintaining the architecture/protocol of 7 Mega Elements could all these 22.5 infra activities of "national security" area be reliably visible, open and "problematic" over there, together with "traditional" green area (5.5 infra activities), to critique and diagnosis among board of directors, executive management, operation managers, strategy, legal, HR, chemical engineers (process safety practitioners), and even beyond. Pragmatically, from designing perspective, chemical engineers should pursue this more matured architecture/protocol/vocabulary, getting inspired from RBPS Core Value, persistently conveying its message and maintaining alive this complex vision of engineering thoughts & practices i.e. this 28 Practices (infra activities)' Diagnosis Map, otherwise the insightful topic "social construction of failure/ accident" [20] would turn out to be a truism. In another words, either RBPS Core Value highest strategically designs safety (and tranquility) through PSCC, or otherwise (chemical engineering/ process safety) failures/ accidents/ catastrophes are socially constructed.

5 Epistemology – Future (Chemical) Engineers' Knowledge Hierarchy

Walter Vincenti's understanding of *engineering knowledge,* as J. C. Pitt ever explained, *is task-specific and aims at the production of an artifact to serve a predetermined purpose* [21], of which, the author believes, CCPS 4 Pillars (20 Elements), as of (chemical) engineering/RBPS knowledge, might have been satisfactorily reflecting the main characters. Bucciarelli discovered, however, that engineering knowledge, of which he described *what engineers do, the reasons they do what they do, and their justification for believing the latter are good and true,* is *historically and culturally contingent, not fixed by the dictates of science and empirical fact alone.* [22]. The author might explain here, echoing the former statement "the landlord … even to buy a new house with new style of design to live within", that the (chemical) engineering knowledge we have here with CCPS 4 Pillars might not be the single unique design/house for the landlord [23], but the landscape (landlord, livelihood & house) of the contingencies, what Bucciarelli calls "a picture of both possibilities and limitations of engineering thinking" [24], is unique. In another words, both the landlord (RBPS Core Value), deeply rooted in most cultural traditions and civilizations, and his livelihood (PSCC), what the author likes to call Constitutional Democracy, should preside the landscape of the contingencies and chair and guide at the highest strategic levels the whole engineering design process, for surety including on accident investigation and diagnosis, so that the objects of design, which "traditionally" have been confined to artifacts e.g. a mega petrochemical project/ operations, might also need to include all processes of decision making and behavior determining (refer to Table 2nd – 28 Infra Activities' Diagnosis Map).

The author might state that *a complete (chemical) engineering design process, including on accident investigation, diagnosis & prevention, shall consist of 3 components: RBPS Core Value (Universal Value), PSCC (Constitutional Democracy) and the application of (chemical) engineering knowledge (e.g. one with CCPS 4 Pillars).* And at same time, we recognize that *future (chemical) engineers need a new epistemology - a new hierarchy of knowledge which consists of Universal Value Knowledge, Constitutional Democratic Knowledge & the relevant (chemical) engineering knowledge.* Edward T. Hall had an earth-breaking saying: *culture is communication and communication is culture* [25], from which the author got inspired and developed the concept of PSCC. When (chemical) engineers verbally (of text) and behaviorally (of practice) abandon the ethical dimension of 7 Mega Elements' vocabulary/ protocol/ narrative, they lose the whole process safety culture or, we could say, both their knowledge hierarchy is deficient, akin to a genocide, and their engineering design process seriously flawed, like building concentration camps.

6 Conclusion

In context of bridging chemical engineering to the Humanities, on accident investigation, diagnosis and prevention, and on engineering design process at large, the author presents 7 Mega Elements' Vocabulary/ Protocol/ Narrative, 28 Infra Cultural Activities' Diagnosis Map and a newly galvanized old Hierarchy of Knowledge, so that some outlooks might be made: a) as one of the largest communities of engineering professions, AIChE would pursue leading roles at dialogues between engineering & the Humanities; b) some small groups of chemical philosopher/anthropologist-engineers would emerge; c) an enlightened era of (chemical/process safety) accident investigation, diagnosis & learning, and engineering design process at large, would arrive; & d) engineering education's future would get brighter than ever.

7 References

[1] Center for Chemical Process Safety, *Guidelines for Risk Based Process Safety*, John Wiley & Sons, Inc., 2007

[2] Zhang, L., An Upgraded Definition of Risk Based Process Safety Core Value and Its Impact in Light of Anthropology & Intercultural Communication on the Future Evolution of Process Safety Culture & Management, 2013 AIChE Spring Meeting and 9th Global Congress on Process Safety, San Antonio, April 2013

[3] Sun-Tzu and Ames, Roger T., Sun-tzu: *The Art of Warfare*, Ballatime Books, 1993

[4] Zhang, L., Benchmarking Risk Based Process Safety Egg (RBPS Temple+2 Universes) against The Art of War, 1st CCPS Asia-Pacific Conference on Process Safety, Qingdao, Sept 2013

[5] Zhang, L., People Means Leadership, NOT Simply mean People – 4 New Dimensions of Process Safety Competency, 2014 AIChE Spring Meeting and 10th Global Congress on Process Safety, New Orleans, April 2014

[6] Zhang, L., How Committed to Process Safety is Committed Enough? – Doing Process Safety Cultural Communication (PSCC) as the Inevitable Ethical Dimension for Vision 20/20, 2016 AIChE Spring Meeting and 12th Global Congress on Process Safety, Houston, April 2016

[7] Zhang, L. and Wasel, W., Corporate Memory Loss Ongoing – Except CCPS 4th Pillar (Learn from Experience) as Espionage, 2018 AIChE Spring Meeting and 14th Global Congress on Process Safety, Orlando, April 2018

[8] Bucciarelli, L. L., *Engineering Philosophy*, Delft: DUP Satellite, 2003, p.2

[9] Ibid. p.99

[10] Ibid. p.6

[11] Ibid. p.4

[12] Ibid. p.4-5

[13] Ibid. p.7

[14] Ibid. p.5

[15] Ibid. p.2

[16] Ibid. p.7

[17] Ibid. p.101

[18] Ibid. p.6

[19] Ibid. p.6

[20] Ibid. p. 101

[21] Pitt, J. C., "What Engineers Know", *Techne* 5:3, Spring, 2001, p.25

[22] Bucciarelli, *Engineering Philosophy*, p.99

[23] Pitt, "What Engineers Know", p.26

[24] Bucciarelli, *Engineering Philosophy*, p.99

[25] Hall, E. T., *The Silent Language*, Anchor Books, 1990, p.186

548354 Applying PSM in Vaca Muerta Shale Oil & Gas Operations: The Experience

How to develop, implement and sustain Process Safety in greenfield shale oil and gas projects? This has been a challenge to many new ventures.

Although conventional oil operations have been in place for decades, the exploration, drilling and fracking operations in unconventional shale fields are relatively new in the World. Most of the projects have kicked off less than eight years ago and there are some aspects of the operations that are unique to unconventional fields.

The objective of this paper is to share with the Process Safety global community, how a fit for purpose Process Safety program was developed and is being implemented in the Upstream Unconventional shale oil and gas YPF & Chevron joint venture in Vaca Muerta / Argentina. The project started in 2013, has currently 14 D&C rigs, 5 fracking sets, several separation and treatment production facilities, 3,500 employees and is expected to grow even more in the next 5 years. The paper describes how the project team developed the idea, planned the steps, engaged all stakeholders and selected the key elements of the Process Safety Management System that would add more value in each stage of the project, including drilling, fracking and oil and gas production. The paper also explains the key challenges, such as the difficulties faced to instill Operational Discipline and other key cultural and technical aspects related to Contractor field Operation crews. The journey to implement Process Safety in the Unconventional shale oil & gas operations is very dynamic and is in progress, however, there are already several positive aspects as well as pitfalls to take away as a lesson learned.

548596 A Survey of Process Safety Workers' Attitudes, Behaviors, and Outcomes: Moving Beyond Human Error and Toward a Systems Model

In an effort to move beyond the "human error" explanation for safety incidents, we explored issues with procedures from the worker's point of view by using an anonymous, holistic survey developed from interviews with then currently employed operators. The current sample ($N = 174$) included individuals employed in the process safety industry (MKO/Advanced Procedures Initiative) and were primarily from the Oil & Gas and Chemical industry. The survey was deployed over the course of approximately four weeks. Twelve distinct constructs emerged from the survey (e.g., perceptions of procedure quality, procedure deviation, attitudes toward the procedure change process, etc.). Results indicated that perceptions of procedure quality was the focal variable in all analyses including positive relationships with attitudes toward the procedure change process and negative relationships with procedure deviations, and safety incidents and near-misses. We conclude that these elements are important factors to consider when evaluating and developing procedure systems and provide additional information beyond what the industry considered paramount.

546680 A Method of the Evacuation Pathway Optimization Under Toxic Gas Releases in Chemical Plants

Toxic gases release in chemical plant is a key factor contributing to major accidents and may cause tremendous human life and economic losses. A scientific planning evacuation pathway is an important guarantee of reducing the losses of accidents. Many studies regarding evacuation planning in chemical plant have been carried out by simulating the accident scenarios. However, at present, a systemic method of evacuation pathway optimization has not been developed, and the effectiveness of emergency preparedness has been compromised. To address this problem, an approach based on the CFD modelling of hazardous gas release consequences to plan evacuation pathways and appropriate shelters is proposed in this paper. The methods stochastic programming and dose-response model are mainly employed in this paper. The toxic gas concentration can be converted into poisoning fatality probabilities by dose-response model. Then, stochastic programming method is used to select all the optimal pathways from the evacuation point to the safe area. Setting the minimal fatality probability as the optimization goal, the optimized evacuation pathways under all the leakage scenarios are finally determined through solving the stochastic programming model. As for places with limited evacuation opportunities, an approach of setting up a safe room is proposed in this paper and relevant design guidance are put forward in this paper considering the detailed situation of the chemical plant. A case study of the optimization of evacuation pathways and safe room setup in a salt chemical plant that contains 88 leakage scenarios of chlorine leakage is investigated. With the proposed methodology, the optimized evacuation pathway and the appropriate safe room can provide scientific references for decision makers and evacuees on reducing losses when toxic gases release in chemical plants.

19 AIChE Spring Meeting
15TH GLOBAL CONGRESS ON PROCESS SAFETY
NEW ORLEANS • 2019

Four Major Gaps That Are Preventing Most Companies Worldwide from Achieving Excellent Process Safety Performance

William G. Bridges
Process Improvement Institute Inc. (PII)
1321 Waterside Lane, Knoxville, TN 37922 USA
wbridges@piii.com

Jeff Thomas
Process Improvement Institute Inc. (PII)
jthomas@piii.com

PROCESS IMPROVEMENT INSTITUTE

Prepared for Presentation at
American Institute of Chemical Engineers
2019 Spring Meeting and 15th Global Congress on Process Safety
New Orleans, LA
March 31 – April 3, 2019

FOUR MAJOR GAPS THAT ARE PREVENTING MOST COMPANIES WORLDWIDE FROM ACHIEVING EXCELLENT PROCESS SAFETY PERFORMANCE

William G. Bridges
Process Improvement Institute Inc. (PII)
wbridges@piii.com

Jeff Thomas
Process Improvement Institute Inc. (PII)
jthomas@piii.com

Keywords: process safety, PSM, human factors, near misses, PHA, HAZOP, procedures

Abstract

Process safety requires implementing many management systems, specific engineered features, and operating and maintenance practices effectively. Most companies believe they have done just that, and yet major accidents continue to occur. Why is that? What is missing?

This paper looks at the statistics of major accidents, combined with results from audits and assessments from more than 50 chemical, petrochemical, oil/gas, and related processing companies world-wide. The paper illustrates the four major gaps that are common to the companies/sites that keep having major accidents, compared to those companies/sites that do not have such accidents:

- Accurate and clear operating and maintenance work instructions (procedures)
- PHA of all modes of operation, especially startup, shutdown, abnormal operations, and online maintenance modes
- Getting Near Misses reported and investigated
- Addressing human factors that are missing from most management systems

The paper also provides guidance for closing each of these gaps.

Introduction

"Process safety management (PSM) is the application of management principles to the identification, understanding, and control of process hazards to prevent process-related incidents"[1]. PSM entails development and implementation of programs or systems to ensure that the practices and equipment used in hazardous processes are adequate and are maintained appropriately. The primary categories of programs or systems have come to be called elements of PSM. However, the basic elements of PSM have been defined by many groups in a number of ways. Table 1 lists the elements of PSM systems from various industry and government groups. Many of the elements with different names have essentially the same meaning. For instance, "maintenance and inspection of facilities," together with some aspects of "personnel" practices, both under American Chemistry Council's (ACC's, formerly CMA's) Process Safety Code of Responsible Care™[2] are essentially the same as the single element, "mechanical integrity," under 29 CFR 1910.1190)[3].

Table 1. Comparison of PSM systems

OSHA 29 CFR 1910.119 EPA 40 CFR 68	AIChE/CCPS Risk-Based Process Safety (RBPS) Standard	Responsible Care© Process Safety Code
Management System	**Commitment to Process Safety**	**Management Leadership**
Employee Participation	Process Safety Culture*	Commitment*
Process Safety Information	Compliance with Standards	Accountability*
Process Hazard Analysis*	Process Safety Competency*	Performance Measurement
Operating Procedures*	Workforce Involvement*	Incident Investigation
Training*	Stakeholder Outreach	Information Sharing
Contractors	**Understand Hazards and Evaluate Risk**	CAER Integration
Pre-Startup Safety Review	Process Knowledge Management	**Technology**
Mechanical Integrity	Hazard Identification and Risk Analysis	Design Documentation
Hot Work Permit	**Manage Risk**	Process Hazard Information
Management of Change	Operating Procedures*	Process Hazard Analysis
Incident Investigation	Training and Performance*	Management of Change
Emergency Planning and Response	Safe Work Practices *	Facilities
Compliance Audits	Asset Integrity and Reliability	Siting
Trade Secrets	Contractor Management	**Codes and Standards**
	Management of Change	Safety Reviews
	Operational Readiness*	Maintenance and Inspection
	Conduct of Operations*	Multiple Safeguards*
	Emergency Management	Emergency Management
	Learn from Experience	**Personnel**
	Incident Investigation	Job Skills*
	Measures and Metrics	Safe Work Practices*
	Auditing	Initial Training*
	Management Review and Continuous Improvement	Employee Proficiency*
		Fitness for Duty*
		Contractors

* Contains some of the 6 elements of a Human Factors element, but all missing elements not covered completely. Mostly covered in CoC/OD element

Note that the newest definition of process safety is CCPS's *Risk Based Process Safety (RBPS)*[4] replaces their earlier process safety definition. In the older definition from CCPS [1], there was an element on Human Factors, which brought strong focus to this necessary element; in RBPS, the human factors subelements are now spread across 6 different elements, though some human factors were inadvertently weakened in the transition to *RBPS*.

Although nearly the entire industry agrees that implementing PSM is the right thing to do, interpreting and converting the PSM requirements into practices is unique to each company, and even unique to each plant site. Not only can the requirements be interpreted differently for each site based on local needs, but each site also starts from a different point when they begin to implement a system that is consistent with this regulation.

This paper looks at the statistics of major accidents, combined with results from audits and assessments from more than 50 chemical, petrochemical, oil/gas, and related processing companies world-wide. The paper illustrates the four major gaps that are common to the companies/sites that keep having major accidents, compared to those companies/sites that do not have such accidents:

- Accurate and clear operating and maintenance work instructions (procedures)
- PHA of all modes of operation, especially startup, shutdown, abnormal operations, and online maintenance modes
- Getting Near Misses reported and investigated
- Addressing human factors that are missing from most management systems

The paper also provides guidance for closing each of these gaps.

1. Procedures Need to be More Accurate and Clear

Operating and maintenance procedures have always been crucial to the safety, quality, and productivity of process systems. With the advent of new safety and quality standards such as OSHA's process safety management (PSM) regulation, OSHA's personal protective equipment (PPE) regulation, EPA's risk management program (RMP) regulation, and the ISO 9000 quality standard, many companies are facing the daunting task of developing or upgrading their procedures to satisfy varied and sometimes complex and conflicting requirements.

Although there are overlapping characteristics that the various regulations and standards share, each approach the procedure-writing process from a somewhat different perspective (e.g., quality or safety, protection of workers or protection of the public/environment), and the required level of detail for documenting the procedures differs greatly between these regulations and standards. For instance, the ISO 9000 quality standard advocates documenting procedures and work instructions that impact quality and suggests keeping the procedures as simple as possible, while the OSHA PSM regulation requires detailed procedures that address: all operating modes (startup, shutdown, etc.), operating limits, consequences of deviations, means to avoid hazards, safety and health considerations, and safety systems and their functions. How can companies develop and maintain procedures that ensure productivity and simultaneously satisfy the different regulations and standards? The key is to remember the ultimate goal of the regulations and standards: to reduce human errors that can impact quality, productivity, and/or safety.

Procedure-related errors are errors that occur because some characteristic of the procedure caused task performance to fail. **This is currently the most critical human factor at most sites since 90% of accidents have at least one root cause related to mistakes within procedures**. Reducing these procedure deficiencies can reduce human error rates by a factor of 2 to 20 [5].

1.1. Identified problems

Deficient Procedures are the most prevalent problem in process industries since procedures have not traditionally been developed from the perspective of optimizing human factors; instead procedures have been traditionally developed to meet a compliance requirement. Examples of procedure deficiencies (inaccuracies) include:

- Incorrect/incomplete/nonexistent (most procedures we have audited have been only 70-75% accurate – the inaccuracies include missing critical steps, steps as written are not what needs to be done, or the steps are out of sequence) [5]
- No/misplaced/incorrect information in warnings (for example, a warning should **never contain** the action to take; it should instead **emphasize** the action to take) and warnings should be located BEFORE the step applicable to the warning
- Poor format and presentation rules

1.2. Objective

The flowchart (Figure 1) nd the procedure quality checklist (Table 2) describes the basic sequence of steps for writing effective procedures. Note that some basic rules are contained in the flowchart, such as "operators write operating procedures," "maintenance craft-persons write maintenance procedures," and "Lab technicians write lab procedures." Procedures are NOT written by engineers or department superintendents. This is a critical to ensure procedures are written in the user's common language and are written at the right level of detail.

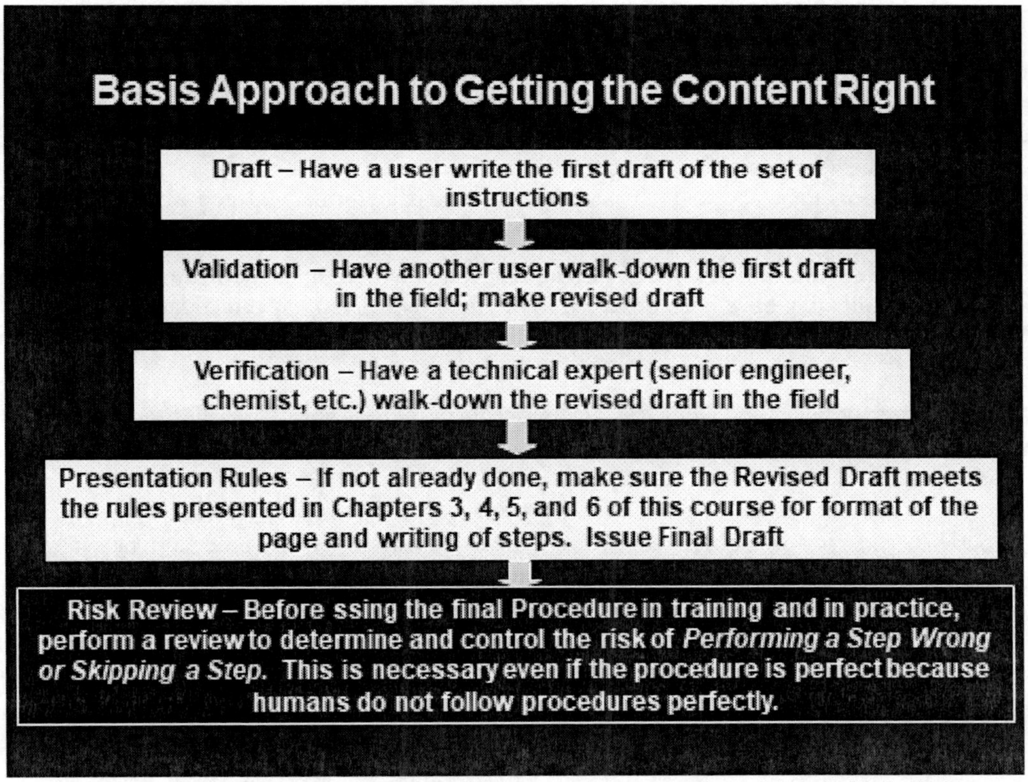

Figure 1. Best Approach to developing written procedures, once the unit has decided which procedures are needed (Extracted from PII Course, Writing Effective Operating and Maintenance Procedures © PII 2003-2017) [6]

The procedure content controls the human error related to "procedure-based" error – if content is wrong, the trainee is very likely to learn the step wrong (trainer uses the procedure as a basis for training or the trainee refers to the procedure step that is wrong when executing the procedure).

Even if the content is correct, not following page formatting best practices can still increase human error rates. If the content of the step is wrong it does not matter if the presentation is clear. If the content is perfect, you can further reduce error rates by 50% to 70% by presenting the information properly.

Rules for developing content[7]

- The procedures should be written at the level of someone who has just completed the basic training for that task. Do not write the procedure for someone just hired or for the 10+ veteran.
- Clearly identify ahead of time what activities need procedures to help ensure error rates are controlled low enough, and identify which activities are considered common "skills" of all of the staff.

> *Example: Starting a pump is typically considered a "skill." This means a procedure step can simply be "Start the benzene recycle pump (P-119)." The procedure does not need to explain how to start the pump. But, starting a pump is a skill that nonetheless must be learned; it requires doing several sub-steps related to positioning of intake and discharge valves, how the pump is throttled at startup, checking local pressure gauge, etc. So, a training module is needed for starting pumps of various types, and the operator needs to learn this necessary skill. Once the skill is learned, it will be applied to so many various cases and so often, that a procedure is no longer needed as a "refresher" on how to "start the pump."*

- The first draft of the procedure must be walked-down in the field by another user. Simply reading through the procedure does not catch enough of the mistakes.
- The revised draft should be reviewed in the field by a technical staff person, such as a process or operations engineer (for operating procedures).
- The procedure needs to be checked to ensure it follow the page format and step writing rules described in the next section. This ensures the accurate steps are clearly presented.

The target is to reach an accuracy of 95% or better[7] (so, no more than one wrong or missing step out of 20 steps). The following have been observed in the field by PII staff by direct data collection at more than 110 sites/plants around the world, based on walk-down review of several to dozens of procedures at each unit/plant:

- At 95% accuracy or better and when the same procedures follow 80% of the rules for procedure clarity (presented next), then most users will follow the written procedures and will try to keep the procedures up-to-date
- At about 85% accuracy or less, about half of the users stop using the procedures
- At about 75% accuracy or less, less than 10% of the users will refer to the procedure or will try to keep it up-to-date. **So, the written procedures are not very effective.**
- Unfortunately, the typical operating procedure walked down by PII staff (accompanied by senior operators, a process engineer, and a shift supervisor) is about 75% accurate (so one step in four is missing or wrong). Usually these inaccurate procedures also follow less than half of the best practices for procedure format, presented next[7].

Procedure Clarity

The importance of the accuracy of the procedure steps were discussed above. The clarity of the procedure steps (how they are written and how the page is formatted) is also important. Following best practices for step and page format reduces human errors by a multiplying factor of 3 to 5.[5],[7]

The best practice rules for writing and validating procedures have been published for many years *(see Bridges & Williams, 1997; Madden & Bridges, 2016, 2017)* [7],[8],[9]. These rules include best practices for formatting of the pages and steps. These have been gradually improved over the past decades and now are incorporated into PII's training materials.[6]

Table 2. Procedure Quality Checklist (courtesy PII, copyright 2003-2017)[6]

#	Issue	Response
	Procedure Content Checklist	
1	Is the procedure drafted by a future user of the written procedure? (Engineers should not author procedures to be used by operators or maintenance staff.)	
2	Is the procedure validated by a walk-down in the field by another future user of the procedures?	
3	Is the procedure reviewed and commented on by technical staff (engineers or vendors)?	
4	Is the procedure checked versus the Page and Step format rules below?	
5	Is a hazard review of step-by-step procedures performed to make sure there are sufficient safeguards (IPLs) against the errors that will occur eventually (when a step is skipped or performed wrong)?	
6	Is the content measured using "newly trained operators" to judge the % of steps that are missing, steps that are confusing or wrong, and steps that are out-of-sequence? (A score of 95% accuracy of content is good; 98% should be the targeted average.)	
	Page Format Checklist	
1	Is the title of the procedure the largest item on the page?	
2	Is the procedure title clear and consistent with other titles, and does it uniquely describes the topic?	
3	Are the document control features the smallest items on the page?	
4	Are temporary procedures clearly identified?	
5	Is white space used effectively? • Is there one blank line between each step? • Does indentation help the user keep their place? • Are the margins large enough to reduce page congestion?	
6	Is type size is 12 pt font or larger?	
7	Is mixed case used for words of steps, with ALL CAPS used only for special cases (such as IF, THEN, AUTO, and WARNING)?	
8	Is the step number very simple (single level of number used)? Only an integer?	
9	Have sections or information not necessary to performing the steps been moved to the back or to another part of the manual or training guide?	
10	Are section titles bold or larger than the text font? Do sections have clear endings?	
11	Is the decision on electronic presentation versus hard copy correct? Are electronic linkages to procedures clear and accurate and easy to use? If paper is chosen, is the color of the paper appropriate?	
12	Is the overall page format (such as Outline format or Two-Column [T-Bar] format) appropriate to the use of the procedure?	
13	Are play script features added for tasks that must be coordinated between two or more users? • Play script is normally used when there are two or more hand-offs of responsibility for steps.	
14	Are rules followed for formatting of Warnings, Cautions, and Notes? (See annotated rules, such as Warnings are for worker safety and Warnings must clearly standout from rest of page.)	
	Step Writing Checklist	
1	Is each step written as a command?	
2	Is the proper level of detail used throughout? This is judged based on: • *Who* will use the procedures • Same level of detail used in similar procedure steps	
3	On average, is there only one implied action per instruction? Best practice is to average 1.2.	
4	Does the procedure indicate when sequence is important? • If sequence matters, each step should be numbered (with an integer or letter)	

	• If sequence does not matter, bullet lists should be used	
5	Are only common words used? Apply "education" level test (5 grade reading level is best)	
6	Do all acronyms, abbreviations, and jargon aid understanding? • Develop a list of such terms for use in procedures *and* communication. • Use terms that users use (within reason)	
7	Is each step specific enough? No room left to guess/interpret: • The meaning of a word or phrase (Check vs. Make sure) • The intent of a step or series of steps • A desired quantity or value • To what equipment the step applies	
8	Is the procedure free of steps that require in-your-head calculations? • Values expressed as ranges rather than targets with error bands • Conversion tables, worksheets, or graphs provided where needed	
9	Are graphics to the user's advantage? • No explanatory paragraphs or lengthy instructions that could be replaced by a picture • No impressive graphics that provide no real advantage	
10	Are references to the user's advantage? • No lengthy explanations or instructions that could be replaced by branching to a reference • No references to a procedure that references still another • No gaps or overlaps between this procedure and a referenced document • If branching, must branch to a procedure, not to a specific step in a procedure	
11	Are rules followed for writing warnings, cautions, and conditional steps? • No more than 2 per page • No actions within a warning or caution (actions must always be numbered steps) • Warnings and Cautions contain descriptions of potential consequences	

For a complete description of best practices, refer to PII's paper *"Best practices for writing operating procedures and trouble-shooting guides"*. [7]

2. PHAs (Process Hazards Analyses) Need to Close Two Major Gaps

Process Hazards is a broad term. The relationship between many components define process hazards. For a PHA team, Process Safety Information (information on the chemicals, technology and equipment) and operating procedures are necessary to identify specific components of and analyze a hazard scenario. These components include: process deviations (parametric and procedural), their causes, the process safety consequence and various safeguards for detecting deviations and causes, preventing causes, detecting and mitigating consequences and intervening safeguards. A thorough review of incident investigations also offers an opportunity to further analyze process hazards.

2.1. Identified problems

Over the last twenty years the CSB has brought needed attention, through its investigations of industry incidences, to two significant gaps in many current PHAs; where the teams did not identify and analyze:

- Process hazards unique to non-routine modes of operations resulting in catastrophic consequences, and
- Process hazards related to Damage Mechanisms.

Though the intent of the US OSHA PSM regulation has always been for PHA teams to analyze the hazards for all modes of operation, OSHA did not give specific requirements for addressing damage mechanisms and only recently have they increased their enforcement attention on PHA of non-routine modes of operation.

Industry has responded by modifying the *Guidelines for Hazard Evaluation Procedures*, 3rd Edition, 2008, CCPS/AIChE[10] to improve the coverage of these two main issues. API in turn issues guidance on addressing damage mechanisms[11]. Industry best practices now exist for addressing both of these weaknesses. Now, with proper approaches, a team with process safety competencies and process knowledge can thoroughly address the issues during a PHA.

Another less significant gap noted by regulators and by the US CSB has been an incomplete consideration, within the PHA, of previous incidents.

This section of the paper lays out the case from the US CSB and US OSHA on these PHA gaps:

- Poor or no coverage of hazards during non-routine modes of operations
- Poor or no coverage of damage mechanisms

2.2. Objectives

2.2.1. Hazard Evaluation of Non-Routine Modes of Operation

Non Routine modes of operation include start-up, shut-down, online maintenance and other abnormal operations. The hazard evaluation of non-routine modes of operation involves reviewing procedures using a HAZOP, simplified HAZOP, or What-if analysis to uncover potential accident scenarios associated with non-routine operations, for continuous or batch operations. Human error is more likely and more critical during non-routine operations. By analyzing procedural steps where human error is more likely, and where human error or component failure could lead to a consequence of interest, risk can be reduced. The hazard evaluation team's objective is to evaluate the risk associated with skipping steps and performing steps wrong.

The purpose of a hazard evaluation of non-routine modes of operation (governed by written procedures) is to make sure an organization has enough safeguards for the inevitable instance when a step is either performed wrong or skipped (inadvertently or due to shortcutting or other reasons).

Industry has found that a HAZOP or what-if analysis, structured to address procedures, can be used effectively for finding the great majority of accident scenarios that can occur during non-routine modes of operation[10],[12],[13],[14]. Experience shows that reviews of non-routine procedures have revealed many more hazards than merely trying to address these modes of operation during the P&ID driven hazard evaluations.

General guidelines for analyzing non-routine modes of operation or batch processes

- Define the assumptions about the system's initial status. "What is assumed to be the starting conditions when the user of the procedure begins with Step 1?"
- Define the complete design intention for each step. "Is the step actually 3 or 5 actions instead of one action? If so, what are the individual actions to accomplish this task?"
- Don't analyze safeguard steps that start with ensure, check, verify, inspect, etc., or where the consequence of skip is "loss of one level of safeguard/protection against…" There is no reason to analyze these steps since they will show up as safeguards of deviations of other steps. This approach is similar to not analyzing a PSV during a HAZOP of continuous mode (i.e., during a parametric deviation analysis); instead the PSV is shown as a safeguard against loss of containment.
- Together with an operator before the review, identify the sections of the procedures that warrant use of:
  - 7-8 Guide Words (extremely large consequences can happen if deviations occur)
  - 2 Guide Words (the system is complex, mistakes are costly, or several consequences could occur)

  - On others, use What-If (no guide words or guide phrases; for use on simpler or lower hazard systems)
- Decompose each written step into a sequence of actions (verbs)
- Apply guide words directly to the intentions of each action

The following preparation steps may also be needed prior to the procedure PHA:
- Walk through procedure in the plant with one or more operators to see the work situation and verify the accuracy of the written procedure.
- Determine if the procedure follows the best practices for "presentation" of the content; the best practices will limit the probability of human error
- Discuss generic issues related to operating procedures, such as:
  - staffing (normal and temporary)
  - human-machine interface
  - worker training, certification, etc.
  - management of change
  - policy enforcement
- Review other related procedures such as lock out/tag out and hot work

- **If the procedures are NOT >90% accurate, then redo procedures before the PHA.**

2.2.2. Damage mechanisms

Recently, more and more regulators, including the US CSB, are making analysis of Damage Mechanisms (DMs) a PHA requirement. This requires that potential process damage mechanisms and their potential consequences of interest be identified. Damage mechanisms can be broken down into three main categories. These categories and examples of each are shown in the following table (Table 3).

Table 3. Types of Damage Mechanisms [15]

Category	Examples
Mechanical	• Mechanical loading failure • Mechanical fatigue • Buckling • Cracking • Embrittlement • Ductile fracture • Brittle fracture
Chemical	• Corrosion – Uniform – Localized – Pitting
Physical	• Thermal failures – Creep – Thermal fatigue – Transformation

The team may use aids such as piping specs, established literature and standards, any applicable MOC documents, etc. to help identify DMs. The PHA leader has several options of how to address damage mechanisms, but PII preferred approach is to discuss during HAZOP under "Loss of Containment" deviation in each node/section.

Damage mechanisms can (and sometimes ***should***) be reviewed prior to the PHA. For example, all MOCs should consider damage mechanisms and their potential consequences of interest before the change is approved. The MOC should also require updates to process safety information which might be relevant to protect against DMs.

Coverage of Damage Mechanism (DM) for each major section [14]

As mentioned above, this is PII's preferred method for thoroughly covering all DM within a PHA/HAZOP of an entire unit. The benefits are that the team can more easily catch changes in DM from section to section and they can more easily identify when unique safeguards, such as Remote Isolation Valves, segregated containments (dike), and unique materials of construction are needed. To facilitate this, a list of generic causes of loss of containment and a generic list of safeguards against loss of containment are covered in the Loss of Containment (LOC) deviation of each section.

For each node (each line, each vessel, each column, etc.), the PHA team should discuss and document each damage mechanism listed in Table 3 (as a cause of loss of containment), consequence of the failure if the damage mechanism occurs, and the safeguards in place to prevent the damage mechanism, detect the mechanism before failure, prevent the release, detect and response to the release, and mitigate or contain the release.

The documentation style varies between PHAs. There are two different styles that have been acceptable in the past.

- One style (Example A, Table 4) uses a reference to a summary table of typical causes (this summary table is not shown but is similar to Table 5 of this paper) instead of listing each individual damage mechanism. In Example A, the same approach is used for safeguards. This approach saves redundant text and some time, but it requires the leader and scribe to be diligent to cover everything in the Typical tables that were referenced in the Cause and Safeguard columns. A modified approach is to relegate some generic causes and safeguards to Typical tables for reference and then to carefully list the specific cause and/or safeguards of interest for LOC for each specific node.

Table 4. Example A – Referencing Generic Tables for Typical Causes and Typical Safeguards [14]

Dev	Deviation	Causes	Consequences	Existing Safeguards	Recommendations
9.10	Loss of containment	Accelerate corrosion External fire High pressure (linked from 9.7) Typical causes of loss of containment (see Table A.1)	Release to atmosphere leading to potential injury of workers and/or community	Pressure Relief Valve's 231A, B, C on Reactor and no valves in line to reactor Most lines / connections are welded construction; only a few flanges Drills conducted each year on evacuation, rescue and isolation Emergency Response personnel are trained at SABIC FTC Generic safeguards protecting against or mitigating process material releases (see Table A.2)	Safety 7. Consider changing the ITPM schedule for managing most PSVs whose inspections are too infrequent based on industry standards and best practice. For instance, the current inspection frequency for PSV-8220 on the Ammonia Receiver (I-2005G) is 9 years, whereas, consensus codes typically recommend testing/inspection every 1-4 years for PSVs in highly toxic services.

- Example B (**Table 5**) does not use or reference a generic or Typical list of causes (damage mechanism) or Typical safeguards, but instead develops a specific listing for the LOC deviation of each node. This style has proven easier to justify to regulators and other outside reviewers but takes more documentation effort.

Table 5. Example B – Specific Listing of Causes and Safeguards [14]

Dev	Deviation	Causes	Consequences	Existing Safeguards	Recommendations
1.9	Loss of containment	Corrosion/erosion External fire and/or flame impingement Gasket, packing, or seal failure Improper maintenance Material defect Operator failing to close or inadvertently opening a valve to the atmosphere (e.g., a valve at a hose connection) Railcar inadvertently derailed Valve leaking to the atmosphere High pressure (linked from 1.5) Acid corrosion caused by high concentration of water (linked from 1.8) High ambient temperature External impact (such as from a mini-engine or another railcar)	Catastrophic release of chlorine from a ruptured railcar Steady release of chlorine from a ruptured connection Steady release of chlorine from a leaky connection High pressure caused by thermal expansion of liquid chlorine if railcar is also over-full	Chlorine repair kit Derailer and warning flag to prevent impact by a mini-engine or another railcar Limited vehicular access to area Maintenance/operator response as required, including isolation if needed Operator periodically monitoring the railcar valves while unloading Personal protective equipment in the area Plugs installed in all chlorine valves to the atmosphere when the valves are not in use Relief valve on each railcar for mitigating releases caused by overpressure Supplier maintenance of railcars (per strictly enforced US DOT requirements) Video monitoring of the unloading area Concrete crossties on rail spur Dike preventing any combustibles spilled nearby from reaching the unloading rack area Concrete railroad ties in chlorine unloading area to prevent fires near railcar	10. Consider installing a chlorine detection system in the unloading and vaporizing area to help detect chlorine releases (especially at likely release points) 11. Verify that periodic maintenance and inspections are being performed in accordance with Chlorine Institute recommendations 12. Review the drainage system for the unloading are, and identify the areas that may be affected by a large chlorine release 13. Consider prohibiting the use of heavy equipment (e.g., cranes) in the unloading and vaporizing area unless special precautions to prevent equipment damage are enacted 33. Consider providing a high pressure alarm for each vaporizer 40. Consider providing a water deluge system in the unloading area to help mitigate chlorine releases from the railcar

Regardless of method, the team leader and scribe must ensure the team rigorously discusses all of the damage mechanisms in Table 5 and adequately documents the results of the damage mechanism review in the PHA/HAZOP analysis tables.

For a complete description of PHA best practices, refer to PII's paper *"Recipe for a Complete Process Hazard Analysis"*. [14]

3. Near Misses Need to be Reported and Enough of Them Investigated

We must learn from accidents and near misses to prevent recurrence. The first step in the learning process is investigation to determine the causes and underlying reasons why accidents and near misses occur. A thorough investigation of root causes will identify the management system weaknesses. Learning which management system weaknesses are leading to near misses and accidents is one of the highest value activities in which a company can invest, and learning from near misses is much cheaper than learning from accidents. Many chemical companies have implemented process safety management systems, and now they are beginning to focus on getting near misses reported and on root cause analysis. This is a very exciting trend. Unfortunately, the chemical industry gets very few near misses reported (the chemical industry is certainly not the only industry with this problem).

To understand more about near misses and getting them reported, it is best to first review the basic definitions.

An incident is either an Accident or a Near Miss.

An Accident is a sequence of unplanned events and conditions that result in harm to people, environment, process, product or image.

A Near Miss is an unplanned sequence of events that could have caused harm if conditions were different or is allowed to progress, but did not in this instance.

Given a consistent understanding of the definition of a near miss, it is possible to estimate how many near misses should be reported for every accident. Studies in several industries indicate that there are between 50 and 100 near misses for every accident. Also, data indicates that there are perhaps 100 erroneous acts or conditions for every near miss. This gives a total population of roughly 10,000 errors for every accident. **Figure 2** illustrates the relationships between accidents, near misses and non-incidents.

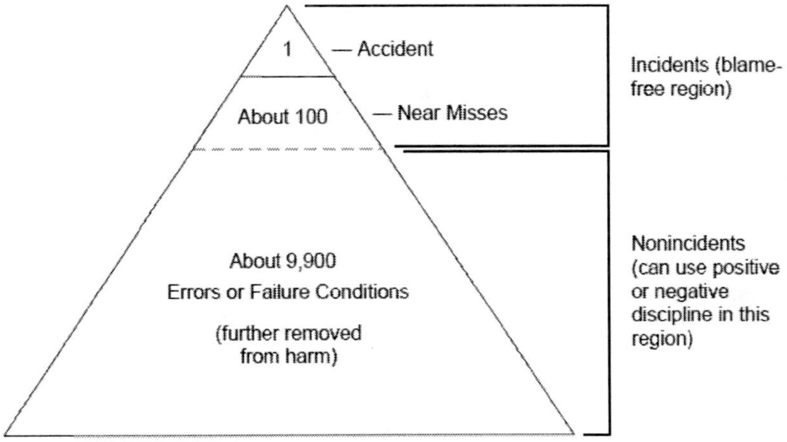

Figure 2. Relationship between Errors and Potential or Actual Impact [16]

Investigating near misses is critical to preventing accidents, because near misses share the causes and root causes of accidents; they are one or two barriers away from the loss/accident. We are very likely preventing many apparently unrelated accidents when we prevent the ones that are obviously related.

3.1. Identified problems

During the past 15 years, we have asked more than 5,000 students of our process safety management (PSM) courses and 3,500 students from our investigator leadership training courses how many near misses they get reported for every accident. The students represented about 400 companies, predominantly in chemical-related process industries (chemical, polymer/plastic, petrochemical, refining, oil and gas exploration, pharmaceutical, pulp and paper, etc.). The answers are quite disturbing. More than 95% said that their ratio of near misses reported to accidents reported falls in the range of 0 to 20. Less than 5% of the individuals said the ratio was greater than 5, and less than 2% said the ratio was higher than 10. Students noted that fear of disciplinary action, lack of management commitment, and lack of understanding of the difference between a near miss and a non-incident were the main reasons why near misses do not get reported. [16]

In conducting about 150 PSM audits and 7000 process hazard analyses (PHAs) during the past 22 years, we have found ratios from 0 to 105. For the first half of the 1990s, more than 90% of the facilities we talked to had ratios in the range of 0 to 0.5, and more than 95% had ratios in the range of 0 to 1. In the last half of the 1990s, more than 90% of the facilities had ratios in the 0 to1 range, and more than 95% had ratios in the 0 to 2 range. In addition, only a few (less than 2% of the facilities) had a ratio higher than 5. By 2005, many facilities had ratios in the 20 to 100 range and each of these had seen great gains (more on this in future articles), but the average across all companies we talked to increased only slightly. By 2011, the ratio of near misses to accidents (any loss event) was about 2 for companies we deal with. But by the end of 2011, many major companies had ratios well above 50! [16]

Our auditors and PHA leaders commented that the primary reasons for lack of reporting were about the same as those found in the survey. The barriers to getting near misses reported, discussed in detail later, are:

- Fear of disciplinary action.
- Fear of teasing by peers (embarrassment).
- Lack of understanding of what constitutes a near miss versus a non-incident.
- Lack of management commitment and lack of follow-through once a near miss is reported.
- An apparently high level of effort is required to report and to investigate near misses compared to low return on this investment.
- There is No Way to investigate the thousands of near misses per month or year!
- Disincentives for reporting near misses (e.g., reporting near misses hurts the department's safety performance).
- Not knowing which accident investigation system to use (or confusing reporting system).
- Company discourages near-miss reporting due to fear of legal liability if these are misused by outsiders.

The good news is that near-miss reporting appears to be improving, and the industry appears to recognize most of the barriers to near-miss reporting. The bad news is that the ratio is still very poor and improvement appears slow!

The number of companies that participated in the informal surveys suggests that the results are a statistically significant representation of the chemical industry.

A formal (written) survey was developed and e-mailed, faxed, and/or mailed to more than 100 companies. Of these, more than 12 replied originally in 1997. Since then, we have reviewed data in detail from more than a dozen major companies. These 25+ companies or affiliates provided data from more than 400 facilities, including more than 150,000 employees in manufacturing and operations. The data were from the chemical industry, polymer industry, refineries, drug/pharmaceutical companies, pulp and paper mills, petrochemical companies, and oil exploration/production.

Table 6 summarizes the results of both the informal and formal surveys.

Table 6. Summary of survey results [16]

Survey Items	Informal surveys		Formal Survey (or actual organization data)
	Classroom polls	Found during Audits and PHAs	
Current Near-miss Reporting Ratio Range (0-5 years prior)	1 to 105	2 to 20	0 to 105
Previous Near-miss Reporting Ratio Range (5-10 years prior)		1 to 5	NA
Current Near-miss Reporting Ratio Average (0-5 years prior)	2-3	3	5
Previous Near-miss Reporting Ratio Average (5-10 years prior)		1	NA
Goal for Near-miss Reporting Ratio (average)	20+	NA	20
Theoretical Upper Limit on Near-miss Reporting Ratio (average)	About 100	NA	100
Number of Participating Companies	400	300+	25+
Number of Participating Facilities	8500+	500+	400+

3.2. Objective

A company should strive to reach a ratio of 50-100 and investigate about 20 near misses per accident. This will provide a statistically significant sample of all incidents (and all important errors) and provide a company with sufficient feedback on which management system weaknesses are causing the errors and component failures. Various companies with different cultures have achieved high ratios with great return on investment.

Table 7 shows a brief description of possible solutions to overcome each one of the barriers identified in the survey.

Table 7. Summary of Survey Results[16]

Barriers	Solutions
Fear of disciplinary action	Implement a policy to NOT punish individuals when their errors lead to accidents and Near Misses.
Fear of teasing by peers (embarrassment)	Ensure that all employees understand the importance of near-miss reporting; demonstrate, through feedback of lessons learned.
Lack of understanding: Near miss vs Non-incident	Develop a list of "in-context" examples that illustrate what you consider to be Near Misses and what you consider to be non-incident
Lack of management commitment and lack of follow-through on reported near misses	Hold management accountable for achieving a Near-Miss reporting ratio
Apparently high level of effort is required to report/investigate Near Misses	Ensure that the data are entered in a database and queried regularly. Share the results with employees so they can see the value of the reported near misses
There is no way to investigate the thousands of Near Misses per month	Let front-line foremen or supervisors decide if a Near Miss or accident needs to be investigated
Disincentives for reporting Near Misses	Ensure that goals and incentives are not tied to lower incident rates (since this discourages reporting), but instead provide incentives for high Near-Miss reporting ratios
Not knowing which accident investigation system to use	Have ONE incident reporting system with ONE approach
Company discourages Near Miss reporting due to fear of legal liability	Involve legal on major Near Misses and accidents to ensure the results are protected as much as possible under attorney/client privilege.

For a complete description of the possible solutions for each barrier, refer to PII's paper *"Gains from Getting Near Misses Reported"*. [16]

4. Key Human Factors are Not being addressed

All accidents (or nearly all, if you consider that there are some natural phenomena that we either cannot guard against or choose not to guard against) **result from human error**. This is because humans govern and accomplish all of the activities necessary to control the risk of accidents. Humans influence other humans in the process – not only do humans cause accidents (unintentionally) by making errors directly related to the process itself, but they also cause errors by creating deficiencies in the design and the implementation of management systems (i.e., we make errors in authorities, accountabilities, procedures, feedback, proof documents and continual improvement provisions). Ultimately, these management systems govern the human error rate by directly causing or indirectly influencing the process. The process-related activities where errors have the most influence include:

- Designing a process
- Engineering a process
- Specifying components

- Receiving and installing equipment
- Commissioning
- Operating a process
- Maintaining, inspecting and repairing a process
- Troubleshooting and shutting down the process
- Managing process, procedure, materials, facility and personnel changes

The 10 human factors categories to be controlled by companies are:

- Available Time (includes staffing Issues) – for responses only
- Stress/Stressors (includes staffing issues)
- Complexity & task design
- Experience/Training
- Procedures
- Human-Machine Interface (includes tools)
- Fitness for Duty
- Work processes & supervision
- Work Environment
- Communication

4.1. Identified problems

Recent major accidents have highlighted the need for increased focus on human factors. The U.S. Chemical Safety Board (CSB) cited (US CSB Video, 2006) human factor deficiencies as one of the main contributors of the catastrophic accident at the BP Texas City Refinery in March 2005[17]. The human factor deficiencies included lack of control of worker fatigue, poor human-system-interface design, poor communication by radio/phone, out-of-date and inaccurate operating procedures, and poor (no) communication between workers at shift handover. The CSB cited similar issues from many other accidents[18] and has urged industry and the U.S. OSHA (the regulator) to pay much more attention to human factors. As a result, the recent U.S. OSHA National Emphasis Program for Refineries included human factors as one of the 12 core elements.

Implementing human factor engineering and policies to prevent accidents is not a new concept. Nearly all (or all, from a more complete perspective) of the causes and root causes of major accidents in the past 30 years have been the result of poor control of human factors. This has been cited in many root causes analysis reports and papers concerning these major accidents.

Process Safety Management (PSM) systems based on OSHA's PSM standard are likely lacking the fundamental human factor elements and implementation guides that, if applied across the applicable PSM elements, would work together to reduce human error. The *Risk Based Process Safety* guideline from the CCPS/AIChE (2007) does contain these human factor requirements.

Many site and company staff do not know there are standards for the control of human factors. However, at last count there were more than 300 non-governmental standards (such as from ANSI, ISO, IEEE, etc.) for the control of human factors, and there are more than 100 government regulations and standards and guidelines. Some of the better government regulations cover aviation, marine operations (shipping) and aerospace. An organization should devote resources to finding and implementing such best practices.

The following is a description of each of the human factors mentioned above and specific ways to control each. Our data from review of more than 2000 incidents reveal that:

- 90% of accidents had at least one root cause related to deficiencies in written work instructions.
- 70% of accidents had at least one root cause related to miscommunication between workers or between workers and supervisors.

- 40% of accidents had at least one root cause related to excessive fatigue of the worker. [5]

4.2. Objective

Not all errors can be prevented. Since the beginning of time, humans have tried to control error rate with more or less success. Human errors have been measured for hundreds of years. Psychologists have studied why humans make mistakes and have gradually put a science around human error probability. Today, the best models for control of human error in the workplace are generally agreed to be related to control of human factors; in turn, these have grown out of what was previously denoted performance shaping factors. What can be done to control each Human factor is discussed in greater detail next.

For a complete description of each human factor, its importance and how to control them; refer to PII's paper *"Human factors and their optimization"* [5].

Time available for the task

Task design is directly linked with this category. Process design, procedure development and worker scheduling need to take into consideration not only the time it takes to complete each task but also the time it takes to manage usual as well as potentially unusual or unexpected situations. True; we do not create norms for exceptions, but exceptions sometimes are more usual than realized.

Stress/Stressors

A well-designed, organized and managed work process helps maintain and promote individual health and well-being, but stresses will arise through no fault of the organization. Supervisors should be trained to recognize stress in workers and, in severe cases, reassign them and help them get assistance to deal with the stress and in other cases, reduce their job pressures for a day by shifting assignments. A wellness program may also be necessary to help cope with stress. Simple approaches, such as training workers to recognize and deal with stress, can help most individuals. Approaches to deal with stress include positive self-talk and practicing relaxation techniques (i.e., breathing control).

Complexity & task design

If roles and responsibilities for accomplishing tasks are not clearly defined, then there will be a risk of serious errors. It is also important that employees and engineers/designers understand the characteristics of the work elements involved, how each element passes information to the other, how each person involved communicates, and that each person has guidelines to adhere to during task design and procedure development.

For optimal performance, task designers often must integrate human and automated equipment in their task analysis.

It is crucial for task designers also to have an understanding of the appropriate allocation of roles and responsibilities to the various participants in a task. Understanding the characteristics and limitations of all of the humans and automated systems involved in the task (particularly critical tasks) will allow for establishing additional controls where needed.

The task analysis process includes *Task definition* which is the initial operational description of the operator tasks required for execution of a given system function; *Performance analysis* of all tasks with respect to human performance requirements; *Error and workload analysis* to be included in the design process, particularly when the likelihood of unacceptably operator error is high; and *Evaluation* since evaluating the design of a new or complex task is important to ensure that the task can be effectively performed as designed. For very high consequence scenarios involved complex human interaction, it is likely prudent to invest in a human reliability analysis (HRA).

Experience/Training

Training to master the Knowledge, Skills, and Abilities (KSAs) required for a job may be obtained from a variety of sources. Companies provide the training or it is obtained from other sources (e.g., trade schools, universities, contractor organizations, etc.) to reach the skills and knowledge established in job descriptions. There are several factors that may result in personnel not mastering the required KSAs for a job. These include course design and delivery methods, course completion, practical skills demonstrations or simulation, and training frequency. Course design begins when the learning objectives have been identified. The design process consists of determining the delivery methods (simulator, on-the-job training, etc.), number of hours required to cover the materials, instructor qualifications, etc. Although some methods, materials and instructors may be more effective or efficient than others, the important issue is that the course content is complete and addresses all of the relevant KSAs, so that the learning objectives are met and the KSAs mastered.

Another determinant of KSA mastery is course completion. Although this factor appears obvious, there are often competing demands on personnel that may pull them out of training at times. As a result, they may miss the instruction related to specific KSAs. Testing may not identify the KSA deficiency because it is impossible to test mastery of all KSAs. Sampling techniques are used to generate examinations. If attendance and participation are not controlled, some personnel may miss training on specific KSAs and testing may not identify the deficiencies before an error is committed.

Another factor affecting KSA mastery is forgetting. An individual's ability to perform a task will degrade over time unless the relevant KSAs are refreshed. Proficiency training (i.e., refresher) will be required for some tasks to maintain the level of mastery that was demonstrated following initial training. One function of training programs is to identify those tasks that require proficiency training.

If certain tasks are performed frequently, proficiency training may be unnecessary. By performing a task, personnel practice the task and obtain feedback on where they have weaknesses. Task performance refreshes the KSAs and successful task performance verifies that proficiency has been maintained. Furthermore, since humans learn by comparison to similar activities, as a worker practices one task, they are learning about similar tasks. Care must be taken here to avoid the practice drifting away from what the procedure requires. Also, if new operators are trained by experienced operators, you must ensure that the new employee is trained according to the procedure, not whatever practice is actually being used. (Of course, ideally, the practice and procedure will be identical.)

Operating and maintenance procedures

Many procedures do not follow best practices for controlling human error, and so the written procedure actually "contributes" to increased error rates. Additionally, many organizations do not have guides on how to troubleshoot (what to do when process deviations occur). The best practice rules for writing and validating procedures have been published for many years. [6], [7], [8], [9]

Procedures have not traditionally been developed from the perspective of optimizing human factors; instead, procedures have been traditionally developed to meet a compliance requirement to have written procedures. Examples of procedure deficiencies (inaccuracies) include:

- Incorrect/incomplete/nonexistent (most procedures we have audited have been only 70-85% accurate – the inaccuracies include missing critical steps, steps as written are not what needs to be done, or the steps are out of sequence)
- No/misplaced warnings or misuse of warnings (for example, a warning should never **state** the action to take; it should instead **emphasize** the action to take an the statement of the action should be in a numbered step)
- Poor format and presentation rules

Human-machine interface

Design for operability refers to designing the Human-System Interface (HSI) to be consistent with the abilities and limitations of the personnel who will be operating it. Weaknesses in the design processes can result in an HSI that is not well suited to the tasks that personnel must perform to ensure plant safety, resulting in increased workload, decreased performance by personnel, and an increased likelihood of errors.

Design for maintainability refers to designing the HSI and associated plant equipment to ensure that personnel are able to perform necessary maintenance activities efficiently. Weaknesses in the design process can result in systems that impose excessive demands on personnel for maintenance and, therefore, are prone to maintenance errors or problems with reliability and availability. *If it is hard to reach, workers will make more errors, including the error of deciding it is not worth it.*

Design for flexibility refers to the way that changes, such as upgrades to the HSI, are planned and put into use. A new HSI component may require the user to perform functions and tasks in new ways. Skills that the user developed for managing workload when using the former design, such as ways for scanning information or executing control actions may no longer be compatible with the new design. The new HSIs must also be compatible with the remaining HSIs so that operators can use them together with limited possibilities for human error. Also, HSI modifications may not be installed or put into service all at one time, causing the user to adapt to temporary configurations that are different from both the original and final configurations. Weaknesses in HSI implementation can increase operator workload and the likelihood of errors.

Fitness for duty

Implement a company fitness-for-duty program with a primary responsibility for detecting and preventing impaired personnel from performing tasks that may affect productivity and safety. Medical evaluations of personnel, behavioral observation programs, employee assistance programs, and drug and alcohol testing are used to detect impairment. Weaknesses in this program may allow impaired personnel to have access to vital areas in a plant where they could commit errors.

Overtime Policies and Practices – Most companies establish limits for work hours to reduce on-the-job fatigue and the potential consequences for poor task performance. Routine authorization for work hours in excess of those recommended may result in fatigued workers. Furthermore, a practice of excluding training or meetings that occur outside of an individual's normal work schedule from work-hour limitations will also contribute to fatigue. But in many cases these have weak enforcement, especially during shutdowns when the company is under stress to re-start soonest.

Shift Scheduling – Shift scheduling may also affect the likelihood that personnel will show performance decrements due to fatigue. A change in the assigned shift or a rotating shift schedule will disrupt circadian rhythms and may increase the likelihood of errors. So, a company must choose the proper shift rotation to allow adjustments to sleep patterns.

Work processes & Supervision

Written plant policies and procedures are meaningful only when they are enforced; otherwise, worker practices are the policy. Once discrepancies are tolerated, individual workers have to use their own judgment to decide what tasks are necessary and/or acceptable. Supervisors and managers should vigorously enforce plant policies/procedures and ensure that plant policies/procedures and practices are revised as necessary to be consistent. To enforce well, a supervisor must (1) lead by example, (2) watch workers often to ensure they are following company policies and procedures, and (3) determine the correct course of action (including recommending changes to policy and procedures) when workers are found deviating from procedures.

Work environment

Programmatic causes of task environment errors are typically found in the company's processes for designing human-system interfaces or in managing maintenance activities. Other programs may also be implicated. Common programmatic causes of task environment errors include:

Industrial Hygiene and Radiation Protection – These programs are responsible for ensuring that task environments have been evaluated to identify hazards and that needed controls are implemented to minimize exposures. Weaknesses in these programs may result in personnel working in task environments that are conducive to errors.

Work Planning and Control – Weaknesses in the work planning and control system may allow work to be planned without consideration of adverse environmental conditions and performed without the necessary compensatory measures. For example, communication devices may not be provided in noisy environments to support task performance. For tasks that involve unusual physical positions or cramped workspace, additional time to complete the task may not be scheduled. Rest breaks for hot and cold environments may not be planned into the work, or additional temporary lighting may not be provided if the work site is not adequately lighted.

Procedures – Weaknesses in the company's procedure development process may result in the design of procedures that are inappropriate for the conditions in which they will be used. For example, procedures that may be used at night, outside and in the rain should be laminated and the type size should be larger to ensure the procedure can be read. Procedures that will be used in vibration conditions may also require larger type size than procedures read in the stationary environment of the control room.

Human Factors Engineering – Weaknesses in the human factors engineering program may result in the installation of new equipment or systems without consideration of task environment characteristics. For example, the impact of control room lighting on the visibility of digital displays or effects of vibration on the legibility of dials or gauges at local control stations should be considered before installation.

Communication

Successful communication requires several steps. The sender first develops the intention to communicate either verbally or in writing. The sender then composes a message that presents the meaning as clearly as possible. The receiver must pay attention to the message and then interpret its meaning. If the communication is successful, the receiver interprets the message consistently with the sender's intended meaning. Table 8 summarizes the rules for successful verbal communication.

The similarity of the meanings given to the message by the sender and receiver can be verified through feedback. An example of feedback verification in verbal communication is when the receiver "repeats back" the message and the sender either agrees with the receiver's repeat back or corrects it. Verification feedback serves an important error-checking function in the communication process. It also allows supervisory oversight of communications to catch errors before they have consequences.

A sender and receiver must both be active for communication to be effective. The sender and receiver share responsibility for ensuring successful communication. However, when companies analyze the causes of events, errors in sending messages are more often identified than errors in receiving. The reasons for the difference are unclear. A company's investigation should consider sending and receiving errors and corrective actions should address both to be effective.

Shift handover is another critical aspect of communication. The status of equipment, repairs that occurred, problems faced, upcoming chores, etc., are all key information the oncoming worker will need to know. One good guide on shift turnover is U.S. DOE 1038-93. [19]

Table 8. Rules for effective verbal (face-to-face, radio, phone) Communication (© PII, 2012)

#	Communication rules
1	Senders MUST require repeat back by the receiver; ask questions to make sure of understanding.
2	Receivers of messages MUST repeat a message they receive.
3	Communicate one task at a time.
4	Sender & receiver must use the standard name and component number (and building name and number) and/or for related reference material (drawing, procedure name/number) in each message.
5	Sender must require receiver to give feedback of foreseeable conflicts. Sender (if message is to subordinate) should follow-up in the field if possible, to make sure message was understood.
6	Receiver must request clarification anytime they believe the message is confusing.
7	Do NOT give verbal instructions to workers who do not have the demonstrated skills to correctly understand and perform the task assignment (unless you plan to supervise/coach the task yourself).
8	Communicate to the most senior member on the work crew.
9	Know your audience & change message accordingly to compensate for lack of knowledge (when appropriate; see other related rules).
10	Use units (and use approved units, such as SI) for process parameters.
11	Always count from left to right when giving instructions related to multiple choices of equipment.
12	Use approved jargon only.
13	Use the words **check**, **make sure**, and **actions** consistently (see procedure writing rules).
14	When possible, write out the task (use a special procedure or work order) rather than communicating verbally. Use formalized templates for consistency and make sure the receiver is provided all information (such as work orders and drawings) necessary for the task.
15	Perform pre-job meeting with work crew (including with contractors or construction company).
16	Perform walkthrough at the location where the work will be completed.
17	Have a backup communication method. Don't rely on one mode of communication. In emergency situations have hand signals as backup to loss of verbal communication.
18	Talk to the operator or technician who is doing the work (no delegation of work); but see Rule 8.
19	Receiver (worker or group) must report back when work is complete.
20	If confusion exists in implementing a task, the shift supervisor (and higher, if supervisor is unavailable) must be contacted to make sure he/she understands the problem.
21	Do NOT use a PA (public address system) for process instructions, since this method of communication does not allow repeat-back.
22	Use Open questions and non-confrontational questioning methods when requesting clarification.
23	When communicating remotely (by phone or radio), if the message is not understood on the second attempt at clarification the Sender must find the Receiver and communicate face-to-face.
24	For large or complicated jobs, in addition to repeat back, the sender must ask the receiver for an assessment of the pre-job briefing to ensure the workers are not confused.
25	Workers and supervisors must keep a shift log to aid in turnover between shifts. Workers and supervisors and support staff (if necessary) must have a minimum or 15 minutes overlap with a relieving shift. *See shift turnover standard for more details.*

Conclusion

Closing the gaps in these 4 areas has a greater than 100:1 payback!

Acronyms and Abbreviations

ACC - American Chemistry Council
AIChE – American Institute of Chemical Engineers
API – American Petroleum Institute
CCPS – Center for Chemical Process Safety (of AIChE)
CMA - Chemical Manufacturers Association
CSB – Chemical Safety Board
DM – Damage Mechanisms
FMEA – Failure Mode and Effects Analysis
GCPS -- Global Congress of Process safety
HAZOP – Hazard and Operability; as in HAZOP Analysis or HAZOP Study
HRA – Human Reliability Analysis
HSI – Human-System Interface
KSA – Knowledge, skills and abilities
LOC – Loss of containment
LOPA – Layer of Protection Analysis
MOC – Management of Change
OSHA - US Occupational and Health Administration
PHA – Process Hazard Analysis
PSM – Process Safety Management
RBPS – Risk Based Process Safety
SOP – Standard Operating Procedure

References

[1] *Guidelines for Technical Management of Chemical Process Safety,* CCPS/AIChE, 1985.

[2] "The Cost and Benefits of Process Safety Management: Industry Survey Results," W. Bridges, *Process Safety Progress*, AIChE, January 1994.

[3] "Necessity of Performing Hazard Evaluations (PHAs) of Non-normal Modes of Operation (Startup, Shutdown, & Online Maintenance)", W. Bridges and Mike Marshall (US OSHA), *18th Annual International Symposium, Mary Kay-O-Connor Process Safety Center, College Station, TX, October 2015.*

[4] *Guidelines for Risk Based Process Safety*, CCPS/AIChE, 2007.

[5] "Human factors and their optimization" W. Bridges and G. Collazo-Ramos, *8th Global Congress on Process Safety*, Houston, AIChE, April 2012.

[6] PII Course 5, *Writing effective operating and maintenance procedures*, © 2003-current

[7] "Best practices for writing operating procedures and trouble-shooting guides", W. Bridges and R. Tew, *13th Global Congress on Process Safety*, San Antonio, AIChE, March 2017.

[8] "Effective Procedure Writing – A Two-Part Series (Part 1: Reducing Risk of Worker Errors; Part 2: Addressing Process Safety Management Requirements for Operating Procedures)," *Chemical Engineering Progress*, 1997, W. Bridges with T. R. Williams (both of JBFA).

[9] "Best Practices for Writing Operating Procedures and Trouble-Shooting Guides", L. Madden and W. Bridges, *12th Global Congress on Process Safety*, Houston, TX, AIChE, March 2016.

[10] *Guidelines for Hazard Evaluation Procedures*, 3rd Edition, 2008, CCPS/AIChE.

[11] "API RP 571 - Damage Mechanisms Affecting Fixed Equipment in the Refining Industry", API, 2nd edition, 2011.

[12] Bridges, W.G., et. al., "Addressing Human Error During Process Hazard Analyses," *Chemical Engineering Progress*, May 1994.

[13] "Necessity of Performing Hazard Evaluations (PHAs) of Non-normal Modes of Operation (Startup, Shutdown, & Online Maintenance)", W. Bridges and Mike Marshall (US OSHA), *18th Annual International Symposium, Mary Kay-O-Connor Process Safety Center, College Station, TX, October 2015 and 12th Global Congress on Process Safety, April, 2016.*

[14] "Recipe for a Complete Process Hazard Analysis – Especially Addressing the Key Demands from US CSB", R. Tew and W. Bridges, *13th Global Congress on Process Safety*, San Antonio, AIChE, March 2017.

[15] PII Course 8, *PHA/HAZOP Leadership training*, © 2003-current.

[16] "Gains from Getting Near Misses Reported", W. Bridges, *8th Global Congress on Process Safety*, Houston, AIChE, April 2012.

[17] US Chemical Safety and Hazard Investigation Board (CSB), *Anatomy of a Disaster: Explosion at BP Texas City Refinery,* published 2006.

[18] US Chemical Safety and Hazard Investigation Board (CSB), *"Pesticide Chemical Runaway Reaction Pressure Vessel Explosion – Bayer CropScience, LP",* published 2008.

[19] "DOE-STD-1038: Guide to Good Practices for Operations Turnover", US DOE, 1998.

546668 Recycle Line from NGL Spheres to Residual Gas Header

Abstract:

HNGL plant is received 1.6 BSCFD of Haradh gas flows into the DGA Gas treating units where CO2 is removed and the produced sweet gas is directed to the NGL recovery units .Hawiyah sales gas is pretreated at HGP for CO2 removal ; hence the produced 2.4 BSCFD flows directly to the NGL units where it is mixed with the Haradh gas. The mixed gas then is processed in the NGL units producing C2+ NGL and sales gas stream. The NGL is pumped to storage spheres and subsequently shipped to NGL piping network through Shedgum Gas Plant. The Residue Gas flows to the sales gas compression area which increase the pressure allowing routing back the gas to the sales gas grid.

The purpose of this project is to install vent line from the two NGL Spheres to Sales Gas (Residue Gas) common header at Hawiyah NGL Plant. The aim of this project is to minimize the continues flaring from the NGL spheres , safe plant resources and protect the environment

Sphere PVs were venting the NGL vapor to the cryogenic burn pit when the pressure of the sphere exceeds 350 psig. In order to lower sphere pressure without losing NGL production, newly installed PVs are set at lower pressure set point of 305 PSIG to recycle the spheres vapors to NGL trains residue gas line, downstream of the brake-compressors..

During summer months July and August (if associated with maximum plant production), spheres are venting about 4.0 MMSCFD of Methane. Based on Methane price of 4.4 $/MMBTU then total saving is 1.05 MM $/year.

547776 Practical Notes on Ragagep Compliance

Process Safety Management (PSM) and Risk Management Plan (RMP) regulations have long required facilities to ensure that they comply with Recognized and Generally Accepted Good Engineering Practice (RAGAGEP). In recent years, authorities having jurisdiction have placed increased emphasis on RAGAGEP compliance. Specifically, emphasis has been placed on documentation and compliance with the industry codes and standards to which a facility is designed and maintained. This poses a unique challenge to older facilities, which have been designed to codes and standards long out of date and especially to owners and operators that do not have access to engineering documents used in the original design of their facility. For facilities using outdated codes, OSHA requires that the operating company document that "equipment is designed, maintained, inspected, tested, and operating in a safe manner."

Over time, the gradual evolution of industry codes and standards may cause facilities to fall out of compliance with the RAGAGEP requirements of the regulations. This paper provides example cases based on authors' experience with various owners and operators of older facilities. Practical methods are discussed for achieving and maintaining RAGAGEP compliance. It includes examples of comparing systems designed to outdated codes and standards against the latest codes, to ensure that they are able to maintain compliance with RAGAGEP.

548482 Propane Safety: Investigation Findings and Lessons Learned in the 2014 Philidelphia FOOD Truck Explosion

On July 1, 2014, a propane cylinder catastrophically failed and exploded on the back of a food truck in Philadelphia, PA. The explosion led to a sudden release of pressurized propane vapor and superheated propane liquid, which rapidly evaporated resulting in a large white cloud that entered and engulfed the rear of the food truck. This cloud ultimately found an ignition source within the food truck and resulted in a significant fireball burning both inside and outside of the food truck. The explosion and ensuing fire fatally injured the food truck owner and her daughter, and also caused injuries to 11 other individuals. This paper will first present the results of our investigation into the cause and origin of the catastrophic failure and explosion event, which included: (1) analyzing video footage of the explosion and ensuing fire; (2) applying propane phase diagrams to determine the expansion and resulting pressures within the cylinder prior to failure; and (3) applying blast techniques to determine the post catastrophic blast and ignition during the event. These techniques included advanced computational fluid dynamic modeling.

The paper will also discuss safety critical design features and processes in the propane industry that must be in place to ensure the safe use and filling of propane cylinders, which include: cylinder requalification; having a fixed maximum liquid level gauge; having a pressure relief valve; and filling cylinders by weight or volume.

548391 Using the Risk Analysis Screening Tool (RAST) in Process Design to Manage Risk

The Risk Analysis Screening Tool (RAST), available through the Center for Chemical Process Safety (CCPS), courtesy of the Dow Chemical Company can enable users to identify hazards and risk scenarios to better manage process safety risks. RAST is a spreadsheet based tool that uses simplified algorithms to identify and evaluate risk scenarios of a unit operation. The risk scenario identification is based on generic HAZOP deviations applied to the unit operation. An engineer can use RAST to identify risks to proposed process designs or plant modifications in the design phase of a project, before P&IDs are developed and a full process hazard analysis study is done. This can save time in process design by weeding out higher risk process alternatives as the process flow diagram is being developed.

The utility of RAST is demonstrated by application to actual case studies. RAST is applied to the runaway reaction at T2 Laboratories in 2007 and to the chlorine release at DPC Enterprises in 2003 showing how the hazardous events and consequences could have been analyzed during the design phase of the projects.

Drug Product vs. Process Safety: Integrating Both into the Design Process from Concept Development to Start-up

Jeffrey Heil, P.E.
Lead Process Engineer
DPS Group Global
959 Concord Street, Suite 100
Framingham, MA 01701
Jeffrey.Heil@dpsgroupglobal.com

Prepared for Presentation at
American Institute of Chemical Engineers
2019 Spring Meeting and 15th Global Congress on Process Safety
New Orleans, LA
March 31 – April 3, 2019

Drug Product vs. Process Safety: Integrating Both into the Design Process from Concept Development to Start-up

Jeffrey Heil, P.E.
Lead Process Engineer
DPS Group Global
959 Concord Street, Suite 100
Framingham, MA 01701
Jeffrey.Heil@dpsgroupglobal.com

Keywords: Process Safety, Process Design, Biotechnology, Inherently Safe Design

Abstract

Process safety has been around for over 350 years but may easily be overlooked or treated as an afterthought in the design process, often leaving hazard reviews and safety valve considerations to the end of detailed design. In the chemical industry, emphasis on safety is generally in respect to process safety, but in the growing biotechnology industry the emphasis is often shifted to product and patient safety. Due to this subtle yet significant difference in focus, it is often challenging for the design team on these projects to secure the necessary project resources to ensure both robust product and process safety in the facility being designed, often on an accelerated schedule. Leveraging experience from working in the oil and petrochemical industry for over 10 years, an in-depth understanding of the project design cycle, and by asking some simple questions along the way; process safety in design can be assured for even the most challenging projects, without a significant impact on project resources.

1 Introduction

1.1 Brief History

The first safety valve, developed in 1681 by Frenchmen Denis Papin, protected the first ever pressure cooker, one of which was used to make meat jelly for England's King Charles II. [1] Papin first invented the steam pressure cooker in 1676 to make the jelly and other food products but then realized no one would be willing to operate the cooker given the risks. Having identified the hazard, he set out to invent a way to mitigate that hazard while ensuring the King got his jellied meat pies.

Following development of the first weighted lever style safety valve invented by Papin, little progress was made until the early 1800's when locomotive steam engines gained notoriety both as a superior means of transportation and a hazardous machine often operated beyond its limits in the name of more power and speed. Safety valves were revised throughout the early 1800's to make these locomotives safer and formally modernized in 1856 with the invention of the first spring actuated pressure relief device that could not be modified to a different pressure, while in service [2].

Modern process safety, as we can still see it in existence today, dates to the early 1900's with the development of modern oil refineries and all the hazards that they include. With these refineries and their complex chemical processes came a wealth of experimentation, safety system design, and industry supported safety initiatives. The goal was the same as it is today, ensuring safety to the employees, the environment, and the facility. The problem with many of these early safety initiatives is that different companies had different ideas of what safe meant and relied on historical data to predict the future. The notion that "it has been running like this for 20 years and will continue for another 20 years" lead to a raising number of deadly process safety incidents, particularly throughout the 1970's to 1980's (i.e. Bhopal, India Methyl Isocyanate release and Piper Alpha explosion in the North Sea.). Plants were operating to the status quo, not improving their practices new technology became available. In parallel, urban sprawl forced industrial plant operation closer to populated areas, increasing risk to the community. The OSHA process safety management (PSM) initiative was introduced into law in the United States to address this very problem [3]. The PSM standard was performance-based meaning that as new technology was developed and safer systems could be designed, those safer designs had to be implemented; i.e., the status quo was no longer good enough. The PSM standard was ratified in 1992: hence, even though the oil and petrochemical industries in the United States have been around for about 100 years, they have only been operating under this standard for 26 years.

Contrast this with today's biotechnology companies. Gentech, one of the earliest, received FDA approval for the insulin drug Humulun in 1982. Like any new technology, investment and growth in biotechnology was slow to start, not taking off until early 2000's. There are many metrics for tracking growth of the Biotechnology industry such as venture capitalist investment and the number of companies in operation, but the high number of mergers and acquisitions over the past 30 years makes such comparisons incomplete. The Nasdaq Biotechnology Index (NBI) was established in 1993 with a baseline index value of 200 and provides a good metric for tracking industry growth over time. After holding steady in the low 200-300s throughout the 1990's, the NBI surged to 1,300 in 2000 as the Biotechnology bubble was forming. The index has steadily risen to a peak of 4,003 on 01-July-2015 and is hovering currently around 3,500 (Q1 2019), exemplifying the continued growth of the industry over the past 20 years. [4]

Pharmaceutical companies (e.g., Aspirin, invented by Bayer in 1899) have been around almost as long the oil companies have and are sometimes confused with biotech. Biotechnology companies are adolescent at best, having typical company life of less than 20 years with little process safety oversight beyond building code requirements as they do not meet the hazard requirements of the OSHA PSM Standard. Not meeting the hazard requirements of the PSM standard does not exempt these companies from a responsibility to ensure a safe working environment. If anything, their proximity to high density population areas and minimal operating experience places them at a higher risk with respect to process safety, but in many cases the operators and designers have only worked in these types of facilities and may not fully appreciate the safety concerns present. Coming from a background where process safety is paramount and the product quality, while important, takes second place, the author has made it his personal mission to increase awareness throughout the Biotechnology industry one project at a time.

1.2 Process vs. Product Safety

In the chemical industry, emphasis on safety is always in respect to process safety; ensuring that the public, employees, environment, and facility (in that order) are sufficiently protected from inherent hazards of a process. "Safety" refers to containment of the process and protection of personnel from harm. If a hazardous material cannot escape the process; it cannot be breathed in, released to the environment, set on fire, or result in damage to equipment that may cause injury to nearby personnel. The previously described process safety management standard formalized the set of tools used by these industries to ensure current best practices are used when it comes to process safety in hazardous environments. The goal of the PSM standard and implications of it being "performance based" is summarized as follows:

> *"... Recognizing the dangers associated with highly hazardous chemicals, Congress mandated (by the Clean Air Act Amendments of 1990) that OSHA set requirements for the management of highly hazardous substances to prevent and mitigate hazards associated with catastrophic releases of flammable, explosive, reactive, and toxic chemicals that may endanger workers. As a result, OSHA developed the Process Safety Management (PSM) standard (issued in 1992), which covers the manufacturing of explosives and processes involving threshold quantities of flammable liquids and flammable gasses (10,000 lbs.), as well as 137 listed highly hazardous chemicals. The PSM standard (29 CFR 1910.119) requires employers to implement safety programs that identify, evaluate, and control these hazards. Unlike some of OSHA's standards, which prescribe precisely what employers must do to comply, PSM is a "performance-based" standard-it outlines key features of safety programs for controlling highly hazardous chemicals, and employers have the flexibility to tailor their safety programs to the unique conditions at their facilities..." [4]*

The biotechnology industry on the other hand often has a more focused emphasis on product safety. Product safety ensures that all current good manufacturing practices (cGMP) are being followed with the goal of protecting the end-user or patient from hazards of a contaminated biological drug product. Product safety was important for the pharmaceutical industry because the medicines being produced were intended for use in the human body and would chemically alter the bodies function. Product cross contamination could result in serious health side effects, especially with highly potent or specialized substances. Biotechnology processes work by growing viruses and manipulating this viral material to deliver highly specialized medicines to the patient. Poor manufacturing conditions could allow unintended bacterial growth alongside the desired product, leaving the future patient with a potentially deadly virus or bacterial infection instead of a cure. The US Food and Drug Administration (FDA) passed the first cGMP regulations in 1969 [6], making them older and with many respects better understood in the industry than the PSM standard is in the oil and petrochemical industry. The practiced engineer will note several similarities in the quoted statements describing the PSM standard and the cGMP standards.

> *"The CGMP requirements were established to be flexible in order to allow each manufacturer to decide individually how to best implement the necessary controls by using scientifically sound design, processing methods, and testing procedures. The flexibility in these regulations allows companies to use modern technologies and innovative approaches to achieve higher quality through continual improvement. Accordingly, the "C" in CGMP stands for "current," requiring companies to use technologies and systems that are up-to-date in order to comply with the regulations. Systems and equipment that may have been "top-of-the-line" to prevent contamination, mix-ups, and errors 10 or 20 years ago may be less than adequate by today's standards." [7]*

Product safety requires process containment too, but in many cases the consequences are less likely to cause bodily harm so long as product quality control procedures ensure quarantine of the affected drug product. Note that this statement applies to the biotechnology industry, the consequences of a loss of containment in the pharmaceutical industry can be as deadly as in a chemical plant, depending on the substance released. Fifty years ago, when the cGMP's were adopted, pharmaceutical operators already knew about process safety and adopted it as readily as oil refineries and industrial chemical plants did. Twenty years ago, biotechnology was in its infancy and there were few large-scale biological processes in operation. Ten years ago, most biological processes ran at atmospheric pressure using water-based chemistry. The risk associated with potential process safety events was minimal in comparison. Today, many biotech processes use advanced solvent-based chemistries needing large volumes of hazardous flammable solvents, a key element in the severity of an incident when a loss of containment occurs. Add to this unique process chemistries such as dichloroacetic acid, methyl chloride, hydrochloric acid, and sodium hydroxide; all of which can easily corrode stainless steel, a common metal alloy used in biological processing, and unique hazards become a reality even at small scale.

2 The Design Process

Designing processes to produce goods takes time, time that the future producer of those goods doesn't always like to part with because "time is money". Each month spent designing or building the facility to manufacturer a product is a month of revenue missed or often in biotech, a month of trials that could be running. A typical design process includes several independent stages or phases. An ideal project would separate each of these phases with well-defined checks creating structure, an opportunity for detailed reviews, and adding time to the project. Initial concept development, feasibility, and design of large-scale chemical or industrial facilities often involves acres of real estate, spans years, and costs millions of dollars. By contrast small-scale bioprocessing facilities, often less than 40,000 ft^2, are fully designed in terms of weeks or months and leverage parallel work paths where possible.

The impact of such a short design schedule is that time for staggered activities, separated by well-defined stage-gates, is often sacrificed in the name of the project schedule. It is possible to crash a project schedule, completing the design early, but doing so limits the amount of design work that can be completed based on previously completed design elements. If a project schedule allots for 6-8 weeks between establishing the basis of design for a facility and documents that are to be issued for vendor bid, scheduling one week for a hazard analysis and another week to implement the resulting design changes (25% of the schedule) is a challenge. By the time the hazard analysis is complete, many aspects of the design have been progressed and resulting changes will result in unbudgeted rework. The same is true for the design of key safety systems such as relief devices or process interlocks that require a complete system design to be finalized. The engineer can anticipate credible overpressure scenarios and safe operating limits for a process but cannot complete these design elements until piping to and from the relief device is complete and the hazard analysis is approved, accepting the level of protection assigned to each process step. It is for these reasons that it is not uncommon to see relief devices for a process being sized and planned for as one of the final project tasks. Explaining to an architect that a storage tank must be moved to meet a flammable liquid set-back identified during a hazard review or that another roof penetration is required for a relief device tail pipe two weeks before the completion of a design are just two examples of real-world challenges that project teams may be faced with.

Understanding the different design phases or stage-gates, the tasks that occur during them, and their impact on the ultimate facility design is important if one is to create a plan for addressing process safety on such a short schedule. Table 1 summarizes one example of a design process, key stage-gates, and design elements to be considered in each phase. Note that jumping a phase and for example combining schematic design and design development (the most common violation of the design process) into one phase can confuse the separation of tasks between these phases and eliminate checks important to the design process.

Table 1 – Key Aspects of the Design Process

Phase	Process Design	Process Safety
Feasibility Study	• Identify product to be manufactured and major unit operations • Establish block flow diagram of process • Identify facility requirements • Confirm business case for proposed facility	• Identify hazardous materials used in the process • Identify hazardous operating conditions (i.e. high pressure) • Establish acceptable risk criteria for operation of the facility

Phase	Process Design	Process Safety
Concept Design	• Breakdown individual processing steps to the equipment level • Develop process flow diagrams for each major process step • Develop preliminary heat and material balance • Develop equipment list and identify long lead equipment (to prioritize design development tasks) • Detailed layout of process equipment • Identify utility requirements and size utility systems • Size and layout major process headers	• Develop a comprehensive hazardous and flammable materials inventory / determine if additional design precautions are required • Identify potential material compatibility concerns and material specifications for each process step • Complete a code review in collaboration with the full design team to ensure all building, fire, and design codes are being followed • Ensure facility citing requirements are met in the layout
Design Development	• Develop initial P&IDs for the process • Identify and document key control loops and automation requirements • Finalize design of all process equipment (size, pressure rating, operating conditions) • Complete detailed sizing and specification of major process headers and utility lines • Finalize process heat and material balance • Prepare equipment and piping specifications and datasheets	• Confirm inherently safe design where possible (pressure profile versus equipment design) • Identify safe operating limits for each unit operation, add control points as required • Review P&IDs for relief devices and confirm their discharge locations • Complete scenario analysis and preliminary sizing for all relief devices. Based on preliminary sizing, define relief device inlet/outlet piping • Confirm piping specifications versus services • Confirm vessel sizing versus hazardous material inventory and update as required • Complete a preliminary hazard analysis on the full process at the end of the design development stage

Phase	Process Design	Process Safety
Detail Design	• Incorporate all recommendations from the design development preliminary hazard analysis • Finalize all equipment and line sizing • Fully specify all process components including equipment, lines, valves and instruments	• Ensure that all actions from the preliminary hazard analysis are incorporated into the design • Review and finalize relief device sizing • Complete sizing and design of effluent handling system for relief device discharges • Double check the set points of key process variables versus the operation's safe operating limits, previously defined. • Review final equipment specifications against material compatibility matrix and equipment specifications

3 Safety Throughout the Design Process

Process safety should be considered in the earliest stages of design. One principle of process safety design is that inherently safe design, creating a system that physically cannot fail, is the only way to prevent an incident from occurring. The easiest example of an inherently safe system is designing a pressure vessel to withstand maximum system pressure, even if all process controls were to fail. While an easy goal on paper, inherently safe design is not always possible and often never fully achievable. A vessel can be designed to the proper pressure, but an unforeseen corrosion issue could reduce the design pressure making the vessel susceptible to premature failure. In anticipation of this and to protect against it, engineers design-in multiple layers of protection to a process starting with inherently safe design at the concept level and ending with proper emergency response procedures developed during commissioning and simulated during plant operation. Continuing with the vessel example, minimizing the quantity of material stored and having an inventory management plan to support smaller on-hand inventories is another way to reduce hazards associated with the process. Additional design principles or layers of protection used are summarized in Figure 1. Different layers of protection against a process safety event are established during concept development and built-up throughout the design process.

Figure 1. Typical Layers of Protection Applied to a Design

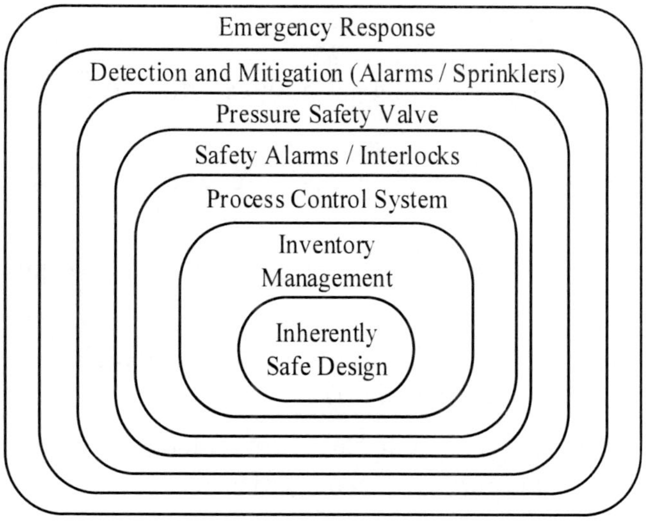

3.1 Concept Design

Concept design is one of the earliest stages of the design process. During concept design types of process equipment are first selected, their location mapped out, and consideration given to how materials will be heated up, cooled down, transferred, and stored when not in use. By asking a few questions as these fundamental design decisions are made, the design predisposes its self to being inherently safer.

- What is the worst-case operating condition, can the system be mechanically designed to handle those conditions?
- Where will hazardous materials be present and can that equipment be installed away from high concentrations of personnel?
- What hazards do the materials pose (toxic, flammable, corrosive)?
- How much of a potentially hazardous raw material must be on-hand? Weekly vs. monthly delivery may reduce stockpiles by 75%.
- Is there a safer way to produce the desired product?
- Is there a safer way to complete this unit operation?

3.2 Detail Design

Concept design develops enough detail to prove that the processing goal is obtainable and business case can be made. Concepts developed such as system location, connectivity, and required unit operations are further developed into a constructible design that can ultimately be operated to produce a finished product. In detailed design, detailed questions are needed as more layers of protection are added. These questions are to be asked informally throughout design and will be formalized in a process hazard analysis (PHA), completed during the detailed design phase.

- Where are the areas of high pressure, temperature, and hazardous material inventories; can they be reduced?
- Are there any specific code requirements or industry best practices for this system?
- What are the expected operating parameters, how can they be exceeded, identified ahead of time, and mitigated?
- Are all systems protected by a mechanical pressure relief device (if needed), do those devices discharge to safe locations?
- How and where could a loss of containment occur, how can it be quickly identified (leak or fire detection)?
- How will the consequences of an event be mitigated once started; how can the event be stopped (process shutdown or fire protection)?
- How can the public, employees, and environment be protected from the consequences of an event if it were to occur (emergency management)?

3.3 Start-Up

Following detailed design and construction, a process must be started, commissioned, qualified, and transitioned to manufacturing. Throughout this process, design assumptions must be confirmed, formal procedures verified, and changes to the design noted so that their impact on the process can be determined. Process safety considerations during start-up must include the full team from engineering to operations and maintenance.

- How does the as-built design compare to the final design, what are the impact of any changes?
- Were changes made to the operating envelope for any of the systems (pressure, temperature, level, concentration)?
- Has a pre-start-up safety review (PSSR) and as-built PHA been completed?
- Do operating procedures match the actual operation of the process?
- Have emergency drills been completed, are there lessons learned from these?

Addressing process safety throughout the design process, versus at the end of design when final design reviews are completed; minimizes late stage design changes, results in a better designed facility, and ensures safe facility operation for years to come.

4 Conclusion

Process safety is a well-established element in the design of oil and petrochemical facilities, taking a front row seat in the design process and being an area of focus once operations begin. Product safety holds a similar role and has been common place for a similar period in pharmaceutical and recently bioprocessing facilities. While the gap between these two elements of design has been closing in recent years as more pharmaceutical companies enter the bioprocessing arena and bioprocessing operations grow, a considerable gap remains. Based on the authors experience designing processes and facilities in both of areas, a few causes for the gaps are easily identifiable:

- Historically, bioprocessing facilities have not contained hazardous processes. Areas of pressure were limited to gas delivery systems (i.e. nitrogen and compressed air) and storage tanks contained aqueous based chemistries (i.e. bioreactors) minimizing the hazards with a loss of containment of these products.
- The shortened design timeline for bioprocess facilities challenges the ability to incorporate process safety into the design at distinct stage gates. The schedule does not allow for a discrete process safety design effort while other design work is on hold.
- There is a general lack of knowledge regarding safety system design in the biotechnology industry. This goes back to the opening statement that in many cases engineers responsible for these systems don't know what they don't know.

Product and patient safety are well defined requirements for biomanufacturing companies and are considered in the earliest stages of the drug discovery process. Applying similar principles to process safety throughout the design lifecycle, and not at distinct stage gates, will ensure a smooth transition from process development to design, validation, and ultimately commercial manufacturing.

5 References

[1] F.M. Editors of Encyclopedia Britannica. "Denis Papin." *Encyclopedia Britannica*, Encyclopedia Britannica, Inc., 18 Aug. 2018, www.britannica.com/biography/Denis-Papin.

[2] Ramsbottom, J. "On an Improved Safety-Valve." *Process Institute of Mechanical Engineers*, vol. 7, 1856, pp. 37–47.

[3] US Department of Labor Final Rule, "Final Rule on Process Safety Management of Highly Hazardous Chemicals; Explosives and Blasting Agents", 1910.109, *Fed Register #: 57.6356*, 24 Fen. 1992.

[4] NASDAQ Biotechnology Index, NBI, https://finance.yahoo.com/quote/%5Enbi/.

[5] *Process Safety Management Introduction*. Occupational Safety and Health Administration, www.osha.gov/dsg/psm/index.html.

[6] US Food and Drug Administration, *Good Manufacturing Practices (GMPs) for the 21st Century - Food Processing*, "GMPs Section One: Current Food Good Manufacturing Practices", 09 Aug. 2004, www.fda.gov/food/guidanceregulation/cgmp/ucm110907.htm.

[7] US Food and Drug Administration, *"Facts About the Current Good Manufacturing Practices (CGMPs)"*, 25 Jun. 2018, www.fda.gov/food/guidanceregulation/cgmp/ucm110907.htm.

546706 LOPA and Sil Risk Analysis As Critical Tools for PSM Culture and Accidents Prevention

Industrial facilities that operate with dangerous products, whether chronic or acute, are increasingly common to be present and part of society at large, often allowing an aggregate of increasingly rich and economically beneficial professions and specialties to the community, be it directly or indirectly. Accidents in these types of units, especially those operating with hazardous products of acute consequence (those that have the capacity to cause fatalities when released to the environment), such as flammable, toxic, explosive products, due to the possibility of occurrence of undesirable events and the behavior of the effects resulting from releases of these types of product, could cause serious damage to the people and / or the installations that are in its vicinity. The various types of accident studies are known as EAR (Risk Analysis Studies), some of which are used more widely and others are specific to specific response needs.

According to BERNSTEIN (1997), the word "risk" derives from the ancient Italian risicare, which means "to dare". While DUARTE (2002) describes that the meaning of the word risk is an expression of thought that has a lot to do with the social, scientific and technological evolution of man. As BILLINTON and ALLAN (1992) note, society has great difficulty distinguishing between a hazard, which can be prioritized in terms of its severity, but does not take into account its probability, and risk, which considers not only the dangerous event, but also their probability of occurrence. KAPLAN (1997) explains that when a person asks "What is the risk?", 3 (three) questions are actually being questioned: What can happen? How can this happen? And if this does happen, what are the consequences? The answers define the scenario, frequency and consequence.

Within the agencies that tend to manage corporate risk, it is noticeable that the risk associated with a particular accident scenario is calculated according to its frequency of occurrence and its severity. Frequency is related to the temporal chance of occurrence of the initiating event, which leads to the release of dangerous product, such as LPP (Liquefied Petroleum Gas) emanation by ball rupture through overpressure of this equipment or, as a second example, the intrinsic rupture of a stretch of line. The consequence, also called severity, considering personal safety, is the capacity that the physical effects, due to the release of dangerous product, associated to the analyzed initiating event, can cause fatality (s), related to the potential of damage related to each effect analysis. The risk, qualitatively or quantitatively observed, and defined in a simplified way, is evaluated as the multiplication of frequency of occurrence and severity. The total social risk related to a particular facility, which can be defined as the risk that a community outside the facility is subject to by its presence in its vicinity, can be calculated as the sum of the risks of all the initiating events arising of the same. In order to do this, for each accident scenario, not only the generating causes, known as Initiating Events, but also the likelihood of ignition, the probability of affecting people, and several other parameters are considered, being one of them very important and permissible of analyzes, which are the failures

safeguards. These, if they act effectively when demanded, can prevent the accident from occurring, since the Initiating Event happens.

Hazardous product risk studies, such as qualitative (Hazard Operational Preliminary Analysis (APPP), Hazard Operability (HAZOP, among others) or those known as quantitative analyzes (Vulnerability Analysis and Quantitative Risk Analysis per se), in addition to being increasingly requested by regulatory agencies, have often been seen as an internal requirement of companies, representing one of the pillars of industrial safety management, in order to keep their risks at low or acceptable levels. However, these analyzes are usually performed in stages

until the final definition of the risk, comparison with acceptability criteria, and later mitigation, if necessary, each of these stages having a minimum time required for its execution, which may make the analysis complete rather time-consuming. Situations similar to these are detrimental to Risk Management when, in fact, what is desired is not a complete risk analysis, but only the achievement of a specific response, such as if my identified risk scenario is effective and properly protected.

The extent of possible damage related to each identified accident is delimited by the intensity of the physical effect causing the damage, and the relationship between the intensity of the physical effect and the corresponding damage is established through the vulnerability models.

On the other hand, semiquantitative risk analyzes such The extent of possible damage related to each identified accident is delimited by the intensity of the physical effect causing the damage, and the relationship between the intensity of the physical effect and the corresponding damage is established through the vulnerability models.

On the other hand, semiquantitative risk analyzes such as LOPA (Layer of Protection Analysis) and SIL (Safety Integrity Level), even though they are less demanded by chemical and petrochemical companies than qualitative or quantitative steps can have beneficial and effective results in terms of knowledge of all the characteristics of the analysis scenarios and of their protection barriers and, in a practical way, much faster to be developed. In addition, they can provide results that do not make up the analysis range of qualitative and / or quantitative studies in depth, such as knowledge and prior review of safeguards, many of which are known as Independent Protection Layers (CIP or IPL - Independent Protection Layers).

Therefore, with the anticipation of an increase in the number of future installations operating with hazardous products and the current installations that already do so, the semi-quantitative studies of LOPA represent an enormous value for the observance of the existing protections in each scenario of process accidents, also serving as the basis for the determination of new Independent Protection Layers or even the reduction of their number, tending to simplify the steps of determining the frequency of the scenario., even though they are less demanded by chemical and petrochemical companies than qualitative or quantitative steps can have beneficial and effective results in terms of knowledge of all the characteristics of the analysis scenarios and of their protection barriers and, in a practical way, much faster to be developed. In addition, they can provide results that do not make up the analysis range of qualitative and / or quantitative studies in depth, such as knowledge and prior review of safeguards, many of which are known as Independent Protection Layers (CIP or IPL - Independent Protection Layers).

Therefore, with the anticipation of an increase in the number of future installations operating with hazardous products and the current installations that already do so, the semi-quantitative studies of LOPA represent an enormous value for the observance of the existing protections in each scenario of process accidents, also serving as the basis for the determination of new Independent Protection Layers or even the reduction of their number, tending to simplify the steps of determining the frequency of the scenario.

546375 Controlling Static Discharge Using Antistatic Additives

Controlling Static Discharge using Antistatic Additives.

Abstract

The presentation will discuss controlling static discharge in non-fuel, non-conductive/low-flash point fluids. The scope of the presentation will concentrate on industrial producers, distributors, consumers of these fluids. Furthermore,furthermore, the presentation covers several industries where these fluids are most commonly found. Common Static-Accumulating Flammable Liquids That May Form Ignitable Vapor-Air Mixtures:

- VM&P naphtha
- Cyclohexane
- n-Heptane
- Benzene
- Toluene
- n-Hexane
- Xylene
- Ethyl benzene
- Styrene

Static discharge has caused catastrophic explosions and fires (U.S. Chemical Safety Board, Static Spark Ignites Explosion Inside Flammable Liquid Storage Tank No. 2007-06-I-KS).

Issues

- Nonconductive flammable liquids can accumulate static electricity during transfer and storage.
- Static sparks can readily ignite flammable vapor-air mixtures inside storage tanks.
- Material Safety Data Sheets (MSDSs) often do not adequately communicate hazard data and precautions.

Key findings:

- Safety publications omit antistat options. (see Process Safety Beacon, August, 2016)
- US affiliates of multinational corporations have different guidelines for antistatic use than their parent companies.
    - NFPA®77 and the American Coatings Association's, *Generation and Control of Static Electricity in Coatings Operation* has guidelines that are different than EU regulations ATEX- Appareils destinés à être utilisés en ATmosphères EXplosibles and ESIG's *Flamability: Safe Working With Industrial Solvents.*
    - The presentation will highlight these differences and how they contribute to the disparity in antistatic additive applications.
- Normal Bonding and Grounding May Not Be Enough! Companies that handle, transfer, and store flammable liquids should contact manufacturers to determine if these liquids can accumulate dangerous levels of static electricity, and if they can form explosive vapor-air mixtures inside storage tanks. If so, *extra precautions*—beyond normal bonding and grounding—may be necessary. Antistatic Additive is an extra precaution.

Recommendations: Knowledge members can use

- Who adds the Antistat?
- How much Antistat is added?
- How is the Antistat added?

546677 Led Your Process Safety Culture, Lead Beyond Hierarchies

The Oil & Gas industry is trying to advance at the next crossroads: "How to achieve that leaders without skills and without training in the management of culture, transform their own culture?"

Our research presents three key ideas that leaders from all over the world

can apply immediately to lead risk management with effectiveness:

frequency, consistency, meaning.

When specific leadership competencies to manage risk have not yet been developed,

Leaders can use this three concepts of easy adoption.

Increasing frequency of interactions with people will allow us to generate safer operations. In full attention economy and contexts of increasing complexity, for risk management, more communication is better communication. But it is not enough to improve the frequency of interactions. That frequency must be accompanied by consistency. The challenge is to achieve possible consistency between what we say and what We do. Leading by example is one of the central keys of risk management. The lack of consistency provides symbolic permissions to normalize deviations. We must maintain a high frequency of face-to-face interactions, but in addition, we must be consistent at every opportunity. And finally, to give meaning to our collaborators. It is necessary to ensure the understanding of risks. Each collaborator need to build his or her own sense of vulnerability and Operational Discipline.

We need workers who operate according to procedures and not adjusted to what the bosses say. This is what meaning allows.

548192 The Importance of Using Color Scale in Process Safety

Everyone knows the importance of color. From physics to fashion going through the automotive industry she will always be present. There is no world without color. Often, they come up on the scale form or not, but the variation in the scenarios are always important. It is so on the traffic sign and the ph tape. From the Diamond of Hommel to across the street. Colors, colors and colors. In addition to not necessarily needing ididomas, such as color of the sky and molten metal have a variety of shades. The main idea of the work is to choose a suitable color to be used in communication within a work area, choosing which and how many elements can evaluated suggest a color scale from the chosen one and present its contribution in the process safety.

550372 Flammable Gas Release Modeling for Real-Time Analysis Using Adversarial Variational Bayes

In this study, we have suggested a method to calculate high-accuracy damage range from flammable material in petrochemical complex. There are many flammable and toxic substances in chemical processes. Computational Fluid Dynamics(CFD) is one of the best way to calculate and simulate scenarios of release range of various fluids or gases. However, it is not used to calculate damage range, because of a lot of time to simulate one scenario. Thus, lots of researchers have enthusiastically studied the surrogate model or reduced order model to get the results of release range in real-time [1-3]. Recently, deep learning based generative model has receive attention because of its superior performance of non-linear manifold estimation [4]. Herein, we propose surrogate model based on adversarial varational Bayes (AVB) [5] for real-time analysis of probability of death. The proposed surrogate model can extract the latent space of high-dimensional flammable and toxic substances dispersion contours. Furthermore, it can generate the estimated contour image from variables of interest much faster and more precise than CFD data memorization with linear interpolation because it does not memorize the data but train the non-linear manifold regarding exact posterior distribution on given observation which basis is predefined in standard variational autoencoder (VAE) [4, 6].

The acrylonitrile was studied about risk assessment for health damage like cancer hazard.[7] But the component has flammable and toxicity, so we should calculate range of damage. The main scenario in this study is the acrylonitrile released from underground model, and the result is shown different damage following hole size, mass flow rate, wind velocity, and wind direction. The superior performance of the proposed model was exemplified by comparing with other surrogate models.

[1] Chen, T., Hadinoto, K., Yan, W., & Ma, Y. (2011). Efficient meta-modelling of complex process simulations with time–space-dependent outputs. *Computers & chemical engineering*, *35*(3), 502-509.

[2] Wang, K., Chen, T., Kwa, S. T., Ma, Y., & Lau, R. (2014). Meta-modelling for fast analysis of CFD-simulated vapour cloud dispersion processes. *Computers & Chemical Engineering*, *69*, 89-97.

[3] Long, K. J., Zajaczkowski, F. J., Haupt, S. E., & Peltier, L. J. (2009). Modeling a Hypothetical Chlorine Release on a College Campus. *JCP*, *4*(9), 881-890.

[4] Na, J., Jeon, K., & Lee, W. B. (2018). Toxic gas release modeling for real-time analysis using variational autoencoder with convolutional neural networks. *Chemical Engineering Science*, *181*, 68-78.

[5] Mescheder, L., Nowozin, S., & Geiger, A. (2017, August). Adversarial Variational Bayes: Unifying Variational Autoencoders and Generative Adversarial Networks. In *International Conference on Machine Learning (ICML)* (pp. 2391-2400). PMLR.

[6] Kingma, D. P., & Welling, M. (2013). Auto-encoding variational bayes. *arXiv preprint arXiv:1312.6114*.

[7] Haber, L. T., & Patterson, J. (2005). Report of an independent peer review of an acrylonitrile risk assessment. Hunam & Experimental Toxicology, 24, 487-527.

546615 Modeling and Analysis of Flammable Gas Dispersion and Deflagration from Offshore Platform Blowout

The operational safety of offshore platform is challenged by a series of adverse factors including harsh environmental conditions and process accidents. Blowout is a type of major accident that may result in catastrophic consequences. This paper uses a computational fluid dynamics (CFD) model to simulate the blowout accident on an offshore drilling platform. The simulation includes the dispersion of flammable gas and the resulting deflagration based on dispersion results. Through the modeling, the blowout gas dispersion processes under different wind speeds and directions are analyzed, and a deflagration scenario due to the ignition of flammable gas is predicted in which the impacts of jet fire, overpressure and heat radiation are assessed. The simulations indicate that wind speed and direction have a great impact on the dispersion behavior of blowout gas. During the deflagration, the maximum overpressure is 280 KPa, the jet fire mainly distributes in the derrick, and the maximum radiation flux of thermal radiation field is 2300 W/m^2. Essentially, the outcomes from simulations could help to conduct a prior risk assessment and make an emergency response plan. Additionally, it could give a reference for the inherent safety design of offshore platform to mitigate the unexpected consequences due to blowout accidents.

548682 Application of Inherently Safer Design Tool (*i*-SDT) for Hazards Assessment and Safer Operation

Inherently ***S***afer ***D***esign ***T***ool (*i*-SDT) is a newly developed semi-quantitative metrics for hazards identification and characterization[1]. It is built on property platform, probabilistic analysis of historical accidents information and characteristic equations of safety parameters (i.e., flammability, toxicity, explosiveness, reactivity, etc.) which are developed in-house to assess potential hazards under various operating conditions. Designers can identify dangerous components and vulnerable equipment of the investigated industrial process and can take necessary corrective actions by applying the inherent safety principles[2] (e.g., minimization, substitution, moderation and simplification). The final output of this *i*-SDT tool is a cluster safety parameter score (C_{SP}) which provides insights regarding the development of safe design and operational parameters to manage the hazards using a very limited amount of process information and safety incidents information.

In this work, the application of the proposed *i*-SDT tool is explained for assessing different alternative routes of an industrial process supply chain. The proposed *i*-SDT safety metric identifies the chemical components with highest risk and the major accident-prone operational units. The detail analysis using *i*-SDT tools shows how different operating conditions can affect the safety performance of the given process. The ultimate object is to find the safer route while considering the techno-economic and environmental performance. It provides Pareto curve to compare alternative routes and gives the freedom to the end-users choosing their safer operating route/technology and process conditions. Therefore, safety can be considered as an important selection criteria for choosing best alternative from the beginning.

Final cluster safety parameter scores (C_{SP}) obtained from the analysis enables designers to compare between available technologies and provides information on various hazardous situations. It can also provide in depth knowledge regarding the cause behind the hazardous behavior. The proposed *i*-SDT metrics allows designer to do the safety analysis simultaneously while doing the initial process synthesis which will reduce the future modification cost of any add-in safety features. This work shows how the unique features of *i*-SDT tool can be applied in a real life scenario while doing inherent safer design during early stage of design.

1. Kazi, M.–K, Eljack F. T., & AlNouss, A. Application of a property-based inherent safety quantification framework for integrating risk assessment into process safety life cycle. *AIChE Spring Meeting, Orlando, Florida, USA, 2018.* (156c), p. 2260.

2. Kletz, T. A., & Amyotte, P. (2010). Process Plants: A Handbook for Inherently Safer Design, Second Edition: Taylor & Francis.

546631 An Approach to Understanding Worker Interactions with Procedures Under Normal Working Conditions

Incidents in high-risk industrial settings are often attributed to the way in which procedures are used. Despite the identification of procedures misuse, little research has been conducted on how workers have interacted with procedures in normal conditions. To understand these interactions a behavior assessment technique has been developed which can describe how a worker interacts with a procedure to complete tasks in everyday work situations. interactions under normal conditions. Using the concept of work-as-imagined/work-as-done, the techniques describes the interactions at a macro whole-of-task level as well as an individual step level. At the macros level, assessment considers if the task was completed correctly and safely (Tier 0). At the micro step-level, assessment considers if the step was completed as it was imagined (Tier 1) or if the step was done in an alternate way (Tier 2). For the alternate approaches, assessment considers if the worker struggled or required assistance, completed the steps out of order or with iterations or if the step was not completed as intended.

Two studies, one involving industrial tasks Second Life® (Study 1), the other at Shells Robert, LA training facility (Study 2), were used to test and develop the behavior assessment technique. In Study 1, the behavior of 13 participants from offshore drilling conducting tasks in either day or night conditions was assessed. In Study 2 the behavior assessment technique was used to test if experience accounted for behavior difference in 25 participants from high-risk industries. Results demonstrate the success of applying the assessment techniques through identifying how the behavior of the participants changes according to their experience and familiarity with the tasks the were performing. Further analysis identified recommendations for design and work flow changes of both written procedures and specific equipment. In conclusion, the initial results suggest the assessment technique is an important contribution to the methods use for understanding early interventions to improve safety in the process industry.

PROCESS HAZARD ANALYSIS (PHA) PREPARATION: A ROLE BASED OPTIMIZATION

Susan S. Frizzell, PE
Adley Services, LLC
Lafayette, LA
Susan@AdleyServices.com

Biography:

Susan Frizzell, PE is the President of Adley Services. Shortly after high school, Susan began her career alongside her father on a drilling rig. Upon graduating from the University of Louisiana with a BS in Mechanical Engineering, she obtained her MBA and began working for a firm performing engineering and process safety reviews on refineries and other land facilities in the New Orleans area. After Susan relocated to Lafayette, LA, she began engineering offshore facilities, complimented with process hazard analysis facilitation. During her almost 20-year career, she has worked upstream, midstream, downstream, and all points in between. She is a licensed professional engineer and an executive member of various organizations.

Abstract

A PHA meeting consumes large resources and can place a strain on departmental manpower. Teams are often faced with the want to be more efficient in PHA conflicting with the need to be thorough and accurate. The key to solving this paradox is efficient preparation. By examining the lessons learned from past PHA's, a role by role analysis of best practices with regards to preparation will allow a PHA team to reduce the costly time spent in the meeting and yet still stay true to the methods and deliver a great, safe result.

When it comes to an optimal PHA, preparation is everything. For example, in a 6-member team PHA, 60 minutes of misused time equates to 6 hours of misused manhours. Whether it is managerial, operations, engineering, or even facilitation, there are lessons learned from previous PHA's that when shared, will not only make the PHA more efficient and the facility safer, but the industry as whole better. The goal being that those reading this be able to take back to their facility these practical preparation tips to share with their present and future PHA team members so that the resources of the facility can be better spent.

1. PHA Preparation Optimized

To fully optimize the Process Hazard Analysis (PHA) preparation, the team needs a clear vision of the end product: "To identify, evaluate, and control the hazards" [1, 2] of an operation or process to minimize the likelihood of occurrence and the severity of consequences of release of a dangerous substance [3]. With that in mind, the team can focus time and energy preparing for the PHA on factors within their control to meet that goal. An efficient, cost-effective PHA begins with including the different roles in the preparation. The PHA planning team should be comprised of management, engineering, operations, the scribe, and the facilitator.

An active commitment to reduce risk begins by assigning tasks to each role prior to the PHA. From the start, management should provide support and direction for the PHA team. They set the tone for obligation and importance. Second, engineering involvement in the preparation ensures accurate and complete process safety information (PSI) is available and accessible. Without correct information, the hazards may not be fully identified. Next, operations should walkdown the facility and identify areas of concern. They have the most intimate knowledge of the workings of the process, which is extremely valuable to the PHA when properly communicated to the team for hazard recognition. The scribe should have the technical skills to ensure equipment is functional and to properly and efficiently document the team's findings. Lastly, the facilitator needs to coach the team before and during the PHA to ensure the methodology is followed.

Successful PHA preparation can improve risk identification and limit cost through appropriate management direction, current and detailed documentation, and clear expectations of the timeline leading up to the PHA. We identify ways that companies can assist in preparation in the months leading up to a PHA and the benefits that preparation has on hazard recognition, team morale, and, most importantly, the bottom line.

2. Management – "*Set the Tone*"

It is important for management to set the tone of the PHA. More specifically, management needs to set the tone before, during, and after. In a PHA, the attitude, the posture, and the dedication of the team is usually a direct reflection of management's attitude, posture, and dedication. In our experience of both facilitating and participating in PHAs, we've seen attitudes of management in a variety of flavors. The following are observations we've made in the past regarding a successful PHA and one not as successful.

The corporate culture and support infrastructure set by management begins well before the first day of the PHA. A typical PHA is done every 5 years [1]. Given that schedule, it is imperative that management starts out on the right foot and begin planning well in advance. A management component that is very serious and very organized will typically induce an attitude of seriousness and organization to the team members. For example, a management team that waits until the last moment and rushes preparation will typically have PHA members that delay and do little preparation, if any at all, for the PHA. PHAs are very expensive and very time consuming. Lack of adequate and efficient preparation will be very damaging to an efficient use of PHA budget and resources. When management in advance sets the tone that a PHA needs to be done correctly and efficiently, the PHA team members tend to take preparation seriously. Just as it is well documented that an effective safety culture of a company starts at the top with management, effective preparation for a PHA also starts at the top with management [4]. The management teams we've seen in the past that do this most effectively set the tone of preparing and doing a PHA as part of your job, not as additional work on top of your daily duties. In other words, management sets the tone that PHAs are in your job description and that we are all PHA team members. For example, a PHA is not additional work given to engineers, it is in an engineer's job description.

The other things we have observed that effective management does by setting the tone is setting up a support team for the PHA team members. These PHA team members will have to transition out of their daily duties, be present at the PHA, and transition back into their regular duties. Often, their project managers and/or functional managers may give some resistance to that. It is imperative for upper management to set the tone with the managers of the team members that they need to have a plan for them to transition into the PHA and transition out of it with regards to their regular duties. Most of these facilities run around the clock all year long and the work never stops. Therefore, it is imperative for management to have a plan for these employees to be present at the PHA while it is going on. A PHA which is constantly stopped for employees dealing with issues outside of the PHA is an unacceptable waste of resources. For example, if a PHA consists of ten team members and one team member must stop the PHA to deal with outside problems unrelated to the PHA, the nine other team members are wasting company resources. When management sets this tone early on that we are dealing with real human lives and significant company resources, the employees and their managers tend to put greater weight on a PHA. And sometimes, the employees just need the help of upper management to persuade lower and mid-level management that their presence, uninterrupted, is necessary to conduct a PHA.

During the PHA, management's role is typically only at the very beginning. It is very effective during the opening remarks of a PHA for a member of management to also make an opening remark. During this address, management will set the tone on several things. It is not uncommon for management to begin opening remarks with a thank you for the team members for their hard work in preparing and committing of their time. Some members of management stress employees' safety; others may stress the importance of safety to the community or environment. Opening remarks is a great time for management to state known worst-case consequences initiated within the limits of the PHA scope or define the assumptions regarding severity. This guides the team towards consistent company-wide risk rankings and avoids the team paranoia over risk ranking too high or too low. It's not management's position to force upon the team a subjective belief; it is management's position to reaffirm the team uses best judgement and not be hesitant of ranking the risks as they believe they should be. Finally, management can give the PHA team members feedback from previous PHAs as to the wording, substance, and ranking of the recommendations to assist with review and completion post-PHA.

Once the PHA begins, management's role is very limited. We've seen effective managers from time to time pop-in, review recommendations, and discuss with team members their thoughts on such. Effective managers will pick a recommendation, look at the supporting data, and ask specific questions of how the team arrived at its conclusion. This serves the purpose of correcting any doubts that the PHA team may be having and serves the purpose of reaffirming to the PHA team what they are doing is correct. Management can also review recommendations with only the facilitator at specific progress intervals of the PHA. The facilitator can then bring suggestions back to the team without the direct watchful eye of management. This minimizes the intimidation factor but gives the team the opportunity to clarify wording or details.

We find in PHAs that the PHA team members want to be very cooperative with management. That is why it's imperative for management to not only set the expectations and tone, but also provide a supportive means and network for a high-quality PHA to be conducted. By management providing a support team and stressing the importance of a PHA during the preparation phase, the PHA team members are motivated to attend the PHA prepared with their full attention. It goes back to the old saying, garbage in, garbage out. Anyone who has sat in a PHA will attest to the opinion that PHAs are grueling and a marathon of being attentive. The PHA team members tend to be very technically sound and understand the PHA process and goal. The biggest hurdle is usually morale and minimizing outside distractions. A great opening statement by management and periodic check-ins to keep morale and encourage the team members goes a long way. Nothing lifts the moral of battle wary troops like a visit to the battlefield by the general. After the PHA is over, feedback and encouragement go a long way in instilling confidence for the PHA team members that will carry over to their next PHA. Although hardly any of this is required by the statutes, these are observations we have noticed with regards to effective management participation in PHAs.

3. Engineering – "The Devil is in the Details"

As an engineer, it is understood that your job is to ask the questions that have not been answered, review the material that has not been presented, and observe that which is not obvious. PHAs are no different. One of the duties of the engineer during a PHA is to gather all the process safety information (PSI). However, we've observed the more effective engineers will take this list, check how current it is, and review the design pressures and temperatures, such as pump discharge pressures. This review helps determine if the PSI is accurate. It also serves the dual purpose of observing any abnormalities that will need to be discussed during the PHA. Looking at this before a PHA is very effective when the engineer's eyes are fresh, and his mind is clear. Once the PHA is ongoing, this task becomes a little more complicated with lower morale and PHA fatigue. This holds true to the majority of the engineer's tasks.

Drawings, drawings, drawings. One of the largest time and money wasters we observe in PHAs is showing up to a PHA with either outdated drawings or needing drawings that are missing. It is imperative for the effective engineer to not only review the drawings to make sure they are up to date and accurately reflect what is present in the field, but it is equally important to review which drawings have been gathered and submitted to ensure that there are no surprises during the PHA. Anyone who has sat in a PHA and heard the words "these are not the most up to date drawings" or "where are all the other drawings at", will attest that it is going to be a long day. The PHA facilitator, the scribe, and pretty much everyone else in the room relies on the drawings being both accurate and complete in their preparation. If the drawings submitted for the PHA are either outdated, incomplete, or just wrong; it will have a ripple effect for the rest of the team both before and during the PHA. If the reader of this paper takes nothing away but one thing, please note that accurate and complete drawings are imperative for an effective PHA team.

PHA team members include those who are present in the room full time and other team members who may be called upon as needed. Effective engineers anticipate ahead of time what possible other resources will be needed during the PHA. For example, engineers for adjacent facilities need to be made aware of an upcoming PHA and given a heads-up as to be ready to answer questions regarding upset conditions and their effects. Likewise, engineers for skid packages or other equipment requiring expertise of vendor engineers need to also be made aware of an upcoming PHA and request their availability if called upon. Similarly, engineers of other disciplines may have better knowledge of failure mechanisms or a specific design review already conducted which is beneficial to the PHA team. It is not uncommon in a PHA to have a question that needs to be answered by a vendor engineer, subject matter experts, or an adjacent facility engineer. Examples of information needed include possible flow rates or pressure to/from receiving/feeding units, adequacy of safe guards in packages provided by vendors, or electrical area classifications. Oftentimes, a simple phone call to these engineers asking if they would be available by phone during the PHA is enough. Other times, effective engineers will require that other personnel be available to come in as requested. PHAs involve a lot of moving parts with each part having its own calendar or schedule. Giving the other possibly necessary personnel a heads up will ensure that should their presence become necessary, the PHA team will not have to recess until they are available.

Effective engineers will typically review MOCs and make phone calls or e-mails to update their current status. The adage that change is constant is very applicable to MOCs. MOCs may have been implemented, postponed, in progress, or completed. It is important to update the status of an MOC. One that has been canceled may require different attention than an MOC that is in progress or completed.

Effective engineers provide the technical expertise of the specific process under review [1]. Before a PHA, effective engineers will take a closer look to evaluate operating trends and upset conditions looking for a pattern that is not readably apparent. Engineers should discuss the facility and how it runs day-to-day with operations prior to the PHA. They should review procedures to ensure they are current and understood. It can be an awkward moment during a PHA when the engineer says, "we've never had a problem with that", and the operator rebuts with "we have a problem with that every day". Engineers know that engineered controls are the strongest defense against hazards. They look for ways to solve problems without relying on administrative controls.

Finally, effective engineers are present during a PHA. This does not mean they are sitting in a chair, it means they are focused, engaged, and completely dedicated to the PHA while they are in that room. Engineering is a life of dealing with problems. To be an effective PHA team member, it is important to make sure the engineer has support to deal with problems while he/she is in the PHA. Effective engineers can have a clear mind because they know that if normal, everyday problems occur, someone else is taking care of them. An engineer that is constantly worrying about what great surprises are awaiting them when they check their email at the end of the day, are not present during a PHA.

4. Operations – "The Ambassador of the field"

The operator is the ambassador of the conditions in the field. The operator, not engineering, management, or the facilitator, is the eyes and ears of the field. As an ambassador, it is important that the operator speak to his constituents (i.e. opposite crews), learn how to effectively communicate, and be present for the meeting of the PHA team. Effective operators speak to their coworkers and get a history of problems, concerns, or other issues they may be having that might not be well documented. Effective operators also walk around and observe conditions of the facility and document such. Documentation being key. Effective operators also review previous incidents not only in their facility, but other similar facilities or neighboring facilities to bring up any concerns that may not be readily apparent when reviewing documentation during a PHA. A great ambassador becomes the eyes and ears of his constituents by listening to what the other operators have to say and physically observing the condition of the facility.

Other things effective operators will do is review operating procedures and note deficiencies. Sometimes operating procedures are written in a way that no one understands them making them ineffective as a safeguard. Other times, there is no written procedure, forcing operations to react to conditions "the way we always do it." If operators are not following the procedures because they are vague, ambiguous, confusing, or out of date, it is up to the operator during the PHA to bring that to the attention of the team members. Frankly, effective operators understand that management and engineers assume that everything they write is perfect, clear, and easy to follow. It is up to the operator to explain why that may not always be the case. Some of the most effective operators we have encountered in a PHA not only describe what the problems are but are very specific about why these are problems.

The next characteristic of the operator as an ambassador for the field conditions is communication. For an operator to communicate with a PHA team, they must speak the same language. Effective operators know that to communicate with a PHA team, they must become a master at reading P&IDs. Being able to understand how the elements of a facility and its attributes are represented on a P&ID are extremely important for an operator. Effective operators can look at a P&ID with little or no guidance and interpret it to see if it is an accurate reflection of what is going on in the field. When the engineer asks a question regarding the P&ID, effective operators are on the same page. If the reader wanted to take away one important thing from this section on operators, please take away that an operator being a master of a P&ID is extremely important to be an effective operator in a PHA.

5. Scribe – "The Show Must Go On"

In American football, a great offensive lineman only receives attention when things do not go as planned. The same can be said about a scribe in a PHA. An effective scribe will give the outward appearance of very little responsibility or having very few complicated tasks. Nothing can be further from the truth. With the goal of quickly and accurately recording information [4], effective scribes understand that a lot of thought and preparation must go into making sure that nothing goes wrong, and if something does go wrong, the PHA suffers minimal interruption. Also, effective scribes are great at being very efficient in the tasks they complete.

At the end of the day, a scribe's responsibility is to ensure that the audio/video equipment works as planned. Although some facilities may be tasked with providing projectors and so forth, the scribe cannot scribe unless all this equipment works. Therefore, an effective scribe will oversee the equipment being used to make sure the parties who are responsible for providing it are communicating and that all the items are compatible. The scribe will ensure that both the scribe's computer and the display have the proper connectors and cables needed. An effective scribe not only brings the connectors and cables needed, but also bring the connectors and cables that were not anticipated as being necessary. An example is the conference room being used for the PHA has HDMI and ethernet and Wi-Fi...or at least that what was told to the PHA facilitator. In the past, we have shown up to conference rooms to find that the HDMI port is in fact a VGA port, or that the HDMI cable is three feet long and the scribe's position is ten feet away. Very effective scribes keep with them a toolbox of supplies such as VGA to HDMI converters, extra batteries for mice and Bluetooth keyboards, and extra cables. Sometimes, the right connectors and cables are present but are malfunctioning. The truly great, effective scribes seamlessly work through these problems and the show continues without interruption. Along those lines, a great practice for scribes to have is to not only back up the PHA file regularly, but also have a written procedure that has been reviewed regarding the backup. We always joke that we PHA the PHA. In other words, what are the risks for file corruption or data loss. Which risk are we willing to accept, and which risk are we going to mitigate? Knock on wood, so far, it's been very effective.

A lesser problem, but a problem none the less, is having to wait on a scribe to catch up. Although not as catastrophic as the PHA having to be stopped completely while technical issues are worked out, a scribe who is not able to keep up with the PHA team stifles the free flow of thought and can leave the door open for other interruptions such as taking phone calls and checking email. Very effective scribes will have experience and practice with the software, know the shortcut keys, and have a vocabulary list prepared for technical terms and names. A typical vocabulary list will be something as simple as a word document open where the scribe has already checked the spelling of names and technical equipment and as the scribe goes along, they can copy and paste the names and equipment labels very quickly without having to correct for spelling. A truly exceptional scribe will have already created templates with standard verbiage for safeguards, recommendations, and comments. For example, when the team decides to make a recommendation, the scribe can copy and paste it into the software and the team can edit the few words that need to be changed to modify the recommendation for the specific application. What this does is it allows the team to make a recommendation and move on to the next node without getting bogged down in the weeds of language choice. We find it is always easier to give a group of people a starting point for a recommendation as opposed to generating the language of a recommendation by consensus of the group. Along those lines, effective scribes review the drawings, MOCs, and other information to become familiar with the vocabulary in the drawings and have an inventory of what PSI they have and what node it corresponds to. Therefore, if a team member erroneously refers to a drawing that does not exist or refers to a drawing or MOC by incorrect name, the scribe will be quick to point out that the item does not exist in the inventory.

Effective scribes will also begin opening vital documents for a specific node as soon as the review of that node begins. Therefore, when the team would like to reference a specific document such as a drawing, MOC, or incident report, the document will be readily available. A common, effective means of doing this is having a second monitor only seen by the scribe, in addition to the screen shared with the projector. The scribe can work on his/her monitor opening documents or correcting language. As the team is ready to view specific information, he/she can drag the window over to the extended screen that is being displayed overhead. This allows the scribe to work without being micromanaged by the team and minimizes the distraction of the team from the scribe flipping through windows or searching for files. Again, the overall theme for an effective scribe is to keep the free flow of thoughts and the team moving. Anytime the team must pause or take a break because of the scribe performing its duties, the team's momentum and productivity can be lost.

Finally, truly effective scribes are great at covering the intangibles. This can include filling all the supplies or doing things to the environment in the room to lift morale. For example, effective scribes will typically have a box of random office supplies, whether it be pens, hi-liters, or post-it notes. At the beginning of each day and after each break, he/she should refill the table or the workspace of the PHA team with extras. So, if a team member would require a specific color pen, they would simply grab it and not have to pause the PHA. This gets back to the overall theme that an effective scribe keeps the evaluation moving forward without interruption. Another intangible that we often see effective scribes will do is they will place motivational posters around the room, such as large check-off list of nodes to visually see progress, to hang inspirational quotes, and so forth. PHAs are really an endurance test of morale and motivation and these little intangibles that the scribes do are of very limited cost, but of great benefit. In all honesty, we often do not realize how great and effective our scribe is until we do a PHA with an ineffective scribe. When you do a PHA with an ineffective scribe, you feel like you're stuck in the mud and going nowhere fast. Although realistically, the delay may add just a few minutes a day, with regards to morale and motivation, it feels like an eternity.

6. Facilitator – "What's Your Plan, Coach?"

Like a good coach, an effective facilitator enables great people to do great things. Effective facilitators setup the team members to be successful and, during the PHA, guide them to produce a high-quality product. Although a facilitator's duty is to ensure the methods are adhered to and the regulations concerning a PHA are complied with [1], a great PHA facilitator is so much more.

A great PHA facilitator, like a coach, ensures that the team is prepared come game day. Effective PHA facilitators will do such things as set deadlines for collection and development of information and suggest further actions to better prepare for game day. Not only do they set the deadlines, an effective facilitator will check in with the team members prior to those deadlines to ensure that progress is proceeding as planned and that the deadline will likely be met. Effective facilitators are great at helping team members overcome obstacles and hurdles in collecting this information. A team being well prepared on game day when they take their chairs and listen to management's opening remarks is no accident. Also, during the preparation, an effective PHA facilitator will ask the hard questions of the team members, such as: Do you have a transition plan for someone to fulfill your role during the PHA? Or, are there any allergies, physical, or dietary restrictions a team member may have so that these can be addressed well ahead of the PHA? To be an effective PHA facilitator, it really comes down to having the team prepare for the work they anticipate doing and addressing as many possible obstacles as one can prior to the PHA commencing. The litmus test for a PHA facilitator is simple - once the PHA commences, no surprises.

In addition to ensuring the rest of the team is preparing for the PHA, the effective facilitator puts in just as much effort as the rest of the team into preparing the PHA file, determining the node segregation, and sharpening his skills as a technical guide. On the day of the PHA, the PHA file should be prepopulated with node descriptions, drawings, common safeguards, etc. The facilitator should have reviewed the node list and noded drawings with other member(s) of the team to ensure the segregations make sense and all parts of the process are included. The effective facilitator should also review recent changes to regulations or process conditions that might affect the hazard identification. He should gain a basic understanding of the process under review along with the potential hazards specific to the facility. Effective facilitators have a list of questions to pose to the team during the appropriate node to ensure all hazards have been addressed.

Again, following the theme of the PHA facilitator being a coach, once the PHA commences, it is game day. It is the PHA facilitator's job to adhere to the methods and regulations regarding PHA facilitation. However, great, effective PHA facilitators understand that the true challenge in a PHA is morale, motivation, and focus. Effective PHA facilitators will typically start off the PHA with a "rules of the road" presentation. They do not assume that everyone in a room regularly participates in PHAs and is familiar with the process. They review the PHA process, the end goal of risk mitigation, the timeline, the daily schedule, acceptable and unacceptable behaviors, and expectations of participation. An effective PHA facilitator will go over these rules of the road and get each team member to commit to their role, to fully participate, and to strive towards the goal. It creates a sense of responsibility and obligation among the members. Once the PHA commences, the effective PHA facilitator will have a sense of when progress is slowing down and when to step in to move the team along. The effective PHA facilitator knows that the team must be given opportunity to discuss causes and consequences, but also has goals and objectives that need to be met within the time allotted.

As the team leader, the facilitator should plan on how to react to differences among team members. The great PHA facilitators have a way of getting conflicts resolved without upsetting or demeaning anyone in the room. Typically, in an argument, somebody must be wrong and somebody must be right. However, the great PHA facilitators have a way of allowing both parties to express their beliefs, guiding them to understand the beliefs of others, and concluding with a common solution. A great way to kill the free flow of ideas in a PHA and stifle productivity is to tell someone, "you're wrong, please be quiet". Great PHA facilitators know how to do that in a way that allows the person to maintain dignity and continue to perform in a PHA.

Another attribute of great PHA facilitators is that they are systematic and thorough. They have a plan of attack for each node. This usually includes the node description and limits review, a review of the appropriate drawings, written list of deviations to encourage the team, and review of MOCs, previous recommendations, and incidents related to the node. The effective facilitator also has a plan in place to limit idle time when critical information necessary to properly review a node is found to be missing. The action is documented as a follow-up for a specific team member to complete prior to the next day allowing the PHA to move forward. Any outstanding to-dos are documented and presented to the team members at the end of each day prior to dismissal. The PHA facilitator is responsible for meticulously documenting the team's findings, which cannot be done without adequate preparation of a game plan.

Finally, great PHA facilitators understand they have an obligation to future PHA facilitator to send their team off with a sense of accomplishment and pride in PHA participation. If the team members of the PHA leave with a bad taste in their mouth or feel discouraged, their participation in the next PHA may not be to the level it could be. Great PHA facilitators compliment their team members and make sure they leave the PHA on a positive note. Therefore, when the team members approach their next PHA, the new facilitator will not be dealing with baggage leftover from the previous PHA. It is a courtesy that great facilitators give to and expect from other facilitators. PHA facilitators are often asked by team members during a PHA, "how can you do this every day?" The response we often use is, "what can we do better to make you volunteer for the next PHA?" Soliciting feedback from team members besides a multiple-choice survey is imperative to be an effective facilitator. The feedback needs to be of substance that can be acted upon and improved upon. We truly love doing PHAs and at the end of every PHA, we express that to the team and solicit their feedback to what we can do better to make them truly love doing it too.

7. Summary

In conclusion, when it comes to an optimal PHA, preparation is everything. Each team member plays an important and specific role. Management sets the tone with their level of commitment to preparation, resources, and action. Engineering ensures the quality of information available to the team and contributes a deep understanding of the process design. Operations represents the field conditions, the procedures, and the most intimate knowledge of the process. The scribe provides accurate documentation through technical skill, proficiency with hardware and software, and team morale support. The facilitator leads the team to identify, evaluate, and control the hazards utilizing a systematic approach to preparation. It has been proven that a small amount of time spent preparing for each role will result in improved hazard identification, higher team morale, and reduced PHA cost.

8. References

1. Occupational Safety and Health Standards, Toxic and Hazardous Substances, Hazard Communication, Available at https://www.osha.gov/laws-regs/regulations/standardnumber/1910/1910.1200 Accessed on December 12, 2018.
2. Bureau of Safety and Environmental Enforcement, Department of the Interior, Offshore, Oil and Gas and Sulphur Operations in the Outer Continental Shelf, Available at https://www.ecfr.gov/cgi-bin/text-idx?tpl=/ecfrbrowse/Title30/30cfr250_main_02.tpl Accessed on December 12, 2018.
3. American Petroleum Institute Recommended Practice 750: Management of Process Hazards, First Edition, American Petroleum Institute, Washington D.C., 1995.
4. CCPS. *Guidelines for Hazard Evaluation Procedures*. Center for Chemical Process Safety, American Institute of Chemical Engineers, New York, NY, 2008.

551204 Building Process Safety Intelligence: What Should Every Graduating Chemical Engineer Know?

Process safety intelligence is derived from derived from a vast corpus of sources - including textbooks, handbooks, journals, conferences, incident reports, safety datasheets, and available materials and thermophysical property data. Dedicated decision-support tools enabling practitioners to harness this intelligence in their daily work, such as Knovel are an important component of chemical engineering practice today. However, many students don't realize the breadth and depth of technical knowledge needed to support process design and safety – they tend to overrely on non-specialist search, and miss pedigreed information available on specialized knowledge platforms, such as Knovel.

The undergraduate engineering space offers an excellent opportunity to plant the foundational information seeking behaviors in future process safety professionals and leaders. Knovel focuses on applied a preview of the types of insight they will be forced to leverage in industry careers, without a direct cost to students and faculty.

Common Process Safety Management Gaps Based on Occupational Safety and Health Administration Process Safety Management of Highly Hazardous Chemicals Violation Trends

Janet Persechino
Engineering Planning and Management, Inc.
959 Concord Street
Framingham, MA 01701
jap@epm-inc.com

Ryan Welch
Engineering Planning and Management, Inc.
959 Concord Street
Framingham, MA 01701
rdw@epm-inc.com

Prepared for Presentation at
American Institute of Chemical Engineers
2019 Spring Meeting and 15th Global Congress on Process Safety
New Orleans, LA
March 31-April 3, 2019

Janet Persechino
Engineering Planning and Management, Inc.
959 Concord Street
Framingham, MA 01701
jap@epm-inc.com
Keywords: OSHA, violations, regulations, process safety

Abstract
On July 17, 1990 the U.S. Occupational Safety and Health Administration (OSHA) published in the Federal Register the proposed standard *Process Safety Management (PSM) of Highly Hazardous Chemicals (HHCs)*. Approximately four months later it was enacted into law via Clean Air Act Amendments (CAAA). These amendments require that the U.S. Secretary of Labor, in coordination with the U.S. Administrator of the Environmental Protection Agency, promulgate, pursuant to the Occupational Safety and Health Act of 1970, a chemical process safety standard to prevent accidental releases of chemicals that could pose a threat to employees [1]. The full text of the final standard was issued in the Federal Register on February 24, 1992.

The OSHA standard is listed under Title 29 (Department of Labor) of the Code of Federal Regulations (CFR) as 1910.119. This standard leans towards a performance-based guidance document that outlines methods and programs to decrease the likelihood of flammable, toxic, reactive and explosive chemical releases. The intent is to reduce the risk to employees, visitors and contractors that may enter or work at chemical processing facilities.

Facilities are subject to the PSM regulation if they hold a threshold quantity of any HHC that is listed in Appendix A of the standard, with some exceptions. The violations that OSHA issues for the 1910.119 standard will cite the specific element and its relevant sub-section. Published data allows for review of the violations, and based on that information an indication of common PSM program gaps can be assessed.

A ten-year span of OSHA PSM violation data was examined (2007-2017). The data shows the top four OSHA cited PSM sections (often referred to as elements) are (highest to lowest):

1) Process Safety Information
2) Mechanical Integrity
3) Operating Procedures
4) Process Hazard Analysis

Each of these elements have known solutions, and specific corrective actions can be easily implemented given the lessons learned, the current level of expertise that exists in the process safety industry and the cost-effective software tools that are now available.

Regulatory compliance is not the only goal of adequate PSM programs, nor is it the goal of this review. The purpose of reviewing the OSHA violation information is not meant to help facilities avoid OSHA citations, but to uncover gaps found by OSHA and bring awareness and attention to some common issues, hopefully leading to focused solutions for safer, more reliable facilities.

1.0 Introduction

There are approximately 24,000 sites that are subject to the OSHA PSM standard nationwide [2]. Facilities are subject to the standard if they process any of the 130+ chemicals listed in Appendix A (of the standard) in excess of the associated threshold quantities, also listed in Appendix A, on-site in one location. They are also subject to the standard if they process flammable liquids and/or gases in quantities equal to or greater than 10,000 pounds. OSHA defines "on site in one location" as areas under the control of an employer, or affiliated employers, that have control over any group of vessels that are interconnected, or in separate vessels that are located in close enough proximity that a HHC could be involved in a catastrophic release.

Exceptions to the applicability include [3]:

- ✓ Retail facilities
- ✓ Oil/gas well drilling or servicing operations
- ✓ Normally unoccupied remote facilities; the standard defines "remote" as a facility that is operated, maintained or serviced by employees who visit the facility periodically to check its operation, no permanent employees are stationed at the facility, and is geographically remote from all other buildings and persons.
- ✓ Hydrocarbon fuel used solely for workplace consumption as a fuel if such fuels are not part of a process containing another HHC covered by the standard
- ✓ Flammable liquids with a flashpoint below 100°F stored in atmospheric storage tanks or transferred, which are kept below their normal boiling point without benefit of chilling or refrigeration

The above information outlines specific threshold quantities that trigger, and exclude, the PSM regulatory requirement. It is strongly recommended that some form of PSM be practiced at plants that may not be subject to PSM, especially those that fall into the following categories:

- Plants that store quantities of HHCs close to the threshold quantities listed in Appendix A of the standard
- Plants that utilize hazardous chemicals not listed in Appendix A of the standard
- Processes exempt from the standard, for example, those listed as exceptions in the standard, plants that have runaway reaction potential, etc.

An overview of "PSM Lite", and related fire code requirements, can be found in these conference proceedings in the paper titled *Fire Code Chemical Process Safety Requirements*. Implementation of an abbreviated form of PSM should be evaluated on a case-by-case basis which will take into account various factors that would clarify the actual risk and the residual risk that would remain after programs are incorporated into the operations.

2.0 Violation and Corrective Action Evaluation

The PSM elements are listed in the 1910.119 standard in the following order/sections:

(c) Employee Participation
(d) Process Safety Information
(e) Process Hazard Analysis
(f) Operating Procedures
(g) Training
(h) Contractors
(i) Pre-startup Safety Review
(j) Mechanical Integrity
(k) Hot Work Permit
(l) Management of Change
(m) Incident Investigation
(n) Emergency Planning and Response
(o) Compliance Audits
(p) Trade Secrets

OSHA has been administering violations for the 1910.119 standard since 1992. For purposes of relevancy, the last 10 years of violation data was examined (2007-2017). These violations were issued when OSHA completed an inspection and/or accident investigation visit to a particular facility.

Each one of the elements listed above has subsections pertaining to the requirements for that element. For example, when OSHA administers a PSM violation during a visit it is listed as: 1910119 E03 I. The nomenclature correlates to a violation for section (e) Process Hazard Analysis (PHA), part 3(i), or (3) the process hazard analysis shall address (i) the hazards of the process. This particular example indicates that the PHA did not properly address any or all of the hazards of the process. The OSHA inspection information also includes other data such as penalty dollar amounts, dates and type of inspection. The information provided in this summary reviews a few segments of this data from years 2007-2017 [4].

3.0 Violation Data Interpretation

All OSHA PSM violations were tallied over the 10-year period (2007-2017), this included:

- ✓ All citation classifications types (serious, repeat, willful etc.)
- ✓ All types of inspections/visits (accident, planned etc.)
- ✓ All types of facilities

This review does not distinguish whether violations were given during the same visit or during separate visits.

The data shows that there are 4 elements, or 4 deficiencies, that clearly stand out as the most prevalent (see Figure 1):

Most Commonly Cited PSM Elements (highest to lowest):

1) Process Safety Information (PSI)
2) Mechanical Integrity (MI)
3) Operating Procedures (OP)
4) Process Hazard Analysis (PHA)

Total Monetary Amount (highest to lowest):

1) Mechanical Integrity
2) Process Safety Information
3) Operating Procedures
4) Process Hazard Analysis

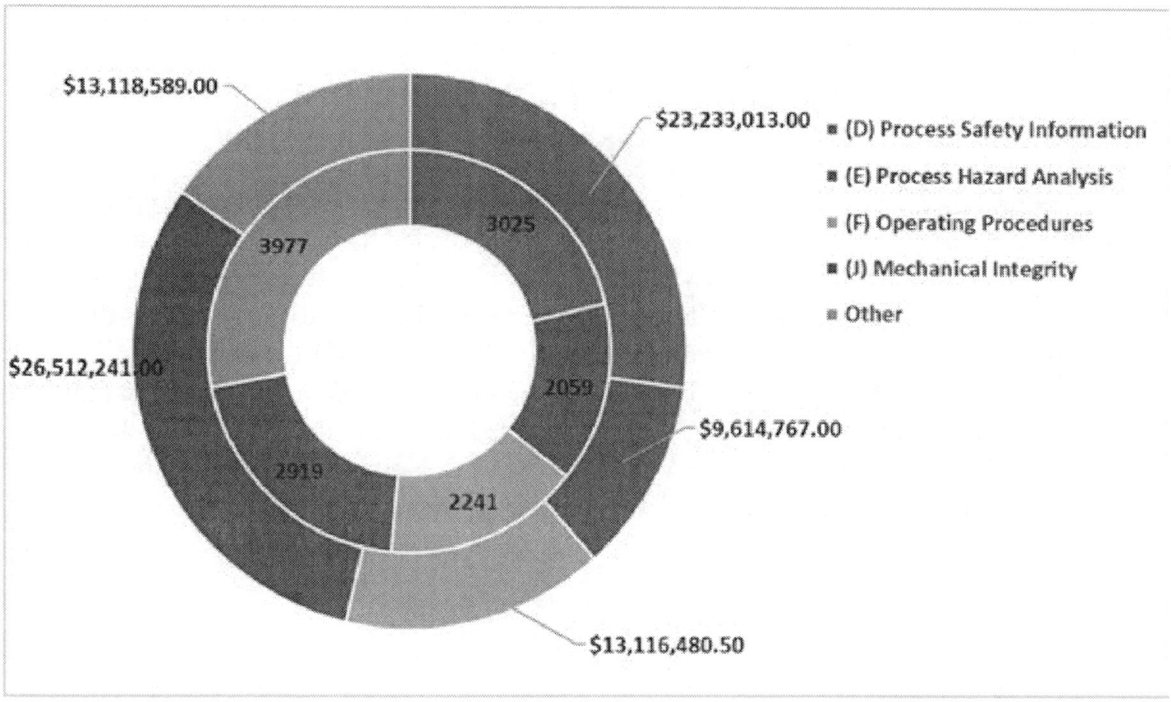

Figure 1: Total Number of OSHA PSM Violations and Total Citation Dollar Amounts by Element: 2007-2017

The top two elements (PSI and MI) have close to the same number of violations and the same goes for the third and fourth most common violations, giving the impression that the struggle to execute PSI and MI is about equal and OP and PHA both come in as close second.

It is alarming that PSI is the most cited element. Having correct and adequate PSI is the foundation for implementing, not only all PSM elements correctly, but any and all chemical safety precautions. PSI is crucial in completing an adequate PHA, the analysis that examines the hazards of the process and ensures that the proper safeguards are in place. The lack of PSI could lead to process hazards being missed, or incorrectly evaluated, during a PHA. Incomplete PSI can also lead to having inappropriate MI programs since some of these programs can be based on the hazards of the process, for example, more frequent inspections than the manufacturer's recommended practices based on the characteristics of the material being processed.

MI is the second most cited element but has the largest monetary amount, this could point to more instances of MI being noted within a single violation. OSHA will increase a fine for a single violation based on the number of "instances", or how many times they see a violation for the same element during the same visit. This review did not take into account the number of instances per violation.

MI can be difficult to implement, but new software solutions and training programs exist to help with proper execution. There are MI software products on the market that include capabilities to capture PSI, PHA and OPs, addressing all of the top-cited elements within one program.

It is helpful to utilize PSM experts for implementation and execution of PSM programs, as they are aware of the depth of information that is needed to meet the intent of the standard and for overall safety. They can often recommend products depending on your particular operation. Individuals dedicated to PSM will have access to resources that might not be available to others who are not embedded in the PSM industry. For instance, access to fee-based databases with previous incident data, one item that must be documented as part of a PHA. The expenditure for bringing in this type of expertise can provide a beneficial return on the investment when examining the average cost per violation and average number of violations per visit (Figures 2 and 3).

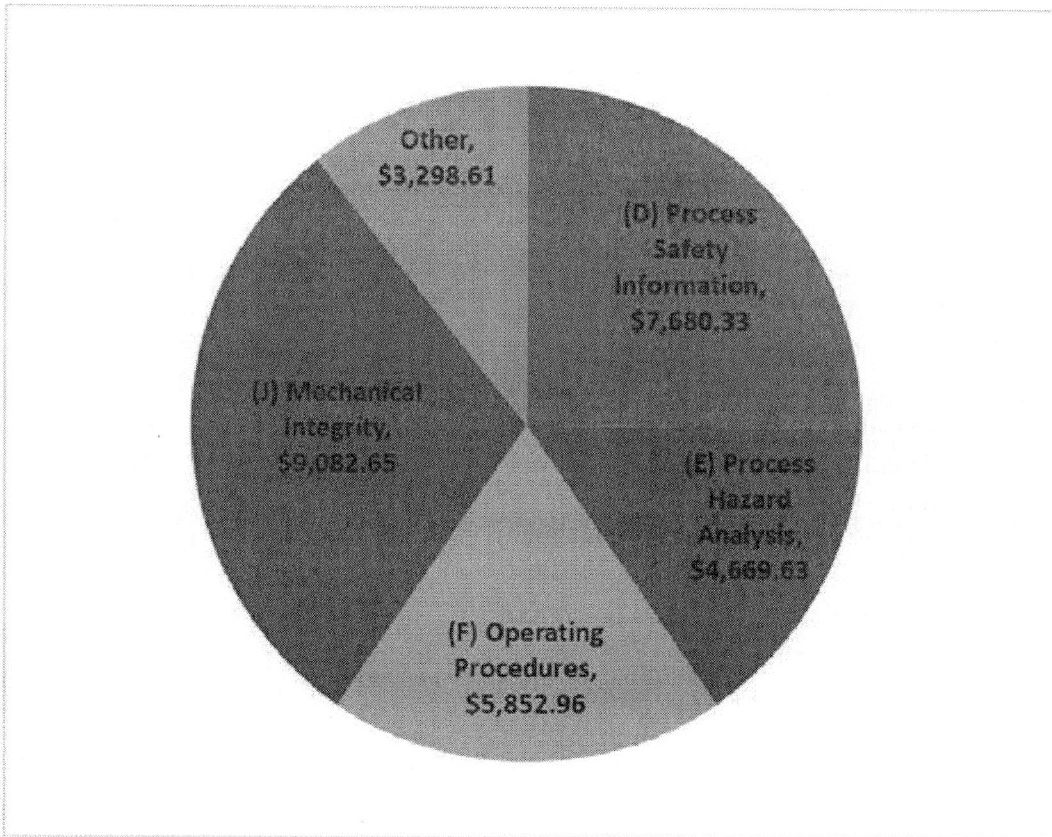

Figure 2: Average Cost per Violation by Element 2007-2017

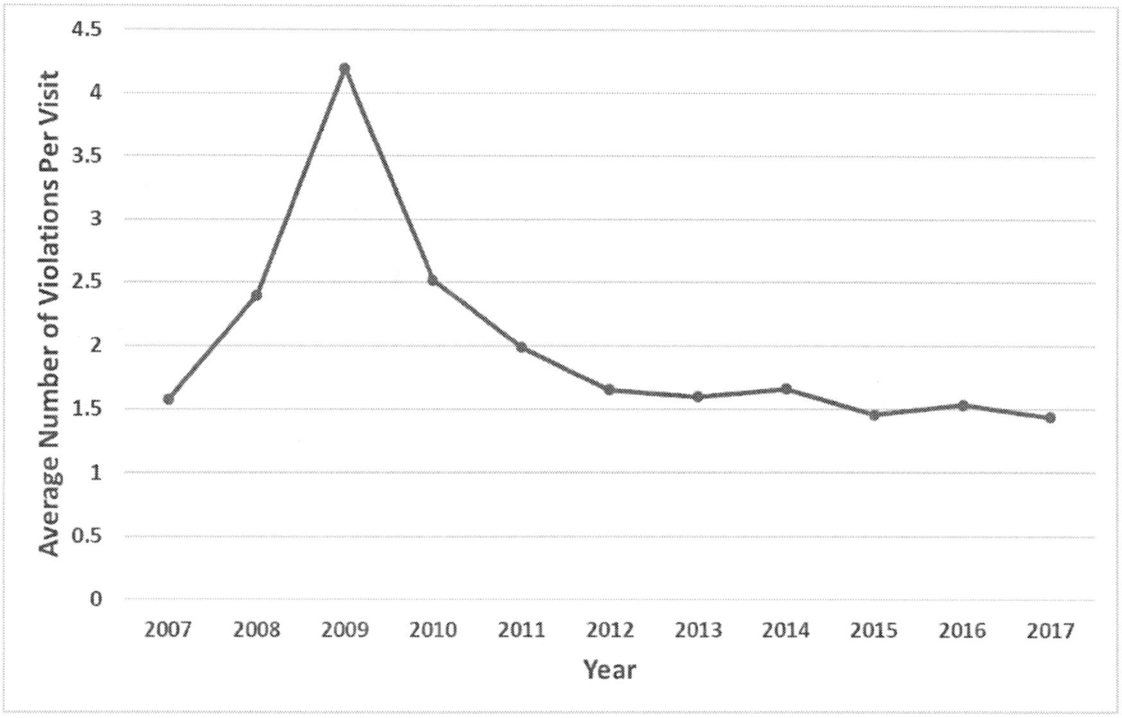

Figure 3: Average Number of Violations per Visit 2007-2017

(To note about Figure 3: the spike in the number of violations in 2009 could be due to the National Emphasis Program (NEP) in which OSHA extended to include a pilot on PSM-covered facilities from the 2007 NEP which only included refineries. The NEP went on from 2011 to only include chemical facilities and not petroleum refineries)

The citation costs are only one aspect to consider when reviewing the return on the investment. Other benefits that can be quickly realized once the programs have been established include but may not be limited to:

- ✓ Improvement in operating efficiencies
- ✓ Increased employee morale
- ✓ Lower turnover rates
- ✓ Safer working conditions
- ✓ Uptime gains
- ✓ Insurance benefits
- ✓ More consistent product quality
- ✓ Decreased likelihood of an incident

A deeper dive into the subsections of the most cited elements yielded the results in Figure 4.

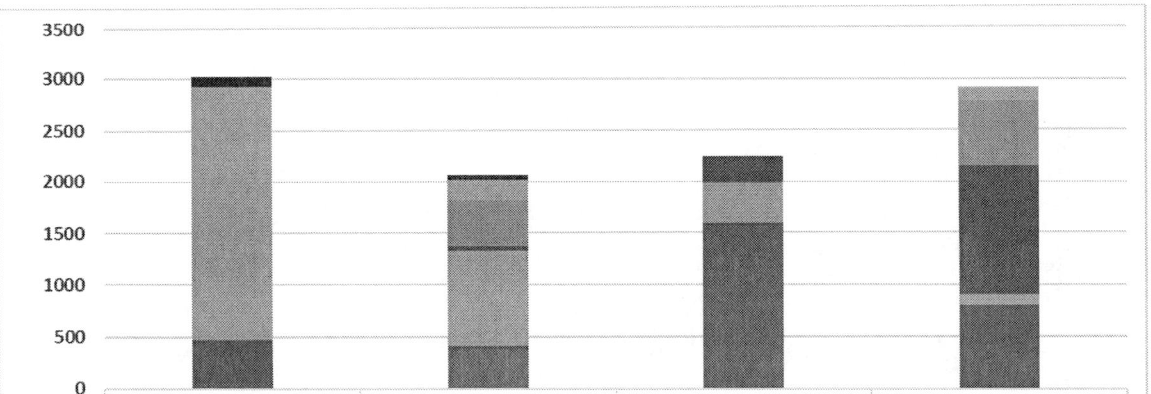

	(D) Process Safety Information	(E) Process Hazard Analysis	(F) Operating Procedures	(J) Mechanical Integrity
■ Blank	101			
■ 07		48		
■ 06		186		135
■ 05		463		640
■ 04		37	255	1235
■ 03	2452	918	391	105
■ 02	418	26	23	798
■ 01	54	381	1572	6

Figure 4: Sub-section Cited in the Top 4 OSHA Most Cited PSM Elements 2007-2017

Figure 4 shows that the biggest gaps over the last 10 years include the following sub-sections of these elements:

1. Process Safety Information (d)
    - Equipment Information (section 3)
2. Mechanical Integrity (j)
    - Written Maintenance Procedures (section 4)
    - Inspection and Testing (section 2)
    - Correcting Equipment Deficiencies (section 5)
3. Process Hazard Analysis (e)
    - Inadequate coverage of the hazards of the process (section 3)
4. Operating Procedures (f)
    - Develop written operating procedures (section 1)

There is a common theme among these four (4) subsection violations. While PSI was the most cited element, the most cited PSI sub-section is Equipment Information, which ties into the MI portion of PSM. Based on this, deficiencies in equipment related information and MI ranks very high on the list. MI related programs help to ensure that chemicals remain inside the vessels, pipes, etc. A review of the OSHA data shows that the industry may be struggling with proper execution of this critical piece of process safety.

As noted above, MI programs are complex and can be difficult to establish, but there are current solutions on the market to help correct these deficiencies, such as economical software solutions that improve compliance by organizing maintenance tasks. The right solution can alert staff to overdue maintenance, and store the maintenance history for each particular piece of equipment. Some tools can also track inventory of spare parts and link Management of Change documents related to the process and/or the equipment. Consultants versed in the regulations and MI techniques can help establish appropriate programs and find solutions useful to your specific operation.

Development of operating procedures for both process and maintenance activities has a high number of citations in the MI and OP elements. It is essential that these procedures are comprehensive, accessible, updated and easy to interpret. Mistakes in operating steps, or failure to recognize abnormal conditions, can quickly lead to catastrophic releases.

There are various writing and auditing techniques that can be used to ensure that these procedures include the correct information and address all stages of the process (start-up, normal conditions, abnormal conditions and shut down). Software tools currently on the market can also assist in organizing this critical information. This approach, while time consuming up-front, is time well spent in providing a safer and more efficient workplace.

Once authored, it is easier to audit the documents for accuracy and audit the operators using a behavior-based system to determine that procedures are being followed correctly.

Vast benefits can be gained using this approach, and many PSM consultants offer services to help author these documents and provide suggestions for document control and review processes.

4.0 Resources

Despite the intent for OSHA PSM to decrease the likelihood and severity of accidents, the OSHA data is showing that PSM gaps still exist in the most fundamental PSM elements. These gaps present an unacceptable risk. This is evidenced by the accidents that continue to occur. The various incident investigations performed by the U.S. Chemical Safety and Hazard Investigation Board often attribute inadequate PSM programs as the root cause.

There are many resources available for assistance with PSM implementation and improvement for plants of all types and sizes. There has been an increase in consulting organizations providing guidance and software solutions for PSM. The list of resources readily available to even the smallest of plants are much more prevalent than they were 10 years ago.

OSHA has published documents with guidelines for PSM for storage facilities and for small businesses. They also have a PSM resource page: https://www.osha.gov/SLTC/processsafetymanagement/. The Center for Chemical Process Safety has many publications, classes and tools to help with all aspects of PSM. Some of these resources are free, or provided at reasonable cost, and CCPS/AIChE (American Institute of Chemical Engineers) membership increases those benefits.

Small businesses often find it difficult to execute PSM. In 2016-2017, in almost the same number of inspections, OSHA administered 57 PSM related violations to companies with under 100 employees and only 18 violations to companies with over 100 employees. OSHA also points to a study that estimates *"employers with 1-25 employees are 47 times more likely to have a release and 17 times more likely to suffer an injury, per employee, than facilities with 1500 or more employees."* [5]

There are a few PSM consulting firms that target small to medium companies, the companies that cannot dedicate resources to PSM, nor have the expertise in-house to meet the intent of the standard. Many of these consulting firms not only focus on OSHA compliance but becoming a partner to help embed PSM concepts into the plant's overall safety culture. Bringing in these third parties is worth the cost because they are embedded in the PSM realm and can bring along expertise from other industries, the latest information and years of experience.

There are cost effective software solutions, some aimed at small businesses from a price point and feature standpoint. These solutions can improve PSM while improving the bottom line through improved efficiency, production uptime gains, insurance benefits and violation and incident avoidance. The smaller chemical processing companies are yet to realize that these solutions exist at very reasonable prices, and the small upfront investments have very beneficial returns that easily justify the expenditure.

It is ideal to use the same consulting firm for various locations to standardize PSM solutions and audits across multiple sites that may have different risk. This type of PSM implementation helps to normalize the oversight and tracking to give a better picture of risk at the corporate level.

5.0 Conclusion

The research objective of this paper was to spot the potential for common gaps and issues in implementing PSM across the United States at all types of chemical processing industries using OSHA violation data, and to use that information to identify opportunities for solutions and preventative/corrective actions. The OSHA violation picture is only one piece of the gap analysis puzzle, but it is one place that collects enough data to increase awareness of PSM weaknesses nationwide regardless of the type of operation. This information can aid in creating education/awareness programs and campaigns aimed at improving PSM execution.

The top four cited violations are PSI, MI, PHA and OPs. Solutions to the deficiencies revealed by the OSHA data are well-known and PSM experts are aware of the many practical and feasible methods that exist to close these deficiency gaps. The return on investment on items outside of the safety aspect are usually realized soon after the solutions are put in place, including: more efficient and consistent work procedures (which can lead to improved product quality), increased up-time and insurance benefits to name only a few.

A review of the OSHA data can help quantify the costs that can be expected from OSHA PSM violations and citations. However, other impacts of violations and/or PSM incidents that are not so easily quantified include the business, legal, liability, company reputation and the impact that the injuries have on families. Sometimes plants are not able to rebuild in the same location or regain the customer share they may have lost due to an incident. Additionally, the OSHA violations are public record and can be accessed by lawyers, local authorities having jurisdiction, insurance companies, activists, etc. Companies may also find it difficult to obtain insurance due to OSHA violation or incident history. While these impacts are difficult to quantify, they must be considered in the overall cost to the business.

PSM implementation is a continuous improvement journey with new lessons learned, new technologies and new resources being created every year. Using the plethora of tools available can help close the deficiency gap.

References

[1] U.S Department of Labor OSHA 3132 Publication *Process Safety Management* 2000
[2] Washington State Department of Labor & Industries July 2010
[3] U.S Department of Labor OSHA 3133 *Publication Process Safety Management Guidelines for Compliance*, 1994
[4] OSHA Violation Data 2007-2017
[5] OSHA *Process Safety Management for Small Businesses* 2017

546725 Safety Performance Evaluation: A Benchmark of Practices across Various High-Risk Systems

The concept of performance has invaded all management spheres and imposed itself as a strong driver of decision making. Following the funding work of Kaplan and Norton (1996), performance appears as a multidimensional concept which evaluation and use require adapted tools and well calibrated decision processes.

Safety performance does not escape these requirements as it englobes various dimensions (workplace safety, process safety, environmental safety), can be approached through various evaluation paradigms (auto evaluation, external assessment through audits and KPIs) and may be used by several stakeholders (risk owner, regional and national authorities, public perceptions). Indeed, different combinations of these three dimensions allow for a large variety of performance governance paradigms carrying various understandings and practices with regard to the way performance should be assessed and used.

By comparing safety performance management in three different risk systems (process, railway and health) in Europe, this works aims at providing safety management actors with a description of the richness and variety of safety performance understandings and practices. Insights gained from this work should enhance safety managers' abilities to grasp the strengths and weaknesses of their approaches and explore the possibilities of a beneficial benchmarking.

19 AIChE Spring Meeting
15TH GLOBAL CONGRESS ON PROCESS SAFETY
NEW ORLEANS • 2019

Lessons learned from analyzing an explosion accident at Shanghai Secco chemical plant

Hanhai Dong
China University of Petroleum (East China)
No. 66, West Changjiang Road, Huangdao District, Qingdao, China
donghanhaiupc@126.com

Yi Liu
China University of Petroleum (East China)
No. 66, West Changjiang Road, Huangdao District, Qingdao, China
liuyi@upc.edu.cn

Dongfeng Zhao
China University of Petroleum (East China)
No. 66, West Changjiang Road, Huangdao District, Qingdao, China
zhaodf@vip.sina.com

Jian Chen
China University of Petroleum (East China)
No. 66, West Changjiang Road, Huangdao District, Qingdao, China
836771528@qq.com

Yanchun Wang
China University of Petroleum (East China)
No. 66, West Changjiang Road, Huangdao District, Qingdao, China
Wychun25@163.com

Wei Wang
China University of Petroleum (East China)
No. 66, West Changjiang Road, Huangdao District, Qingdao, China
15621097585@163.com

Prepared for Presentation at
American Institute of Chemical Engineers
2019 Spring Meeting and 15th Global Congress on Process Safety
New Orleans, LA
March 31 – April 3, 2019

Lessons learned from analyzing an explosion accident at Shanghai Secco chemical plant

Hanhai Dong
China University of Petroleum (East China)
No. 66, West Changjiang Road, Huangdao District, Qingdao, China
donghanhaiupc@126.com

Yi Liu
China University of Petroleum (East China)

Dongfeng Zhao
China University of Petroleum (East China)

Jian Chen
China University of Petroleum (East China)

Yanchun Wang
China University of Petroleum (East China)

Wei Wang
China University of Petroleum (East China)

Keywords: event sequence diagram; Multilayer Bow-tie Analysis; FLACS

Abstract

On May 12, 2018, six workers from Shanghai Aijinke Engineering Construction Services Co., Ltd. died on the spot of an explosion when they were maintaining a benzene tank in the tank farm at Shanghai Secco Petrochemical Co., Ltd. The accident was characterized as a disastrous accident caused by hot work in a confined space, which resulted in massive casualties and a direct economic loss of 11.66 million yuan. This paper aims to analyze the accident occurrence process in a chronological order firstly by using the event sequence diagram (ESD). Multilayer Bow-tie Analysis was adopted to explore and classify the causes of the explosion of the benzene tank, along with the corresponding consequences, and the preventive measures. The analyses show that the primary inducements were the unsafe behavior of on-site workers, hazardous conditions of in-tank substances, risky factors in the construction environment, and deficiencies in management. Casualties, fire and explosion, equipment damage, and property losses were all included as consequences. The FLACS software was used to simulate the variation of explosion pressure and explosion temperature varying with time when benzene vapor was ignited at the tank bottom during the accident, and a flow field diagram for the explosion was achieved correspondingly. The results obtained are of direct significance for investigation of similar accidents in the future and are instructive to equipment safety evaluation as well. Based on the qualitative and quantitative analyses, we emphasize three considerable reasons leading to the accident, which includes inapposite flammable gases detection points before the hot work, failing to test the flammable gases during the hot work, and deficiency of the Management of Change (MOC). Finally, the paper concludes with a summary of lessons learned from the accident as well as suggestions for accident prevention and safety improvement.

1 Introduction

The term confined space refers to closed or semi-closed facilities and places with limited entrance and exit, and poor ventilation, which may lack oxygen or contain flammable, explosive, toxic, or harmful substances. The main causes of accidents in a confined space include fire, explosion, poisoning, asphyxiation, and physical injury [1]. According to the statistics of the Ministry of Emergency Management of the People's Republic of China, from January 2012 to April 2018, a total of 64 accidents involving three or more deaths in a confined space occurred in China, resulting in 225 deaths and 82 injuries [2]. Among them, on April 8, 2014, three people were killed and two injured by an explosion during the repair and welding of the desulfurization tank at a factory in Wuhai, China. On March 18, 2015, the operators of a chemical company in Binzhou, China, suffered an explosion accident during an operation in a confined space, resulting in four deaths and two injuries. Hot work in a confined space is the common cause of the two accidents above.

Due to the poor ventilation, a large amount of flammable gas may be accumulated in a confined space. When construction workers perform hot work inside, they will face potential safety risks and problems, such as difficulty in escaping, delayed personnel rescue, and inconvenient communication. Hot work in a confined space can easily lead to fire, explosion, and personal injury accidents, which threaten not only the personal health of operators but also the property of enterprises.

Researchers have done a lot of risk analysis and exploration for the particular operating environment in a confined space. Selman et al. [3] analyzed the mortality rate of confined space operations in similarly industrialized countries from 0.05 to 0.08 per 100,000 workers and pointed out that the main risks are toxic atmospheric hazards and physical hazards. Selman et al. [4] divided the emergency rescue of confined space accidents into three levels: individual self-rescue, non-entry rescue, and progressive rescue. Botti et al. [5] applied the Internet of Things (IoT) technology and proposed a holistic approach to analyzing the procedure of confined space operations. Meanwhile, many studies adopted experimental and numerical simulation methods to observe the evolution of characteristic parameters during the diffusion and detonation of flammable gas in a confined space. The results obtained are helpful to the quantitative analysis of confined space accidents. Bang et al. [6] established a theoretical model to predict the pressure rise caused by the deflagration in a confined space. He et al. [7] presented an LES turbulence model to simulate the diffusion process of LPG in the underground confined space.

This paper mainly analyzes a disastrous accident caused by hot work in a confined space at Shanghai Secco Petrochemical Co., Ltd (Secco). On May 12, 2018, the workers from Shanghai Aijinke Engineering Construction Services Co., Ltd. (Aijinke) triggered the explosion of flammable gas when they were performing inspection and maintenance inside a benzene tank. The serious accident resulted in the immediate death of six on-site workers and direct economic loss of 11.66 million yuan. Through a series of qualitative and quantitative analysis methods, we reveal the root causes of the accident, simulate the development process of the explosion, and propose corresponding rectification suggestions. Here we want to emphasize that the reason why we write this paper is not to pursue responsibility and punishment but to reflect on how to avoid the recurrence of similar accidents.

2 The occurrence of the explosion accident

On the morning of May 12, 2018, Aijinke workers arrived at the utility tank farm of Secco to prepare for maintaining TK-75-0201 benzene tank (0201 benzene tank). Before the start of the operation, a Secco field operator used a hand-held gas detector to measure oxygen and flammable gas at the manhole outside the tank. The on-site supervisor of Aijinke, the on-site supervisor of Secco, and the on-duty monitor of Secco utility tank farm signed the work ticket successively without verifying the measurement data and checking the personal protective equipment (PPE) and explosion-proof tools of the on-site workers according to the requirements specified in the operating permit.

At 1:15 p.m., six workers of Aijinke continued to dismantle the floating pontoon in the tank and two workers worked outside for monitoring. At the scene, another Secco field operator was in charge of regular oxygen and flammable gas measurement. At 3:25 p.m., a sudden explosion occurred in the 0201 benzene tank.

After the explosion, Secco immediately launched their emergency response plan and reported the accident information to the response center of Shanghai Chemical Industry Zone Management Committee (SCIZMC), the fire brigade, the medical center, and the production operations monitoring center of Sinopec. At 3:50 p.m., the fire brigade put out the fire by spraying water. At 5:50 p.m., the on-site rescue work ended and a total of six Aijinke workers in the tank died on the spot. Figure 1a shows the scene of the 0201 benzene tank after the explosion.

Figure 1. The 0201 benzene tank and floating pontoons

2.1 Description of the timeline

The occurrence of accidents is a complex system problem, and it gives useful information on how accidents arise in practice [8]. In March 2018, Secco found that the breather valve of the 0201 benzene tank leaked slightly, which caused the VOC concentration to exceed the standard. After inspection, it was judged that the floating pontoon leaked. In order to find the root causes, we describe in detail the fundamental behaviors of the primary responsible party and the contractor prior to the explosion.

- On April 16, Secco established a maintenance order and a work list in the SAP system.
- On April 19, Secco began to empty the 0201 benzene tank, performing four days of steam dilution and eight days of nitrogen replacement.
- On April 24, Secco informed Aijinke to formulate maintenance plan.
- On April 26, Secco carried out the Job Hazard Analysis (JHA) around the maintenance work.
- From May 1 to 7, Secco opened the manhole of the tank for natural ventilation and inspection, and found that more than half of the floating pontoons had effusion.
- On May 8, the relevant departments of Secco held a special meeting and decided to change the original plan to replace all the floating pontoons.
- On May 9, Secco asked Aijinke workers to enter the tank and destructively dismantle all floating pontoons at the regular OSBL equipment maintenance meeting.
- From May 9 to 11, Aijinke workers entered the tank and dismantled the floating pontoons. A total of 38 floating pontoons were dismantled, as shown in Figure 1b.

2.2 Analysis by event sequence diagram

Event Sequence Diagram (ESD) is an accident scenario-oriented simulation method, which treats events as the basic research object, and takes into account the domino effect and risk transfer effect [9]. ESD is also an intuitive reflection of the accident scenario. It provides a more literal representation of the system state, indicating the behavior of crucial process variables, and even operator and hardware state changes [10].

In order to further illustrate and point out the critical links of the accident, this paper makes an intuitive description of events by using ESD. As shown in Figure 2, the basic events of accidents are divided into initial events, judge events, action state events, and termination events. At the same time, the path indicated by the dashed line represents the potential events that may occur in this accident, and helps to make a comprehensive and logical analysis of the whole process. The graphic symbols in the ESD are explained in the Appendix.

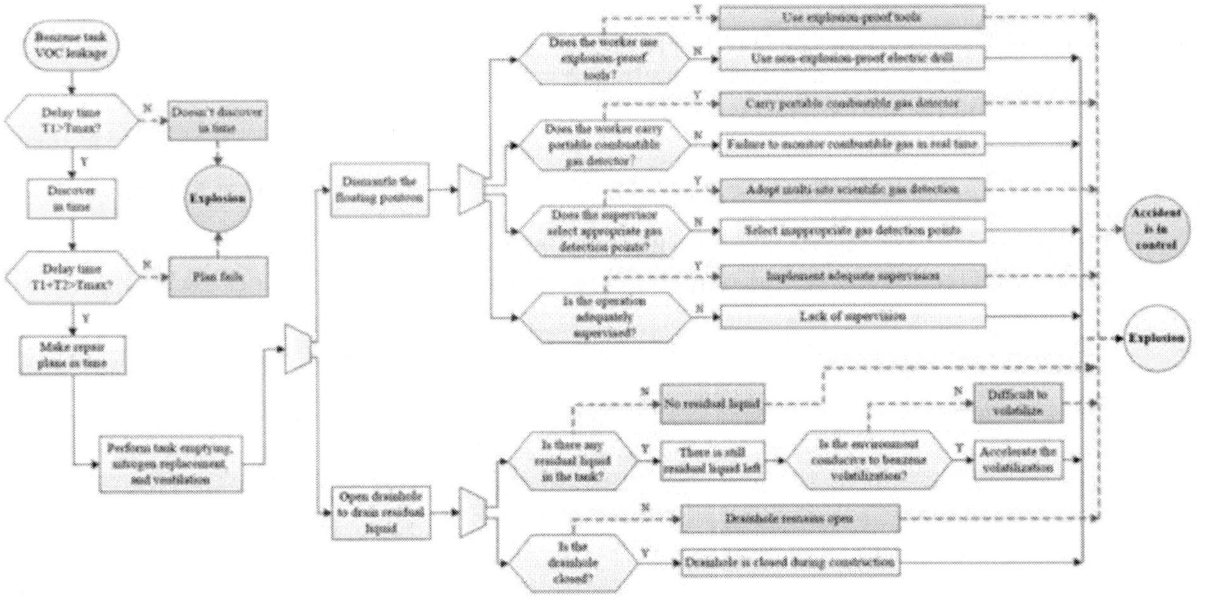

Figure 2. The ESD of the accident

As can be seen from the ESD, there are nine pivotal judgment nodes on the occurrence of the event chain, which can determine the direction of the incident. Among them, whether to discover leakage in time and whether to make a repair plan in time are the nodes at the time aspect. When the detection instrument and worker cannot find leakage and formulate the response plan in the maximum allowable delay time, they will directly lead to the explosion. Enterprises need to enhance equipment reliability and inspection frequency, so as to reduce emergency response time. Whether there are residues in the tank and whether environmental conditions are conducive to benzene volatilization are two nodes at the environmental aspect, which are also the leading causes of accidents. They require the workers to make a detailed hazard identification of the construction environment, and improve scientific detection. The critical nodes regarding human factors include whether the drainhole is closed, whether the on-site workers use explosion-proof tools, whether they carry portable combustible gas detector, whether they select appropriate detection points, and whether the supervisors take adequate monitoring [11]. Strengthening safety training before the operation, improving individual risk identification ability, perfecting risk control measures, and enhancing safety supervision level are essential means to avoid accidents caused by human factors.

All of these factors, organizational and operational, contributed to the accident, with a hierarchy of casualties among them. Through the comprehensive description, ESD can not only find out the nodes where the problem lies, but also guide the prevention of similar accidents. Therefore, enterprises should pay attention to the basic events expressed by nine judgment nodes and take appropriate interventions to block the development path of dangerous accidents.

3 Multilayer bow-tie analysis

3.1 The process of MBTA

Among many techniques available, bow-tie analysis is prevalent because it organically combines the two sophisticated risk analysis methods of Event Tree Analysis (ETA) and Fault Tree Analysis (FTA) [12], systematically summarizes the causes and consequences of accidents, and further integrates the prevention and mitigation of accidents or risks [13]. Bow-tie analysis has been proven to be a reliable and efficient tool, mainly used for process safety analysis, risk management, and accident risk analysis. It helps to understand which significant event combinations will lead to the top events, and which security barrier failures would escalate the top event to a certain consequence.

Traditional bow-tie analysis mainly includes top event, causes, consequences, preventive barriers, and mitigation barriers, which shows a lot of information. However, the single-level structure lacks logic and is not conducive to people's deep understanding. Thus, this paper establishes a Multilayer Bow-tie Analysis (MBTA), which resolves the top-event-related ETA and FTA individually into hierarchical structures, to deeply analyze the root causes, specify the severe consequences, and propose the corresponding safety barriers and scientific disposal methods. MBTA uses graphics to visually display the relationship between different levels and avoid repeated analysis effectively. Figure 3 shows the MBTA model, in which the code explanations are listed in Table 1.

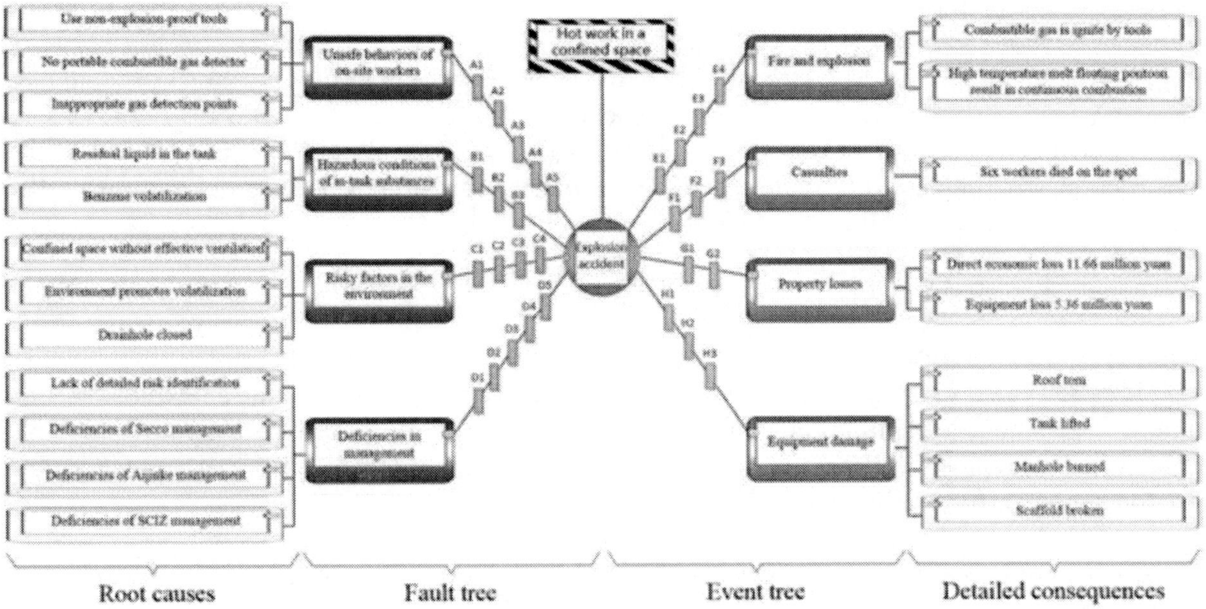

Figure 3. The MBTA model

Table 1. The explanations of MBTA

Code	Safety barrier	Code	Safety barrier
A1	Use explosion-proof tools	D4	Implement the main responsibility
A2	Carry portable combustible gas detector	D5	Improve contractor management
A3	Strengthen operation training	E1	Call the fire brigade in time

A4	Enhance supervision and assessment	E2	Spray water to extinguish fire
A5	Set up warning signs	E3	Fire isolation
B1	Remove the residual liquid	E4	Launch the emergency plan
B2	Hazardous material isolation	F1	Call for medical treatment
B3	Combustible gas detection in the tank	F2	Search and rescue the wounded
C1	Forced ventilation	F3	Launch the emergency plan
C2	Open drainhole	G1	Insurance claim
C3	Continuous detection during operation	G2	Launch the emergency plan
C4	Take environmental risk assessment	H1	Dangerous equipment isolation
D1	Compliance with rules and regulations	H2	Equipment repair
D2	Increase the intensity of punishment	H3	Equipment replacement
D3	Strengthen process safety management		

In the above MBTA model, we adopt the analytic hierarchy process (AHP) to extract the common characteristics of 12 causes of accidents and classify the causes into four categories: unsafe behaviors of on-site workers, hazardous conditions of in-tank substances, risk factors in the construction environment, and deficiencies in management. Moreover, we could conclude the direct and indirect reasons for the accident.

3.1.1 Direct reasons

Benzene is a colorless and transparent liquid with a strong aromatic odor. It has a boiling point of 80.1 ℃, a saturated vapor pressure of 10 kPa at 20 ℃, a flash point of -11 ℃, an explosion limit range of 1.2%-8.0%, and minimum ignition energy of only 0.20 mJ. Benzene is a highly flammable substance. The explosive mixture of benzene vapor and air can burn and explode when exposed to ignition source or high temperature.

At the construction site, the temperature is 24.7 ℃, and the pressure is 100.68 kPa. Under these conditions, benzene is easy to volatilize and form a gas explosion environment in a confined space. During the process of dismantling, residual benzene in the floating pontoons leaked onto the bottom of the tank and was not cleaned up in time. Because there was no effective ventilation in the tank, benzene vapor mixed with air to form an explosive environment and benzene concentration reached the explosion limit. The non-explosion-proof tools used by the operators produced ignition energy, which brought about the deflagration of the flammable gas. The high temperatures generated by the combustion melt the other aluminum alloy pontoons, resulting in sustained combustion for a short period.

3.1.2 Indirect reasons

Aijinke failed to abide by the relevant safety production rules and regulations. Managers did not fulfill their duties of safety management and supervision. Aijinke failed to give safety technical information to workers following the operating rules before the construction. Without implementing and supervising PPE, carrying portable gas detector, and using explosion-proof tools, operators were arranged to work in a confined space. When significant changes occurred in the job conditions, Aijinke did not promptly request workers to stop working and implement the Management of Change (MOC).

Secco had some omissions in safety risk management, professional management, and special operation management. In addition, Secco did not follow the safety production rules and regulations. On-site inspectors failed to carry out gas detection in accordance with specifications. After the change of operation, the manager did not ask the contractor to stop the process in time and revise the construction plan according to the regulations. Secco failed to report the maintenance plan to SCIZMC as required.

SCIZMC failed to implement the safety management system fully and did not discover that Secco performed maintenance operations without permission.

3.2 Lessons learned from MBTA

This paper puts forward a series of preventive measures, such as strengthening operation training, removing residual benzene in the tank, improving risk assessment of the environment, and enhancing process safety management. It is noticed that the environment in a confined space is harsh and there are many potential unsafe factors. Furthermore, the contractor's maintenance work cycle spans the statutory weekend. Based on the two reasons above, the enterprise should upgrade the level of hot work and adopt a broader system of operation approval and management. On-site workers need to stop working immediately when safety problems are found in the construction process and report to their superiors by relevant regulations.

Through refining the consequences of accidents, we summarized that detailed safety emergency plans and energy isolation methods are universally applicable in various measures to control and mitigate the impacts of accidents. They require Secco and Aijinke to formulate a comprehensive emergency response plan before, including fire extinguishing, personnel search and rescue, equipment maintenance, property damage remediation and so on. When an accident occurs, workers have to stay away from dangerous substances and places immediately.

As with many disasters, what ended up in a sequence of equipment damage and human errors started mostly as a management problem. Many unsafe practices were tolerated and even encouraged. The risk management of confined space operations should be divided into ten steps according to relevant regulations, namely training, pre-operation planning, job approval, hazardous energy isolation, ventilation, gas detection, personal protection and tool safety, warning signs, monitoring, and emergency response, so as to strengthen process safety management comprehensively. It is necessary to enhance the issuance of management of work tickets for hazardous operations such as hot work and confined space operations. Enterprises must strictly inspect illegal operations and supervise safety precaution measures to ensure that the operation process is safe and controllable. For changes occurring in the construction process, enterprises ought to strictly implement the MOC, analyze the potential risk and possible accidents, formulate corresponding safety measures, and carry out safety education for all operators.

MBTA provided a detailed analysis of the accident and established the corresponding safety barriers. However, there are still some certain subjective problems, and its content should also be flexible. Therefore, we encourage employees, accident investigators, case learners, and other relevant personnel to improve and develop the content of MBTA continually, thereby effectively expanding the application scope and effectiveness of MBTA.

4 Quantitative simulation of the explosion

The analysis of the accident should include not only qualitative analysis but also quantitative analysis [14]. We use FLACS software to simulate and analyze the occurrence of the benzene explosion. According to the relevant information, the 0201 benzene tank is an internal floating roof tank with a volume of 10,000 m^3. Its diameter is 30 m and height is 19 m. The design temperature range and pressure range are -10 – 65 ℃ and -0.49 –1.96 kPa, respectively. The tank adopts box-type aluminum alloy internal floating pontoons, and a manhole with a diameter of 0.6 m is set at the position of 0.75 m above the ground. Figure 4 shows the 0201 benzene tank model.

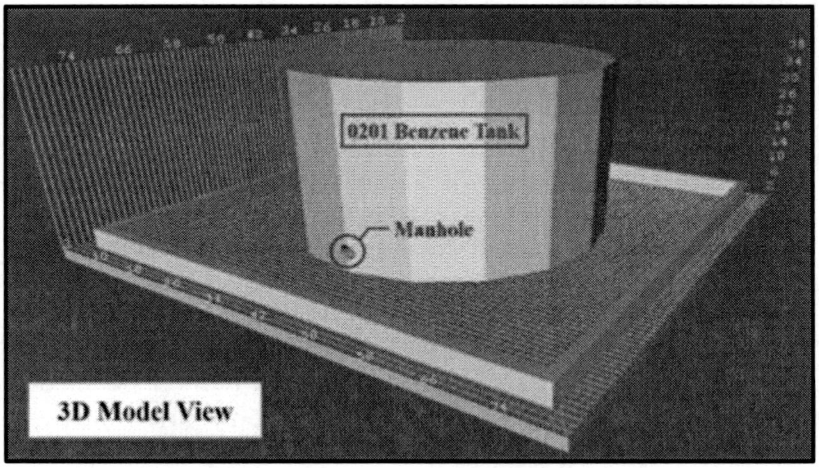

Figure 4. The 0201 benzene tank model

The simulated environmental conditions are fixed at the temperature of 24.7 ℃ and the pressure of 100.68 kPa, which are the same as the construction environment. We make reasonable assumptions: (a) The residual benzene in the tank evaporates continuously into the confined space and forms a uniform mixture of flammable gas. When the volatilization concentration of benzene reaches the lower explosion limit (LEL), the ignition source that persists during the construction operation detonates the flammable gas and results in an explosion. (b) Six on-site workers dismantled the float pontoons at the bottom of the tank, so the ignition source is set at the height of 0.75 meters from the geometric plane center, as shown in Figure 5 point I. There are several detection points of pressure and other characteristic parameters in the model of benzene tank, as shown in Figure 5, points M1-M5. (c) The benzene tank is windless, and the flammable gas accords with the real gas state equation. (d) The gas explosion process satisfies a series of fundamental equations, such as the continuity equation, momentum equation, and energy equation.

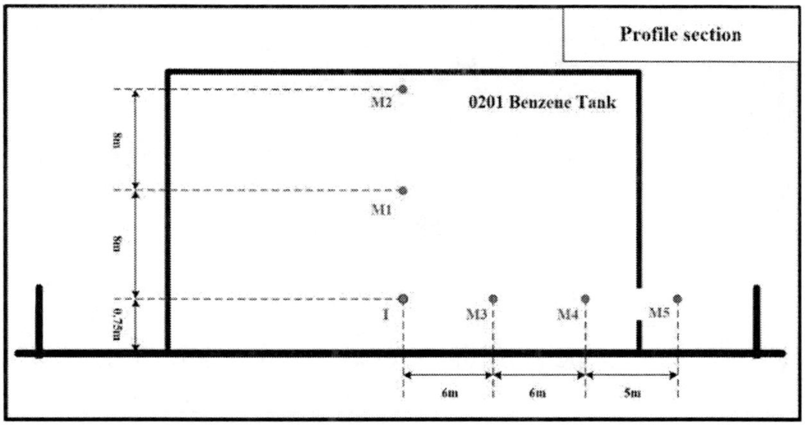

Figure 5. The profile section of the model

We selected the maximum profile section of the tank as the research object and intercepted the temperature field and pressure field at four different moments in the explosion process, as shown in Figure 6 and 7. The simulation results intuitively describe the development process of the explosion. When the flammable gas in the tank was ignited, the flame diffused around in the form of a spherical wave and reached the bottom of the tank at 0.35 s, as shown in Figure 6a and 7a. Since the radial distance of the tank is less than the vertical axial distance, the flame propagated horizontally to the wall, as shown in Figure 6b and 7b. At 1.52 s, the flame reached the manhole and sprayed out to the outside of the tank in the form of jet flow. During this period, the pressure and temperature increased continuously, which are the leading causes of the manhole being burned, as shown in Figure 8a. When the flame propagated to the top of the tank at 2.11 s, the uncombusted gas was recirculated by the viscous effect and blocking effect of the wall, which further caused the recirculation vortex. The recirculating vortex enlarged the contact area of the flame front, resulting in flame acceleration and pressure increase, as shown in Figure 6c and 7c. Figure 6d and 7d illustrate the flammable gas burned out at 3.24 s. Simultaneously, the explosion temperature and pressure reached the peak value contributing to the tearing of the roof and the lifting of the tank, as shown in Figure 8b and 7c.

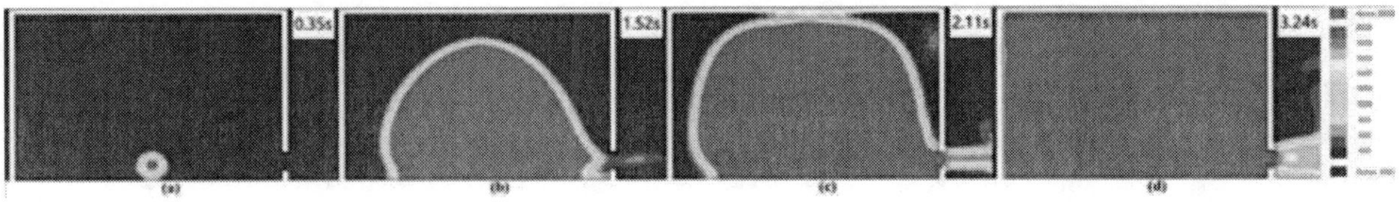

Figure 6. Temperature simulation of the explosion

Figure 7. Pressure simulation of the explosion

Figure 8. The results of the accident

Figure 9 reveals the change of temperature and pressure recorded by the detection points M1-M5 during the explosion. Table 2 describes the data of the detection points, including maximum and minimum explosion pressures, maximum and minimum explosion temperatures, peak data arrival times and so on. Combining the data of detection points and the pressure field diagram in Figure 7, it is shown that the pressure in a confined space will be uniform immediately for a certain time and further proves the accuracy of the analysis of the flame propagation process. In particular, the external detection point M5 was affected by the propagation of explosion flame in the tank at 1.6 s, resulting in a peak explosion overpressure of 0.031 bar and a peak temperature of 1895 ℃. Subsequently, due to the pressure rise in the tank, the pressure at the detection point M5 dropped to negative pressure until the flammable gas burned out.

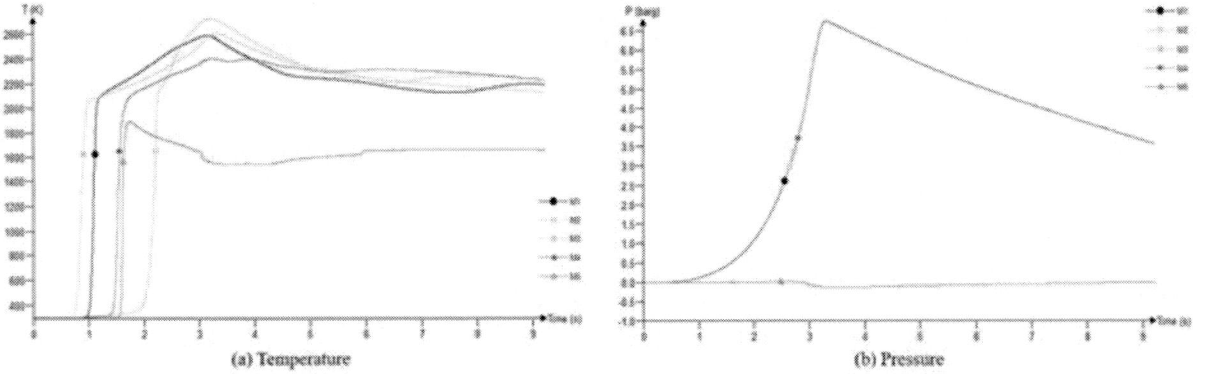

Figure 9. Parameters change of the detection points

Table 2. Data from the detection points

No.	Temperature				Pressure			
	Tmax/K	**Time/s**	**Tmin/K**	**Time/s**	**Pmax/bar**	**Time/s**	**Pmin/bar**	**Time/s**
M1	2589.4	3.11	298.1	0.04	6.75	3.24	0.00	0.04
M2	2732.2	3.24	298.1	0.04	6.76	3.24	0.00	0,04
M3	2609.3	3.28	298.1	0.00	6.76	3.24	0.00	0.00
M4	2306.2	3.25	298.1	0.00	6.75	3.24	0.00	0.00

M5	1895.5	1.68	298.0	0.69	0.05	1.65	-0.15	3.29

Although there is only a small amount of residual benzene, the benzene volatilization forms a flammable gas with the air inside the tank, which generates a substantial destructive power when ignited. The simulation results show that the flame propagation velocity of the flammable gas increases constantly, with the peak value reaching up to 41 m/s. Since the flame is propagated in the form of subsonic velocity, it can be judged that there are mainly the phenomena of deflagration occurred in the process of explosion [15]. Soloviova-Sokolova et al. [16] calculated the flame burning rate of the benzene vapor was 0.6 m/s at 298 K, 0.1 MPa and 0.6 eq., which clearly illustrated that the peak propagation velocity of the benzene vapor in a confined space would be several times higher. Meanwhile, the explosion pressure in a confined space is usually greater than that in an open area due to lack of heat loss and volume expansion.

When workers are engaged in hot work in a confined space, they should carry out comprehensive risk identification, select several representative flammable gas detection points, and detect the flammable gas in real time. Workers ought to stop work immediately and withdraw in an orderly manner once the flammable gas exceeds the standard. Through the analysis of the explosion process, the flame propagates to the wall in a very short time. Thermal radiation and explosive shock wave are the primary causes of casualties in the tank. Owing to the special working environment in a confined space, the phenomenon of mass casualties and injuries often occurs under the effect of high temperature and high pressure, which needs people's sufficient attention.

5 The subsequent event

The explosion accident of Secco has aroused widespread concern, and 22 responsible persons have also been punished. But unfortunately, lessons from the accident did not prevent a subsequent similar event. On November 26, 2018, three contractor workers of the Ningbo Engineering Company (NEC) and one worker of Secco inspected the discharge tank of cracking gas compressor in the olefin separation unit. Without ascertaining the concentration of oxygen in the tank, a contractor worker of NEC entered the tank for construction and fainted unexpectedly. A Secco worker immediately went into the tank to rescue him, but he also fell into the tank. The accident resulted in the death of two workers who had been rescued after asphyxia.

It is the second confined space operation accident in the same company within six months. The main causes of the accident are still the deficiency of contractor management and risk identification in a confined space. While expressing sadness, the public should profoundly reflect on why accidents continue to occur, what is the role of punishment and accountability, and how relevant regulations and reform measures can be truly implemented. Drawing lessons from the previous accidents is necessary. Enterprises ought to raise the safety awareness of their employees and use strict supervision to ensure safe production of chemical plants.

6 Conclusions

This paper provides an introduction of an accident caused by hot work in a confined space, together with the detailed ESD-based description of nine pivotal judgment nodes leading to this disastrous accident. It concludes that enterprises should take suitable intervention measures to block the pathway of the accident. The MBTA, covering the causes and consequences of the accident, offers proactive and reactive measures based on an intuitive analysis. Meanwhile, it emphasizes the importance in selection of appropriate detection points, real-time detection of in-tank operation, and implementation of MOC. Also, we apply the FLACS software into quantitative simulation of the explosion process in the tank. The results contribute to the guidance of accident investigation and safety measures selection. Finally, it suggests that enterprises adopt technical remedies and better safety management practices, since they are the essential basics for effective implementation of risk control measures.

Appendix

Symbol	Annotation
	Initial event: Beginning event of the ESD.
	Judge event: The event has two possible outcomes that determine whether subsequent events occur.
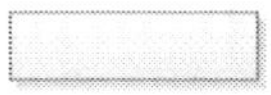	Action state event: Describe the action state of the event.
	Termination event: Ending event of the ESD.
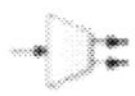	Output AND gate: Input event lead to multiple output events occurring.

7 References

[1] D.Burlet-Vienney, Y.Chinniah, A.Bahloul and B.Roberge, Design and application of a 5 step risk assessment tool for confined space entries, Safety science 80 (2015), 144-155.

[2] The Office of the State Council Security Committee Notified 8 Cases of Blindly Rescuing in a Comfined Space Leading to Casualties Enlargement, Available at http://www.chinasafety.gov.cn/default.htm Accessed on June 8, 2018.

[3] J.Selman, J.Spickett, J.Jansz and B.Mullins, Safety science 109 (2018), 333-343.

[4] J.Selman, J.Spickett, J.Jansz and B.Mullins, Safety science 113 (2019), 78-90.

[5] L.Botti, V.Duraccio,M.G.Gnoni and C.Mora, A framework for preventing and managing risks in confined spaces through IOT technologies, ESREL (2015), 3209-3217.

[6] B.Bang, C.Ahn, J.Lee, Y.Kim, M.Lee, B.Horn, D.Malik, K.Thomas, S.James, A.Yarin and S.Yoon, Theoretical, numerical, and experimental investigation of pressure rise due to deflagration in confined spaces, International Journal of Thermal Sciences 120 (2017), 469-480.

[7] B.He, X.Jiang, G.Yang and J.Xu, A numerical simulation study on the formation and dispersion of flammable vapor cloud in underground confined space. Process Safety and Environmental Protection 107 (2017), 1-11.

[8] I.M.Shaluf, F.Ahmadun and A.M.Said, Fire incident at a refinery in West Malaysia: the causes and lessons learned, Journal of Loss prevention in the Process Industries 16.4 (2003), 297-303.

[9] S.Swaminathan and C.Smidts, The event sequence diagram framework for dynamic probabilistic risk assessment, Reliability Engineering & System Safety 63.1 (1999), 73-90.

[10] J.Zhou, G.Reniers and N.Khakzad, Application of event sequence diagram to evaluate emergency response actions during fire-induced domino effects, Reliability Engineering & System Safety 150 (2016), 202-209.

[11] E.Zarei, M.Yazdi, R.Abbassi and F.Khan, A hybrid model for human factor analysis in process accidents: FBN-HFACS, Journal of Loss Prevention in the Process Industries 57 (2019), 142-155.

[12] N.Khakzad, F.Khan, P.Amyotte, Dynamic risk analysis using bow-tie approach, Reliability Engineering & System Safety 104 (2012), 36-44.

[13] M.B.Mulcahy, C.Boylan, S.Sigmann and R.Stuart, Using bowtie methodology to support laboratory hazard identification, risk management, and incident analysis, Journal of Chemical Health and Safety 24.3 (2017), 14-20.

[14] C.Zhu, J.Zhu, L.Wang and M.S.Mannan, Lessons learned from analyzing a VCE accident at a chemical plant, Journal of Loss Prevention in the Process Industries 50 (2017), 397-402.

[15] I.C.Tolias and A.G.Venetsanos, An improved CFD model for vented deflagration simulations–Analysis of a medium-scale hydrogen experiment, International Journal of Hydrogen Energy 43.52 (2018), 23568-23584.

[16] J.V.Soloviov-Sokolova, V.A.Alekseev, S.S.Matveev, I.V.Chechet, S.G.Matveev and A.A.Konnov, Fuel 175 (2016), 302-309.

Strengthening Process Safety Culture throughout Learning from Each Other´s Vulnerabilities

Luciana Poli SIlva
PETROBRAS
Av. Republica do Chile, 65, 4th floor, aisle 402.
lucianapoli@petrobras.com.br

Prepared for Presentation at
American Institute of Chemical Engineers
2019 Spring Meeting and 15th Global Congress on Process Safety
New Orleans, LA
March 31 – April 3, 2019

Strengthening Process Safety Culture throughout Learning from Each Other´s Vulnerabilities

Luciana Poli Silva
PETROBRAS
Av. República do Chile, 65, 4th floor, aisle 402.
lucianapoli@petrobras.com.br

Keywords: andragogy, dialogue, leadership, safety leadership culture, trust, vulnerabilities.

Abstract

A fundamental approach to any program that targets the implementation of a process safety culture is to work within reality, going beyond procedures and compliance. To facilitate the emergence of the unspoken is essential to stimulate the dialogue among leaders that brings about their vulnerabilities and opens new perspectives to address them effectively. This paper presents a pilot program designed to increase the Process Safety Leadership Culture in a Brazilian Energy Company. Based on a benchmark made within the industry and in andragogic principles, the program created a trust ambiance among peers and inspired leaders to recognize and share their vulnerabilities, as well as to take actions towards the strengthen the role of being a safety culture leader.

1 Introduction

A lot has been said and written about Leadership Process Safety Culture, but there is a gap between what the compliance procedures dictate and what managers really do. When it comes to the Brazilian Energy Industry, the pressure leaders are under in order to achieve the results they are expected to can be considered significant. How to balance the production challenges with a safety culture that must be among the core values of the operation? Are these factors compatible?

The pilot program entitled '*Strengthening Process Safety Culture*' dealt with reality, using andragogic approaches to bring about the complex factors forty managers face in their routine, such as the existence of several levels separating top managers from the operational team, privileging safety over production goals, and being an effective representative of the Company´s policies. All these factors can jeopardize the safety culture from becoming stronger every day.

This paper describes the implementation of a methodology designed to create a safe environment, in order to build trust among participants and at the same time, make them reflect and dialogue about how they can evolve from compliance to commitment.

In the coming sections, we will present the central concepts used to develop the program – Culture, Safety Culture and Leadership - exploring its implication on an actual work environment.

In the sequence, we will talk about what it takes to make the unspoken, unwritten rules emerge. From there, how we stimulated the increase in the self-conscious level as well as in the collective one, building a matrix entitled 'Personnel Influence on Safety' to diagnose how one really sees himself concerning his leadership in safety culture.

We will then present the method used to address the issues managers pointed out, in order to transform the discussions into actions to strengthen the safety process leadership culture.

In the conclusion, we will present our findings and discuss possible future approaches to amplify the safety leadership culture further.

2 Safety Culture and Leadership: definitions and implications

There are many possible definitions for culture. Taylor (2015) says culture has to do with the behaviour patterns that are encouraged, discouraged or tolerated over time in an organization. Schein (2015) says that culture is "the total sum of everything an organization has learned in its history in dealing with the external problems – which would be goals, strategy, how we do things – and how it organizes itself internally."

Since we humans are gregarious social beings, we are naturally wired to fit in with our community. We use our perceptions to read what the tribe accepts and we adapt by behaving accordingly, because we want to fit in and be rewarded. Therefore, we reinforce the existing norms. This process is perpetuated over time; established behavior influences the behavior of new employees and so on.

Most cultural sources are non-verbal. Taylor (2015) says it is a myth to believe that culture reflects the values statements that are declared in the company's annual reports. Most commonly, the statement will be the organization´s intents, not the culture as it actually is.

If we want to change culture, it is necessary to have a clear direction and a focused, effort-driven leadership to lead people towards the goal culture. Culture has to do with what is encouraged and allowed. For Taylor (2015), three major sources send cultural messages: behaviours - of other people, especially important, recognized ones; symbols - artefacts, events and decisions that are meaningful to the community; and systems - mechanisms that manage people and tasks.

In this sense, culture has to do with the manifestation of what is valued in the organization.

According to Lafraia (2014), culture shapes reality. Cultural lenses structure the perception of the people involved in the organization.

When it comes to safety culture, its meaning may vary from one organization to another, or even in different departments of the same company. It is perfectly possible to have diverse lenses that translate the meaning and the importance of safety practices in diverse ways.

OGP, in its 452 Report (2013), says that a positive Safety Culture has to do with a culture in which safety is a core value for everyone who works in the organization, and it is pursued in an everyday basis. In order to build a Safety Culture, there are factors that must be considered, such as "leadership commitment to safety; employee involvement and motivation; responsibilities and accountability; production and bottom line pressures versus quality issues; and actions, or lack of action, to correct unsafe behaviours and unsafe conditions"; among others.

In order to build a consistent safety culture, leaders must educate their teams by demonstrating, by their practices, that some policies are fundamental. Safety culture should use this lens, rather than a command and control opaque one. Leaders must educate people to understand why safety is a core value that concerns the most precious gift: one´s life.

Safety culture is not about only following the rules, but making the team understand that the rules serve a greater principle. This approach can influence values and beliefs and change the individual and team´s behavior. The challenge leaders face is to conciliate the educational role with the policeman role, guaranteeing that compliance is required and everyone in the organization, starting by himself, values that as an unnegotiable principle.

In order to develop a strong safety culture, some elements must be considered (OGP, 2013):

Elements of a Strong Safety Culture

- an informed culture: the company has quality data concerning safety and uses it well;
- a reporting culture: people feel they can report safety concerns without worrying about being punished;
- a learning culture: the organisation actually learns from its unsafe conditions and changes them, banning any equipment, practice or whatever takes place that hazards safety;
- a flexible culture: the company changes its command chain if necessary to adapt;
- a just culture: people understand what is acceptable and unacceptable. What is not tolerated is dealt fairly.

For Peter Drucker (Cohen, 2018), leadership can be defined as the lifting of a man's vision to higher sights. Throughout leadership, people perform in higher standards, they are inspired and evolve in multiple levels.

When it comes to safety leadership, we can quote Ram Charan (2007) when he says that "a leader who does not produce other leaders is not a great leader." A safety leader should act in a way to enhance in everybody´s minds and actions a safety leadership commitment.

In order to make it possible, it is crucial that the leader himself is aware of his beliefs and constantly monitoring his behaviours. It takes time and a lot of effort to build a strong safety culture. Despite that, a safety culture is fragile and easily threatened by wrong messages. (OGP, 2013)

In the next section we explore the method used to address self-consciousness about safety.

3 Creating a trust environment that favors Safety Process Leadership Culture

According to Malcon Knowles, who develop the paradigm of andragogy, adults have a learning style characterized, among other factors, by changing one´s concepts according to new experiences one lives; the pursuit for autonomy; increasing one´s own motivational level to match the social role assumed; and acquiring knowledge to immediately solve practical issues (Muresan, 2013).

Considering this practical approach, we began the program to strengthen safety leadership culture by making each participant reflect on its daily practices concerning management, as well as the safety implications involved in routine decisions.

To start this reflection, we did a benchmark and decided to invite a senior top manager from another company in the Energy Industry to give a testimony. He made an honest statement about his team perception of his management practices. Concerning safety process, he mentioned that he realized that, in order to influence people, he must first change himself. He then shared with an amazed group of participants some practical examples of safety interventions he changed in his auditing procedures, for example, in order to be coherent with the message he wanted to pass forward.

The testimony had an impressive impact on the managers. A remarkable phrase that was repeated by managers was: "*Simplify. Be the Powepoint. Less Presentations, More You.*"

After the testimony, there was a presentation on the invited company's safety philosophy, about its journey into safety leadership. The practical steps mentioned were followed by videos with managers from different parts of the world talking about the way they see safety culture benefits.

With that background environment set, we forwarded to a self-perception diagnosis. The goal was to stimulate the increase in the individual conscious level as well as in the collective one. We used a matrix entitled 'Personnel Influence on Safety' that was obtained in the benchmark process with the above mentioned company.

It aimed to diagnose how one really sees himself concerning its leadership in safety culture. It involved two dimensions: the personal influence towards oneself; and the personal influence concerning other people´s safety.

The figure below illustrates the matrix:

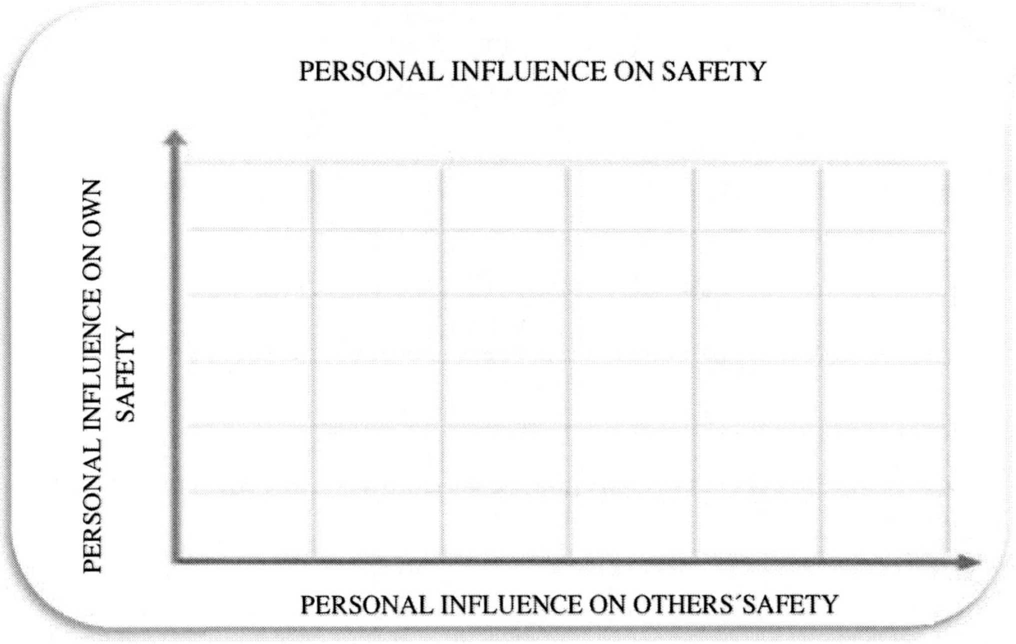

Figure 1. Personal Influence on Safety Matrix

We used the Self-perception as a tool to promote reflection and consciousness. We asked the participants to fill in the matrix in two different papers: one in his scrapbook (explained below) and another one on a plain sheet to hand over to the program facilitator. All the matrices were compiled and the result is shown in figure 2. As indicated, most participants initially evaluated themselves as influencing more themselves than others.

The coordination staff evaluated the matrix result as requiring attention from the participants. The staff observed the fact that some managers attending the program were obese, others worked extra hours, and most of them had reported having trouble balancing personal and professional life. Therefore, to promote deeper self-reflection, the consultant who was facilitating the program posed some questions to the participants regarding personal care. The remarks made were an invitation to honestly look at oneself and feel if the level of personal influence corresponds to what one would recommend to a loved person. Quoting Brené Brown (2015), the consultant said "Treat yourself like someone you love. Do you do it on a regular basis? Do you put yourself first?" Those remarks made some participants change their self-evaluation.

The intention was to call leaders ´attention to the fact that genuine influence on safety starts with their own safety, and goes beyond process or operational safety. To be legit and truly be a safety leader, one should take good care of oneself.

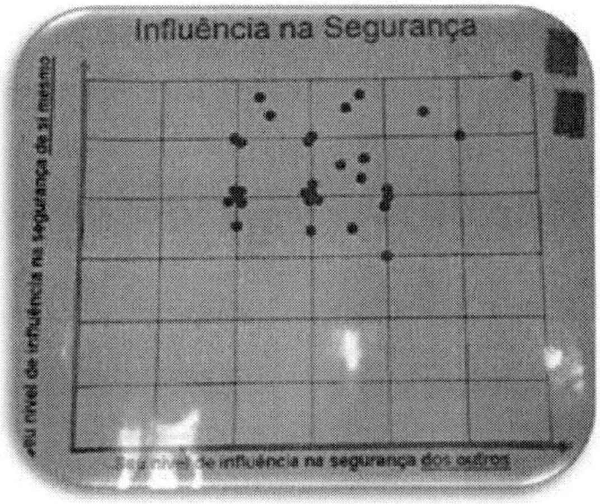

Figure 2. Compiled Personal Influence on Safety Matrix with the results from all the participants

We also developed a scrapbook that was given to each participant as a job-aid material to be used both during the program and after. The content of the scrapbook followed the program plan and was an invitation to self-analysis. The participants had special pages to do some journaling about what they had just listened to or dialogued about. There was also a special page dedicated to register insights. Figure 3 illustrates the scrapbook.

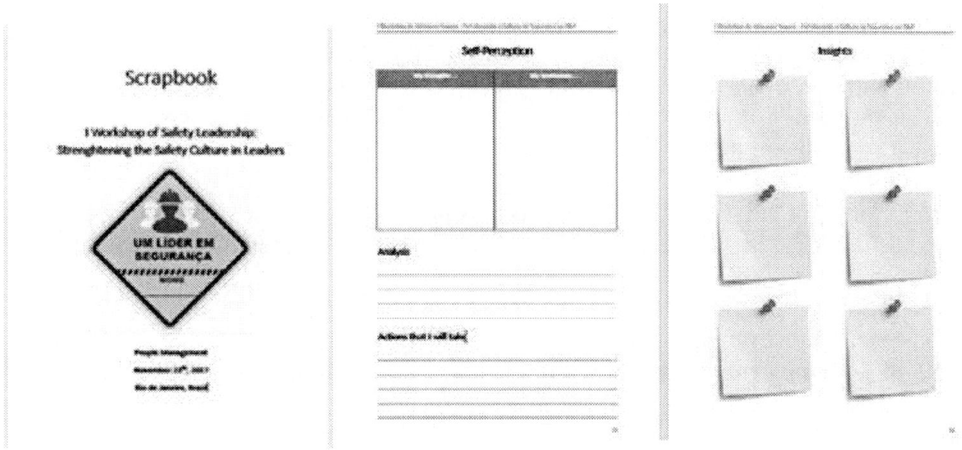

Figure 3. Pages from the Scrapbook used to stimulate self-consciousness and to plan actions

4 Exposing and Addressing Vulnerabilities

4.1 Control Room

We will now present the method used to address the issues managers pointed out, in order to transform the discussions into actions to strengthen the process safety leadership culture.

We prepared a special control room, where we designed a pathway decorated with images of accidents from the company and from the industry, and the number of fatalities. We also put an alarm sound playing during the route and videos showing accidents. These resources were intended to create an emotional ambiance to take the analysis to another level.

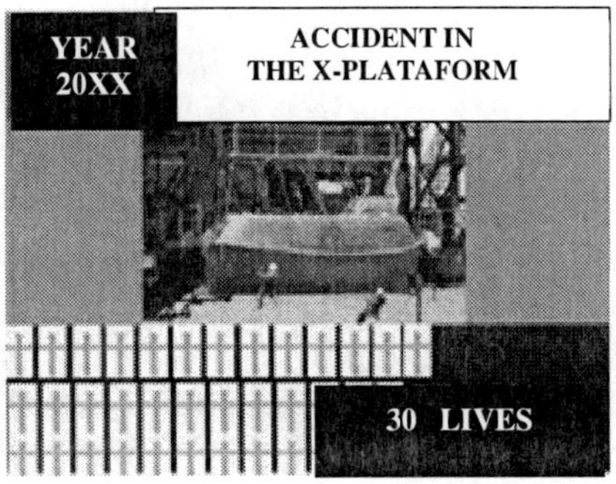

Figure 4. Picture of the Control Room and example of poster from accident used to decorate it.

Besides the planned journey passing several industry accidents, the Control Room was decorated with red material, to create an atmosphere of urgency. After going through the posters and watching the videos, the participants were invited to give their testimony of accidents they´ve been through.

The group tone got emotional as some of them shared unpleasant stories they had lived. The goal of inducing the reflection to an in-depth level was successfully achieved.

After that point, the participants went back to the regular room and dialogued about some questions that were in their scrapbook. The questions were related to the *Elements of a Strong Safety Culture* preconized by OGP (2013), mentioned in section 2 of this paper.

The methodology privileged social interaction among participants and the active listening of each other's considerations. Below is the list of the questions addressed to the participants.

Questions to Reflect and Dialogue:

1. Am I able to influence my operational team?
2. How many hierarchical levels are there between the operational team and me? To what extent does my speech actually makes sense to them?
3. How much of what is said by the operational team do I believe genuinely applies to me? Do I believe I receive good quality feedback?
4. Of the security actions implemented in my management, do I receive direct feedback on the effectiveness or otherwise of its implementation?
5. What degree of tolerance do I accept as a leader, regarding violations of safety standards?
6. Do I, as a leader, make it mandatory that the urgent inspection recommendations are executed immediately?
7. Do I, as a leader, take every opportunity to motivate, mentor and develop my team to enhance individual safety awareness and safety leadership?
8. Do I praise when I see a safe behavior?
9. Do I perform Behaviour Audits in a protocolled way, to meet the Scorecard, or do I do it because I believe in the effectiveness of what can arise from it?
10. Is Safety a value or is it a Priority for me as a leader?
11. Do I have the courage to step in and stop an activity when I see someone at risk? Do I always intervene?
12. Do my behaviours as a leader have safety as a foundation?
13. When I find that the environment is unsafe, do I privilege safety over other goals?
14. Safety comes first in my management?
15. How do I, as a leader, ensure that safety in my management is dynamic enough to ensure continuous improvement in operational procedures?
16. Thinking Safety as a chain, how can I, as a leader, be an effective link between operations and the company policy?

4.2 Results

In order to capture the precious discussions and share insights among the participants, in the final part of the workshop there was a collective presentation based on what each small group discussed about the questions above. Each group had to have at least one delegate, and others could participate if they wanted to.

Some of the findings reported were, for example: "*I have to go to the platform more often*"; "*Behaviour audits can be very useful, I should profit more from them*"; "*There are too many levels between my operational team and I. I have to get together with them on a regular basis.*"

After sharing insights, all managers filled in the Personal Influence on Safety Matrix again. The group reported that, this time, the self-perception was lower, since many managers realized they need to invest more efforts and more quality time in Safety, in both dimensions – with oneself and towards others.

After that, each participant revised and complemented his individual action plan (figure 3). They also wrote a personal commitment as a Safety Process Culture Leader.

At the very end of the program, we asked everyone to do a check-out by sharing with the entire group one action of the action plan. This intended to create a deeper commitment, as well as possible insights to others. Also, it was a practice that could be used with the managers team.

According to the group evaluation, presented in the word cloud below, the program succeeded in strengthening the safety culture in leaders, who mentioned as the top benefits from the program: (*Safety as a*) Value; Communication; Life; Commitment; Example; Integration; Consciousness; and Synergy.

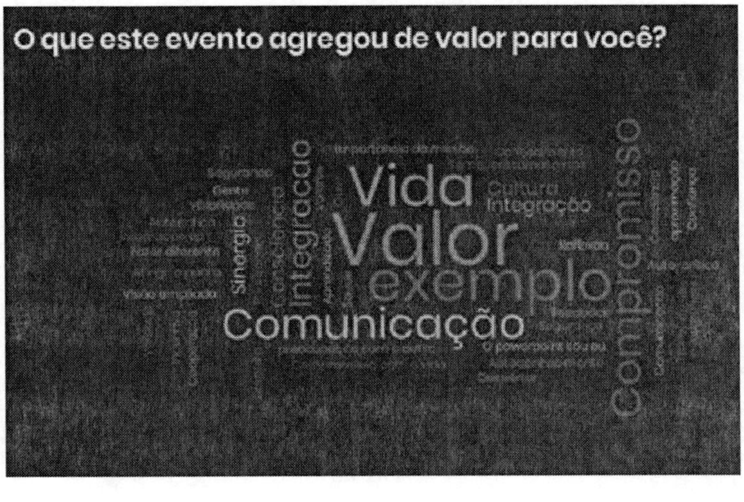

Figure 5. Evaluation: Word cloud containing the managers´ perception about the contribution brought by the program to participants.

5 Conclusions

The feedback given by the program´s participants was encouraging. It indicated that testimonies from partners, combined with relevant questions based on the daily problems managers face concerning safety, created a trust ambiance among managers.

The activities planned based on andragogic approaches favored dialogue and in-depth self-analysis. The use of the same instrument (Personal Influence on Safety Matrix) twice, as well as the individual scrapbook, facilitated the self-consciousness process and lead managers to review their initial assessment and to elaborate an action plan.

Leaders play a decisive role in strengthening the safety culture in the organizational world. Bringing about their vulnerabilities and having honest conversations with peers definitely helps managers to find ways to evolve from compliance to commitment.

Strengthening the safety leadership culture is a process that does not come to an end. It is an on-going process that must be continuously taken care of by leaders who value life and put safety first. Future projects that aim to implement effective actions in Safety Culture should take into consideration the company´s culture, as well as daily operational problems and safety as a core value to preserve life and business.

6 References

B. Brown, "*Daring Greatly: How the Courage to Be Vulnerable Transforms the Way We Live, Love, Parent, and Lead*", Penguin Random House, NY, 2015.

C. Migueles, J.R.B. Lafraia, G. C. Souza, "Criando o Hábito da Excelência: Compreendendo a Força da Cultura na Formação da Excelência em SMS", Quality Mark, Rio de Janeiro, 2014.

C. Taylor "*Walking the Talk: Building a Culture for Success*" Random House Business Books, London, 2015.

M. Muresan, "*Using Cybergogy and Andragogy Paradigms in Lifelong Learning*", 5th World Conference on Educational Sciences – WCES. Romania, 2013

OGP, *Shaping Safety Culture through Safety Leadership.* Report N° 452. International Association of Oil & gas Producers. October, 2013.

R. Charan, *Ram Charan on Leadership*,

Available at https://hbr.org/ideacast/2007/01/harvard-business-ideacast-27-r.html. Jan, 2007 (accessed Dec 4, 2018)

T. Kuppler, *Edgar Schein on Culture*,
Available at https://www.leadershipandchangemagazine.com/edgar-schein-on-culture/ Accessed on January 2, 2019.

W. Cohen, *Drucker's Top Tips to Successful Leadership*
Available at https://www.managementmattersnetwork.com/strategic-leadership/columns/druckers-top-tips-to-successful-leadership. Jun, 2018 (accessed Dec. 4, 2018)

7 Acknowledgments

I would like to thank Petrobras who fully supported this project, as well as Shell Brasil, for a very profitable partnership. A special thanks to my Human Resources, HSE and E&P work colleagues, who had a decisive role in making this project a reality.

548424 Development of Human Reliability Analysis Method for Oil and Gas Operations

The oil industry has grown in recent decades in terms of quantity of facilities and process complexity. However, human and material losses still occur due to major accidents at the facility. The analysis of these accidents reveals that many involve human failures that, if prevented, could avoid such accidents. These failures, in turn, can be identified, modeled and quantified through Human Reliability Analysis (HRA), which forms a basis for prioritization and development of safeguards for preventing or reducing the frequency of accidents. The most advanced and reliable HRA methods have been developed and applied in nuclear power plant operations, while the petroleum industry has usually applied Quantitative Risk Analysis (QRA) focusing on process safety in terms of technical aspects of the operation and equipment. A joint project from the B. John Garrick Institute for the Risk Sciences and a big company from the oil and gas industry is focusing on developing a HRA method focused on this industry operations, to bridge the existing gap between HRA and the Petroleum industry. The Phoenix HRA methodology was used as a basis, which has three layers i) a crew response tree (CRT), which models the interaction between the crew and the plant; ii) a human response model, modeled through fault trees, that identifies the possible crew failures modes (CFMs); and (iii) "contextual factors" known as performance influencing factors (PIFs), modeled through Bayesian networks. In addition to building on such a structure, the method development will rely on interviews with key personnel, visitations to installations, and analysis of past incidents data.

547327 Studying the Dependence of Operator Performance on Step Level Complexity

In the chemical, oil, and power industry, procedures are used by operators to perform their work correctly without compromising safety. Unfortunately, many procedures fail in adequately conveying information to their users. Heuristics tell us that technical jargon, vague language, and operator training all play a role in determining the effectiveness of procedures. To what extent these variables play a role is still not fully understood. Knowing this information will allow plants to more effectively write procedures. To counter this, researchers at Texas A&M's School of Public Health and Mary Kay O'Connor Process Safety Center are investigating the underlying factors that cause operators to deviate from procedures.

One possible factor may be that complex steps hinder an operator's ability to perform a task. This work aims to understand why certain steps are more complex than others. To do this, a set of procedures on how to operate a gas chromatography mass spectrometry (GCMS) machine is written. A study is performed in which participants use a set of procedures to analyze samples through gas chromatography. Using a scoring rubric, the complexity of each step is determined. The procedures have been designed to obtain an even distribution of complexity. Participants for this study are chemical engineering students with little to no prior experience in handling a GCMS machine. Thus, the experience, education, expertise, and training of the participants is controlled. The performance of each participant is reviewed and is grouped into one of four quadrants. For this study, a decrease in operator performance is expected as the complexity of each step increases.

548400 Prevent Them All or Stop Them Small: Tripod Best Practices in Reducing Chemical Accidents and Enhancing Response Preparedness

Prevent Them All or Stop Them Small: Tripod Best Practices in Reducing Chemical Accidents and Enhancing Response Preparedness

The connection between industry, government, and public safety, which we call the Tripod, is vital when implementing the 3-P's - Prevention, Protection, and Preparation. As identified by AICHE's Vision 2020 program, there are significant opportunities for organizations that address industrial process safety to work together. Responsible Collaboration between industry, regulatory and investigatory authorities, labor organizations, communities, and technical organizations promotes mutual understanding of risks, effects process safety and can reduce chemical accident risk in both small and large communities. Community based safety partnerships and leadership are fundamental to the success of such efforts.

For the past five years, a series of geographically focused Chemical and Ammonia Safety Days have cultivated and showcased chemical emergency prevention and planning efforts between Industry, NGOs such as the Ammonia Safety Training Institute (ASTI), and a myriad of public agencies including fire departments, health departments, LEPCs, SERCs, local regulators, and state and federal agencies such as EPA, CSB, DHS and OSHA. Within the context of Chemical Accidents and Releases, the focus of these efforts has been to Prevent Them All or Stop Them Small.

These Safety Day programs are designed for and with local community stakeholders from industry and first response agencies with a goal of seeding a permanent, engaged geographic collaboration for accident prevention and response readiness ownership and partnership. Agenda and curriculum elements include technical training focused on local risk, first responder field practicums and case studies, accident prevention program prioritization, practice and accountability, bowtie analyses of event propagation and management, and emergency response planning centered on the first 30 minutes and one-plan approach. The programs often include field demonstrations and exercises for fire departments and facility responders.

As a best practices demonstration, the presenters for this session will share the developed programs, recruitment methodologies and the demonstrated results of these efforts. In addition to Responsible Collaboration, other Vision 2020 principles embraced and promoted by the Safety Day Approach include Disciplined Adherence to Standards, Intentional Competency Development, Enhanced Application & Sharing of Lessons Learned, and Enhanced Stakeholder Knowledge. As a bonus, the societal theme of Harmonization of Standards is enhanced by the community of practice element of engaging federal, state and local regulators as well as industry stakeholders and process safety professionals.

548394 Risk Analysis Screening Tool – CCPS Project 280

A spreadsheet based Risk Analysis Screening Tool (RAST) is available through the Center for Chemical Process Safety (CCPS), courtesy of the Dow Chemical Company. The tool allows users to identify hazards and understand risks including causes, consequences and probabilities. RAST can be used for existing or proposed installations to facilitate MOC reviews, assess compliance with risk criteria, or evaluate proposed designs.

The users can input the type of equipment being used along with the equipment characteristics, chemical data, process conditions, and the plant layout. Using these inputs, the tool is used to help identify hazard scenarios and evaluate the consequences of representative loss of containment scenarios. RAST also allows the users to specify their own hazard scenarios based on their own knowledge of the process. Users can also input their own risk ranking matrix. With the hazard scenarios and the risk matrix, the user can select those scenarios of most concern and use the tool to create a Layer of Protection Analysis (LOPA) spreadsheet for the process.

As an example a comparison of RAST calculated scenario identification and consequence results to actual case studies from the CSB will be presented. Courses on how to use RAST, and how to adjust the pre-loaded default parameters are available.

550863 Impacts of Climate Change on Process Safety Management in Oil and Gas Operations

Impacts of Climate Change on Process Safety Management in Oil and Gas Operations

Chizaram Nwankwo

Coventry University

Project objectives and goals

The aim of this study is to analyse the impacts of climate change on process safety management in oil and gas operations. The objectives for this study include the following: -

- To evaluate past, present and future trends of climate change around various global regions
- To establish the relationship between climate change and process safety management in oil and gas operations
- To proffer adaptation and mitigation measures for climate change impacts on process safety management in oil and gas operations

In the last two decades, climate change has emerged as one of the most controversial topics that has been argued on several fronts (Hulme 2009). The idea of global warming and the negative impact of human-driven activity on the global weather has been fiercely debated (Cook *et al.* 2016). However, Field *et al.* (2012) argue that the facts and figures tend to reiterate the continuous occurrence and proliferation of climate change, as well as its increasing impacts in different industrial sectors. Indeed, the warming of the climate is unequivocal (Panteli and Mancarella 2015). According to IIGCC *et al.* (2017), 70% of global greenhouse gas (GHG) emissions are attributed to energy-related activities. The oil and gas industry contributes to 60% of these energy-related emissions due to production of fossil fuels as their main operational activity (Puig *et al.* 2015). Evidence from NASA (2018) suggests that there has been 1^{0}C increase in temperature which is close to the set mark in accordance to the Paris Agreement.

There are several case studies such as the flood in San Jacinto River, Houston Texas in 1984 and Coffeyville Refinery, Kansas in 2007. In the case of the San Jacinto flooding event in 1984, 8 pipelines were ruptured and 29 were undermined, 35,000 barrels of oil were spilled and ignited, while 547 people were injured and $23 million was incurred in damage costs (Matek 2017).

Hence, adaptive measures need to be applied to manage the severe consequences of climate change on oil and gas operations (Smith 2013). Such adaptive measures include more robust design specifications for oil and gas assets exposed to harsh climatic conditions (Cruz 2010). Ramasamy and Yusof (2015) suggest that this could be vital in offsetting the impacts of climate change on assets such as offshore rig platforms and equipment, transport pipelines, storage facilities etc. However, there is still little attention accorded to design and adaptation of oil and gas assets to offset the effects of climate change on them (Skjaerseth and Skodvin 2010). Therefore, this study seeks to analyse the impacts of climate change on deterioration of oil and gas assets.

Description of methods and results

Firstly, this study adopts a pragmatic paradigm since it explores the consequences of actions, is problem-centred, pluralistic, real-world practice oriented and allows for use of mixed models (Creswell and Creswell 2018). The area of investigation for this study is focused on impacts of climate change on oil and gas operations. The predominant approach taken for this study is a correlational research approach which tries to establish a relationship between climate change and oil and gas operations. The selected approach is due to the continuous occurrence of accidents in the oil and gas industry due to asset deterioration.

After thorough literature review of climate change trends and asset integrity issues in the oil and gas industry, this study identified a gap in knowledge that the oil and gas industry has not duly considered climate change during design of assets; hence posing a risk to oil and gas operations conducted under harsh climatic conditions. Therefore, the data collection for this study employed "concurrent triangulation" method of quantitative and qualitative data to ensure robust analysis of key variables such as climate change and oil and gas operations. Consequently, data collection tools for this study include questionnaires, interviews, archive data and documentary analysis. The data collection for this study was conducted over a period of 3 months.

Firstly, it involved piloting the questionnaire among academic and industry professionals to identify any errors, as well as testing and refining the data collection instrument. Next, the study was approved as a medium-risk project by Coventry University Ethics Committee, as primary and secondary data were collected from archive data, documents, questionnaires and interviews. The population sample for the study includes oil and gas industry professionals such as Mechanical Engineers, Facility Managers, Production and Well Engineers, Process Engineers, Health and Safety Officers, as well as Safety and Reliability Engineers. The questionnaires used for this study were designed using Bristol Online Survey (BOS) tool and covered questions around climate change and oil and gas operations. These data were predominantly collated and analysed using various software such as Zotero, Microsoft Excel and SPSS software. Statistical analysis conducted in this study include trend analysis, descriptive statistics, Pearson's correlation, Cronbach's alpha reliability analysis and Relative Importance Index (RII). Lastly, the findings of this study were benchmarked and justified with peer-reviewed academic literature.

Potential for application of results

This research could aid in curbing several accidents caused by asset deterioration due to increasing number of climate change events. It will help the oil and gas industry to avoid the enormous financial, legal and moral costs associated with accidents. Also, this study will provide a deeper insight on current climate change statistics and proffer adaptation and mitigation measures on how to deal with these problems. Similarly, it could broaden the scope of researchers to start incorporating impacts of climate change on various industrial sectors and subsectors asides the oil and gas industry.

References

Cook, J., Oreskes, N., Doran, P.T., Anderegg, W.R., Verheggen, B., Maibach, E.W., Carlton, J.S., Lewandowsky, S., Skuce, A.G., and Green, S.A. (2016) 'Consensus on Consensus: A Synthesis of Consensus Estimates on Human-Caused Global Warming'. *Environmental Research Letters* 11 (4), 048002

Creswell, J.W. and Creswell, D.J. (2018) *Research Design: Qualitative, Quantitative, and Mixed Methods Approaches.* 5th Edition (International Student Edition). Sage Publications

Cruz, A.M. (2010) *Vulnerability of Oil and Gas Infrastructure to Climate Change and Extreme Weather Events* [online] available from <http://indico.ictp.it/event/a09141/session/35/contribution/25/material/0/0.pdf> [5 March 2018]

Field, C.B., Barros, V., Stocker, T.F., and Dahe, Q. (2012) *Managing the Risks of Extreme Events and Disasters to Advance Climate Change Adaptation: Special Report of the Intergovernmental Panel on Climate Change*. Cambridge University Press

Hulme, M. (2009) *Why We Disagree about Climate Change: Understanding Controversy, Inaction and Opportunity*. Cambridge University Press

IIGCC, CERES, and IGCC (2017) *Global Climate Disclosure Framework for Oil & Gas Companies* [online] available from <http://www.iigcc.org/files/publication-files/Global_Climate_OG_DisclosureFramework.pdf> [14 November 2017]

Matek, B. (2017) *CHARACTERISTICS OF OIL AND GAS PIPELINE ACCIDENTS CAUSED BY CLIMATE CHANGE INTENSIFIED HURRICANES* [online] Baltimore, Maryland: Johns Hopkins University. available from <https://jscholarship.library.jhu.edu/bitstream/handle/1774.2/46088/Benjamin%20Matek.pdf?sequence=1&isAllowed=y> [26 June 2018]

NASA (2018) *Global Climate Change* [online] available from <https://climate.nasa.gov/> [9 May 2018]

Panteli, M. and Mancarella, P. (2015) 'Influence of Extreme Weather and Climate Change on the Resilience of Power Systems: Impacts and Possible Mitigation Strategies'. *Electric Power Systems Research* 127, 259–270

Puig, D., Haselip, J., and Naswa, P. (2015) *Adaptation to Climate Change in Colombia's Oil and Gas Industry: Recommendations to Promote Risk Management* [online] UDP-CRD-OLADE-01(EN). Denmark: UNEP DTU Partnership. available from <www.unepdtu.org> [19 September 2018]

Ramasamy, J. and Yusof, S.M. (2015) 'A Literature Review of Subsea Asset Integrity Framework for Project Execution Phase'. *Procedia Manufacturing* [online] 4, 79–88. available from <http://linkinghub.elsevier.com/retrieve/pii/S2351978915011336> [14 May 2017]

Skjaerseth, J. and Skodvin, T. (2010) *Climate Change and the Oil Industry: Common Problem, Different Strategies*. Manchester University Press

Smith, M.H. (2013) *Assessing Climate Change Risks and Opportunities for Investors: Oil and Gas Sector* [online] Analysis of Climate Change Risks and Opportunities. Technical Report. ANU and Investor Group on Climate Change. available here. [5 April 2018]

Further Information

Chizaram Nwankwo (E: nwankwo4@uni.coventry.ac.uk)

546246 Study of the Effects of Wind Speed and Surface Roughness Under Stable Atmospheric Condition Using Aloha Model and CFD Model

Study of the effects of wind speed and surface roughness under stable atmospheric condition using ALOHA model and CFD model

Many substances released in chemical processes are toxic, flammable, or even both. When these substances accidentally release to the atmosphere, human health and the environment could be threatened. As results, a reliable quantitative risk analysis tool is crucial to provide decision makers useful information on downwind concentration and hazards. ALOHA consequence model and Computational Fluid Dynamics (CFD) numerical method are widely used in chemical industries. Validation of these models under different scenarios is a necessary preliminary step before applying them to risk analysis. CFD models generally have acceptable concentration predictions on dense gas dispersion. ALOHA model is verified with DEGADIS model, but there is lack of study on effects of wind speed and surface roughness on dense gas dispersion. This paper compared field data from the Kit Fox experiments, which were CO_2 gas releases under stable atmospheric conditions with increased roughness, with CFD and ALOHA models' predictions. Also, statistical performance indicators were utilized to evaluate the accuracy of ALOHA model. Statistical performance indicators showed that ALOHA had systematic errors for dense gas dispersion with increased roughness under stable atmospheric conditions. Comparing effects of wind velocity profile and surface roughness values in CFD model and ALOHA model, this paper provided a better understanding of effects of key parameters in both ALOHA and CFD models. Furthermore, engineers would have better quantitative risk assessments under stable atmospheric conditions when they are using ALOHA consequence model and CFD model.

548222 Consequence Assessments and Collaboration to Streamline Project Siting

Collaboration is often sought after across disciplines, but all too often disregarded in favour of linear project development. For siting and consequence, as with other areas where design interacts with environment, health and safety constraints, this often leads to more costly design due to increased project iterations.

Stantec recently completed a consequence assessment for a compressor station to assess flammable extents during an emergency blowdown. At the same time, the client was experiencing siting issues due to sound emissions from the same process. The facility was grappling with competing factors: blowdown the facility in time to avoid catastrophic failure of their infrastructure, avoid excessive flammability extents from the blowdown stack location, and avoid extraneous noise exposure to nearby receptors. Using transient consequence modelling tools to model the blowdown sequence, Stantec was able to simultaneously and efficiently assess the flammable extents and sound emissions from the blowdown operations.

Consequence modelling for flammability was completed using a transient pipeline blowdown modelling tool to simulate the mass, energy, and momentum throughout the pipeline system. The model first simulated normal operating conditions, and then simulated the blowdown to estimate mass flow rate, liquid mass fraction, and release temperature from the blowdown stack. Several sizes of reducers upstream of the blowdown stack were introduced to estimate the pressure drops in the piping that would be required to appropriately reduce noise levels but also achieve full depressurization of the facility in time to meet safety criteria.

Consequence modelling of flammability was completed using SLAB, while the predicted change in sound pressure level from the blowdown stack was completed using Cadna acoustic modelling software. By combining these two analyses, the assessment costs streamlined for both required studies. In addition, the acoustics study benefited from additional insight provided by the consequence modelling tool.

548348 CFD Simulation of the Effect of Wind Speed and Release Elevation on Dispersion of Flammable Fluids at LNG Export Facilities

Facility siting studies for LNG export terminals are required by federal regulations to assess the potential hazard from accidental releases of flammable fluids. Federal regulations require under 49 CFR 193.2059 the evaluation of flammable vapor-gas dispersion distances from releases from flammable fluids in the facility for a range of representative wind conditions based on historical weather data expected to produce the maximum dispersion distances. A parametric CFD study will be conducted to evaluate not only the effect of wind speed but also the effect of the release elevation on the dispersion distance from flashing and jetting fluids. Hazard distances will be compared between the 2D software PHAST and the 3D software FLACS for a set of exemplar fluids found at LNG export facilities. The ability of the FLACS software to handle detailed 3D geometry provides for the evaluation of the effect of obstructions, congestion, confinement and mitigation measures on the dispersion distances. Release scenarios typical of LNG export facilities will be presented, with a focus on jetting and flashing releases of LNG, ethylene, propane, mixed refrigerants and heavy hydrocarbons.

546616 A Comparative Study of the Optimization Models for the Gas Detector Placement in Process Facilities

Gas detector network is an important layer of protection in process facilities for prevention gas release accidents of turning into fire and explosion disasters. But traditional standards just provide basic principles of the installation of gas detectors. Recently mathematical programming gradually became an important method to solve the problem. When using the mathematical programming method to optimize the detector layout, different optimization strategies will result in different optimization target models and detector layouts. In this study, three models have been developed and contrasted, namely minimal detection time P-Median model (MITP), minimal leakage concentration P-Median model (MICP), and minimal individual risk P-median model (MIRP), for the optimization of the gas detector placement based on stochastic programming. And as a basis, possible release scenarios were identified. In order to achieve the goal of using finite discrete scenarios to represent all scenarios effectively and quantitatively, clustering analysis was used to filter similar scenarios and select representative leak scenarios. The input data of these models was obtained from computational fluid dynamics. Furthermore, an adaptive particle swarm optimization (APSO) algorithm was applied to solve these models. The gas detector network optimization in a diesel hydrogenation unit was carried out as a case study. Results demonstrate that the MICP model is superior to the MITP model and MIRP model in describing the integrity of hazardous gas leakage and diffusion distribution. But the cumulative detection time of its optimized layout is slightly longer than that of MITP. Moreover, the control effect of MIRP model on personal risk caused by hazardous gas leakage was increased by 24.2%, although the minimum number of detectors was slightly higher than that of MICP model.

549411 Reducing Industrial Risk in a Converging It/OT World: 20 Experts Share Their Strategies

Rapid adoption of Industrial IoT requires tight integration between IT and OT. CISOs and corporate OT leaders are now scrambling to develop strategies, secure budget, and implement solutions to secure their mission-critical industrial control systems against burgeoning cyber threats.

Effective integration of IT and OT cybersecurity strategies is essential to address risks successfully. To secure industrial facilities and ensure safe, reliable production, IT and OT security teams – traditionally two separate disciplines with different priorities – must collaborate and share cybersecurity and risk management best practices.

In a series of interviews in late 2017 and early 2018, more than 20 different experts on the front lines of OT cybersecurity risk mitigation shared practical advice for making control systems more secure. These experts represented a diverse range of industries including oil and gas, chemicals and refining, and power generation.

In their interviews, they shared their insights into how to leverage IT/OT convergence and develop an effective OT cybersecurity program that meets challenges in real-world ICS environments. Technology plays an important role, but as these experts also point out, more technology isn't always the solution. Just as important is the need to establish effective communication and trust between cybersecurity professionals and OT engineers.

Security project teams must make a concerted effort to create alignment between business management, the cybersecurity community, and OT professionals. Without alignment, security initiatives will fail to attain the support required to ensure success.

In this presentation, attendees will learn:

- Practical, real-world recommendations from OT cybersecurity experts
- Best practices for breaking down barriers and building trust between cybersecurity professionals and OT engineers
- How to align business management and plant operations on cybersecurity strategies

549417 The Intersection of OT Security and Process Safety

The rise of cyberattacks on critical infrastructure creates troubling concerns for security, economic stability, and personnel and public safety. However, many companies are still relatively unprepared to successfully defend their facilities from cyberattacks launched by increasingly sophisticated threat actors. Safe production is at risk.

Safety focuses on incident prevention and measures risks to personnel. Cybersecurity focuses on protecting systems from threat actors and measures risk based on the threat landscape. The increased network connectivity that exists within industrial facilities today means that cybersecurity is now closely linked to the safe operation of industrial facilities and their processes. Successful defense of cyber-physical systems from nation-state attacks requires close coordination and collaboration between cybersecurity and process safety. Every safety initiative must now consider cyber risk. Every cybersecurity initiative must now consider safety.

Considerable overlap exists between risk analysis in safety and cybersecurity. Safety and cybersecurity disciplines can learn much from each other. However, safety and cybersecurity standards and corporate organizational structures remain siloed in far too many companies. This session focuses on the importance of executing a cohesive strategy for both safety and cybersecurity to ensure safe and reliable operations.

In this session, we will:

- Examine the existing cyber threat landscape
- Discuss how established process safety best practices can be applied to OT cybersecurity
- Provide practical guidance for improving cyber defenses and reducing safety and environmental cybersecurity risks

551385 Process Safety at Offshore Aerodromes (Helidecks)New Application for Process Safety

Offshore air transport is an importance activate of the oil & gas industry. In this way, the management of air operations is essential for maintaining the safety and reliability. The present work has the objectives of divulging the methods used by
Petrobras to audit maritime units with helideck and to demonstrate the contribution of such methods to the safety of process in the air operations. To achieve these objectives, the paper was based on the Center for Chemical Process Safety (CCPS) publication "Guidelines for the Safety of Risk-Based Processes", and aims to share the Bow tie methodology for the qualitative analyzes of risks inherent in Air Operations, focusing on the barriers typically faced by oil and gas industry, in order to contribute to the process by providing a graphical and structured approach to the required barriers, both to avoid catastrophic consequences and to prevent events recognized as precursors.

Thermodynamics and Computational Methods for Relief Valves

By Umesh Mathur, P.E.
umesh.mathur@aggreko.com , mathuru@gmail.com
Aggreko Process Services
3732 Magnolia Street, Pearland, Texas 77584, USA

15th Global Congress on Process Safety (AIChE, March, 2019, New Orleans)

1. INTRODUCTION

The computation of flow through a relief valve is one of the most important aspects of safety system design. This paper discusses the fundamental question of choosing the proper thermodynamic basis for such calculations. We also discuss the discrepancies we have observed in the literature on this crucial question and attempt to dispel some of the misconceptions we have found.

Our concern with this issue arose from a study of the relevant literature on this subject, which shows that the flow of a compressed gas (or gas-liquid mixture) through a rigidly held valve or hole in a pipeline must be treated as an irreversible, isenthalpic process. This reasoning flows simply from the First Law of Thermodynamics as applied to a flow process.

For example, as early as 1947, Hougen and Watson[1] discussed "Free Expansion" of gases (p. 551) and noted as follows:

"The unrestrained expansion of a gas is known as free expansion. Under conditions of no restraint no work is done, and under adiabatic conditions no heat is added. Free expansion under flow conditions is commonly known as throttling or as the Joule-Thompson effect referred to in Chapter XII.

The Joule-Thompson effect is measured experimentally by expanding the gas slowly and in steady flow through a well insulated porous plug; in this way, potential work is lost, and no heat is allowed to enter or leave the system through the walls. The fluid flows reversibly into and out of the process, but the expansion step is completely irreversible. Since no heat is added or lost, the process takes place under conditions of constant enthalpy, according to Equation (VII-11), page 207. However a change in internal energy results if any change in flow energy pV occurs."

Similarly, in a discussion of valves and throttles, Elliott and Lira[2] observe as follows:

"A throttling device is used to reduce the pressure of a flowing fluid without extracting any shaft work and with negligible fluid acceleration. Throttling is also known as ***Joule-Thompson expansion*** *in honor of the scientists who originally studied the thermodynamics. An example of a throttle is the kitchen faucet. Industrial valves are modeled as throttles....Changes in kinetic and potential energy are small relative to changes in enthalpy as we just discussed. When in doubt, the impacts of changes in velocity can be evaluated as described in Example 2.9. The amount of heat transfer is negligible in a throttle. The boundaries are not expanding, and there is also no mechanical device for transfer of work, so the work terms vanish. Therefore, a throttle is isenthalpic."*

Finally, in a discussion of nozzles, Klein and Nellis[3] observe that:

"...The potential energy terms of the inlet and outlet flows are not usually significant. Nozzles neither require nor produce power ($W_{in} = W_{out} = 0$) and are very nearly adiabatic. Note that the kinetic energy terms associated with the flows, particularly the outlet flow, cannot be neglected as the purpose of the nozzle is to increase the kinetic energy of the fluid. With these simplifications, an energy balance on the nozzle becomes:

$$H_{in} + \tfrac{1}{2}V_{in}^2 = H_{out} + \tfrac{1}{2}V_{out}^2 \qquad (4\text{-}54)\text{"}$$

In the light of these assertions by these noted authorities, and others[6][10], it is hard to understand how the flow through a relief valve, or a hole in a vessel that is open to the atmosphere, can be treated as isentropic (***reversible*** adiabatic) in nature, as opposed to the isenthalpic (***irreversible*** adiabatic) basis that is consistent with free expansion. This is because an isentropic expansion process would necessarily show an outlet enthalpy for the gas from the nozzle or valve (or an orifice in a punctured vessel) to have a lower enthalpy than the inlet, the difference in such enthalpies being equal to the work done, for example on a rocket or airplane. However, a relief valve or vessel remains stationary and absolutely no work is done during a free expansion by the gas flowing through.

Nevertheless, several standard references have used isentropic formulas for free expansion[4],[5] in stationary valves, or nozzles, from the inlet to the throat. We believe that this methodology is flawed as it violates the steady flow energy equation (i.e., the first law of thermodynamics written for an open system). This procedure would result in an erroneous prediction of outlet gas temperature when entropy is held constant. This, in turn, would predict an erroneous outlet gas density and thus, for a desired gas relief rate, the required cross-sectional area for the valve throat and also any conduit or pipe downstream of the valve.

The practical consequence of this anomaly in predicting outlet gas properties (especially the gas density at the throat) would be that the relief valve sizing could be incorrect. Therefore, it is important that the proper thermodynamic basis (isenthalpic expansion) be used in all such computations.

Conversely, for a given relief valve, errors in throat density based on an isentropic path would also affect the relief rate. Both types of errors could lead to potentially serious process safety concerns.

For these reasons, when calculating the release rate for gas at a high pressure through a relief valve, it is important to be able to use the proper basis for gas enthalpy (and therefore accurate computations of fluid thermodynamic properties). The same considerations apply to the gas release rate calculations resulting from a rupture in a pipeline or vessel containing a gaseous mixture under pressure.

To help clarify the proper methodology for a relief valve - or a hole in a pressurized vessel - the computations that should be performed are described below. We also show a few simple examples. Note that the possibility of sonic velocity at the throat, or choking, of the valve (or at the orifice in a ruptured vessel) is explicitly considered. In both such cases, the procedure described below is subject to the following assumptions:

- The opening can be represented by a convergent nozzle of exit area equal to the area of the throat of the relief valve, or hole in the case of pipeline rupture

- The compressed gas pipeline, or leaking container, is a large reservoir in which upstream pressure and temperature are maintained constant
- The compressed gas undergoes an irreversible, adiabatic process of expansion through the nozzle.

The flow through such a device from either (a) the inlet to the throat, or (b) from the throat to the outlet follows the steady flow energy balance equation (Equation 9.21, Himmelblau & Riggs[6]):

$$\Delta E = Q + W - \Delta(H + KE + PE) \qquad (1)$$

Where:
ΔE = energy accumulation (zero at steady state)
Q = net heat transfer into the system (very close to zero for a relief valve)
W = net work done by the system (zero for a relief valve)
ΔH = change in enthalpy of the system
ΔKE = change in kinetic energy of the system
ΔPE = change in potential energy of the system (zero for a relief valve)

Clearly, for steady flow through an orifice, valve (or relief valve), ΔE, Q, W, and ΔPE are all zero. Therefore, we are left with:

$\Delta(H + KE) = 0$, or

$\Delta H = - \Delta KE$, or

$$H_{out} = H_{in} - \Delta KE \qquad (2)$$

In other words, the outlet enthalpy is equal to the inlet enthalpy minus the change in kinetic energy in going through the throat or orifice. (It should be noted that the same logic would apply for discharge through an ordinary valve, a relief valve, or an orifice plate).

If there is no change in velocity, the outlet enthalpy will be the same as that of the inlet gas. This phenomenon is generally referred to as a Joule-Thompson (J-T) expansion. For an ideal gas, enthalpy is a function of temperature only; therefore, a J-T expansion would result in no change in temperature. However, for most real gases, a J-T expansion results in a decrease in temperature. There are a few notable exceptions, however, such as hydrogen or helium gas for which the J-T coefficient has the opposite sign[10]. On the other hand, if the gas speeds up, its kinetic energy will increase and, therefore, its enthalpy will decrease in accordance with equation (2) above. This kinetic energy increase would result in a decrease in outlet gas enthalpy, and therefore temperature, even for an ideal gas.

For ideal gases, the computation of enthalpy for ideal gases is straightforward as it is a function of temperature only. For real gases, however, enthalpy is also a function of pressure, temperature, and composition and use of an equation of state (EOS) becomes necessary[2] to calculate the enthalpy departure from the ideal, as discussed below.

COMPRESSIBILITY FACTOR AND ENTHALPY FOR REAL GASES

For real gases, enthalpy is found using a suitable equation of state (EOS), such as the Soave-Redlich-Kwong (SRK), Peng-Robinson (PR), or Lee-Kesler (LK). This procedure may be summarized as follows:

At the given upstream pressure (P), temperature (T), and composition, solve the EOS for the compressibility factor (z). This requires solving a cubic polynomial in z when using any cubic EOS, such as the SRK or PR, using either an explicit analytical or numerical trial and error method. For the LK, however, a trial and error procedure is always required as the LK EOS requires solving the 8-constant Benedict-Webb-Rubin[2] (BWR) equation, which is not a simple cubic in z, once for the simple fluid and again for the reference fluid.

Using the appropriate thermodynamic expressions for the EOS, use this value of z to calculate ΔH_d, the enthalpy departure from ideal.

Calculate the ideal gas enthalpy, H_0 (function of temperature T_0 only).

Find the real gas enthalpy using:

$$H_r = H_i - \Delta H_d \qquad (3)$$

Where:

H_r = real gas enthalpy
H_i = ideal gas enthalpy
ΔH_d = enthalpy departure from ideal

This procedure must be followed for calculating enthalpy at both upstream and downstream conditions.

The gas velocity at the throat may be sonic or sub-sonic. Also, as the gas speeds up in passing through the valve throat or hole (since the cross-sectional area is lower than that of the pipe), its kinetic energy increases and this results in a reduction of the enthalpy (Elliott & Lira[2], Eq. 2.53):

$$H_t = H_0 - \Delta KE \qquad (4)$$

Where:
H_0 = upstream gas enthalpy[J/kg]
H_t = exit gas enthalpy at the throat or hole[J/kg], and
ΔKE = change in kinetic energy[J/kg]

ΔKE is found as follows (Elliott & Lira's Eq. 2.53):

$$\Delta KE = (u_t^2 - u_0^2)/2 \qquad (5)$$

Where:

ut = throat (or hole) velocity[m/s], and

u_0 = upstream velocity[m/s].

Clearly, the outlet and inlet enthalpies are equal only if the change in velocity is negligible.

RELEASE RATE CALCULATION

The computation of the release rate from a pipe rupture should be done subject to the following equations:

Bernoulli's Equation

This equation relates upstream and downstream conditions for an arbitrary flow geometry, as follows:

$P_0 + \rho_0.u_0^2/2 + \rho_0.g.z_0 = P_e + \rho_e.u_t^2/2 + \rho_t.g.z_t + \Delta P_{friction}$ (6)

Where the subscripts o and e refer to the inlet and exit planes, respectively, and

P	= pressure[Pa]
ρ	= density[kg/m^3]
u	= velocity[m/s]
g	= gravitational constant[m/s^2]
z	= elevation above datum[m]
$\Delta P_{friction}$	= friction losses[Pa]

The friction losses have been ignored for simplicity in this presentation.

Obviously, for a hole, or throttle (or relief) valve, the elevations z_0 and z_t are the same.

Sonic Velocity

The sonic velocity in any medium[2] may be expressed as:

$u_s = (\gamma . z . R . T / M)^{0.5}$ (7)

Where:

u_s	= sonic velocity[m/s]
γ	= gas C_p/C_v ratio[]
z	= gas compressibility factor[]
R	= gas constant = 8314.46[J/(kgmol.K)]
T	= temperature[K]
M	= gas molecular weight[kg/kgmol]

In Equation (7), properties at the throat must be used for choked flow calculations. For ideal gases, the compressibility factor z is equal to 1. For real gases, however, the compressibility factor z is generally not equal to 1 and can be found only using an appropriate EOS.

C_p, C_v, and γ=C_p/C_v ratio

Similarly, for ideal gases, the value of γ is found from the value of C_p, which is a function of temperature only, and the value of C_v is found from the following:

$C_v = C_p - R$ (8)

For real gases, however, the value of γ again requires use of an equation of state to calculate the ΔC_{pd} and ΔC_{vd} departures from ideality. The real gas C_p and C_v values are found using the ideal values and these calculated departures from ideality, by subtraction, in a manner similar to that described above for enthalpy:

$C_{p1} = C_{p0} - \Delta C_{pd}$ (9)

Where

C_{p1} = real gas C_p[J/(kg.K)]

C_{p0} = ideal gas C_p[J/(kg.K)]

ΔC_{pd} = departure in C_p from ideal gas value[J/(kg.K)]

Similarly, $C_{v1} = C_{v0} - \Delta C_{vd}$ (10)

In Equation (10),

C_{v1} = real gas C_v[J/(kg.K)]

C_{v0} = ideal gas C_v[J/(kg.K)]

ΔC_{vd} = departure in C_v from ideal gas value[J/(kg.K)]

In a conduit, orifice, or valve, the actual gas velocity, u_e, can never exceed the sonic limit, u_s. Note that u_s, in general, is different at the upstream conditions compared to the value at throat conditions. The upstream and throat temperatures and pressures of the gas may be different owing to the kinetic energy effect, Equation (4).

At a sufficiently low upstream pipeline or vessel pressure, the hole velocity would be lower than the sonic velocity limit. Above some critical limit for the upstream pressure, however, the maximum discharge flow is limited by the sonic velocity criterion at the hole.

Density

Gas density is computed using:

$\rho = P . M / (z . R . T)$ (11)

Where:

ρ = density[kg/m^3]

P = pressure[Pa]

M = gas molecular weight[kg/kgmol]

z = gas compressibility factor[]

R = gas constant = 8314.46[J/(kgmol.K]

T = temperature[K]

The gas compressibility factor is 1 for ideal gases; it is calculated from the EOS for real gases.

1.2 Velocity

The velocity in a pipe or hole (or valve throat) is found using:

$u = W / (\rho . A)$ (12)

Where:

u = velocity[m/s]

W = mass flow rate[kg/s]

ρ = gas density[kg/m^3]

A = cross sectional area[m^2]

For a circular pipe, the cross-sectional area is defined as:

$A = \pi . D^2/4,$ (13)

Where:

D = inner diameter[m]

For non-circular cross sections, the equivalent diameter D_e is defined using the concept of the hydraulic radius which is defined as:

D_e = 4 (Cross-sectional-area/Wetted-perimeter) (14)

It can readily be seen that this definition holds for a circular pipe also: $4 . \pi . D^2 / 4 / (\pi . D) = D$

As noted earlier, for ideal gases, the compressibility factor z is equal to 1. For real gases, however, finding the gas density ρ requires first calculating z from the EOS. This is done using the procedure as described above.

It should be noted that conditions at the throat of a valve (or at the hole in a pipe) are not known. Therefore, the computation of the gas flow rate must be done using an iterative procedure. When using Microsoft Excel, this is best performed using the Solver Add-in. Solver is a non-linear optimization technique that can simultaneously adjust the unknown quantities and solve the problem by enforcing the constraints imposed by the Bernoulli Equation (6) for throat velocity, and the Steady-Flow Energy Equation (1) for hole enthalpy. Also, the calculated throat velocity from the Bernoulli equation can never exceed the sonic velocity limit, Equation (7), and this constraint must be added explicitly when the upstream pressure exceeds a threshold value.

When solving for an unknown gas flow rate, if the upstream pressure is sufficiently low, hole velocity could remain sub-sonic. However, if the upstream pressure is sufficiently high, the pressure at the hole and also the gas mass flow rate would vary in a manner that obeys the sonic velocity limit at the hole.

Example

CASE (A): Methane is flowing through a 300[mm] diameter, above-ground pipeline at an absolute pressure of 46[bar] and a temperature of 300[K]. The pipeline develops a hole with an equivalent diameter of 50[mm]. Calculate the mass rate of leakage to atmosphere and also the velocity of the gas jet emerging from the hole. The pressure inside the pipeline may be assumed to be constant owing to continuous supply from the source.

CASE (B): Rework Case (A) assuming that the pressure in the pipeline is 1.5[bar] instead of 46[bar].

Case (A) P_o = 46[bar]

PROBLEM: Solve for release rate from a hole in a pipeline - sonic velocity at throat.

WE SHOW THE FOLLOWING DETAILED SOLUTION FOR THE IDEAL GAS CASE ONLY. Note that the solution of the real gas case requires use of complex computer programs to solve the chosen equation of state (e.g., Lee-Kesler) iteratively for the required thermodynamic quantities. Since these routines are imbedded in an outer loop for the unknowns (T and P at the throat, and the mass flow rate), the computational burden is too great for hand calculations.

The POLYMATH software[7] was used for solving this problem. All variables are defined in Table 1. The problem statement for Case (A) for the ideal gas case is as follows:

 # Initial guesses

$T_t(0) = 300$	# Initial guess for T_t[K]
$W(0) = 20$	# Initial guess for W[Kg/s]
$P_t(0) = 46$	# Initial guess for P_t[bar]

 # Aly-Lee ideal gas enthalpy constants:

A = 33298	# Ideal gas enthalpy constant A
B = 79933	# Ideal gas enthalpy constant B
C = 2086.9	# Ideal gas enthalpy constant C
D = 41602	# Ideal gas enthalpy constant D
E = 991.96	# Ideal gas enthalpy constant E
F = -120280000	# Ideal gas enthalpy constant F

 # The following three functions must be made equal to zero by adjusting the three unknown variables W[kg/s], T_t[K], and P_t[bar]:

$f(T_t) = h_t - h_{t2}$

$f(W) = u_t - u_{t2}$

$f(P_t) = u_t - u_s$

 # Initial guess ranges for non-linear equations

W(min) = 1

W(max) = 50

$T_t(min) = 200$

$T_t(max) = 400$

$P_t(min) = 10$

$P_t(max) = 46$

This problem requires solving the three simultaneous, non-linear algebraic equations (defined above) by varying T_t (the hole temperature), W (the mass flow rate), and P_t (the hole pressure):

$f(T_t) = 0$

$f(W) = 0$

$f(P_t) = 0$

In this case, the upstream pressure is high enough to cause the velocity at the hole to reach the sonic limit. This problem requires solving three simultaneous non-linear algebraic equations. The POLYMATH solution shown below for Example A was obtained using a software package called POLYMATH[7]. Note that any other similar package can also be used. We verified this numerical solution using the MATHCAD[8] and EXCEL[9] software packages also.

NOTE: The numerical solutions for Cases (A) and (B) are shown in Table 1, following the problem statement for Case (B).

Case (B) $P_0 = 1.5$[bar]

PROBLEM: Solve for release rate from a hole in a pipeline. Assume that the throat velocity is sub-sonic. This means that the outlet pressure is fixed at 1[atm] = 1.01325[bar].

When the outlet pressure is fixed, we are left with two non-linear equations. Also, for consistency with the methodology in (A) above, we use a very large upstream diameter to make the inlet velocity very small.

NOTE: If the calculated outlet velocity exceeds the sonic limit, we would revert to the problem formulation under (A) above. (This would add an equation requiring outlet velocity to equal sonic velocity, and thus again require solving three simultaneous non-linear equations.)

As noted earlier, the solution of the real gas case requires use of complex computer programs to solve the chosen equation of state (Lee-Kesler, in these examples) for the required thermodynamic quantities.

In this case, the upstream pressure is low enough so that the velocity at the orifice does not reach the sonic limit. Accordingly, this problem requires solving only two simultaneous non-linear algebraic equations. The solution shown below was obtained using a software package called POLYMATH. Note that any other similar package can also be used. As with Case (A), we verified this numerical solution using the MATHCAD and EXCEL software packages also.

The POLYMATH problem statement for Case (B) is as follows:

$f(T_t) = h_t - h_{t2}$

$f(W) = u_t - u_{t2}$

 # Initial guesses

$T_t(0) = 300$ # Initial guess for T_t[K]

$W(0) = 20$ # Initial guess for W[Kg/s]

Using POLYMATH, the solution for Case (A) shows that the gas release rate, W, for an ideal gas is 17.16[kg/s]. As noted above, identical results were obtained using the Mathcad and Excel software packages.

TABLE 1: SONIC OR SUB-SONIC FLOW THROUGH A VALVE OR HOLE

VARIABLES AND EQUATIONS

Variable	Description & Units	Explicit equations
A	Ideal gas enthalpy constant A	A = 33298.
B	Ideal gas enthalpy constant B	B = 79933.
C	Ideal gas enthalpy constant C	C = 2086.9
D	Ideal gas enthalpy constant D	D = 41602.
E	Ideal gas enthalpy constant E	E = 991.96
F	Ideal gas enthalpy constant F	F = -120280000.
P_t	Throat pressure [bar]	$P_t = 1.01325$
kcal_j	Conversion: 1[kcal] = 4184[J]	kcal_j = 4184
M	Gas molecular weight[kg/kgmol]	M = 16.0
T_0	Inlet gas temperature[K]	$T_0 = 300$
P_0	Inlet gas pressure[bar]	$P_0 = 1.5$
z_0	Inlet gas compressibility factor[]	$z_0 = f(T_0,P_0,x)$
$(c_p/c_v)_0$	Inlet gas C_p/C_v ratio[]	$(c_p/c_v)_0 = 1.32$
R	Gas constant[J/(kgmol.K)]	R = 8314.46
D_0	Diameter of pipe[m]	$D_0 = 0.3$
D_t	Diameter of Throat[m]	$D_t = 0.05$
h_0	Inlet enthalpy[kcal/kg]	$h_0 = (A*T_0 + B*C/\tanh(C/T_0) - D*E/\tanh(E/T_0) + F)/M/kcal\_j$
rho_0	Inlet density[kg/m³]	$rho_0 = P_0*M*1.e5/R/T_0$
A_0	Pipeline cross-sectional area[m²]	$A_0 = 3.14159/4*D_0^2$
A_t	Throat flow area[m²]	$A_t = 3.14159/4*D_t^2$
h_t	Throat enthalpy[kcal/kg]	$h_t = (A*T_t + B*C/\tanh(C/T_t) - D*E/\tanh(E/T_t) + F)/M/kcal\_j$
u_0	Upstream velocity[m/s]	$u_0 = W/rho_0/A_0$
rho_t	Throat density[kg/m³]	$rho_t = P_t*M*1.e5/R/T_t$
u_t	Throat velocity[m/s]	$u_t = W/rho_t/A_t$
u_s	Sonic velocity at throat[m/s]	$u_s = (z_0*(c_p/c_v)_0*R*T_t/M)^{0.5}$
u_{t2}	Throat velocity from Bernoulli Eq.[m/s]	$u_{t2} = (2/rho_t*(P_0*1.e5 + rho_0*u_0^2/2 - P_t*1.e5))^{0.5}$
dh_{KE}	ΔKinetic Energy = Δ(u²)/2[kcal/kg]	$dh_{KE} = (u_{t2}^2 - u_0^2)/2/kcal\_j$
h_{t2}	Calc. enthalpy at throat = inlet enthalpy - Δ(u²)/2[kcal/kg]	$h_{t2} = h_0 - dh_{KE}$
T_t	Temperature at throat[K]	$f(T_t) = 0$
W	Mass flow rate[kg/s]	f(W) = 0
P_t	Pressure at throat[bar]	$f(P_t) = 0$

TABLE 1: SONIC OR SUB-SONIC FLOW THROUGH A VALVE OR HOLE (CONTINUED)

CASE (A) AND CASE (B) SOLUTIONS (USING POLYMATH)

Variable	CASE (A) Value (SONIC)	CASE (B) Value (SUB-SONIC)
A	33300.0	33300.0
B	79930.0	79930.0
C	2086.9	2086.9
D	41600.0	41600.0
E	991.96	991.96
F	-120300000.0	-120300000.0
P_t	27.71686	1.01325
kcal_j	4184.0	4184.0
M	16.0	16.0
T_0	300.0	300.0
P_0	46.00	1.50
z_0	1.00	1.00
$(c_p/c_v)_0$	1.32	1.32
R	8314.46	8314.46
D_0	0.30	0.30
D_t	0.05	0.05
h_0	226.2098	226.2098
rho_0	29.50683	0.96218
A_0	0.07069	0.07069
A_t	0.00196	0.00196
h_t	205.2641	210.3263
u_0	8.22527	7.71172
rho_t	20.86568	0.73254
u_t	418.7383	364.6543
u_s	418.7383	427.2981
u_{t2}	418.7383	364.6543
dh_{KE}	20.94576	15.88352
h_{t2}	205.2641	210.3263
T_t	255.6215	266.1833
W	17.15555	0.52433
P_t	27.71686	1.01325

CASE (A) RESULTS

This problem was also solved using the Lee-Kesler equation of state to calculate non-ideal gas properties, using a suitable computer program. This solution is provided below, with the ideal gas results quoted alongside for comparison:

	CASE (A)	
	IDEAL GAS	REAL GAS
Inlet pressure[bar]	46.00	46.00
Inlet temperature[K]	300.00	300.00
Calculated mass flow[kg/s]	17.28	18.25
Inlet velocity[m/s]	8.27	8.09
Throat velocity[m/s]	424.65	421.48
Sonic velocity at throat[m/s]	424.65	421.48
Inlet Enthalpy[kcal/kg]	225.61	214.87
Delta Kin. Energy[kcal/kg]	21.54	21.22
Outlet Enthalpy[kcal/kg]	204.07	193.64
Throat pressure[bar]	27.32	25.73
Throat temperature[K]	254.27	250.83

It should be noted that these results are quite similar for this case because the behavior of the gas (methane) is close to ideal at the conditions specified. However, these results would very likely differ significantly if, for example, the inlet temperature was considerably lower and the pressure much higher in which case the gas behavior would be markedly non-ideal.

CASE (B) RESUTS

Case (B) was also solved using the Lee-Kesler equation of state to calculate non-ideal gas properties. This solution is provided below, with the ideal gas results quoted alongside for comparison:

	CASE (B)	
	IDEAL GAS	REAL GAS
Inlet pressure[bar]	1.5	1.5
Inlet temperature[K]	300.0	300.0
Calculated mass flow[kg/s]	0.524	0.525
Inlet velocity[m/s]	7.71	7.70
Throat velocity[m/s]	364.7	364.2
Sonic velocity at throat[m/s]	435.06	435.3
Inlet enthalpy[kcal/kg]	226.21	226.21
Delta Kin. Energy[kcal/kg]	15.88	15.85
Outlet Enthalpy[kcal/kg]	210.33	210.36
Throat pressure[bar]	1.01325	1.01325
Throat temperature[K]	266.18	266.25

As might be expected, the gas mass flow rate is far lower for Case (B) compared to Case (A). Also, for Case (B), it was found that the conditions at the throat are virtually identical for the ideal versus the non-ideal gas.

2. ADDITIONAL EXAMPLES

To illustrate the importance of the issues discussed in this paper, we show a few additional examples for the differences in our numerical results compared to others.

2.1 Gas Release From A Hole In A Vessel

A 0.1[in] hole forms in a tank containing nitrogen at 200[psig] and 80[°F]. Determine the mass flow rate through this leak (Crowl[4], Example 4-4, p. 143).

The solution for this problem using the Lee-Kesler EOS was 4.26e-2[lb/s], compared to 3.86e-2[lb/s] in the reference (gas velocity at the throat was sonic). This difference is over 10% and not negligible.

2.2 Flow Through A Nozzle

Air at p_0=7.0e5[Pa] and T_0=293[K] discharges to the atmosphere. Compute the discharge mass flux, G, the pressure, Mach number, and velocity at the exit[5].

Using the Lee-Kesler EOS, the results are as follows:

Quantity	Lee-Kesler (this article)	Ideal gas (isentropic)[5]	Difference
Mass flux[kg/(m^2.s)]	1589	1650	-3.68%
Exit temperature[K]	238.5	244	-5.5 [K]
Exit pressure[Pa]	212041	370000	-42.7%
Mach number	1.0	1.0	-
Sonic velocity[m/s]	338	313	+8%
Velocity[m/s]	338	313	+8%

Note that the nozzle is not predicted to be choked when the downstream pressure is above 2.12[bar] (kinetic energy effects must be included), as opposed to 3.7[bar] in the original reference.

If a relief valve is to be designed for this scenario, the size would be under-predicted somewhat using the isentropic formulas used in the original reference. However, this discrepancy could be significantly larger in cases where an isentropic expansion incorrectly predicts partial condensation of the gas at the throat. In that situation, the density of the two-phase mixture at the throat could be far lower than that of the uncondensed gas (for the isenthalpic case) and the valve could be under-sized significantly.

3. CONCLUDING REMARKS

It has been shown that calculation of the release rate from a relief valve (or a hole in a vessel) should be done using an isenthalpic (i.e., irreversible adiabatic) path for the flow process. Using an isentropic thermodynamic path for this problem violates the First Law of Thermodynamics, as expressed for steady flow in a stationary relief valve or a hole in a pipeline. This could lead to an incorrect design both when sizing relief valves and estimating release rate from an existing valve or hole in a vessel.

In addition, it is important to use the proper equation of state to estimate real gas properties reliably when compared to the ideal gas assumption, especially when gas conditions are far from ideal. The Bernoulli equation (for throat pressure), and the sonic velocity limit (when flow is choked at the throat) must also be included for a complete definition of the problem. If the upstream pressure is low enough and the throat is un-choked, the isenthalpic methodology - as described in this paper - should again be followed to ensure a theoretically sound and consistent basis. Iterative computations for non-ideal cases (real gases) generally are too intensive for hand calculations and require use of computerized chemical engineering software.

REFERENCES

1. Hougen, O.A., K. M. Watson: "Chemical Process Principles" (John Wiley & Sons, New York, and Chapman and Hall, London, 1st Edition, Part II, p. 551, 1947, and 2nd Edition with R.A. Ragatz, Vol. II, p. 653, 1959)
2. Elliott, J.R., and C. T. Lira: "Introductory Chemical Engineering Thermodynamics" (Pearson Education, Inc. publishing as Prentice Hall, pp. 68-69, 2nd Edition, 2012)
3. Klein, S. and G. Nellis: "Thermodynamics" (Cambridge Univ. Press, 1st Edition, p. 168, 2011)
4. Crowl, D. A., and Louvar, J.F.: "Chemical Process Safety" (Pearson Education, 3rd Edition, pp. 130-131, 2011)
5. Green, D. W., and R.H. Perry: "Perry's Chemical Engineers' Handbook" (McGraw-Hill, 8th Edition, Section 6, pp. 23-24, 2008)
6. Himmelblau, D.M., and J.B. Riggs: "Basic Principles and Calculations in Chemical Engineering" (Pearson Education, 8th Edition, Fig. 9.18, Item 3, p. 525, 2012)
7. Cutlip, M.B., and M. Shacham: "Problem Solving in Chemical and Biochemical Engineering with POLYMATH, Excel, and MATLAB" (Prentice Hall, 2nd Edition, 2007)
8. Mathcad 2001 Professional: "Mathcad User's Guide With Reference manual", (Mathsoft Engineering and Education, Inc., 2001)
9. Freund, S.M., M. Jones, and J.L. Starks: "Microsoft Excel 2013: Complete (Shelly Cashman Series)", (Course Technology, 1st Edition, 2013)
10. Lewis, G.N., and M. Randall: "Thermodynamics", (McGraw-Hill, 2nd Edition, Revised by K.E. Pitzer and L. Brewer, p. 48, 1961)

548434 Engineered Solutions for Effective Flare System Analysis

Relief and flare system documentation is an important part of Occupational Safety and Health Administration (OSHA) Process Safety Management (PSM) 29 CFR 1910.119. Flare system analysis (sizing or rating) can be performed using commercially available software. Effective flare system analysis requires reliable Process Safety Information (PSI) such as process and instrumentation diagrams (P&IDs), relief system evaluation, Heat and Material Balance (H&MB), relief valve data sheets, isometrics, and disposal system loads. Good engineering methodologies can be implemented to effectively design header systems and rate existing header for any change. Some methodologies include use of dynamic simulation to predict realistic relief loads, reboiler pinch analysis for applicable systems, and selection of global relief scenarios. This paper will describes some methodologies and a combination of instrumented systems for effective flare system analysis. Effective flare analysis helps save capital expenditure spent on flare headers and possible flare load reduction yields reduced radiation footprint on a facility.

548364 Flares Are the End of the Line; How to Permit Correctly

INTRODUCTION

-

Flare is one of the key equipment in any refinery or LNG facility. Many relief valves relieve to flare, and if the flare is not permitted corrected – and functional - the plant is down. You cannot operate the plant without an active flare or flare field. The permitting of a flare is one of the prime activities in the permitting process. But the permit is dependent on the type of flare. This presentation covers various Federal Regulations involved in the permitting process, whether general or specific to each type of flare.

FLARE TYPES AND THEIR DESIGN FUCTURES

-

There are mainly three types of flares. The most common and most economical one is an elevated flare. There are elevated flares as high as 660 ft. The can be classified many ways; self-supported, guy wire supported, derrick supported, demountable type etc. One of the most important is the type of burner, whether a pipe tip or "furnace" burners. The second type is a ground flare with rows of furnace burners arranged in arrays at grade and surrounded with a wind fence. The last type is an enclosed flare with furnace burner enclosed with a tall stack. This type can be large as 70 ft in diameter and 150 ft tall. The flare type on a given project is selected based on the plot availability, location of the plant, sparing philosophy etc.

PERMITTING REGULATIONS

This presentation will discuss cover various Federal regulations that are applicable to all types of flares and the various parameter that the regulators are needing when apply for the permit. The presentation will cover the required destruction efficiency of the flaring gases, exit velocity for an elevated flare, alternate means of emission limitation for ground flares. In particular, the tip velocity, burner type, backpressure, emission testing and smokeless capacity are specific to each type of flare.

WHAT YOU WILL LEARN

-

- This paper will provide sufficient information to ensure the correct type of flare is selected and permitted
- This paper will cover various flares types employed in refinery and in LNG plants
- The limitations of these flare
- Importance of flare in any facility
- Application Federal Regulations
- Parameter required in the permitting applications
- Ground flare testing to establish minimum destruction efficiency
- Alternate means of emission limitation requirements in lieu of testing

548232 A Case Study on Application of HAZOP for a Brewery

Abstract

Beer is the most consumed alcoholic beverage in the world, and most popular beverage after water and tea. It's an alcoholic beverage produced by the saccharification of starch, often derived from the malted grains such as barley and wheat, and the fermentation of the resulting sugar. The brewing process can be broken into four main constituents before the bottling line, Mashing, Lautering, Wort boiling and Fermenting. Breweries commonly rely on ammonia refrigeration to keep their supplies cold. However, some brewers fail to realize that ammonia is one of the dangerous chemicals covered under 29 CFR 1910.119, OSHA's regulation for preventing or minimizing the release of toxic, reactive, flammable or explosive chemicals. "Safety matters at all levels", even for companies that may not be PSM level facilities.

A HAZOP study identifies hazards and operability problems. The study involves investigating how the plant might deviate from the design intent. It is based on the principle that several experts with different backgrounds can interact and identify more problems and failure scenarios working together and combining their results. This paper presents a review of consequences identified in brewery HAZOP. It also outlines some of the recommendations and action items identified from the case study of a brewery.

Keywords: Process Safety Management (PSM), PHA, HAZOP, Brewery, Ammonia, OSHA.

548410 A Basic Review of Fundamental of Thermodynamics and Its Application in the Analysis of Consequences and Accident Investigation

Once the stage of hazard identification has been concluded in a risk analysis, the next stage corresponds to select and specify the risk scenarios that will be modeled their effects or damages associated. This stage correspond to the consequences analysis and usually takes place in three basic steps: a) source term definition, b) dispersion model and d) effect or damage model.

Depending on the risk scenario to be modeled, several thermodynamic fundamentals or principles are applied in each of these steps supporting or sustaining the behavior and expected results of the consequences analysis model. In this work we propose to review the main thermodynamic fundamentals that support each of the models applied in each of the three steps that define a consequence analysis. This review of basic thermodynamic principles defines a set of different thermodynamic basic principles to apply to representative and different cases of consequence analysis. These basic thermodynamic principles are applied in the consequence modeling of a pair of cases of accident inventions to corroborate their validity.

548331 Evaluating Safe Location Criteria for Vents and Relief Devices Discharging to Atmosphere

In some process technologies, it may not be feasible to dispose the effluents from safety valves and vents to a closed system and discharging directly to the atmosphere can be the only alternative. Some examples include High Pressure Polyethylene Processes, Olefins and Storage units. The discharge can be directed to the atmosphere if:

- Local regulations permits it;
- Does not harm the environment;
- Does not harm the neighborhood;
- And is directed to a Safe Location.

The criteria for defining what a safe location is can vary, but it must include at least a consequence assessment of the flammable, toxic and thermal radiation effects. Some guidance can be found on industry standards and engineering references such as API 521 and the CCPS Guidelines for Pressure Relief and Effluent Handling Systems

One aspect to discharge hydrocarbons to a Safe Location is to evaluate the dispersion of the discharged hydrocarbons and ensure that it is diluted bellow the lower flammable before reaching the ground level or elevated structures. These areas are typically with people circulation and ignition sources. Another aspect that must also be evaluated is the thermal radiation generated due to an ignition of the discharge affecting personnel or damaging equipment or other assets. Lastly, if there is a possibility of the flammable mixture reaching confined spaces, the explosion possibility must also be assessed.

API 521 gives some guidance on chapter 5.8 for estimating the dispersion and concentration profile from a typical relief device discharge. It also describes that if the vapor relief velocity is high (> 30 m/s), the air entrance rate is also high and the vapors typically would be diluted below lower flammable limit in a very short distance.

This phenomenon explains the standard recommendation of locating the discharge of the relief device at least 3 m above any adjacent equipment, building or structure located within 15 m horizontally from the discharge point.

Experience has shown, however, that for some special cases these typical recommendations does not apply, such as:

- Heavy gas releases;
- Cryogenic releases;
- Control valve releases;
- Discharge piping with unconventional design (rain cap, horizontal, etc.);
- Discharge location within a building wake.

Adding to the complexity of the analysis, the relief devices can also be fitted with steam or other purge gas injection that could positively affect the dispersion.

After some case studies, which included the analysis of High Pressure Polyethylene plants Vent Stacks and Olefins unit relief and venting devices, Braskem developed an internal engineering standard with guidance and criteria for evaluating existing projects and new designs in order to ensure they comply with safe location. The guidance includes:

- Relief cases selection;
- Simplified and rigorous dispersion analysis;
- Consequence assessment endpoints;
- Thermal radiation effects to assets;
- Liquid, vapor and two-phase releases;
- Process Hazard Analysis;
- Diluting/snuffing steam;
- New systems and retrofitting criteria;

Applying the criteria in the developed standard allows the evaluation if the relief is in a safe location or not. If the discharge is not in a safe location, a risk ranking and recommendations to change the system must be made. One kind of a typical recommendation is to install the relief in a higher location. However, for some cases, achieving Safe Location is not possible and changes in the system are needed, such as relocating to a closed system or design changes to eliminate the relief scenarios associated with that relief device.

The development of an internal engineering guideline ensures a standard approach throughout the company for evaluating the Safe Location of hydrocarbon discharges and improves the quality of the risk evaluation during the PHA cycles.

548407 Experimental Study of Electrostatic Hazard inside Scrubber Column Using Response Surface Methodology

Experimental Study of Electrostatic Hazard inside Scrubber Column using Response Surface Methodology

Jingyao Wang[a,b], Yue Sun[a,b], Xiaodan Gao[a,b], Benjamin Wilhite[a,b], M. Sam Mannan [a,b]

[a]Mary Kay O'Connor Process Safety Center, Artie McFerrin Department of Chemical Engineering, Texas A&M University, College Station, Texas 77843-3122, USA

[b] Artie McFerrin Department of Chemical Engineering, Texas A&M University, 3122 TAMU, College Station, Texas 77843, United States

Abstract

Electrostatic ignition has triggered significant fire/explosion incidents in liquid processing industries including chemical, oil/gas, food, pharmaceutical and transportation. In this work, the impact of operating flowrate and temperature on the electrostatic energy accumulation in scrubber columns was studied. An *in-situ* charge measurement apparatus combined with a particle scattering laser function was installed inside a pilot-scale scrubber column, which can measure the aerosol particle size distribution, concentration and charge simultaneously. Four electrostatic ignition scenarios were considered including discharge from a single aerosol droplet, aerosol clouds, ungrounded conductor charged by the aerosol and ungrounded conductor charged by the liquid. Response surface methodology was applied to construct a quadratic empirical model correlating the electrostatic energy response with the flowrate and temperature. The obtained electrostatic response shows that corona from a charged single droplet was too small for safety concern. However, the ungrounded conductor polarized by charged aerosol or liquid can accumulate significant energy that is higher than the minimum ignition energy of some flammable chemicals. Temperature has two-sided effect where high temperature generates more concentrated charged aerosol but also facilitates the charge relaxation in the liquid phase. In all scenarios, increasing gas/liquid flowrate can cause significant electrostatic energy accumulation and can increase the hazards of many incidents. To reduce the electrostatic hazards, it is recommended to reduce the gas/liquid interaction flowrate and eliminate the presence of ungrounded conductors in liquid process operations.

Keywords: Scrubber column design, Electrostatic hazards, Response surface methodology, Process safety, Hazard identification, Aerosol hazards.

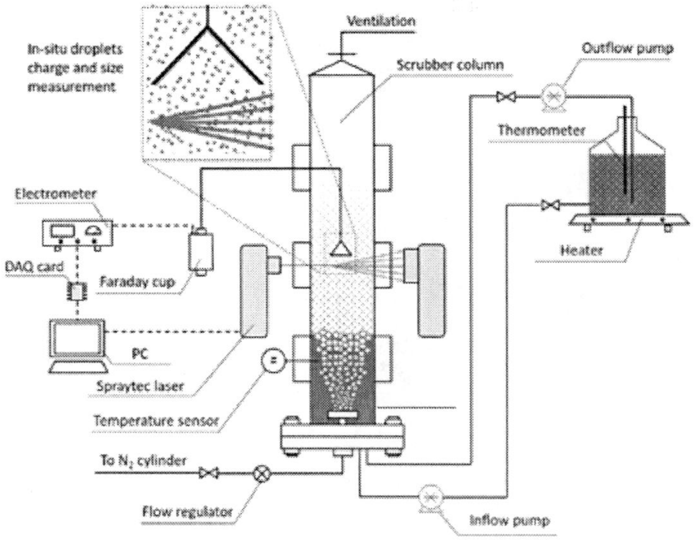

550852 Process Safety in Oil Refineries: Vacuum Distillation Column Hazop Study

Errors that can occur in the identification studies for hazards and risks which is one of the elements of a process safety management cause resulting major accident having undesirable long - term effects on human and environment. For preventing major accidents, identification and analysis of process hazards due to equipment failures, design or montage defects, improper maintenance, misuse, material defects, metal fatigue, corrosion, welding defects, human failure etc. are the main principles of process safety management. The aim of present study is to investigate possible accidental risks occurring in Vacuum Distillation Column of oil refining within contributing to the process safety frame.

In this cross - sectional study, semi quantitative HAZOP analysis was used for identifying accidental risks. By means of HAZOP studies 66 deviations, 238 deviation causes, 287 consequences, 651 barriers and 113 recommendation were identified. In the determination of accidental risk in the vacuum distillation column by applying semi quantitative HAZOP analysis method, effective results were obtained. The most common causes of deviations in the column have been found as control loop failure, leakage and mechanical failure. It is stated that the most common accidental risks are as a portion of 66.9 % operational deterioration, fire, explosion, release known as major accident risk. It is identified that the accidental risks mostly cause existence damage and the highest accidental risk occurs in the "vacuum column overhead system" and "bottom of the vacuum column system" 87.7 % of 113 recommendations are modification in the design, 5.3% is about modifications in the process safety management system.

551179 API Process Safety Site Assessment Program - Trends and Good Practices in Process Safety

The Process Safety Site Assessment Program is a collaborative industry effort managed by API and is focused on improving process safety in petroleum refining and petrochemical facilities. The program has conducted over 65 process safety assessments and 38 HF RP-751 assessments since its inception in 2012. The assessments are conducted by independent and credible third-party teams of industry-qualified process safety expert assessors using industry-developed protocols.

The presentation will review trends, both positive and low scoring, seen in the extensive benchmarking data collected. Good practices observed in the assessments will be identified and discussed. The emphasis will be to have the attendees leave with an understanding that typically some site(s) has developed a way to handle challenging areas of process safety.

545445 Emerging Trends in Process Safety Management Relating to Contractor Workforce Compliance

Following a Six Sigma Black Belt process improvement session conducted 2016 to 2018 for petrochemical complexes along the Gulf Coast the presentation identifies several emerging trends for Best Practices in industrial workforce compliance.

Safety Councils work for petrochemical plants and their contractors to facilitate innovative ways to reduce incident rates resulting in a 95% reduction in recordables. 2017 marked an evolution in Safety Council deliverables, slowly migrating away from site specific training to develop new workforce compliance and risk management processes. Working within the industry we developed several internationally recognized Best Practices for contractor compliance, risk management, labor cost management. ISTC collaborated with our partners bringing years of quantitative data to identify problems that are developing, have occurred in the workplace and formulating solutions, cost structures and most importantly compliance management to combat workforce issues. The presentation, best practices in contractor compliance highlights the journey from identifying the problems, and the associated issues with hiring, compliance, auditing, accident response and even Department of Homeland Security, CFATS inspections. Beyond the logistics of building solutions for a particular problem, the presentation reflects on how companies were able to share their issues among peers without violating confidentiality or company procedures to solve what was to be discovered as an industry trend. Using Six Sigma tools, the units were able to identify root problems, collaborate on a solution, model cost recovery and implement the solutions to the contractor workforce with a minimum of disruption to the business at hand.

19 AIChE Spring Meeting
15TH GLOBAL CONGRESS ON PROCESS SAFETY
NEW ORLEANS • 2019

Análisis Cuantitativo De Riesgo (QRA) Como Herramienta Para La Toma De Decisión Sobre La Infraestructura En Sistemas De Transporte De Fluidos Peligrosos

Daniel Mauricio Moncada Saavedra
CENIT -Transporte y Logística de Hidrocarburos
Cll 113 #7-80 Piso 13 Bogotá – Colombia
daniel.moncada@cenit-transporte.com

Maria Isabel Montoya Rodriguez
CENIT -Transporte y Logística de Hidrocarburos
Cll 113 #7-80 Piso 13 Bogotá – Colombia
maria.montoya@cenit-transporte.com

Prepared for Presentation at
American Institute of Chemical Engineers
2019 Spring Meeting and 15th Global Congress on Process Safety
New Orleans, LA
March 31 – April 3, 2019

Análisis Cuantitativo De Riesgo (QRA) Como Herramienta Para La Toma De Decisión Sobre La Infraestructura En Sistemas De Transporte De Fluidos Peligrosos"

Daniel Mauricio Moncada Saavedra
CENIT -Transporte y Logística de Hidrocarburos
Cll 113 #7-80 Piso 13 Bogotá – Colombia
daniel.moncada@cenit-transporte.com

Maria Isabel Montoya Rodriguez
CENIT -Transporte y Logística de Hidrocarburos
Cll 113 #7-80 Piso 13 Bogotá – Colombia
maria.montoya@cenit-transporte.com

Palabras clave:

Abstract

La pérdida de contención en activos utilizados para el transporte de hidrocarburos puede ocasionar impactos considerables a personas, medio ambiente, económicos y de reputación. Con el fin de mitigar la ocurrencia este tipo de eventos, se implementan estrategias sobre los activos, siendo la identificación y análisis de riesgo un elemento clave durante el proceso.

El presente análisis utilizó un enfoque cuantitativo, tanto para probabilidad como para consecuencia, con el fin de poder estimar el riesgo que genera la operación de una línea de transporte en un área de alta consecuencia clasificado como sitio identificado, planteando posibles alternativas de control y mitigación del riesgo.

Como resultado del estudio realizado se pudo establecer el nivel de riesgo actual asociado a la operación de la línea en el sitio de interés. Como actividades de mitigación para reducir el riesgo asociado se identificó: la reparación de los defectos, así como el realineamiento y reubicación de la tubería.

1 Introducción

Para el presente análisis se implementó un cálculo de la probabilidad de falla a partir del análisis de integridad estructural del sector de interés. Adicionalmente, se consideró el análisis de la susceptibilidad que se presenta a partir de las amenazas presentes, la cuales son identificadas a partir de las actividades de monitoreo e inspección desarrolladas sobre la infraestructura. El cálculo de la interacción entre las amenazas a la integridad; así como el resultado final de probabilidad de falla, se puede realizar mediante diferentes enfoques, el presente estudio utilizo cálculo de probabilidad de excedencia y aplicación de redes bayesianas para la estimación.

Para el cálculo de la consecuencia se buscó verificar la posible afectación a personas, producto de la ignición del producto transportado. Dentro de los sucesos finales se consideró: incendio de piscina, incendio por chorro de fuego, llamarada y la sobrepresión debida a la explosión. El cálculo de estos efectos se realizó en el sitio donde se presenta la perdida de contención, así como también para el caso en el que el hidrocarburo es transportado por un cuerpo de agua o se desplaza en tierra.

Una vez determinada tanto la probabilidad de que se desarrollen los sucesos finales identificados como la consecuencia que cada uno de ellos puede generar, se procedió a realizar el cálculo del riesgo expresando el resultado en términos de riesgo social e individual, comparando el resultado contra criterios de tolerabilidad referentes en la industria. Para la condición actual establecida se identificaron alternativas las cuales buscaban disminuir el nivel de riesgo, implementando actividades de mitigación del nivel de riesgo.

2 Análisis de Riesgo Cuantitativo

3 Frecuencia Final de Ocurrencia de los Sucesos Finales

En el sector bajo interés se tienen que la línea es susceptible a las siguientes amenazas:

3.1 Análisis de Condición.

En este aparte se relacionan los resultados de análisis de probabilidad de falla, obtenidos en el sector de interés, PK 43+000 al PK 45+000 del poliducto, para las amenazas de corrosión externa, corrosión interna, geotecnia y daños por terceros involuntarios:

3.1.1 Corrosión Externa (CE).

A continuación, se describe el estado de salud de las barreras de CE del poliducto en el sector solicitado (ver Ilustración 1), así como el análisis de probabilidad de excedencia de las anomalías ILI identificadas con la corrida del 2014. En la Ilustración 2 se observa la disponibilidad de la URPC y se reporta que la tubería cuenta con recubrimiento en buen estado. Bajo las condiciones y premisas consideradas para el cálculo y proyección de la POE (Probabilidad de excedencia; entre ellas una rata de corrosión de 2-3 mpy aproximadamente; no hay riesgo de excedencia de los criterios de profundidad y ERF (Factor Estimado de Reparación) de las anomalías ILI hasta el año 2021 en el sector 43+590 y 42+746 (Anomalías ILI con DR(Distancia de registro): 2646.76, 3490.08, 3492.38, 3492.71) (Ilustración 2).

Ilustración 1. Perfil de riesgo y salud de barreras CE.

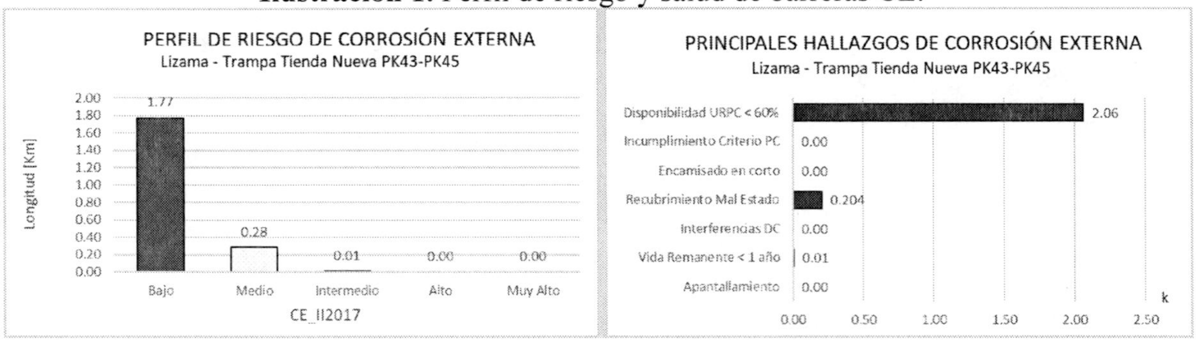

Ilustración 2. Análisis de POE de anomalías de CE en el tiempo.

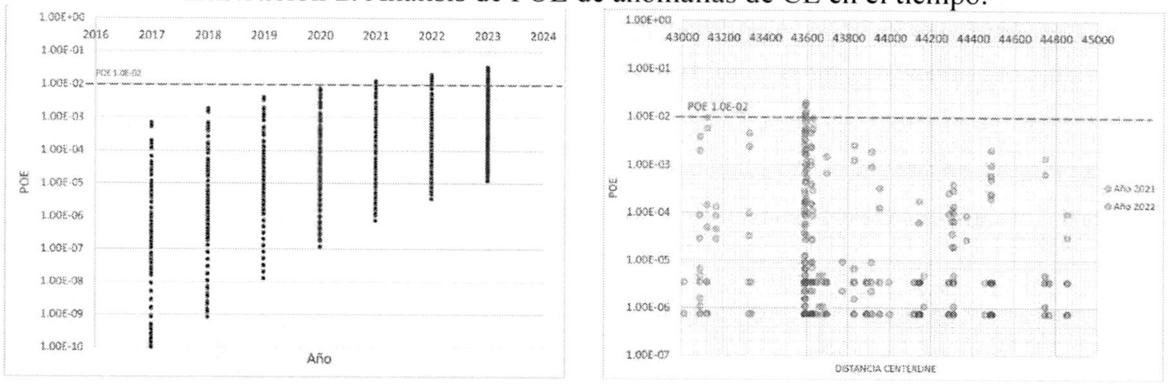

3.1.2 *Corrosión interna (CI).*

A continuación, se describe el estado de salud de las barreras de CI de la tubería en el sector solicitado (ver Ilustración 3), así como el análisis de probabilidad de excedencia de las anomalías ILI identificadas en la corrida del 2014. De acuerdo con los resultados de valoración de riesgo, se presenta un correcto estado de salud de las barreras de corrosión interna. Adicionalmente, de acuerdo con las condiciones y premisas consideradas para el cálculo y proyección de la POE, no hay riesgo de excedencia de los criterios de profundidad y ERF de las anomalías ILI en el mediano plazo (Ver Ilustración 4).

Ilustración 3. Perfil de riesgo y salud de barreras CI.

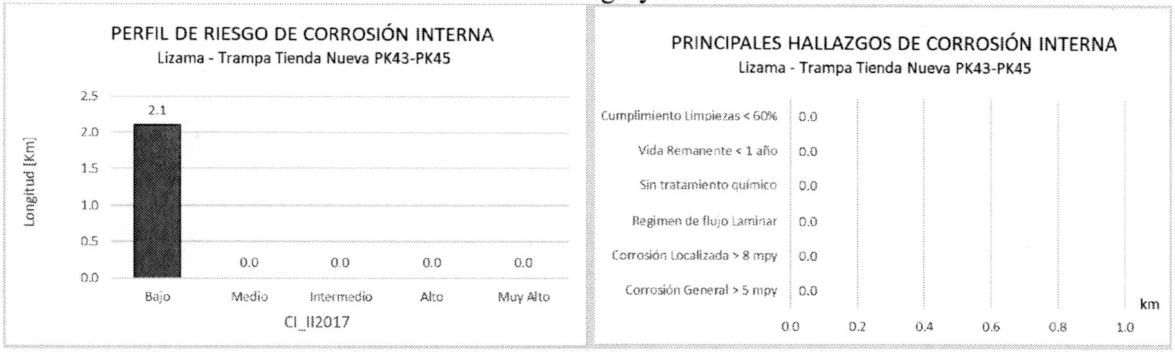

Ilustración 4. Análisis de POE de anomalías de CI en el tiempo.

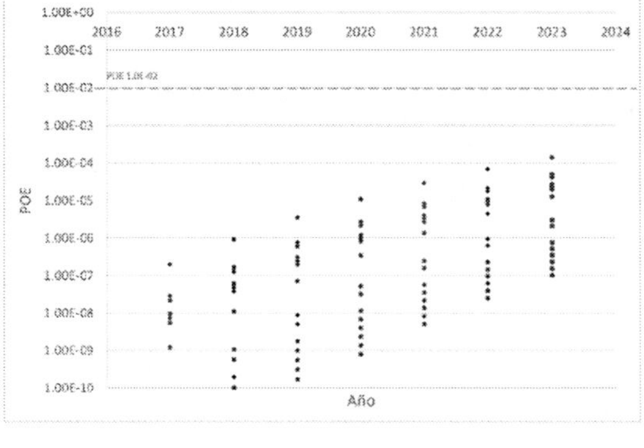

3.1.3 Geotecnia.

En el sector de interés se reporta hallazgo de geotecnia en la distancia centerline 44643 – 44703 (PK 44+400), donde se presenta proceso erosivo incipiente. Actualmente, no se identifica avance de la condición, razón por la cual, se continuará seguimiento a través de la inspección visual del derecho de vía (IVDDV).

Ilustración 5. Perfil de riesgo y salud de barreras de Geotecnia.

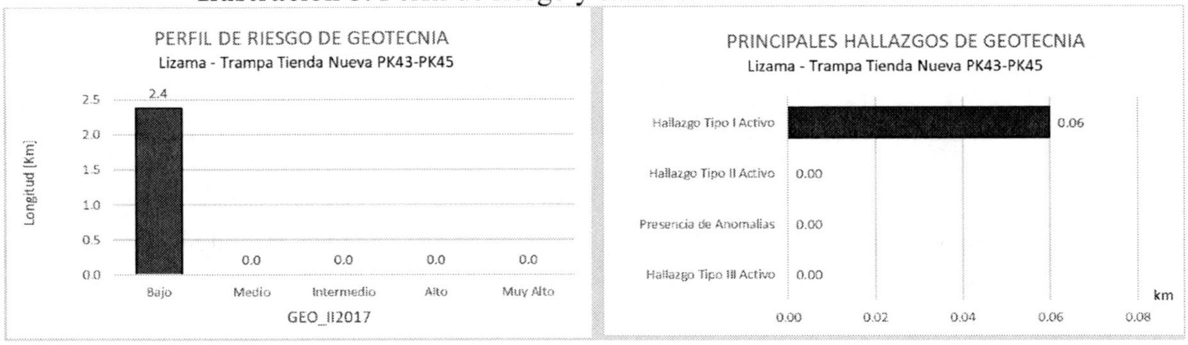

3.1.4 ***Daños por Terceros Involuntarios (DPT).***

Debido a la condición de la tubería y a la gestión realizada por el operador para el control de la amenaza, ésta presenta una susceptibilidad baja de ser golpeada por un tercero sobre el derecho de vía. A continuación, se relaciona el perfil de riesgo de la amenaza de DPT en el sector analizado:

Ilustración 6. Perfil de riesgo y salud de barreras de Daños por Terceros Involuntarios.

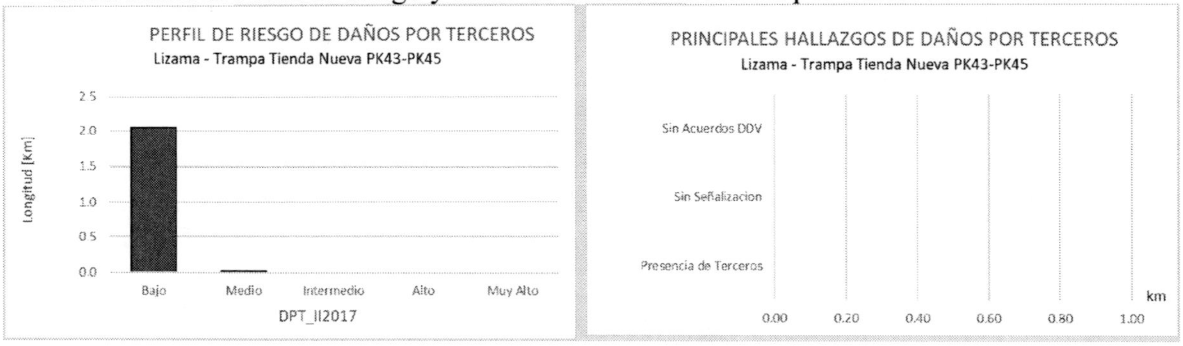

De acuerdo con el análisis ejecutado, el evento más probable obedece a una pérdida de contención generada por la amenaza de corrosión externa; donde el análisis estructural identifica que el modo de falla se daría por fuga. Sin embargo, las anomalías encontradas no presentan riesgo de excedencia de los criterios de profundidad y ERF hasta el año 2021.

Con el fin de poder estimar la probabilidad de falla de la línea en el sector de interés, se realizó el análisis estructural considerando principalmente defectos por perdida de espesor, bajo estas premisas se obtuvo un valor de probabilidad de excedencia de los criterios de profundidad y ERF de 7.09E-4. Del análisis de condición se concluye que el iniciador más probable en este sector es la falla de la tubería por fuga asociado a corrosión externa, para el cual a continuación se presenta el análisis de riesgo.

4 Análisis de consecuencia

A continuación, se presenta el análisis de consecuencia para los sitios en los cuales se ha identificado que se puede llegar a presentar una pérdida de contención.

4.1 Identificación de escenarios

4.1.1 Selección de sustancias.

En el poliducto se transportan las siguientes sustancias: Gasolina, GLP, Jet A1 y B2

4.1.2 Identificación de iniciadores.

Los iniciadores representan las situaciones iniciales de pérdidas de contención o exposición que se pueden presentar en los das diferentes equipos. Los iniciadores pueden ser de dos tipos, genéricos o específicos. Los iniciadores genéricos, involucran una pérdida de contención debida a causas no específicas del proceso u operación, como pueden ser corrosión, errores de construcción, fallas de la soldadura, etc. Los iniciadores específicos se caracterizan por tener causas concretas debido a las condiciones de proceso, el diseño, los materiales utilizados, etc.

A continuación, se describen los iniciadores tenidos en cuenta para cada uno de los sitios en los cuales se identificó la inestabilidad de taludes.

- **Sector PK 43+590.** Este sitio tiene la menor distancia desde la infraestructura a la escuela bajo análisis, 30 metros aproximadamente, sumado a esto en las cercanías de este se ha identificado la presencia de anomalías las cuales, bajo las condiciones actuales de operación y deterioro hacia el año 2021, pueden exceder los límites de profundidad critica a la falla.

- **Sector PK 43+805 – PK 44+110.** En este sitio se tiene que ante la posible pérdida de contención del producto se puede llegar a presentar la formación de una ruta de derrame en tierra, por la cual se desplazaría el producto hacia el interior de la Escuela.

4.1.3 Identificación de sucesos finales.

La identificación de los sucesos finales que pueden tener lugar después de que se da el iniciador, se debe realizar utilizando los árboles de eventos o sucesos. Dichos árboles permiten ver el comportamiento de la sustancia tras la descarga y estimar la probabilidad de ocurrencia de cada suceso, teniendo en cuenta los sucesos intermedios que condicionan la evolución del suceso final.

Esta identificación de escenarios se lista en Tabla 11 para los escenarios que involucran sustancias inflamables tales como Gasolina o Nafta. Los posibles sucesos finales identificados en el poliducto son: derrame continental, incendio de piscina, incendio de chorro y llamarada.

Tabla 1. Identificación de escenarios.

Escenarios	Descripción	Iniciadores	Sustancia	Tipo de pérdida
Afectación directa				
E1	Poliducto	L3	Gasolina	Incendio de piscina
				Incendio de chorro
				Llamarada
E2	Poliducto	L3	GLP	Incendio de piscina
				Incendio de chorro
				Llamarada

Escenarios	Descripción	Iniciadores	Sustancia	Tipo de pérdida
Afectación directa				
Afectación indirecta				
E3	Poliducto	L3	Gasolina	Incendio de piscina
				Llamarada

4.2 Estimación del área de afectación por los efectos que involucran producto

4.2.1 Condiciones para la simulación.

Para determinar las consecuencias, es necesario definir las condiciones propias del entorno de la infraestructura que servirán como base para la simulación de los efectos. Estos parámetros son presentados en la Tabla 2. Las estabilidades atmosféricas fueron determinadas a partir de la rosa de vientos basada en datos estadísticos y modelos meteorológicos, la cual es presentada en la **¡Error! No se encuentra el origen de la referencia.**.

Tabla 2. Condiciones atmosféricas

Parámetro	Valor	Estación de Referencia
Temperatura promedio (°C)	31	Colosal
Humedad relativa promedio (%)	83	
Rugosidad el terreno (m)	0,08	
Estabilidad atmosférica más probable y velocidad del viento (m/s)	2F	
Estabilidad atmosférica F y velocidad del viento (m/s)	2F, 5D	

4.2.2 Distancias de afectación directa

Las distancias de afectación representan las longitudes dentro de las cuales se pueden manifestar los efectos correspondientes a los niveles de afectación, para cada suceso final (incendios de piscina o llamaradas). Se denominan distancias de afectación directas, ya que los sucesos finales de que dependen se originan y desarrollan su efecto directamente en el sitio en el que se produce la pérdida de contención del producto

4.2.3 Estimación de Descargas.

Para la simulación de los posibles efectos producto de una pérdida de contención, es necesario conocer cuánto producto podría derramarse en cada punto de estudio sobre la infraestructura. El volumen derramado corresponde al estimado teniendo en cuenta el llenado de la tubería, las condiciones de operación y el tiempo de respuesta operativo, es decir, el tiempo de detección más el tiempo de actuación (generalmente el seccionamiento del sistema o parada de bombeo).

Tabla 3. Tiempos de actuación

Equipo	Tiempo de Detección (min)	Tiempo de Actuación (min)
	Fuga	**Fuga**
Poliducto	30	180

Tabla 4. Volumen y Tasa de descarga

Volumen (US bls)	Tasa de descarga (bls/hr)
59	22

4.2.4 *Simulación de Efectos de los Sucesos Finales.*

En este numeral se presentan los resultados de las distancias de afectación directa, obtenidas para cada uno de los niveles de afectación asociados a los sucesos finales. En la **¡Error! No se encuentra el origen de la referencia.** se presentan las distancias de afectación obtenidas para el evento incendio de piscina, es importante resaltar que este depende en gran medida del área de empozamiento que se forma una vez se produce la descarga del producto.

En la **¡Error! No se encuentra el origen de la referencia.** se presentan las distancias de afectación para incendio de chorro obtenidas durante las simulaciones, es importante resaltar que la distancia obtenida depende de las propiedades químicas de la sustancia y de las condiciones en las cuales se produce la descarga.

Tabla 5. Distancia de afectación incendio de piscina.

Equipo / Abscisado	Iniciador	Sustancia	Volumen Derrame	Niveles de Afectación por Incendio						
				Incendio de Piscina						
				1.6 Kw/m²	5.0 Kw/m²	7.3 Kw/m²	9.5 Kw/m²	14.5 Kw/m²	20.9 Kw/m²	37.5 Kw/m²
			(Bbl)	Distancia (m)						
K43+590	Rotura mínima	GLP	59	46,94	28,63	24,58	21,98	18,01	14,55	8,38
K43+590	Rotura mínima	Gasolina	59	94,63	49,69	35,17	27,20	20,98	20,98	-

Tabla 6. Distancia de afectación incendio JET FIRE.

Equipo / Abscisado	Iniciador	Sustancia	Volumen Derrame	Niveles de Afectación por Incendio						
				Chorro de fuego						
				1.6 Kw/m²	5.0 Kw/m²	7.3 Kw/m²	9.5 Kw/m²	14.5 Kw/m²	20.9 Kw/m²	37.5 Kw/m²
			(Bbl)	Distancia (m)						
K43+590	Rotura mínima	GLP	59	45,9	33,61	30,93	29,33	27,13	25,23	23,42
K43+590	Rotura mínima	Gasolina	59	44,43	31,20	28,46	26,82	24,58	22,96	20,84

En la **¡Error! No se encuentra el origen de la referencia.** se presentan las distancias de afectación por llamarada; las cuales están relacionadas con la volatilidad del producto junto con las condiciones climáticas de las zonas del evento.

Tabla 7. Distancia de afectación Llamarada.

Equipo / Abscisado	Iniciador	Sustancia	Volumen de Derrame	Niveles de Afectación por Llamarada más probable		Niveles de Afectación por Llamarada más estable	
				LII	LII/2	LII	LII/2
			(Bbl)	Distancia (m)		Distancia (m)	
K43+590	Rotura mínima	GLP	59	24,12	58,53	26,48	66,7
K43+590	Rotura mínima	Gasolina	59	-	-	-	-

4.2.5 Distancias de Afectación Indirecta

Las distancias de afectación indirecta representan las longitudes de potencial afectación por sucesos finales que se originan en una trayectoria o ruta de derrame (en tierra o en agua) o, a partir de estancamientos o acumulaciones de producto. En este numeral se presentan los resultados de las distancias de afectación obtenidas para cada uno de los niveles de afectación asociados a los sucesos finales que se pueden presentar en la ruta de derrame. Estos resultados corresponden a la radiación generada por piscina de fuego, evento que en los cuerpos de agua depende de la sección del cuerpo de agua, así como por el frente de llama generado por el volumen derramado.

En la Tabla **88 y ¡Error! No se encuentra el origen de la referencia.**9 se referencian los corredores de afectación generados por incendio de piscina sobre las rutas de derrame en tierra identificadas en el poliducto, en el sector comprendido entre los PK 43+000 - PK 45+000.

Tabla 8. Distancias de afectación indirectas POOL FIRE.

Abscisado	Medio de Transporte	Sustancia	Incendio						
			Incendio en piscina						
			1.6 Kw/m^2	5.0 Kw/m^2	7.3 Kw/m^2	9 Kw/m^2	14.5 Kw/m^2	20.9 Kw/m^2	37.5 Kw/m^2
			Distancia (m)						
Km 43 - 45	Ruta en tierra	Gasolina	82,02	54,68	27,34	24,61	19,14	13,67	8,20

Tabla 9. Distancias de afectación indirecta FLASH FIRE

Abscisado	Medio de Transporte	Sustancia	Niveles de afectación por llamarada más probable		Niveles de afectación por llamarada más estable	
			LII	LII/2	LII	LII/2
			Distancia (m)		Distancia (m)	
Km 43- 45	Ruta en tierra	Gasolina	2,14	4,74	8,64	32,24

4.2.6 Áreas de Afectación

En sistemas de ductos, el área de afectación directa se define por un corredor medido desde el eje central del ducto. Este corredor está dado por la mayor distancia probable de afectación obtenida en la simulación de efectos para todas las sustancias analizadas. Por otro lado, el área de afectación indirecta está dada por varios corredores a lo largo de las posibles rutas de derrame tanto en suelo como en agua. En la Ilustración 77 se presenta el área de afectación identificada.

Ilustración 7. Área de afectación identificada.

4.2.7 Identificación de Elementos Vulnerables

Para calcular el riesgo de la población respecto de la posible afectación producida por la pérdida de contención descrita, se requiere el número y localización de las personas, en la Tabla 100 se presenta la población identificada al interior de la escuela.

Tabla 10. Presencia de población Escuela

Presencia de Población							
Escuela - Dia				Escuela - Noche			
Niños	63	Profesores	6	Niños	0	Profesores	2

5 Evaluación del Riesgo.

A continuación, se presentan los resultados del análisis de riesgo el cual tuvo en cuenta la información presentada en los numerales anteriores asociada a las posibles causas de falla, los modos en que puede darse cada falla identificada, la frecuencia de ocurrencia de la falla, la probabilidad de ocurrencia de los sucesos finales, y finalmente, la estimación de la magnitud de las consecuencias de tales sucesos finales en el entorno en términos cuantitativos.

5.1 Riesgo Individual.

Los resultados de riesgo individual expresados en número de fatalidades por año (o año-1), sí permiten establecer zonas de aceptabilidad del riesgo según los diferentes criterios que se utilizan en la industria. En la Ilustración 8 se presentan los resultados de riesgo individual en el sector comprendido entre los PK 43+000 y 45+000.

Ilustración 8. Riesgo Individual.

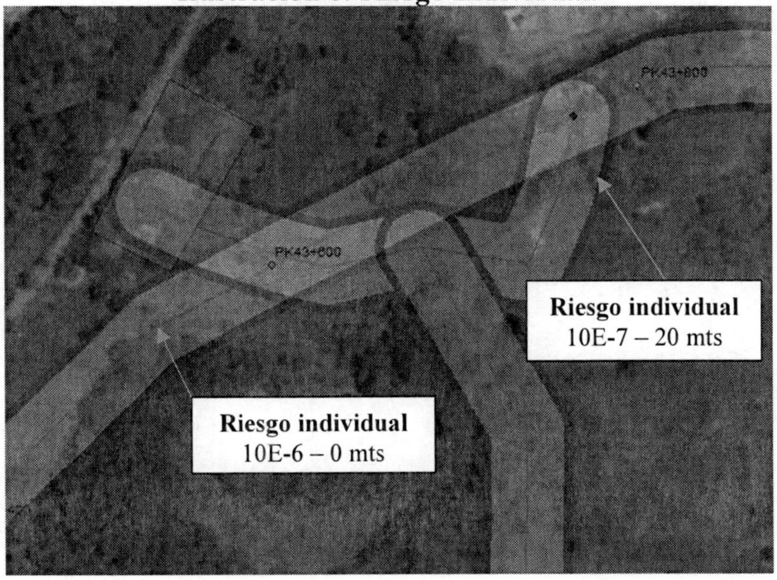

5.2 Riesgo Social

El riesgo social representa el número de fatalidades que pueden llegar a ocurrir en los sitios evaluados en el poliducto en el PK 43+000 – PK 45+000, al revisar la información contenida en la Ilustración 9 se puede apreciar como el riesgo se mantiene en la región tolerable, muy cercana al límite inferior ALARP. El evento dimensionante corresponde a una llamarada en la ruta de derrame, con el ducto transportando gasolina, la cual puede llegar a afectar a ocho (8) personas.

Ilustración 9. Riesgo social Escuela PK 43+000 – PK 45+000.

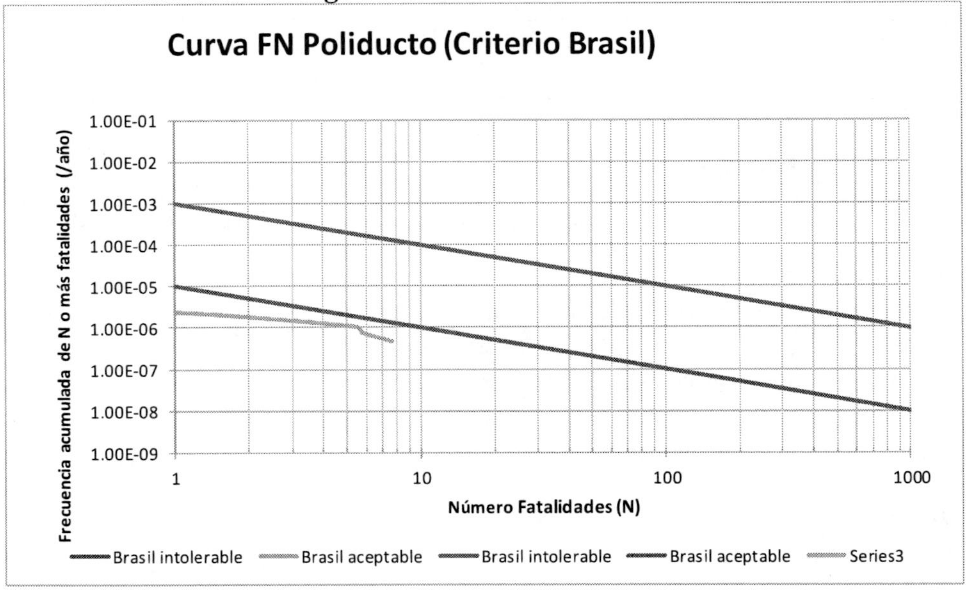

6 Alternativas de mitigación de Riesgo

Considerando los resultados del análisis de riesgo cuantitativo sobre el Poliducto para el sector PK 43+000 - PK 45+000, se tiene que ante una rotura total el evento dimensionaste corresponde a una llamarada con una distancia de afectación de 135 metros, para el caso de una fuga por corrosión el evento dimensionaste corresponde a una llamarada con una distancia de afectación de 32,24 metros, finalmente como resultado de una perdida de contención se puede llegar presentar la formación de una ruta de derrame en tierra la cual ingresaría a la escuela, con una distancia de afectación de 8.6 metros generada por una llamarada, tal como se presenta en la Ilustración 10.

Ilustración 70. Distancias de afectación

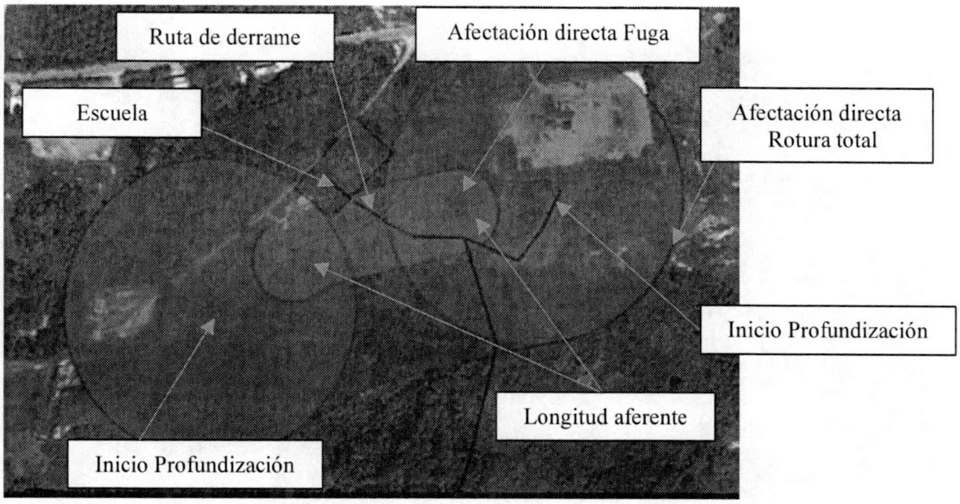

Teniendo en cuenta que el evento corresponde al presentado en la ruta de derrame, se hace necesario considerar la evaluación de alternativas de mitigación de riesgo. Con base en lo anterior, se consideró que la actividad de realineación y profundización cubra la longitud aferente de esta ruta, la cual es de 155 metros e inicia en la coordenada 74°33'35.466"O - 6°9'8.213"N y finaliza la coordenada 74°38'15.025"O - 6°9'8.737"N. Lo anterior, tiene como objeto de eliminar la ruta de derrame actual que conduce a la escuela en caso de una falla y perdida de contención, dado que el vertimiento se genere en sentido contrario al actual.

Sumado a lo anterior es necesario considerar una distancia de protección adicional para el dado caso en el que se llegue a presentar la rotura total de la línea, esta distancia se establece a partir del radio de impacto del evento llamarada, localizando los puntos sobre la línea que no presentan afectación a la escuela. A partir del análisis se estableció que esta distancia es de 316 metros e inicia en la coordenada 74°31'33,57146"O - 6°9'8,22442"N y finaliza en la coordenada 74°32'28,98241"O - 6°9'9,70756"N, esta distancia cubre la longitud aferente descrita anteriormente.

Dado que la alternativa de mitigación tiene como fin eliminar la afectación indirecta y el impacto directo sobre la escuela a continuación se presentan dos alternativas de trazados propuestos, que aseguran que las rutas de derrame en caso de una perdida de contención tendrán dirección opuesta a la ubicación de la escuela.

Ilustración 80. Opción 1 realineamiento entre PK 43 + 695 a PK 44 + 205.

Ilustración 91. Opción 2 realineamiento entre PK 43 + 625 a PK 44 + 460 de 752 metros.

En la ilustración 12 se presenta las dos alternativas propuestas y la distancia de separación del trazado actual de tubería, así como de la escuela.

Ilustración 102. Alternativas propuestas para mitigación de riesgo

7 Conclusiones

Del análisis realizado se tienen las siguientes conclusiones.

- El máximo número de fatalidades a generar es de ocho (8) y es generado por una llamarada en la ruta de derrame en tierra la cual, en el dado caso de que se presente el evento, ingresaría a las instalaciones de la escuela.
- La mayor susceptibilidad de la línea es por la amenaza de corrosión externa, especialmente en algunos sectores en los cuales el recubrimiento se encuentra en mal estado, esta condición junto con las anomalías encontradas en la línea, impactaron la frecuencia de falla establecida para el sector.
- En términos generales el nivel de riesgo se encuentra en la zona tolerable, muy cercana al límite inferior ALARP.

8 Recomendaciones.

- Es recomendable realizar actividades tendientes a disminuir el volumen disponible a desplazar en la ruta de derrame, para ello se puede realizar la profundización de la línea lo cual además de disminuir el efecto de los eventos por JET FIRE y FLASH FIRE disminuye la cantidad de producto disponible a desplazar al favorecer la evaporación de este.

- A partir del análisis realizado se tiene que la distancia a realinear y profundizar es:

  Opción 1 realineamiento entre PK 43 + 695 a PK 44 + 205 con una longitud de 516 metros
  Opción 2 realineamiento entre PK 43 + 625 a PK 44 + 460 con una longitud de 752 metros.

- Con ambas alternativas se elimina la afectación indirecta y la afectación directa, sobre la escuela, estas distancias cubren todos los posibles eventos que se pueden presentar y tienen afectación sobre la escuela en mención.

- Se pueden construir cortacorrientes los cuales eviten que la ruta de derrame ingrese o haga su tránsito en las cercanías de la escuela, junto con la construcción de estructuras de retención las cuales detengan el avance del producto mientras se realizan actividades de recuperación de este.

- Es recomendable rehabilitar el recubrimiento en aquellos sectores en los cuales se encuentra en mal estado, junto con la atención de aquellas anomalías que sean críticas.

9 Referencias

[1] Fingas, M. (2011). Oil Spill Science and Technology. Kidlington, UK: Gulf Professional Publishing Elsevier.

[2] D.O.T. Cornell Law School. Obtenido de 49 CFR 192.905 - How does an operator identify a high consequence area?: https://www.law.cornell.edu/cfr/text/49/192.905 Accesado 15 Enero, 2019

[3] D.O.T. Cornell Law School. Obtenido de 49 CFR 192.5 - Class locations: https://www.law.cornell.edu/cfr/text/49/192.5 Accesado 15 Enero, 2019

[4] Yellow Book. TNO. Methods for the calculation of Physical Effects Due to releases of hazardous materials (liquids and gases) – Third edition Second revised print 2005

[5] Purple Book. TNO. Guidelines for quantitative risk assessment – First edition 1999/2005

Referencias no citadas:

[6] ASME.ASME B31.8S. Managing System Integrity of Gas.New York, USA.2005.

[7] ASME.PIPELINE TRANSPORTATION SYSTEMS FOR LIQUID HYDROCARBONS AND OTHER LIQUIDS ASME B31.4-2002. New York, USA.2002.

[8] API.API Std 1160. Managing System Integrity for Hazardous Liquid Pipelines. Washington, USA.2008.132 Pags.

548036 Best Practice in MOC and Pssr Automation for Opex

Management of Change (MOC) and Pre Start-up Safety Review (PSSR) are still the most challenging elements of OSHA's Process Safety Management (PSM) standard and the EPA's Risk Management Program (RMP) rule. Now MOC and PSSR are requirements of the US Department of Interior's Bureau of Safety and Environmental Enforcement's Safety and Environmental Management Systems (SEMS). Even though the PSM standard has been around since 1992 and the industry has been managing change for several decades, we still can get it wrong, sometimes with disastrous results. The mindful or diligent efforts of many working in concert are necessary to ensure that change is identified, analyzed and executed in a quality way. But just when we get the process on paper and get the workforce up-to-speed, we have employee turnover, complacency sets in and sometimes there is regulatory change.

Many high-profile companies have addressed this challenge by creating and automating standard processes across their manufacturing or production facilities. MOC and PSSR are two of these major processes that have been successfully automated or **digitized**. Technology has been used to **automate** each step of the MOC workflow, enabling the collection of change ideas, formalizing the analysis of the ideas, and documenting the approval and execution of the change. Reports and dashboards can provide visibility of the health or current state of your MOC and PSSR processes. The use of **software** can enable the standardization of best practices.

On a well-meaning quest to standardize on a MOC/PSSR software solution and process, many companies are stifling creativity and innovation. Operational Excellence cannot be achieved in such ways. This paper discusses how the chemical process industry has defined the MOC and PSSR best management practices and how they should be automated for operational excellence.

Chemical Process Safety Requirements in the Fire Code

Ronald Melucci
Engineering Planning and Management, Inc.
959 Concord Street
Framingham, MA 01701
rbm@epm-inc.com

Prepared for Presentation at
American Institute of Chemical Engineers
2019 Spring Meeting and 15th Global Congress on Process Safety
New Orleans, LA
March 31-April 3, 2019

Ronald Melucci
Engineering Planning and Management, Inc.
959 Concord Street
Framingham, MA 01701
rbm@epm-inc.com
Keywords: NFPA Fire Code, International Fire Code, Process Safety

Chemical Process Safety Requirements of the Fire Code

Abstract

When most Chemical Process Safety professionals think about regulation and enforcement, they think about the applicable Occupational Safety and Health Administration (OSHA) and Environmental Protection Agency (EPA) regulations. However, the majority of US states adopt NFPA 1 or the International Fire Code (IFC) as their state fire code which contain requirements for chemical process safety. These requirements are applicable for processes of all sizes and levels of complexity below PSM and RMP thresholds and they address both fire and non-fire related chemical hazards. These requirements are enforced by municipal or state fire officials/marshals, who are becoming more aware of these requirements and chemical process safety in general. This paper will review some of the code requirements with a focus on requirements that may be commonly overlooked in the industry.

Introduction

The OSHA Process Safety Management (PSM) and EPA Risk Management Program (RMP) regulations, enforced by OSHA and EPA, respectively, are familiar to most safety professionals who work in industries that include processing of large quantities of chemicals. However, the applicable fire code requirements related to chemical process safety are often more restrictive than the federal regulations and are generally lesser known to such safety professionals.

The two model fire codes adopted by states in the US, are: *NFPA 1, Fire Code,* published by The National Fire Protection Association (NFPA), and the *International Fire Code* (IFC) published by the International Code Council (ICC). The states that do not adopt the IFC will typically use NFPA 1 and each state may also institute their own NFPA 1 amendments. Local amendments are almost always more restrictive than the model code verbiage, and many states have legislation that prohibits fire code amendments that reduce the level of safety/standard of care intended by the model fire code.

Both of these model fire codes contain many requirements that are similar to PSM and RMP regulations and contain references to them for additional guidance. Generally speaking, the model fire codes do not regulate all hazardous materials, but focus on hazards that are acutely dangerous to people or property. The PSM and RMP regulations address these hazards as well but additionally include materials that present a risk of chronic health or environmental hazards.

However, many states and municipalities have local fire code amendments that often include lower chemical quantity thresholds, such as those of the Commonwealth of Massachusetts as described in the following sections. The enforcement of these requirements is therefore not limited to OSHA and the EPA, as is commonly believed. The requirements are also enforceable by the local (municipal) fire authorities through the fire code references to these regulations.

Fire Code Requirements

The Commonwealth of Massachusetts currently adopts NFPA 1 – 2015 edition, with amendments, as 527 CMR 1.00: *Massachusetts Comprehensive Fire Safety Code*. Since 2006, 527 CMR has included chemical process safety requirements that are more restrictive than the federal regulations. The impetus for this was several accidents that occurred involving chemical process quantities below the OSHA and EPA thresholds. These Massachusetts (MA) amendments have been added to Chapter 60 of NFPA 1 – 2015 edition and include specific reference to OSHA PSM and EPA RMP regulations. Chemical processes with quantities above the PSM/RMP thresholds are defined in MA as "Category 5" processes. However, the 527 CMR amendments have also defined processes with lower quantities, i.e. Category 1, 2, 3 and 4, each with specific and unique requirements as described herein.

The NFPA 1 committee is currently reviewing proposals to adopt requirements with a "tiered" quantity threshold approach similar to MA. This is in recognition of the fact that severe hazards often exist, and severe accidents have occurred, in chemical processes with quantities below the PSM/RMP thresholds. It is likely that such a "tiered" approach will be incorporated into NFPA 1 and subsequently adopted by many US states. A preview of the applicable NFPA 1 and 527 CMR requirements are summarized herein.

Massachusetts Amendments to NFPA 1, Section 1.12.8.34.2 introduces the requirements as: *Process or Processing of Any Hazardous Material at any Facility* and it states the following:

(1) "A permit shall be required for the process or processing of any hazardous material at any facility identified in this *Code* as Category 2 through Category 5, as described in Chapter 60.
(2) The Authority Having Jurisdiction (AHJ) may require technical assistance in accordance with *1.15* to evaluate the adequacy of a Category 3 or Category 4 facility process safety conditions, programs, procedures, and practices undertaken at the facility, but only after a notice of denial has been properly served upon the person making application."

As part of the MA amendments to NFPA 1 – 2015 edition, Section 60.8, ***Hazardous Material Process or Processing***, was added to 527 CMR. Section 60.8 applies to both new and existing facilities that process hazardous materials, with a list of exceptions such as:

- Motor vehicle service stations
- Refrigeration systems using a refrigerant other than ammonia or LPG,
- Certain wastewater treatment operations
- Storage of hazardous materials in atmospheric vessels.

Definitions:

NFPA 1 defines "**Hazardous Materials**" as "A chemical or substance that is classified as a physical hazard material or a health hazard material, whether the chemical or substance is in usable or waste condition."

OSHA and EPA define "**Process**" as "Any use, storage, manufacturing, handling, or the on-site movement of hazardous chemicals, or combination of these activities." The MA state fire code defines "Process" as "A sequence of operations in which the sequence can be inclusive of physical operations such as heating, cooling, mixing, distilling, compressing, and pressurizing, and chemical operations, such as polymerization, oxidation, reduction, and other chemical reaction processes. The sequence can involve but is not limited to: preparation, separation, combination, purification, or any actions that cause a change in state, energy content, or chemical composition."

MA Amendments to NFPA 1 Section 60.8 categorizes hazardous material processes into five (5) categories, as follows:

Table 1: Hazardous Material Threshold Quantities in the MA Amendments to NFPA 1

Category	**Quantity (gallons)**
Category 1	< 2.5
Category 2	2.5 – 60
Category 3	60 – 300 (or Group H)
Category 4	300 and not Category 5
Category 5	≥ the threshold quantities stated in 29 CFR 1910.119

The MA Amendments also add Section 60.8.5.1 to NFPA 1, which requires that facilities operating hazardous material processes must maintain the required documents and procedures for periodic inspection and review by the head of the Fire Department to remain in compliance with the state fire code. Requirements for each process category are as follows:

Table 2: Outline of Regulatory Requirements in the MA Amendments to NFPA 1

Category	Requirements
Category 1 (Vessel Capacity ≤ 2.5 gal)	• OSHA 1910.1450 (Occupational exposure to hazardous chemicals) • OSHA 1910.1200 (Hazard communication to employees) • Policy per NFPA 1, Ch. 66 (Flammable liquids) and Ch. 67 (Flammable solids) • Requirements of NFPA 1, Section 60.1 – 60.5
Category 2 (Vessel Capacity ≤ 60 gal)	• All requirements of Category 1 • Permit required per NFPA 1, Sections 1.12.8.34.2 and 60.8.1.1
Category 3 (Vessel Capacity ≤ 300 gal OR Group H Occupancy)	• All requirements of Category 1 & 2 • Category 3 Hazard Evaluation • Requirements of NFPA 1, Section 60.6
Category 4 (Vessel Capacity ≥ 300 gals AND not Category 5)	• All requirements of Category 1-3 • Limited Safety Program (PSM "lite")
Category 5 (Vessel Capacity ≥ the threshold quantities stated in 29 CFR 1910.119)	• All requirements of Category 1-4 • Process Safety Management (PSM) Program 29 CFR 1910.119 (OSHA) • Risk Management Program (RMP) 40 CFR Part 68 (EPA)

Category 1 – Requirement Details

Category 1 processes, as stated in Table 2 above, must possess documentation of compliance with OSHA 1910.1450, *Occupational Exposure to Hazardous Chemicals in Laboratories*, OSHA 1910.1200, *Hazard Communication,* maintain and implement a policy in compliance with NFPA 1, Ch. 66, *Flammable and Combustible Liquids*, and Ch. 67, *Flammable Solids*, demonstrate compliance with NFPA 1, Section 60.1.5.1, *Emergency Action Plan*, and Sections 60.5.1.4.3.2(5) through 60.5.1.4.3.2(9) and comply with permitting requirements in Sections 1.12 and 60.8.4 of the MA Amendments to NFPA 1.

OSHA 1910.1450 and 1910.1200

Compliance with OSHA 1910.1450 requires facilities to set and adhere to exposure limits, monitoring plans, hygiene plans, precautions, training, medical plans, and other plans, as found to be acceptable by the AHJ, to react to and minimize potential exposure to hazardous chemicals. OSHA 1910.1200 requires hazard communication methods for employees such as inventory documentation, hazard programs, labels, safety data sheets, and training.

NFPA Chapter 66 and 67

In addition to demonstrating compliance with OSHA standards, facilities meeting Category 1 requirements must implement and maintain a written policy in compliance NFPA 1, Ch. 66 and Ch. 67.Chapter 66 is divided into 29 sections, most of which contain multiple requirements. Section 66.6 contains the prescriptive requirements for fire and explosion prevention and risk control, and Section 66.6.4 goes on to list requirements related to a facility's hazards analysis, which is required to be reviewed to ensure that fire and explosion hazards are addressed by fire prevention, fire control, and emergency action plans. The AHJ should be consulted with for the development of the hazards analysis to ensure that the extent of the fire prevention and control is deemed acceptable or an engineering evaluation of the operation and application of fire protection and process engineering principles is soundly applied. The hazards analysis must include, but not be limited to, analysis of the following:

(1) Fire and explosion hazards
(2) Emergency relief from process vessels
(3) Applicable facility design requirements in Chapters 17, 18, 19, 28, and 29 of NFPA 30
(4) Applicable requirements for liquid handlings, transfer, and use, as covered in Chapters 17, 18, 19, 28, and 29 of NFPA 30
(5) Local conditions/exposures
(6) Emergency response capabilities of the local emergency services

Additional guidance for conducting a fire risk assessment can be found in NFPA 551, *Guide for the Evaluation of Fire Risk Assessments.* Facility management must repeat the hazards analysis when any of the following occur:

(1) Changes occur in the materials in process
(2) Changes occur in process equipment
(3) Changes occur in process control
(4) Changes occur in operating procedures or assignments

This is consistent with the requirements of OSHA 1910.119 Sections (e) and (l). Not all sections of Chapter 66 may be related to processes; however, Section 66.17 is for processing facilities. Section 66.17 includes prescriptive requirements for processing facilities such as location of process vessels and equipment (Section 66.17.4), containment, drainage, and spill control (Section 66.17.10), and management of operations hazards (Section 66.17.15). Later sections of Chapter 66 include additional requirements for dispensing, handling, transfer, and use of liquids (processes) (Section 66.18) and specific operations such as recirculating heat transfer system, vapor recovery and vapor processing system, solvent distillation units, and cooking oil storage tank systems in commercial kitchens (Section 66.19).

NFPA 1 Chapter 60

In addition to Chapters 66 and 67 of NFPA 1, Category 1 processes must also adhere to the requirements of 60.1 – 60.5 of NFPA 1 and the MA Amendments to NFPA 1. Depending on whether the quantities of materials at the facility exceed the MAQ or not, determines which requirements of Section 60.1 are applicable (Section 60.1.3).

Section 60.1.5 of NFPA 1 requires that an emergency action plan, consistent with the available equipment and personnel, shall be established to respond to fire and other emergencies in accordance with requirements set forth in NFPA 1. Section 10.8 of NFPA 1 includes additional requirements and guidance on generating emergency action plans, such as what must be included in the plans, purpose, responsibilities, etc. Section 60.4 of NFPA 1 contains requirements related to permissible storage and use locations and sets prescriptive requirements on the maximum allowable quantities for each individual control areas based on the area's occupancy type.

Section 60.5 contains the fundamental requirements for hazardous materials, including requirements for personnel training, ignition sources, systems, equipment, and processes (Section 60.5.1.4, 60.5.1.5, and 60.5.1.6 respectively). In Massachusetts, Section 60.5 includes multiple requirements that come from both NFPA 1 and the MA Amendments to NFPA 1. Despite Category 1 processes having a vessel capacity much less than Category 2-5 processes (Category 1 processes have a vessel capacity ≤ 2.5 gallons), there are numerous NFPA and state fire code requirements that must be followed. Therefore, the importance of fire protection engineers and their understanding of the NFPA and state fire codes cannot be understated for processing facilities. Each subsequent category process contains additional fire code fire protection/chemical process safety requirements of increasing complexity.

Category 2: Requirement Details

Category 2 processes must comply with all Category 1 requirements and additionally submit for a permit to the AHJ, as required per 1.12.8.34.2 and 60.8.1.1 of MA Amendments to NFPA 1. Permits submitted to the AHJ must comply with the requirements of Section 1.12 of NFPA 1 – 2015 edition. The nature of the operations and materials at the facility determine the permit requirements. Depending on the nature of the process, the AHJ may require technical assistance to evaluate the adequacy of facility process safety conditions, programs, procedures, and practices undertaken at the facility, but only after a notice of denial has been properly served upon the person making application.

Category 3: Requirement Details

While meeting all the requirements of Category 1 and 2 processes, Category 3 processes must also provide the following documentation and evaluations:

- Compliance with the requirements of Category 1 and 2 processes;
- Category 3 hazard evaluation; - per Section 60.8.2
- Implementation of process safety controls to mitigate hazards of normal and abnormal conditions identified in the hazard evaluation;
- Maintain documentation for review for a minimum of two (2) years;
- Third-party review (if required by the fire department/AHJ);
- Post-Incident Analysis per 60.8.6 (to be initiated within 48 hours of incident);

MA Amendment to NFPA 1, Section 60.8.2.2 provides requirements and details for the hazard evaluation that is required for Category 3 processes. **Hazard evaluations** are meant to identify hazards, including adjacent vessels that contain hazardous materials, and determine the required preventative, protective, and safety control measures in conformance with **r**ecognized **a**nd **g**enerally **a**ccepted **g**ood **e**ngineering (and safe work) **p**ractices (RAGAGEP) associated with a particular process or condition. Section 60.8.2.5 requires that a person with *specialized knowledge* beyond that of an average person, about risk assessment, process hazard analysis, and/or process safety management principles, for the process or processes being evaluated *conduct the necessary evaluations*. Section 60.8.2.5 further emphasizes the importance of both chemical process safety specialists and fire protection engineers for analysis of hazardous processing facilities.

In addition to generating a hazard evaluation, Category 3 processing facilities (vessel capacity ≤ 300 gal OR Group H Occupancy) must also meet the requirements of NFPA 1, Section 60.6, *Emergency Action Planning, Fire Risk Control and Chemical Hazard Requirements for Industrial Processes.* Where quantities of materials in use require compliance with Protection Level 1, Protection Level 2, Protection Level 3, or Protection Level 4 based on materials exceeding the Maximum Allowable Quantity (MAQ) in the following categories shall comply with the requirements of Chapter 7 of NFPA 400, *Hazardous Materials Code*:

(1) Unpackaged organic peroxide formulations that are capable of explosive decomposition in their unpackaged state
(2) Oxidizer Class 3 and Class 4: solids and liquids
(3) Pyrophoric solids, liquids, and gases
(4) Unstable reactive Class 3 and Class 4: solids, liquids, and gases
(5) Highly toxic solids, liquids, and gases
(6) Water-reactive liquids, Class 3

NFPA 400, Chapter 7, *Emergency Planning, Fire Risk Control, and Chemical Hazard Requirements for Industrial Processes,* contains requirements and guidance on topics such as:

- Process review and plan preparation
- Operating and maintenance procedures
- Safety reviews
- Incident investigation plans

Section 7.2.3 contains specific requirements for Emergency Action Planning that facilities must comply with. Some requirements of Section 7.2.3 include:

- Procedures for notifying and coordinating with the fire department and other applicable agencies
- Procedures and schedules for conducting drills
- Alternative measures for occupant safety
- Maintenance of fire protection equipment

Lastly, in lieu of complying with Chapter 60 in its entirety, occupancies containing high hazard Level 1 to high hazard Level 5 contents shall be permitted to comply with Chapter 10 of NFPA 400, as stated in NFPA 1, Section 60.7. MA Amendments to NFPA 1 further states that complying with NFPA 1, Section 60.7, is subject to an independent review in accordance with Section 1.15 and a copy, including its recommendations, shall be submitted to the Building Official.

Category 4: Requirement Details

Category 4 processing facilities (vessel capacity ≥ 300 gal) must meet:

All the requirements of Category 1, 2, and 3 processes

And also create and maintain a **Limited Safety Program (PSM "lite")**. The Limited Safety Program documents evaluation, policy and/or required procedures to ensure compliance with the following requirements of the MA Amendment to NFPA 1:

(1) Process information including, but not limited to, MSDS for the chemicals and products being processed, process chemistry, piping and instrumentation diagram, safety relief design, process control safety alarms and interlocks;

(2) Facility suitability including, but not limited to, the *Building Code* compliance, electrical hazard classification, ventilation design, fire alarm and fire protection, spill containment and control;

(3) A process hazard safety analysis including, but not limited to, effects in the event of failure, suitable administrative and engineering controls to minimize failure and to control unanticipated releases, and emergency responses to safeguard life and property;

(4) Written procedures, including routine operating and maintenance, as well as precautionary, shut-down and emergency response measures;

(5) A written training program for operating and maintenance personnel and outside contractors whose work or activity may affect process safety;

(6) A written records management protocol, which tracks any changes, including but not limited to changes to chemicals, equipment, operating procedures training program. Such records shall include the date of such change and the name of the manager responsible for such change; and an internal review at a maximum every three years.

These requirements are similar to those found in OSHA 29 CFR 1910.119, *Process Safety Management of Highly Hazardous Chemicals*.

Category 5: Requirement Details

Category 5 processing facilities must provide evaluation and documentation per 29 CFR 1910.119 OSHA, *Process Safety Management of Highly Hazardous Chemicals*, and 40 CFR Part 68 EPA, *Risk Management Program Rule*, all while meeting the requirements for Categories 1-4. OSHA's *Process Safety Management* (PSM) *of Highly Hazardous Chemicals* was issued in February 1992 and contains 14 elements that focus mainly on worker safety and catastrophic loss of containment. The Process Hazard Analysis (PHA) is the "heart" of the standard and is the most time consuming. The PHA must be completed every five (5) years and *is required to be completed by someone knowledgeable with the PHA methodology*. Additionally, compliance audits for the standard must be completed every three (3) years and may be completed in-house or by a third party (preferred). EPA's *Risk Management Program* (RMP) *Rule* consists of 12 elements and is focused on public safety and the environment. The program contains Chemical Accident Prevention Provisions (CAPP) and Accidental Release Prevention/Risk Management Plan Rule. The RMP must be submitted to the EPA every fire (5) years or if there are any changes.

Conclusion

Some fire code requirements for chemical process safety apply to processes with quantities much lower than the PSM/RPM thresholds. Failure to comply with these requirements can result in fire department citations, fines, remedial work and in extreme cases, shutdown of operations. Between 2013 and 2018, 214 OSHA PSM violations were recorded in Massachusetts alone, which resulted in over $485,000 in penalties. There are far more non-PSM facilities than PSM facilities in MA, so the magnitude of the problem can be seen by simple extrapolation. Failing to comply with the fire code and the federal requirements can also result in significant property loss, injuries and fatalities with much higher costs than the violation penalties. Compliance with the local fire code and federal regulations should be determined by experienced chemical process safety specialists and fire protection engineers to minimize the risk of significant loss.

Fire protection engineering and science is its own discipline and chemical process safety is its own discipline usually practiced by the chemical engineering field. Finding ways to create awareness and bridge the gap between the two disciplines, and sharing chemical process safety management concepts with those in the fire protection arena, will likely improve the oversight, understanding and enforcement of process safety resulting in better overall safety. There may be a way of joining these disciplines into one overarching safety category and therefore giving a better sense of how interrelated they really are.

546769 Poly-Butadiene Formation in Olefins Debutanizer Condenser

Abstract Title : Poly-butadiene formation in Olefins Debutanizer condenser

Discipline : Process Safety

Topic : Poly-butadiene formation in Debutanizer condenser

Author Name : Ali Bazroun

Participating team: Fahad AlShammari, and Eldridge Monette, Bertrand, Da, Cruz, Fahad AlSaad, Abdullah Al-Ismail

ABSTRACT

In one of Sabic olefins plant Plant, the condenser of the debutanizer tower was fouling. This was evident through the pressure increase in the tower and the cooling water control valve reaching the maximum opening (69%) limited by the mechanical stopper. The condenser process service is butadiene, which has the tendency of polymerizing and formation of popcorn, which could plug relief piping or accumulate in the condenser until the vessel shell is over stressed causing loss of containment and leading to fire or explosion.

The risk assessment team evaluated that the risk as significant, meaning that it could result into fatality and multiple lost time injury with a likelihood of 10E-3. With the propagating shell deformation, the likelihood could increase to L3 (10E-2) over a longer period. The risk could become C2-L3 and remain Significant. To ensure effective risk mitigation, a major plant modification was executed by constructing a new process line that bypasses the debutanizer tower and sends the material to the flare. The project execution was remarkable as it was completed in about a week. 15 tons of C4 is being flared per. This, however, allowed the olefins unit to mitigate the risk effectively without any interruption of the production. The change was not done before in the local industry, and it reflected how effective measures can be taken to mitigate risk through strong leadership and excellent engineering. Interim mitigations, moreover, were in place until the modification was completed: for example, inhibitor injection in the exchanger to inhibit the popcorn formation as soon as possible, and an emergency procedure in case loss of containment. Without the effective mitigation, Sabic operation team would have had to shutdown the olefins unit for over 2 months. This olefins unit feeds two poly ethylene that would shutdown consequently, causing significant losses over the mentioned period.

After isolating the exchanger, borescope test proved the formation of polymer. A root cause analysis was initiated, along with cleaning the tower and fabricating a new exchanger. The details of the RCA will be shared in the full paper.

19 AIChE Spring Meeting
15TH GLOBAL CONGRESS ON PROCESS SAFETY
NEW ORLEANS • 2019

Industrial process plant safety by collaboration of a Safety Instrumented System and a Dynamic Simulator

Tadaaki Ando
Yokogawa Corporation of America
12530 West Airport Blvd, Sugar Land, TX 77478, USA
Tadaaki.Ando@us.yokogawa.com
(Tadaaki.Andou@jp.yokogawa.com)

Yasuhiko Yamashiro
Yokogawa Electric Corporation
2-9-32 Nakacho, Musashino-shi, Tokyo, 180-8750 Japan
Yasuhiko.Yamashiro@jp.yokogawa.com

Prepared for Presentation at
American Institute of Chemical Engineers
2019 Spring Meeting and 15th Global Congress on Process Safety
New Orleans, LA
March 31 – April 3, 2019

Industrial process plant safety by collaboration of a Safety Instrumented System and a Dynamic Simulator

Tadaaki Ando
Yokogawa Corporation of America
12530 West Airport Blvd, Sugar Land, TX 77478, USA
Tadaaki.Ando@us.yokogawa.com
(Tadaaki.Andou@jp.yokogawa.com)

Yasuhiko Yamashiro
Yokogawa Electric Corporation

Keywords: Industrial process plant, SIS (Safety Instrumented System), Dynamic Simulator

Abstract

Safer HMI design for operators' quick and appropriate response is very important. Allowed maximum delay for operator's action from the alarm is not same all the cases because it is depends on many factors, i.e. ambient temperature, density of material, O2, and etc. So design of SIS includes HMI should be based "worst case". However, the 'worst case' is difficult to be found.

A "Dynamic Plant Simulator" can precisely simulate behaviors of the real process by visualizing how a hazard is diffused at various stages of a chemical reaction. And combination of DCS/SIS/HMI simulator with "Dynamic Plant Simulator", safety logic design is optimized about trip point and alarm threshold, and HMI design is modified safer manner.

The effective result of application of "cyber model by Dynamic Process Simulator" to a vinyl acetate monomer (VAM) plant and its value for operator training will be discussed.

1 Introduction

Industrial process plants in Japan have been seeking for the highest productivity and optimization of their processes, same with any process plant all over the world. These plants were renovated and expanded as time passed for the purpose, the plants themselves became complicated and troublesome as a result. Few years ago, severe process plant incidents occurred in Japan. The most probable causes of these severe incidents are out of "incorrect judgement and operation" and "shortage of prediction for technical aspects" during exceptional manual work that has been done by incidents. Understanding the hazardous situation by risk analysis and designing safety measures utilizing a SIS (Safety Instrumented System) are important and effective for resolving "incorrect judgement and operation" and "shortage of prediction of technical aspect". Safety measure includes human intervention via SIS man-machine interface which showing alarm, event, and process data.

The safety measure and its function are developed on the desk by brainstorming or so-called "hazard and operability study (HAZOP)." The verification of the safety function hardware could be established by reliability based analysis, SIL calculation in other word. However, the verification of safety logic is the issue because it has only been verified "on the desk". Especially the determination of the criteria for the emergency shutdown is the most important, but it is not possible to be confirmed with "try and error" on a running plant.

2 Safe HMI Design

Safer HMI design with considering different conditions and situations in the real process for operators' quick and appropriate response is very important. To design it, knowing the behavior of the process itself and related factors which affect the process is necessary.

For example, a reaction speed varies on the ambient temperature, density of material, O_2, catalyst, and etc., so allowed maximum delay for operator's action from the alarm is not same all the cases. So design of SIS includes HMI should consider "worst case". However, finding the worst-case scenario with how much is the allowed operator reaction is difficult because of many factors.

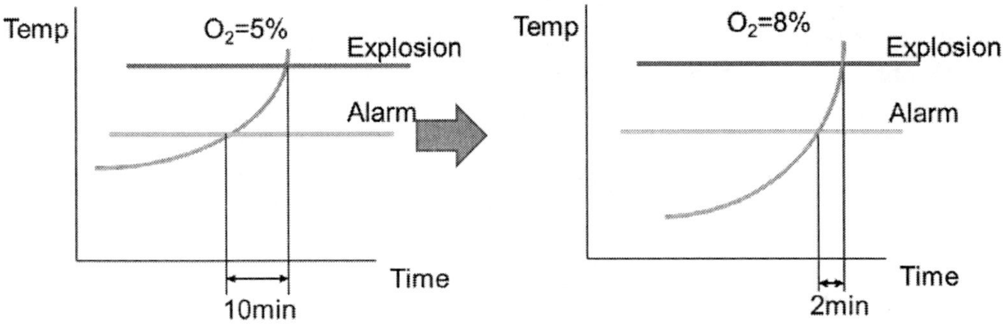

Fig.1 Example: Difference of O_2 density vs. allowed reaction time

Another example is the 'thermal runaway'. Even if feeding of material, heat, and catalyst were stopped by ESD because of a high temperature of a reactor, reaction continues due to a feed-forward loop of heating, means reaction generates heat to continue the reaction itself.

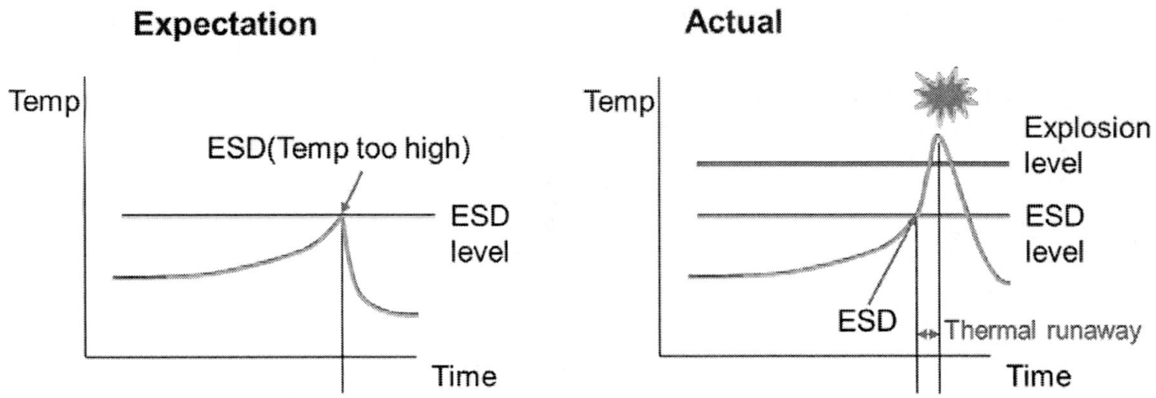

Fig.2 Even after ESD, thermal runaway leads the process to explosion

Thus, considered 'worst case' could not be a 'worst', or could be 'too much pessimistic'.

A "Dynamic Plant Simulator" can precisely simulate behaviors of the real process by visualizing how a hazard is diffused at various stages of a chemical reaction. And combination of DCS/SIS/HMI simulator with "Dynamic Plant Simulator", safety logic design is optimized about trip point and alarm threshold, and HMI design is modified safer manner.

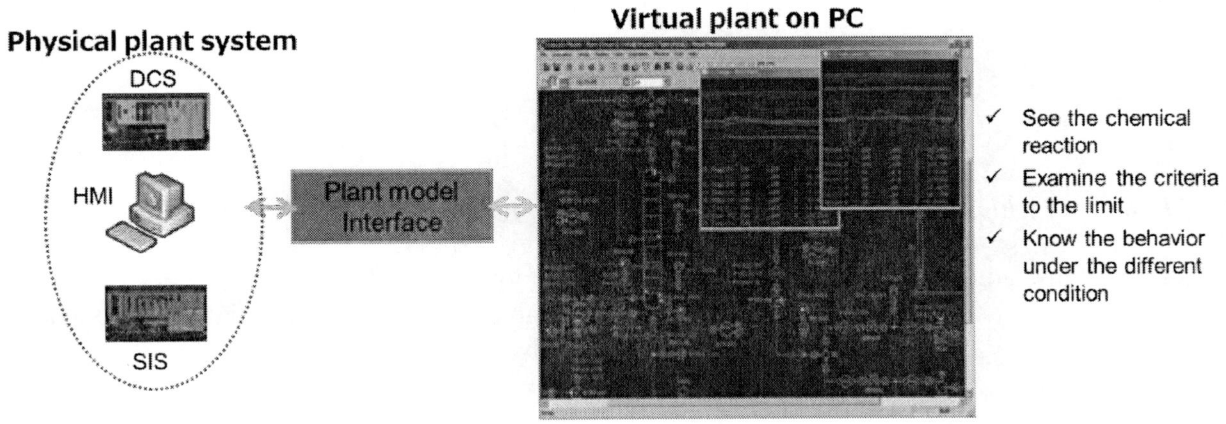

Fig.3 Dynamic Plant Simulator

3 "Dynamic Plant Simulator" at VAM (Vinyl Acetate Monomer) process

"Dynamic Plant Simulator" with "DCS/SIS/HMI simulator" was applied for several shutdown conditions of VAM plant. "Dynamic Plant Simulator" can simulate a variation of factors like ambient temperature and thermal runaway, because is based on a theory of chemical reaction. And "DCS/SIS/HMI simulator" is copying a real function and operation of DCS/SIS/HMI.

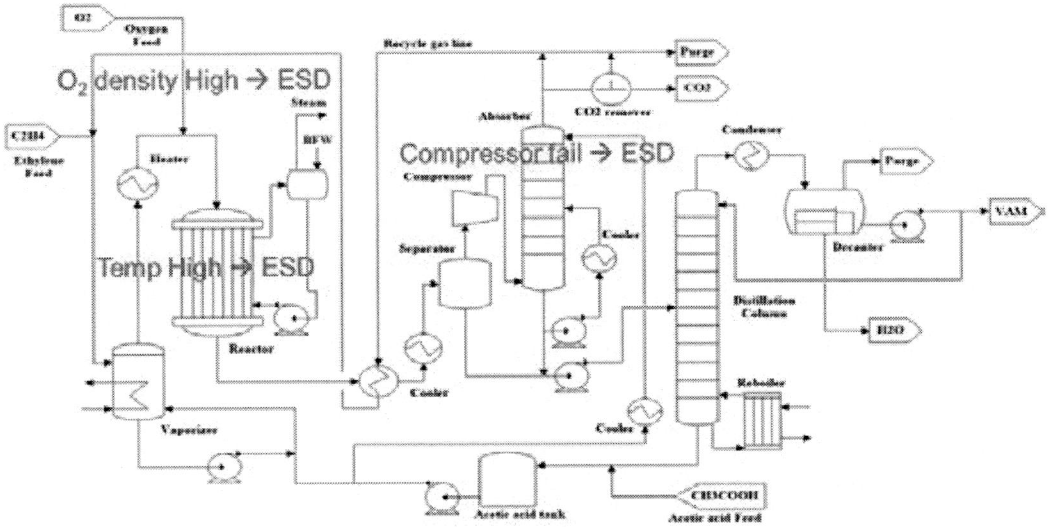

(Main Reaction) $C_2H_4 + CH_3COOH + 1/2O_2 \rightarrow CH_2{=}CHOCOCH_3 + H_2O$

(Side Reaction) $C_2H_4 + 3O_2 \rightarrow 2CO_2 + 2H_2O$

Fig.4 VAM process model

3.1 Shutdown by high temperature to avoid a damage to reactor

Reactor temperature must be monitored and controlled lower than 200 Degrees C (392 Degrees Fahrenheit) to avoid a damage of the reactor. If the temperature reached to a set point for emergency shutdown, emergency shutdown action stops feeding of Ethylene and O_2.

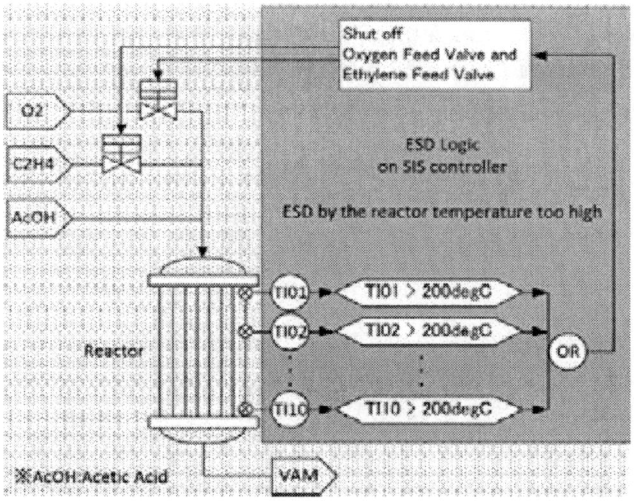

Fig.5 ESD logic for high temperature shutdown

When 190 Degrees C (374 Degrees Fahrenheit) is selected as the set point to shutdown, the result of simulation is, the reactor temperature must become over 200 Degrees C due to a runaway reaction in the reactor. It means reactor must be damaged. When 180 Degrees C (356 Degrees Fahrenheit) is selected, the simulation shows the reactor temperature can be kept lower than 200 Degrees C, so no damage to the reactor.

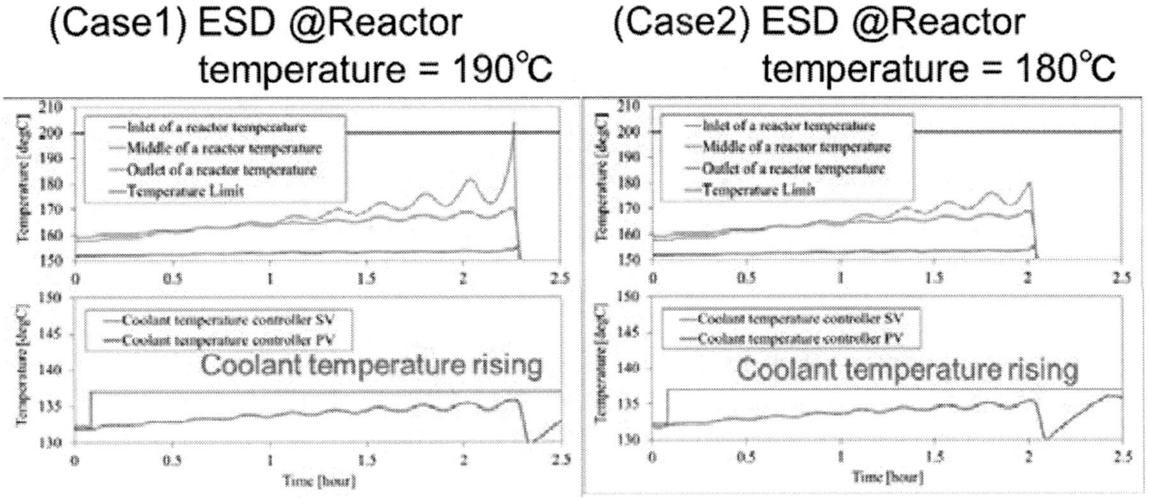

Fig.6 Simulation results

And simulation suggested the reactor temperature must be monitored in multipoint, middle of the reactor and top of the reactor. When the reaction temperature is lower than 160 Degrees C (320 Degrees Fahrenheit), temperature of a middle of the reactor is higher than other points, but when the reaction temperature is higher than 160 Degrees C, the temperature of top of the reactor is higher than other points.

3.2 *Shutdown by high O_2 density to avoid explosion*

If the density of O_2 in the reactor exceeds 8%, there is a risk of explosion. However, the O_2 density varies depends on other factors, i.e. feed pressure of ethylene to the reactor. Simulation has been done about O_2 density when the feed pressure of ethylene is suddenly changed. Simulation of some cases of dumping the feed pressure to lower side which increase the density of O_2 in the reactor were done.

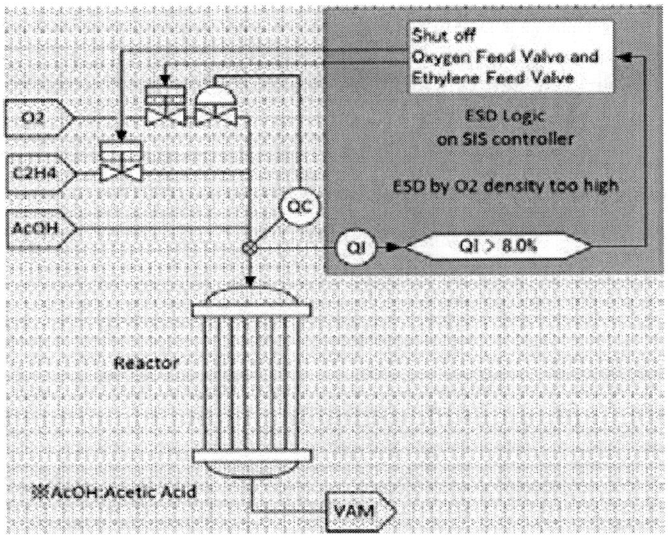

Fig.7 ESD logic for high O_2 density shutdown

When the feed pressure of ethylene dumped 10%, density of O_2 in the reactor doesn't reach to 8%. So there is no risk of explosion. However, the dumping of ethylene feed pressure is 20%, O_2 density will be higher than 8%, then an emergency shutdown must be activated. Shutdown does affect to business.

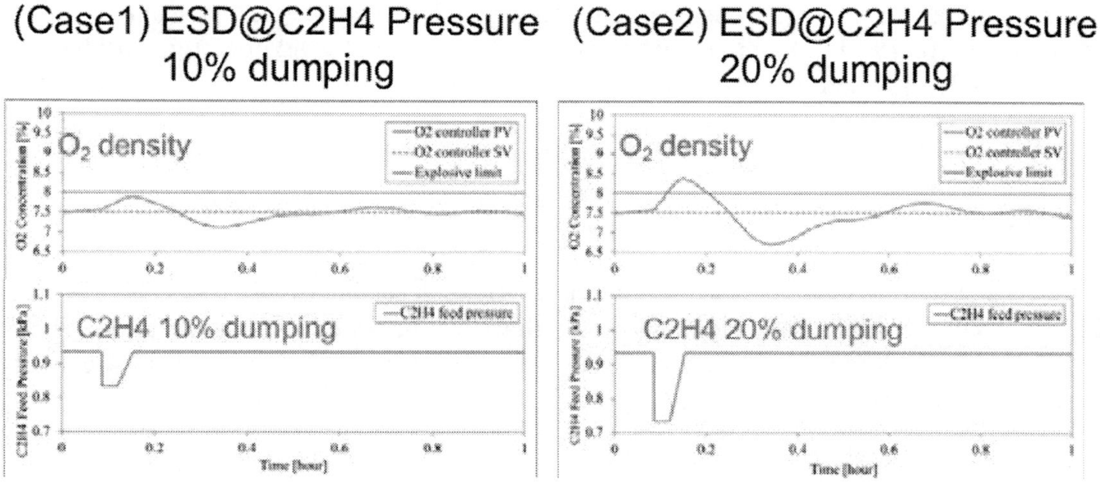

Fig.8 Simulation results

If there is a possibility of dumping ethylene feeding pressure, feeding O_2 in normal operation should be lower to avoid the risk of reactor explosion. On the other hand, lower O_2 density leads lower reaction speed, less production. Finding a best balance with safety and productivity is the key.

Using dynamic process simulator, many factors of process can be manipulated not only within normal operation range, but also close to abnormal, emergency situation or more un-realistic range, and results can be confirmed. This type of "try-and-error" approach is not possible on the real process, but possible in the "cyber" world to find a best balance of setting for maximize production with keeping or improving safety.

4 "Dynamic Plant Simulator" for Operator training

The training of operators is important in many reasons. Especially training for response to critical / emergency situation is indispensable, because operator rarely or never face such a situation and consequent secondary action to DCS/SIS or response to alarm flooding in a real plant. Therefore operator are impossible to familiarize with their required response in right procedure to critical / emergency situation. If operators are not capable or not knowing how to response such a situation, risk must propagate real hazard or accident. Furthermore, operators need to re-start the process after confirming the cause of emergency doesn't exist anymore and checking all the conditions of materials are steady, status of equipment are ready to re-start. This procedures are also not normal for operators.

"Dynamic Plant Simulator" realizes operator training with "equal to real" environment. Operators will be trained for correct and timely manner response until proficient. And operators will have more understanding for process itself by playing on the simulator.

5 Conclusion

The effectiveness of "Dynamic Process Simulator" for safer design of DCS/SIS/HMI is confirmed on VAM plant. The best balance of safety and productivity can be found and optimization of safety action is created by simulation. The created new design of DCS/SIS/HMI is possible to use an actual system directly. And operator training with "Dynamic Process Simulator" will increase a total safety of the plant.

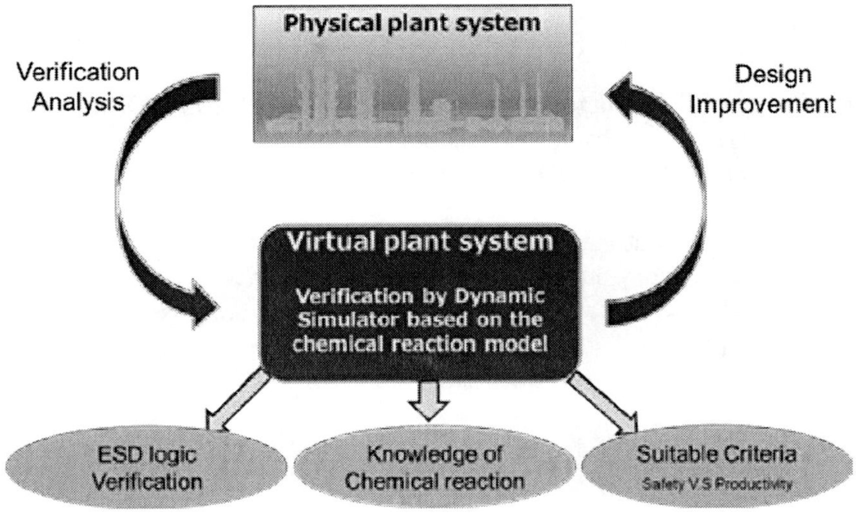

Fig.9:Virtual plant system realizes well balanced plant operation

Analysis of Chemical Effluent Discharge through the Emergency Relief System using Air-Water Properties

Basar Ozar
Jensen Hughes
600 W Fulton St. Suite 500
Chicago, IL 60661
bozar@jensenhughes.com

Jens Conzen
Jensen Hughes
600 W Fulton St. Suite 500
Chicago, IL 60661
jconzen@jensenhughes.com

Damian D. Stefanczyk
Jensen Hughes
3610 Commerce Dr. Suite 817
Baltimore, MD 21227
dstefanczyk@jensenhughes.com

Prepared for Presentation at
American Institute of Chemical Engineers
2019 Spring Meeting and 15th Global Congress on Process Safety
New Orleans, LA
March 31 – April 3, 2019

Analysis of Chemical Effluent Discharge through the Emergency Relief System using Air-Water Properties

Basar Ozar
Jensen Hughes
600 W Fulton St. Suite 500
Chicago, IL 60661
bozar@jensenhughes.com

Jens Conzen
Jensen Hughes
600 W Fulton St. Suite 500
Chicago, IL 60661
jconzen@jensenhughes.com

Damian D. Stefanczyk
Jensen Hughes
3610 Commerce Dr. Suite 817
Baltimore, MD 21227
dstefanczyk@jensenhughes.com

Keywords: two-phase flow, hydrodynamic load, emergency relief system, chemical effluent

Abstract

A sudden opening of a pressure relief device (typically a pressure relief valve or a rupture disk) during an overpressure event will cause the chemical effluent to accelerate through the emergency relief system (ERS) piping. The rapid change of momentum of the effluent creates transient hydrodynamic loads on the piping. ASME B31 in the United States and European Directive 2014/68/EU requires these transient loads to be considered along with other applicable loads (i.e., dead weight, thermal) in the system design. Additionally, the pressure distribution along the piping may be of importance, particularly when other vessels are connected to the same ERS. The back pressures generated on these vessels should be evaluated and compared with the vessels' MAWP (maximum allowable working pressure) as well as the set pressure of any isolation devices. As such, a detailed time history of the effluent pressure and hydrodynamic loads are required to perform a proper evaluation. The effluent may be composed of proprietary chemical mixtures containing various materials. Therefore, the exact material properties of the effluent during the transient might not always be known. In this study, a previously published methodology is used for calculating the dynamics of the effluent and the resultant hydrodynamic loads for a typical ERS by using the RELAP5 (Reactor Excursion and Leak Analysis Program) software. Air-water properties are assumed for the effluent. Key output parameters, such as pressure drop and hydrodynamic loads, are then compared with established hand calculation methods using the actual effluent properties. Given the mechanical nature of the hydrodynamic load event, it was found that the use of air-water properties provides a reasonable representation of the important characteristics of the actual effluent discharge in mechanical design space.

1 Introduction

Pressure build up in a system may occur as a result of a process upset, instrument failure or fire. These upset scenarios can result in either, or both, vaporization due to exothermic runaway reactions or generation of gases due to the decomposition process, which are of particular interest in terms of system and public safety. The extent of the potential hazards caused by such incidents are generally determined by the rate and amount of the energy being generated, the chemical substance makeup, and the quantity of the effluent.

The vessels or systems, which contain such chemicals (e.g. chemical reactors or storage tanks), are typically protected by an ERS that prevents from over-pressurization of the equipment as a consequence of the vaporization of chemicals and/or generation of gases. The process for designing of an ERS was previously discussed by Conzen and Wood (Ref. [1]) and Ozar (Ref. [2]). The design process consists of a process hazard analysis (PHA), defining credible and limiting upset scenarios, adiabatic calorimetry testing, vent sizing calculations, fluid transient analysis, pipe stress analysis, and a dispersion analysis. The central task in the design process is the fluid transient analysis (Refs. [1] and [2]), which is the main focus of this study.

The ERS is typically equipped with either a rupture disk or a pressure relief valve (PRV). Some systems may have both pieces of equipment. The sudden opening of the relief device during an overpressure event will typically result in a two-phase discharge and cause the chemical effluent to accelerate through the ERS piping. The rapid change of momentum of the effluent creates transient hydrodynamic loads on the piping. ASME B31 (Ref. [3]) in the United States and European Directive 2014/68/EU (Ref. [4]) requires these transient loads to be considered along with other applicable loads (i.e., dead weight, thermal) in the system design. Additionally, the pressure distribution along the piping may be of importance, particularly when other vessels are connected to the same ERS. The back pressures generated on these vessels should be evaluated and compared with the vessels' MAWP as well as the set pressure of any isolation devices. As such, a detailed time history of the effluent pressure and hydrodynamic loads are required to perform a proper evaluation.

The two-phase behavior of the chemical effluent has been studied extensively. Following earlier developments by Epstein et al. (Ref. [5]) and Grolmes and Leung (Ref. [6]), a simplified correlation was suggested by Leung (Ref. [7]) to predict flashing choked flow in a frictionless duct or nozzle. The correlating parameter, ω, was based on an isenthalpic (constant enthalpy) assumption.

Leung and Grolmes (Ref. [8]) integrated the momentum equation for homogenous two-phase flow in a horizontal duct using "ω" as a closure relationship for two-phase mixture specific density. Leung and Epstein (Ref. [9]) extended the derivation to an inclined duct configuration. Leung (Ref. [10]) provided the derivation of the reaction forces during a two-phase discharge from ERS piping.

Computational methods have also been used by various industries to investigate two-phase flow behavior in pipes and the hydrodynamic loads due to a sudden acceleration or deceleration of the fluid and the propagation of pressure waves. RELAP5 is a tool for analyzing small-break loss of coolant accidents and system transients in nuclear power plants. The program has the capability to model one dimensional single and two-phase fluid flow and heat transfer problems. The software has undergone extensive review and validation by the US Nuclear Regulatory Commission (Ref. [11]). It has also been benchmarked (Refs. [12], [13] and [14]) for predicting pressure wave propagation and hydrodynamic loads in piping systems. Refs. [15] and [16] provide a discussion about some of the nuclear power plant applications. However, RELAP5 is limited to non-condensable gas-water and steam-water two-phase flow applications.

This study provides a discussion about using the analytical methods developed by Leung and Grolmes (Ref. [8]) and Leung (Ref. [10]) to demonstrate the similarity between the two-phase mixture of common industrial chemical effluents and air/steam-water two-phase flow. Computational and analytical calculations for three different discharge scenarios are also provided. The comparisons between the computational and analytical results are used to demonstrate the applicability of using air/steam-water properties to simulate chemical effluent discharge to reasonably predict the hydrodynamic loads and pressure distribution in the ERS following the actuation of a pressure relief device.

2 Theoretical Background

The governing steady-state conservation equations for one-dimensional homogenous two-phase flow in a horizontal duct with a constant flow area are (Ref. [8]):

Mass

$$G = \text{constant} \tag{1}$$

Momentum

$$vdP + G^2\left(vdv + \frac{4fv^2dL}{2D}\right) = 0 \tag{2}$$

Energy

$$h_0 = h + \frac{G^2v^2}{2} \tag{3}$$

G, v, P, f, L, D and h are the mass flux, specific volume, pressure, two-phase friction factor, length of the duct, hydraulic diameter of the duct, and specific enthalpy, respectively. The specific enthalpy and specific volume are expressed in terms of quality (x) and phasic properties, i.e., $h_0 = h_{f0} + x_o h_{fg0}, h = h_f + xh_{fg}$, and $v = v_f + xv_{fg}$. Subscript 0 denotes stagnation conditions and fg the difference between vapor and liquid properties.

Epstein et al. (Ref. [5]) and Grolmes and Leung (Ref. [6]) suggested a simple relation between the two-phase specific volume and pressure as given by:

$$\frac{v}{v_0} - 1 = \omega\left(\frac{P_0}{P} - 1\right), \tag{4}$$

where ω is a non-dimensional parameter. Leung (Ref. [7]) correlated ω for flashing choked flow in a frictionless duct or nozzle. The correlating parameter was shown to be in the following simplified form:

$$\omega = \frac{x_0 v_{fg0}}{v_0} + \frac{C_{f0} T_0 P_0}{v_0}\left(\frac{v_{fg0}}{h_{fg0}}\right)^2 \tag{5}$$

C_{f0} is the heat capacity of the liquid. The last term on the right-hand side is zero for non-flashing gas-liquid two-phase flow and the non-dimensional parameter, ω, is provided by:

$$\omega = \frac{x_0 v_{fg0}}{v_0} \tag{6}$$

Leung and Grolmes (Ref. [8]) considered a two-phase horizontal discharge from a large reservoir (Figure 1). They integrated Equation 2 over the length, L, by substituting Equations 4 and 5 and using the isenthalpic (adiabatic and reversible) flow approximation. The momentum equations after integration yields to:

$$N \equiv 4f\frac{L}{D} = \frac{2}{G^{*2}}\left[\frac{\eta_1 - \eta_2}{1-\omega} + \frac{\omega}{(1-\omega)^2} ln\left[\frac{(1-\omega)\eta_2+\omega}{(1-\omega)\eta_1+\omega}\right]\right] - 2ln\left[\frac{(1-\omega)\eta_2+\omega}{(1-\omega)\eta_1+\omega}\left(\frac{\eta_1}{\eta_2}\right)\right] \tag{7}$$

The non-dimensional parameters are defined by:

$$\eta_1 \equiv \frac{P_1}{P_0}, \eta_2 \equiv \frac{P_2}{P_0}, \eta_3 \equiv \frac{P_3}{P_0}, \; G^* \equiv G/\sqrt{\frac{P_0}{v_0}}, N \equiv 4f\frac{L}{D} \tag{8}$$

η, $G^*$and f are the dimensionless pressure, dimensionless mass flux and two-phase friction factor, respectively. Leung and Grolmes (Ref. [8]) recommended a range between 0.003 and 0.005 for the friction factor.

G^* and η_1 are related via Bernoulli's equation for compressible fluid (Ref. [8]):

$$G^* = \frac{\sqrt{-2[\omega ln\eta_1 + (1-\eta_1)\omega - (1-\eta_1)]}}{\omega\left(\frac{1-\eta_1}{\eta_1}\right)+1} \tag{9}$$

The exit condition can be either subcritical, in which case $\eta_2 = \eta_3$, or critical where the local choking statement would require (Leung and Grolmes, Ref. [8]):

$$G^* = \frac{\eta_2}{\sqrt{\omega}} \tag{10}$$

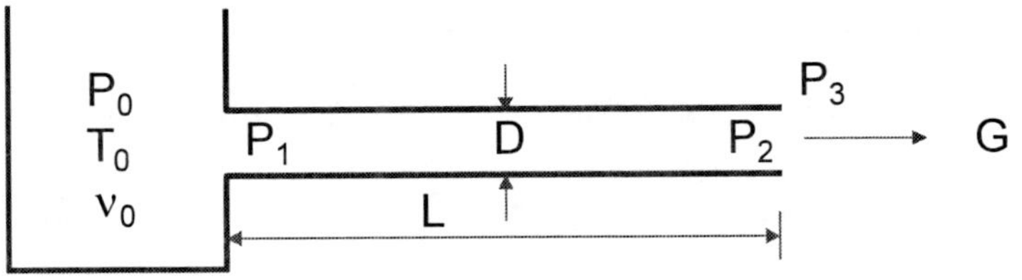

Figure 1. Two-phase horizontal discharge from a large reservoir.

With $\eta_1 = \eta_2 = \eta$, the above two equations satisfy the following transcendental equation (Leung and Grolmes, Ref. [8]):

$$\eta^2 + (\omega^2 - 2\omega)(1-\eta)^2 + 2\omega^2 ln\eta + 2\omega^2(1-\eta) = 0 \tag{11}$$

Similarly, the generalized thrust expression (T_R) is provided by Leung (Ref. [10]):

$$\frac{T_R}{P_0 A_e} = \frac{G^2}{P_0 \rho_0}\left(\frac{v_2}{v_0}\right) + \left(\frac{P_2}{P_0} - \frac{P_3}{P_0}\right) \tag{12}$$

The first term represents the fluid momentum expulsion rate and the second term represents the exit plane pressure (imbalance) force. For the case of unchoked exit, the exit pressure imbalance term vanishes, and the trust is only dependent on the exit momentum rate (Ref. [10]):

$$\frac{T_R}{P_0 A_e} = \frac{2[-\omega ln\eta_2 - (\omega-1)(1-\eta_2)]}{\omega\left(\frac{1}{\eta_2}-1\right)+1} \tag{13}$$

This approximate analysis shows that for two-phase flow in ducts (e.g. piping systems) ω is the sole independent variable and hence is the scaling parameter.

3 Transient Fluid Analysis using RELAP5

3.1 Description of the benchmark scenarios

A chemical reactor with a volume of 2.7 m^3 is considered for the benchmark scenarios. The reactor vessel is rated for 3 bars-g. The reactor is equipped with an emergency relief nozzle and a rupture disk of 41.1 mm inner diameter (DN40, 3.6 mm thickness). The nozzle is located near the top of the vessel. The piping downstream of the rupture disk has an inner diameter of 80.9 mm (4 mm thickness) and is 20 m long. It is a straight and horizontal section of piping.

Three different relief scenarios are used for the benchmarks. A summary of the postulated scenarios is provided below:

- Scenario 1: Polyurethane (PUR) and toluene diisocyanate mixture at 60 °C.
- Scenario 2: N-methyl pyrrolidone, formic acid, trimellitic anhydride and methylene diphenyl diisocyanate (MDI) at 90 °C.
- Scenario 3: Acetone under-fire scenario with vessel initially filled to 20% of its volume.

The chemical mixtures for Scenarios 1 and 2 are gas (CO_2, carbon dioxide) generating reactions. Alternatively, the scenarios are also called gassy reactions. A foamy mixture is expected inside the reactor vessel for these scenarios. Therefore, the chemical effluent discharging through the ERS is a two-phase gas-liquid mixture.

Scenario 3 is the boiling of a solvent due to a fire load applied on the vessel. The level of the boiling mixture is low, and liquid acetone is not entrained into the ERS. Only acetone vapor discharges through the ERS for this scenario.

3.2 *Description of the computer model*

The calculation was performed using the RELAP5/MOD3.3 (Patch 05) computer code, which is a light water reactor transient analysis code developed for the United States Nuclear Regulatory Commission (US NRC). It is suitable for the simulation of a wide variety of hydraulic and thermal transients in both nuclear and nonnuclear systems involving mixtures of steam, water, non-condensable gases and solutes under single phase and two-phase conditions (Ref. [17]).

A RELAP5 model was developed for the simulation of the venting scenarios from the reactor vessel into the ambient. The model was used to determine the hydrodynamic loads resulting from the sudden opening of the rupture disk. The mass flow rate of the two-phase effluent and the pressure distribution along the ERS piping were also calculated. Figure 2 provides an overview of the nodalization diagram of the model.

Time dependent volumes (TDVs) were used for the pressure boundary conditions and the thermo-physical condition of the fluid at those boundaries. Pipe components (P) were used to model the piping. Single junctions (SJs) connected the components such as pipes with pipes or pipes with TDVs. The valve junctions (V) was similar to the SJs but with the additional ability of the junction to open and close.

The length-to-diameter ratios (L/D) were set close to 1.0 for each pipe node and the time step was selected to be below the acoustic Courant limit. This is recommended for simulations involving propagation of acoustic waves and more generally for models where rapid fluid momentum changes are anticipated. The rupture disc was assumed to open linearly over a period of 10 msec, which is consistent with typical data (Ref. [18]). A steady-state calculation was performed for the first 5 seconds of each simulation to initialize the model. The transient portion of the simulation was initiated at 5 seconds, when each rupture disc started to open and was fully open at 5.01 sec. No heat transfer through the pipe walls was considered for the hydrodynamic fluid analysis. This is often a reasonable assumption for industrial processes due to the short duration of the relief event.

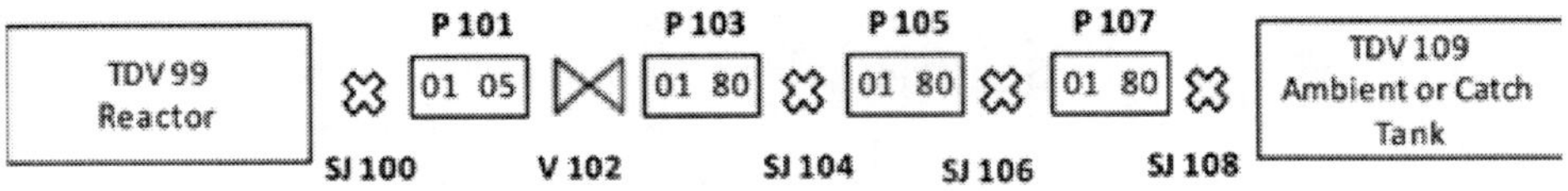

Figure 2. Nodalization diagram for the RELAP5 model.

The hydrodynamic loads (also commonly referred to as fluid force) on the modeled piping were computed through command files by using the RELAP5 output data. The generalized force equation in one-dimensional form can be resolved by a pipe segment bounded by two elbows as shown in Equation (14) (Ref. [19]). Note that the sign convention for the force is considered positive in the nominal direction of the flow as defined within the physical system.

$$F_W = -\frac{\mathrm{d}}{\mathrm{dt}} \int \rho A V dx \tag{14}$$

F_W is the hydrodynamic load, ρ is the fluid density, A if the inner flow area of the pipe, V is the fluid velocity, and x is the distance along the pipe. For pipe segments with an open end an additional force component needs to be considered, which is regarded as the thrust load. Examples of open ended pipe segments are pipes that connect to pressure vessels, large manifolds, or a catch tank. The total thrust load is calculated as shown in Equation 15, which is made up of the momentum flux (due to fluid momentum expulsion) and pressure terms:

$$F_T = -(P_E - P_A)A - \dot{m}_E V_E, \tag{15}$$

where P_E, P_A, $\dot{m}_E$ and V_E are the pressure at the pipe exit, pressure of the reservoir the open piping is connected to, and the mass flow rate and fluid velocity through the pipe exit, respectively.

3.3 Boundary conditions for the RELAP5 model

A summary of the upstream boundary (reactor vessel) conditions for the RELAP5 models of the three scenarios are provided in Table 1. 10% over-pressure (10% greater than the set pressure) was used for Scenarios 1 and 2, which is typical for relief system sizing of reactive chemicals. 21% over-pressure was used for Scenario 3 per API 520's (Ref. [20]) recommendation for under-fire scenarios. Two-phase gas liquid discharge was assumed for Scenario 1 and 2, where foamy mixtures inside the reactor vessel were established. Single-phase gas discharge was assumed for Scenario 3. The gas-liquid mixture of chemical effluent was simulated by using the air-water properties. The quality values of the air-water mixtures were adjusted so that the ω value of the gas-liquid mixture of the chemical and that of the air-water mixture would match. The quality and the ω value of chemical mixtures were determined based on the initial fill ratio and the homogenous two-phase mixture assumptions. A single-phase acetone vapor release scenario (Scenario 3) was simulated by using air. The determination of the choking condition at the rupture disk was made based on the analytical model, which was described in Section 2. The downstream boundary conditions were set to be at typical atmospheric conditions (air at 20 °C and 1 bar-a).

Table 1. Upstream Boundary Conditions for the RELAP5 Model

Upstream Boundary Conditions	Scenario 1	Scenario 2	Scenario 3
Pressure	4.4 bar-a	4.4 bar-a	4.8 bar-a
Flow regime	Two-phase gas-liquid	Two-phase gas-liquid	Single phase vapor
Simulant	Air-water	Air-water	Air
ω	0.085	0.336	Not calculated
Choking at the burst disk	No	Yes	Yes

4 Results and discussion

The time history of the mass flow rate at the rupture disk calculated using RELAP5 for Scenario 1 is provided in Figure 3. The flow quickly accelerates following the opening of the rupture disk. The effluent experiences the frictional and form losses only for the sections of the piping, which it has already occupied at a given time. Therefore, the transient mass flow rate is slightly greater than the steady-state mass flow rate while the effluent is filling up the ERS piping. Once the entire ERS piping is filled, the mass flow rate achieves a steady-state value.

Figure 4 shows the pressure-time histories downstream of the rupture disk and at the end of the ERS piping for Scenario 1. The pressure downstream of the rupture disk gradually increases to its steady-state value as the effluent fills the piping. The pressure at the exit of the piping remains near atmospheric pressure (downstream boundary condition).

Figure 5 shows the dynamic load calculated on the ERS piping. The total load and its constituents are provided. There is a spike for the wave load as the rupture disk opens. This spike is caused by the depressurization wave traveling from the rupture disk to the reactor vessel. Typically, this section of the piping is filled with gas or vapor for which the speed of sound is well known. Therefore, the initial spike is captured accurately and the speed of sound of the two-phase mixture (which is typically unknown) does not have a significant role on the results. The wave load continues to be the dominant mechanism of the load as the effluent continues to fill up the ERS piping. Once the piping is filled up, the flow momentum term replaces the wave load. The flow momentum is due to the discharge of the effluent from the ERS piping into the ambient or a catch tank and it is the first term on the right-hand side of Equation 15. The pressure force is zero since the exit of the ERS piping is at the same pressure as the ambient; there is not choking at the exit of the piping.

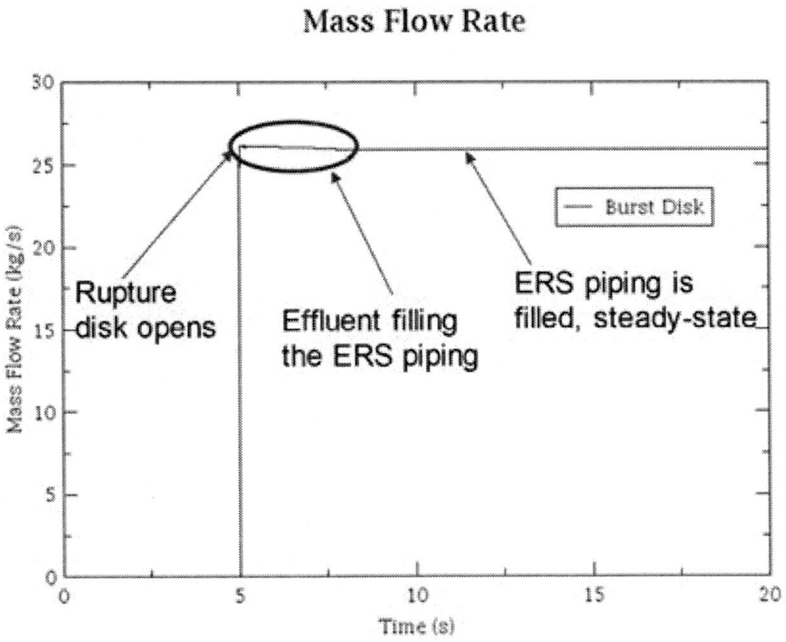

Figure 3. Mass flow rate for Scenario 1.

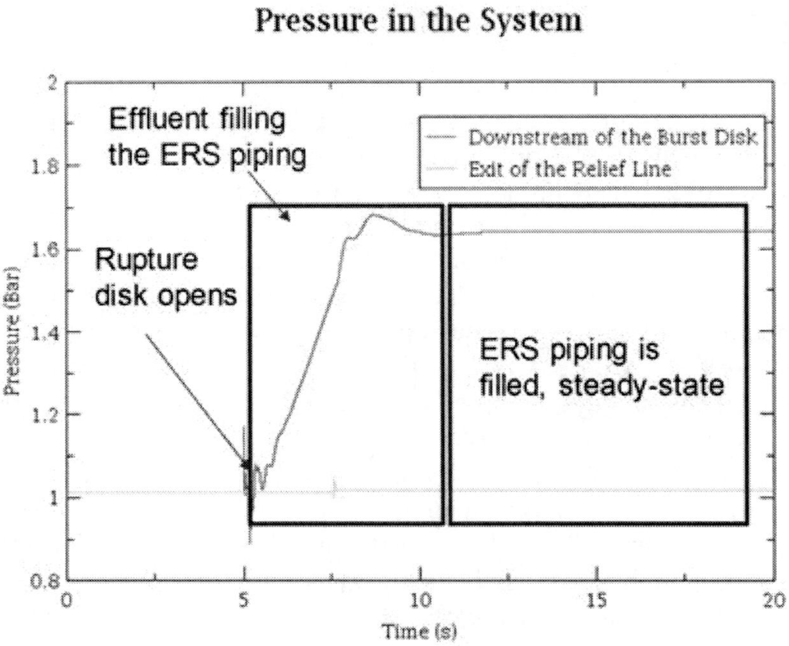

Figure 4. Pressure in the ERS piping for Scenario 1.

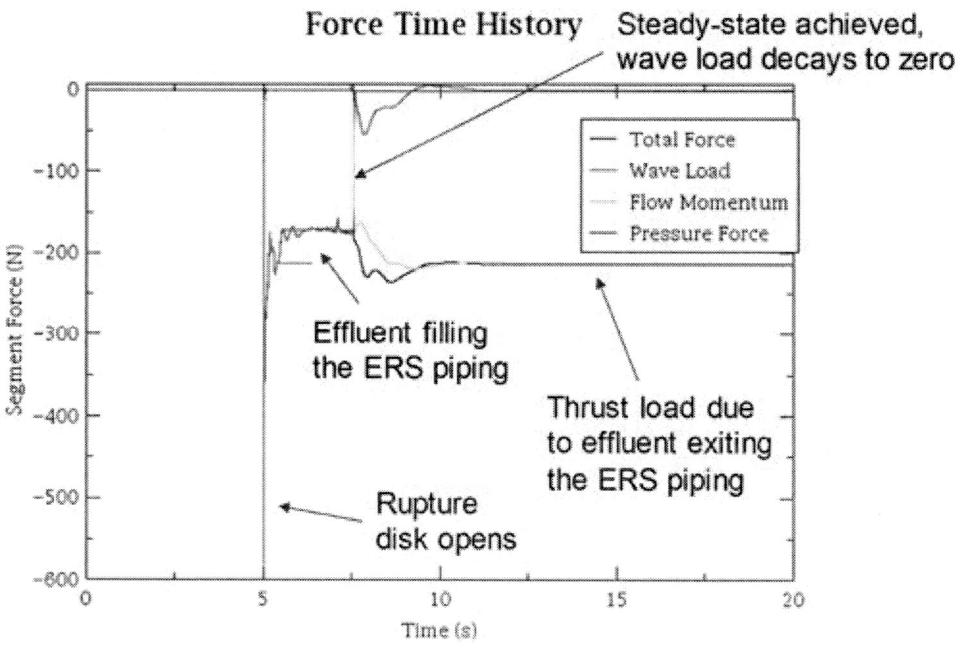

Figure 5. Dynamic loads for Scenario 1.

It is interesting to note that the wave load (while the ERS piping is filling up) and the load due to flow momentum (at steady-state) are almost similar. This is explained using the mathematical relations for the wave and thrust loads. The wave load for an effluent while filling up an empty piping (using Equation 14) is mathematically same as the flow momentum term in Equation 15.

The steady-state pressure, mass flow and thrust load values were compared between RELAP5 calculations and the analytical model, which was provided in Section 2. Actual chemical properties were used for the theoretical model. The comparisons between the two methods are provided in Table 2. The results demonstrate that the two-phase chemical effluent can be simulated with air/steam-water properties as long as the ω values are matched for these mixtures.

Table 2. Comparison between calculations using air-water (RELAP5) and actual two-phase chemical properties

Parameter	Scenario 1	Scenario 2	Scenario 3
Pressure downstream of the burst disk (bar-a)	1.6\1.7	1.6\1.7	1.9\1.6
Mass flow rate (kg/s)	25.9\26.1	16.3\16.4	1.2\1.2
Steady-state thrust load (N)	212.5\191.4	176.3\199.9	219.3\235.0

Black: RELAP5 results   Red: Theoretical model

In the chemical manufacturing industries, the effluent can be composed of proprietary chemical mixtures containing various materials. Therefore, it can be challenging to know the exact material properties at any given time during an emergency relief event. The method presented in this study requires only the stagnation properties of the chemicals at the venting conditions for predicting key parameters, such as the pressure distribution and wave loads during the transient. These properties are typically established by adiabatic calorimetric testing or similar characterization of the chemicals. The method presented in this study may also be used with other commercially available software, which can predict transient fluid dynamics and wave loads.

5 Conclusions

In this study, a previously published methodology is used for calculating the dynamics of the effluent and the resultant hydrodynamic loads for a typical ERS by using the RELAP5 software and assuming air-water properties for the effluent. Key output parameters, such as pressure drop and hydrodynamic loads, are then compared with established analytical methods using the actual effluent properties. The following conclusions were observed:

- The fluid transient analysis is a key aspect of the design of the emergency relief systems. It provides the wave loads required for the dynamic pipe stress analysis. It also provides key parameters such as the pressure distribution, mass and volumetric flow rates through the system, and wave loads. These key parameters are often used to determine back pressure effects on other vessels, for chemical dispersion analysis and dynamic pipe stress analysis.

- The analytical approach shows that there is a similarity between the air/steam-water and the gas/vapor-liquid chemical mixtures if the "ω" values of the two mixtures are matched. This allows for an emergency relief event to be simulated with air/steam-water properties.

- Similar pressure distribution in the piping system and loads can be achieved through the similitude of "ω". Therefore, "ω" of the air/steam-water mixture should be matched with the "ω" of the chemical effluent at the stagnation conditions. The quality of the air/steam-water mixture can be adjusted until a similar "ω" is achieved.

- The sonic velocity of the effluent is typically not known. Also, it is practically difficult to match the sonic velocity of the air/steam-water mixture with the sonic velocity of the chemical effluent, which is a major requirement for tracking pressure waves and wave loads. This study demonstrates that the emergency relief scenarios are pipe filling events, which only require the momentum flux to be calculated for a reasonable prediction of the initial wave load. Therefore, the sonic velocity of the effluent is not required to capture the initial wave load.

- The effluent can be composed of proprietary chemical mixtures containing various materials. Therefore, it can be challenging to know the exact material properties at any given time during the transient. The method presented in this study may be used with most commercially available software and air/steam-water properties for a reasonable representation of the important characteristics of the actual effluent discharge in mechanical design space.

- The study needs to be extended further to identify applicability and limiting conditions for this method.

6 References

[1] Conzen, J., Wood. R.G., (2015), "Modeling and Computation of Reaction Forces on Relief Piping During Depressurization", AIChE 11th Global Congress of Process Safety.

[2] Ozar, B., (2018), "Analysis of Chemical Effluent Discharge through the Emergency Relief System using Stream-Water Properties," DIERS Fall Meeting, 15-17 October, Houston, USA.

[3] ASME. *ASME Code for Pressure Piping, B31.* The American Society of Mechanical Engineers, New York, NY, 2016.

[4] "Directive 2014/68/EU of the European Parliament and of the Council," Official Journal of the European Communities, May 2014.

[5] Epstein, M., Henry, R.E., Midvidy, W., Pauls, R., (1983), "One-dimensional Modeling of Two-phase Jet Expansion and Impingement," Thermal-hydraulics of Nuclear Reactors, 2nd International Topical Meeting Nuclear Reactor Thermal-hydraulics, Santa Barbara, CA.

[6] Grolmes, M.A., Leung, J.C., (1984), "Scaling Considerations for Two-phase Critical Flow," Multi-phase Flow Heat Transfer III, Part A: Fundamentals, T.N. Veziroglu and A.E. Bergles, eds., 549, Elsevier, Amsterdam.

[7] Leung, J.C., (1986), “A Generalized Correlation for One-Component Homogenous Equilibrium Flashing Choked Flow,” American Institute of Chemical Engineers, Vol. 23, pp. 1743-1746, October.

[8] Leung, J.C., Grolmes, M.A., (1987), “The Discharge of Two-phase Flashing Flow in a Horizontal Duct,” American Institute of Chemical Engineers Journal, Vol. (33), p. 524.

[9] Leung J.C., Epstein, M., (1990) “The Discharge of Two-phase Flashing Flow from an Inclined Duct,” Transactions of the ASME, Vol. 112, pp. 524-527, May.

[10] Leung, J.C., (1992), “Reaction Forces During Two-phase Discharges,” American Society of Mechanical Engineers Journal of Fluids Engineering, Vol. (114), p. 689-692.

[11] Information Systems Laboratories, (2016), “RELAP5/MOD3.3 (Patch03) code manual volume I to VIII, NUREG/CR-5535/Rev P5-Vol I to VIII”, prepared for USNRC.

[12] International Agreement Report, (2007), “Simulation of the Propagation of Pressure Waves in Piping Systems with RELAP5/MOD 3.2.2: Comparison of Computed and Measured Results”, TÜV SÜD - NUREG/IA-0206, June.

[13] Sokolowski, L., Koszela, Z., (2012), “RELAP5 Capability to Predict Pressure Wave Propagation Phenomena in Single- and Two-Phase Flow Conditions”, Journal of Power Technologies 92 (3), page 150 – 165.

[14] Miller, J.S., Ramsden, K.B., (2005), “Validation of a Thermal Hydraulic Computer Code to Perform Two-Phase Multi-Component Force Calculations for Structural Evaluations”, Proceedings of 13th International Conference on Nuclear Engineering (ICONE13), 16-20 May 2005, Beijing, China.

[15] Ozar, B., Henry, C.E., Harvill, R., Norton, D.A., (2012), “Analysis for Low Pressure Cooling Injection System Suction Hydrodynamics for a Boiling Water Reactor,” Proceedings of 20th International Conference on Nuclear Engineering (ICONE20), July 30-August 3, 2012, Anaheim, CA.

[16] Henry, C.E., Lim, J., Ozar, B., (2012), “Investigation of Hydrodynamic Loads Associated with Pyrotechnic Valve Actuation,” Proceedings of 20th International Conference on Nuclear Engineering (ICONE20), July 30-August 3, 2012, Anaheim, CA.

[17] Riemke, R.A., Davis, C.B., Schultz, R.R., (2009), “Recent Hydrodynamics Improvements to the RELAP5-3D Code”, Proceedings of the 17th International Conference on Nuclear Engineering, ICONE17-75380, July 12-16.

[18] Health and Safety Executive, “Offshore Technology Report 2002/023: Testing and Analysis of Relief Device Opening Times,” Pipeline Simulation and Integrity Ltd., Yorkshire/UK.

[19] Moody, F.J., (1990), “Introduction to Unsteady Thermofluid Mechanics”, John Wiley and Sons, New York, Section 7.4, p. 365-366.

[20] API Standard 520, (2008), “Sizing, Selection, and Installation of Pressure-relieving Devices in Refineries,” 8th edition.

547437 Evaluation of the Thermal Stability of Styrene in Contact with Impurities

Safe handling and storage of reactive monomers have long been an issue for polymer manufactures. The unsaturated double bond may self-react to initiate unintended polymerization reactions under many scenarios, for example, completely depletion of inhibitors, exposure to fire or external heat, and in contact with impurities. Usually, a trace amount of incompatible species is able to trigger unintended reactions and lead to catastrophic consequences. Thus, understanding how the impurities promote self-reaction is expected to provide valuable guidance to safe monomer-handling practices. Our goal in this study is to quantify the contamination effects of some most common contamination chemicals on styrene self-polymerization. Thermal analysis tools including DSC and ARSST are applied to examine the exothermic characteristics of styrene mixed with contaminating substances at different concentration levels and mixing ratios. From the preliminary results, we observed the onset of self-polymerization shifted to a lower temperature when in contacted with water and also alkaline. A trace amount of concentrated strong acid initiated unwanted exothermic activity even at room temperature. Self-heating rate was also elevated when diluted acid was present. Additional research is needed to understand what is the impact in of other types and concentrations of contaminants such as rust, metal ions. Our study demonstrates significant implications in the prevention of runaway incidents during transportation and storage of styrene.

549038 A Mixed-Integer Programming Framework for Placement of Fire Detectors in Chemical Processing Facilities

In chemical processing plants, gas and fire detectors provide an important layer of protection for personnel. However, given the large variety of chemical and physical sensors and number of factors influencing their capabilities to successfully detect hazards, the decision of where to place sensors is challenging. Gas detectors (e.g. ultrasonic, catalytic bead) must be in contact with gas leaks to be effective, while fire detectors (e.g. point infrared, open path infrared, optical) require a visual path to the fire, free of obstructions, for successful detection. Numerical optimization techniques provide an opportunity to overcome these challenges and determine effective sensor placement.

The optimal gas detector placement problem was formulated by [1] as a stochastic program based on the p-median facility location problem and relied on scenarios generated from CFD simulations within a petrochemical facility. Their results showed that coupling coverage with expected detection time using a scenario-based method outperformed the existing coverage-only approaches and has inspired formulations based on MINLP approaches for non-uniform failure probabilities [2].

But while gas detectors must be placed in the path of gas leaks for effective detection, fire detectors must be placed within unobstructed visual distance of flames. The optimal placement of visual fire detectors in processing plants can be particularly difficult due to the large number of physical obstructions (e.g. piping, valves, tanks).

This work focuses on an optimization formulation for placement of fire (visual) detectors. The scenario-based, p-median approach demonstrated by [1] for gas detectors is translated to a maximum coverage model for fire detectors in a 3D space with consideration to failure probabilities. We show that this approach allows for the placement of fire detectors for a fixed sensor budget while under uncertainty. We demonstrate the effectiveness of this formulation on several case studies. We further discuss how the framework is readily scalable for larger problem sizes and amenable to various objectives.

We would like to thank the Purdue Process Safety and Assurance Center for their support in this work.

References:

[1] Legg, S. W., Benavides-Serrano, A. J., Siirola, J. D., Watson, J. P., Davis, S. G., Bratteteig, A., and Laird, C. D. (2012). A stochastic programming approach for gas detector placement using CFD-based dispersion simulations. Computers and Chemical Engineering, 47:194–201.

[2] Liu, J., and Laird, C.D., A Global Stochastic Programming Approach for the Optimal Placement of Gas Detectors with Nonuniform Unavailabilities, to appear in Journal of Loss Prevention in the Process Industries, 2017.

546855 Advance Alarm Management Applied to DCS Improvement in a Petrochemical Plant: A Case of Study

Alarms use in process industry is fundamental for avoiding process accidents in which people, the environment or process equipment could be damage. Currently, Carbon Black production plants are equipped with distributed control systems (DCS) for setting alarms values and to provide indication of abnormal events to the control rooms operators. However, the basic control system could not meet the needs against the different production conditions and the presence of improper alarms could send a bad signal to control rooms and it could be the start of a process safety event.

In the present study, an alarm system assessment was performed in a petrochemical plant, located at Mamonal industrial park, in Cartagena – Colombia; based on data analysis, it was observed that when process conditions changes occured, the DCS was not be able to adjust alarms settings automatically, according to the load fluctuation and/or process conditions. In order to solve this problem, in first place the DCS performance was studied and it was reported an alarms list, which provided information regarding to the alarm flooding phenomena, alarm chattering and others alarm system issues.

On the other hand, the information earned and alarm rationalization methodology were used for establishing an action plan in order to improve DCS performance. Results showed the utility of these tools and it was identified the need to perform betters in the alarm system logic to avoid flooding alarm phenomena and to make operations safer.

548361 Overcoming Motivational Biases in Hazops

Hazard and Operability Studies (HAZOPs) are one method of assessing the risk in a facility. Based on the risk ranking of a scenario, recommendations may be required to reduce the risk. Often, these recommendations require additional capital expenditure and/or operational resources. In aging plants where resources are not readily available or in greenfield projects where there is a tight timeline to start up, there is sometimes a bias to risk rank scenarios such that there is no need to add additional safeguards. Motivational biases can outweigh common sense such that scenarios are not ranked according to their real risk. We have developed a method to overcome some of these motivational biases with the use of a polling system to engage all participants in the HAZOP. Participants are asked to vote anonymously on the severity and likelihood for a given scenario, then the results are discussed. During this discussion time, arguments for all the different risk rankings are discussed, including those voted on by the minority in the room. This facilitates an open discourse about the real risks for a given scenario rather than shifting the attention to the most vocal person in the room, who may be strongly influenced by a desire to reduce spending or minimize the effort required for maintenance. Following this discussion a second poll is performed to reach a consensus. We have found this to be an effective method for reducing motivational biases and ensuring that the right decisions are made based on the most logical argument instead of based on budget or time constraints. Tracking polling results increases the accountability of participants and makes them less likely to skew the risk ranking. The requirement for consensus ensures that all opinions are heard and considered, even if they are not popular. Together these make the polling system an effective method for overcoming motivational biases in HAZOPs and ensuring that safe decisions are made.

546515 Integrating Culture into Process Safety Assurance

Purpose: PETRONAS Chemicals Group Berhad (PCG) has embarked on a group-wide culture transformation, introducing the concept of the Results Pyramid® and a Culture of Accountability® i.e. a workplace culture where people take accountability to think and act in the manner necessary to achieve results.

Methodology: This paper will discuss two such activities anchoring on the Results Pyramid® that have been deployed across several plants and units within PCG.

The first activity is an enhancement of existing assurance programmes, where culture subject matter experts are assigned to observe plant leadership teams and front-liners during interviews and site verifications conducted by the technical assessors. The observations are synthesised into belief statements, which are presented together with the technical findings.

The second activity is Power Up, a session designed to incorporate powerful and specific feedback exchange between various teams, culminating in a specific list of actions to improve delivery of each team's activities in order to achieve the shared key results. This workshop can be done either as a follow-up to an assurance activity, in preparation for a project/turnaround activity or as a part of a periodic performance review.

Results: Examples of observations, feedback exchanges and specific actions from these activities will be shared, together with the impact on process safety leading and lagging indicators. Recommended practices and lessons learnt for continuous improvement will also be included.

Conclusion: Applying the Results Pyramid® concept to the design and implementation of Process Safety-related programmes has helped to identify the underlying leadership, mindset and behaviour issues impacting their success. By doing so, more specific interventions can be put in place to ensure full compliance to safety requirements and maximise effectiveness of these programmes.

548442 Emergency Plans Integration By Diferent Companies

In Brazil, different companies with pollution potential must have an emergency plan. However, each company analyzes its risks and contingency measures separately, based on the Conama 398 law. With the establishment of the Area Plan, companies from the same territory with pollution potential caused by oil in waters began to analyze how to combat a major accident quickly and effective together. This work shows how the integration of the individual contingency plans of each company and the cooperative result of the area plan in territory of significant scope and considerable environmental sensitivity.

No Brasil, diferentes empresas com potencial de poluição devem ter um plano de emergência. No entanto cada empresa analisa seus riscos e medidas de contingência separadamente, com base na lei Conama 398. Com a instituição do Plano de Área, empresas do mesmo território com potencial de poluição causadas por óleo em águas passaram a analisar como atender conjuntamente de forma rápida e eficaz a um grande acidente em seu território de abrangência. Este trabalho mostra como foi estruturada e organizada a integração dos planos de emergência individuais de cada empresa e o resultado cooperativo do plano de área em território de abrangência significativa e considerável sensibilidade ambiental.

548111 Moving Back to Flammable Refrigerants - Facility Siting and Personnel Safety Considerations

"Everything old is new again – Paul Allen"

Not just a catchy tune from the 1970's, this jingle summarizes a current technological shift in the world of refrigeration. Historically, refrigeration cycles used toxic and/or flammable materials as the working fluid. As the technology matured and made its way into transportation and consumer applications, the working fluid also evolved. Halogenated hydrocarbons, such as chlorofluorocarbons (CFCs), were developed as replacements to conventional refrigerants such as ammonia and propane. CFCs, which may have toxic effects, are desired for their tunable working range and for being non-flammable. However, the environmental drawbacks to the use of CFCs has been well researched and documented. As we embark into the future, where climate change and environmental quality are at the forefront of concern to consumers and industry alike, we are seeing a move back to flammable refrigerants. Everything old is, indeed, new again.

New construction and expansion projects can certainly consider and plan for the use of flammable refrigerants in commercial and industry facility design and siting, but what about retrofit applications? What are the risks and hazards associated with converting a CFC refrigeration system, designed and sited for the use of non-flammable refrigerants, to a system now utilizing flammable refrigerants? Beyond the need for additional fire protection systems, what risks are posed to the facility and personnel? In this paper, we seek to answer these questions by comparing the hazards and risks of mid-scale refrigeration systems using either non-flammable or flammable refrigerants. We explore how facility siting constraints may change with respect to the use of flammable refrigerants, and perform a quantitative risk assessment (QRA) for an example retrofit project in order to assess the potential for increased risk that should be considered in such projects.

548210 Making Sense of Risk Results

Risk based studies are very useful, but at the same time, the results are extremely complicated. While we have a basic set of results that can be used to illustrate the risks to people, these are relatively simple for risk practitioners and sometimes too complicated for site managers.

FN curves show the risk to all people from all sources, and individual risk contours illustrate the risk from all scenarios to people across specific locations, but how do we translate these results into actions. Generally, the action is to relocate people to lower risk areas, or upgrade buildings to mitigate the impacts identified. However, this is not always feasible. What if the site footprint is limited; there are public receptors who cannot be relocated; or building upgrades are cost-prohibitive or infeasible?

We must better quantify and evaluate the **sources** of risk rather than focusing on mitigating the impacts from the events. We need to evaluate the type of risks, the potential initiating events, and then we can assess the potential methods to effectively mitigate the risks. Existing risk result presentations are often very poor at illustrating the sources of risk. To effectively identify the sources of risk requires that we carefully dissect the risk results, while maintaining their validity and intent.

This paper will focus on how we find these risk drivers and make enough sense of them to create actionable recommendations that protect people.

551205 Ammonia Dispersion Modeling for Ammonia Storage with Multi-Scenario Failure System

PT Pupuk Kalimantan Timur is one of the largest fertilizer companies in Southeast Asia where located in Bontang City, East Kalimantan Province, Indonesia. The total production capacity of PT Pupuk Kalimantan Timur are 2.7 million Tons / year Ammonia, 3.43 million Tons / year Urea and 350 thousand tons / year NPK.

To accommodate ammonia production, PT Pupuk Kalimantan Timur currently operates 3 ammonia tanks with a capacity of 2 x 25,000 tons and 1 x 50,000 tons. In order to manage the hazard risk contained in the ammonia tank, it is necessary to do a modeling to map the potential and ammonia leakage potential areas and the impact of their exposure to the workforce and the community around the factory area.

The results of this modeling are expected to be used by Pupuk Kalimantan Timur's management in carrying out corrective actions for occupational health and safety performance, especially in controlling the potential hazard of ammonia, especially in ammonia tanks.

In this paper modeling is done using the AERMOD® Modeling System software. AERMOD Modeling System was chosen because it can be used on:

1. U.S. EPA regulatory Gaussian plume air dispersion model.
2. For near-field impacts (50 km max).
3. Prefered model for complex terrain.
4. Improved treatment of urban effects.

For modeling, we choose the AS-T-101 ammonia tank which has the largest capacity, which is 50,000 tons. Ammonia distribution models will be simulated based on weather conditions and ammonia leak source scenarios that can occur in the AS-T-101 ammonia storage tank system. Ammonia leakage sources will be displayed in 5 scenarios, i.e:

1. 1 Vent opens
2. 1 PSV poping
3. 2 PSV poping
4. 3 PSV poping
5. 3 PSV and 1 vent poping

The AERMOD Modeling System program is used to make ammonia distribution models for each event scenario assuming an exposure radius of up to 10 km. Based on the results of the calculation and modeling of the potential distribution of ammonia from AS-T-101 using the AERMOD Modeling System, the highest exposure of ammonia is in the 5th scenario, which is up to 36611 μg / M3.

546560 De-Mystifying the Concepts of Hazardous Area Classification

Purpose

The purpose of this topic is to address and clarify major items of confusion associated with hazardous area classification assessments and compliance auditing in classified locations. The ultimate objective is developing documentation that provides a framework for the control of electrical energy ignition sources in environments made hazardous by the potential presence of flammable vapors.

Background

In February of 1992 Occupational Safety and Health Administration (OSHA) 29 CFR 1910.119 Process Safety Management of Highly Hazardous Chemicals became law in the United States and confusion reigned supreme. Over the next several years industry professionals began the process of sorting through the responsibilities, regulations and impending deadlines that had been placed squarely in their laps. Acronyms such as PSM (Process Safety Management), PSI (Process Safety Information), RAGAGEP (Recognized and Generally Accepted Good Engineering Practices), MOC (Management of Change), and PHA (Process Hazards Analysis) became commonplace in the vocabulary of process safety professionals nationwide.

Employers now were required to meet stringent deadlines regarding the creation and management of a plethora of process information associated with equipment in the process covered by this law, i.e. any covered process. This information would be used to support the exhaustive efforts needed to create a baseline PHA for each covered process by May of 1997. In addition OSHA required all PHA's be placed on a five (5) year life cycle requiring their review and revalidation. Along with the revalidation of PHA's came the task of also revalidating all process safety information on the same five (5) year cycle.

As a result, every willing and able bodied employee was thrust into the quagmire of supporting, participating in, facilitating, or conducting PHA's. The required information used to support PHA's included the materials of construction, Piping and Instrument Diagrams (P&ID's), Relief System design and design basis, ventilation system design, material and energy balances, safety shutdown systems, and last but not least **electrical classification**. Yes, "electrical classification" was and still is legally considered an integral part of required process safety information.

This abstract addresses the following and provides an interactive discussion regarding the three (3) major myths surrounding electrical classification:

- assigning a classification to a location
- proper use of the classification information
- keeping the documentation evergreen

The following topics will be addressed and clarified:

- What Is Hazardous / Electrical Classification?
- The Process of a Hazardous Area Classification Assessment
- The Concept of Electrical vs Hazardous Area Classification
- The Compliance Assessment (where does it fit in the process of controlling electrical energy ignition sources)
- Common Compliance Assessment Issues

546108 Maintaining Safety Excellence By the Implementation of Process Safety Inspection in Badak LNG

Badak LNG as one of the most experienced LNG Producers in the world, has undergone safety culture change throughout its operation. The goal of this change is to achieve excellent safety performances, indicated by the lower rate of Lost Time Injury Frequency (LTIF), lower number of Loss of Primary Containment (LOPC) Tier-1, High Potential Incident and multiple Train trip. This change was triggered by the occurrence of a fatal incident in 2005 as well as lost time injury incident in 2006. Since then, Badak LNG has integrated Process Safety Management (PSM) into its Company Management System, strengthened the roles and responsibilities of Managements, and pursued for interdependent safety behaviour of each personnel. One of PSM implementations is through process safety inspection.

Process safety inspection is an activity to assess the current equipment integrity based on visual observations compared with equipment's standard condition. The standard condition checklist for process safety inspection in Badak LNG is based on Physical Barrier Assessment tools, owned by Det Norske Veritas (DNV). The assessed items are divided into 5 categories: Inherent Safety, Hazard Prevention Measure, Hazard Control, Mitigations, and Support System. The overall score is determined by total number of standard items in all categories compared to the total number of verified items in all categories. The minimum acceptable score is at 80%.

During the year of 2012-2017, the average score of process safety inspections in Badak LNG is 90%. As a result of these inspections, Badak LNG had also identified the need of improvement in some items such as blinds, pipework and flanges, passive fire protection, fire & gas detection system and lock-open/lock-close block valves. In overall, by the implementation of PSM, Badak LNG has aligned on the right path to maintain its safety excellence achievement in LNG business, such as zero LTIF since 2007, zero multiple Train trip since 2012, zero LOPC Tier-1 since 2016, and zero High Potential Incident in 2017.

548411 Using HAZOP/LOPA to Create an Effective Mechanical Integrity Program

Many people view the conduct of a HAZOP/LOPA to address regulatory requirements as a chore, and stop there. However, the implementation of a quality HAZOP/LOPA has the potential to provide a framework for addressing numerous safety and operational optimization issues at plants, including the formulation/refinement of the Mechanical Integrity Program. The purpose of this paper is to focus on the mechanical integrity program, illustrate how a quality HAZOP/LOPA can support the effective implementation of some of the new requirements for California Refineries, and optimize key elements of an effective Mechanical Integrity Program, e.g.:

- Inspection/testing methods

- Testing intervals

- Maintenance outage periods

- Repair prioritization and allowable outage

- Identification of low priority equipment

544678 *"Identifying Critical Process Parameter to Ensure Asset Integrity" or "Pulse of Mechanical Integrity Management System"*

The management of equipment begins with an awareness that ageing is not about how old the equipment is, but is about what is known about its condition and the factors that influence the onset, evolution and mitigation of its degradation. The presentation provides an overview of Integrity Operating Window (IOW) parameters of any process that are key to understand the asset degradation and prevent Process Safety Incidents.

IOW are the situations in the process industries where a process parameters (temp., pressure, dosing rate, conc. etc...) when gets out of operating windows could result in significant loss of primary containment leading to potential severe consequences.

Arrazi - SABIC developed an IOW system that covers IOW identification, associated risk, mitigation (interim – long term), stewardship/management of IOWs and elimination.

The presentation shall cover Insight concept of the Topic, major Process Safety accidents related to breeching IOWs, actual example, IOW Management Process, its influence on PM plans etc.

three key bullet points that we can include in the agenda:

- Knowing operating limits with associated mechanical integrity or process safety risk to ensuring proper actions to be in operating envelope
- Developing the engineering leadership to systematically guide the operators, superintendents and engineers to be in safe zones
- Dynamic inspection frequency based on plant conditions and sharing some industrial major incidents for this critical item

19 AIChE Spring Meeting
15TH GLOBAL CONGRESS ON PROCESS SAFETY
NEW ORLEANS • 2019

The Practice of Process Safety Analysis and Management System Based on Safety Reliability
(THE NEW TITLE: The PHAMS system in SINOPEC)

Jiang Xue
SINOPEC Research Institute of Safety Engineering,
QingDao,China
Jiangxue.qday@sinopec.com

The PHAMS system in SINOPEC

Jiang Xue
SINOPEC Research Institute of Safety Engineering,
QingDao,China
Jiangxue.qday@sinopec.com

Keywords: Safety Barriers Reliability, PHAMS, HALOPA, SIL, Safety Management Online.

Abstract

The big data technique has been in trend for the process safety assessment. In China, it's the very need to transfer the qualitative safety management to quantitative safety management, which should rely on the professional risk analysis technique, as well as the standard safety reliability database. Based on the above, it's very important to integrate several professional risk analysis technique, as HAZOP, LOPA, SIL etc. to realize the integral risk identification and evaluation, grading and management plan making, supporting by the proprietary SINOPEC risk metrics and grading standards, as well as safety equipment reliability database,which is developed as the Process Safety Analysis and Management System(PHAMS) for SINOPEC, to support all SINOPEC branches to make dynamic risk analysis and management online

1 Introduction

With the smart factory's development, the static, periodic risk assessment and management can not satisfy the identification and real-time safety control of plant's risk events. The PHAMS which establishes the plant-wide dynamic risk analysis, can effectively solve the risk rating and monitoring of the unit and realize the plant-wide or corporate-wide management optimization.

The key data for HALOPA, SIL and Management module in PHAMS is the dependent protection barriers, which we rely on RBI correction, Personnel exposure probability and ignition probability to optimize the six mainly safety barriers' failure probability, simultaneously, do use of the Bayesian Correction Modelling to verify the instant failure probability on site, as well as the SIL identification, which can validly save the grading cost on time and personnel, and all of these are towards to the deeper research on dynamic risk analysis for PHAMS.

2 HALOPA Analysis System

2.1 SINOPEC RISK MARTIX

According to international common matrix criteria, as well as statistics from the SINOPEC accident data and the maintenance data, the frequency of risk occurrence and severity classes are coupled to the 8×7 matrix by a linear exponential model, and be assigned risk index to each risk level.

SINOPEC RISK MARTIX		Occurrence Frequency							
		1	2	3	4	5	6	7	8
		$\leq 10^{-6}/y$	10^{-6}-$10^{-5}/y$	10^{-5}-$10^{-4}/y$	10^{-4}-$10^{-3}/y$	10^{-3}-$10^{-2}/y$	10^{-2}-$10^{-1}/y$	10^{-1}-$1/y$	$>1/y$
Severity Classes	A	1	1	2	3	5	7	10	15
	B	2	2	3	5	7	10	15	23
	C	2	3	5	7	11	16	23	35
	D	5	8	12	17	25	37	55	81
	E	7	10	15	22	32	46	68	100
	F	10	15	20	30	43	64	94	138
	G	15	20	29	43	63	93	136	200

Figure 1 SINOPEC Risk Matrixp[1]

Based on this matrix, SINOPEC has established four-level risk list for establishing targeted management of branch network. For example, the executive board of the group is in charge of the red-level risk list, the branch directors is in charge of the orange-level risk list, etc.

2.2 HALOPA Analysis Modelling

2.2.1 System risk analysis based on equipment operation cycle

The HALOPA analysis model can be used to identify the hazard of the device or process unit, clarifying the initial event, process deviation, independent protective layer and consequence events, combined with the event tree analysis method to analysis the reduction influence of the occurrence frequency of initial events. The protection layer performs a failure frequency evaluation role to determine the final frequency of the on-site device risk events during each operating cycle under the coupling reduction of the existing protective layers at each level [2].

An important assumption of the model is that each protection layer is independent, that is, the failure of a single protection layer will not affect the risk reduction impact of other existing protection layers [3]. Each time a protective layer takes effect, the frequency of accidents is reduced by the risk reduction of the protective layer.

When the frequency of the initial event does not exceed 1 time/year or is less than twice the test frequency of the first protective layer, the evaluation of the consequence frequency can be referred to the following formula [4]:

$$f_I^C = f_I \times P_{enable} \times \prod_{j=1}^{J} PFD_{Ij} \quad (1)$$

where

f_I=Occurrence frequency of initial event I annually;

Penable=Moderation factors by enabling conditions and other conditional probabilities.

PFDIj=Failure probability of dependent protection layer j of initial event I.

When the frequency of the initial event exceeds 1 time/year or exceeds the test frequency of the first protective layer twice, the first protection layer failure shall be used as the initial event, and the calculation formula of the consequence frequency is as follows:

$$f_I^C = f_{\mathrm{IPL\_first}} \times P_{enable} \times \prod_{j=2}^{J} PFD_{Ij} \qquad (2)$$

where

$\mathbf{f_{IPL\_first}}$= The failure probability of 1st dependent protection layer

The HALOPA analysis model in PHAMS designed a modified calculation model for equipment failure frequency and personnel exposure probability, and realized BPCS, SIS, physical protection, leakage storage and collection, fire and explosion suppression for equipment operation cycle and device in service. The system's six major types of protection layer failure frequency have the correction factors of risk reduction, and realize system risk assessment based on the device production cycle. For example, according to the system risk assessment of a diesel hydrogenation unit based on the PHAMS system, on-site data collection of various protective layers and key correction conditions under one cycle of the driving operation to date, online preliminary analysis of the risk overall value and grading of the feeding unit The results are as follows:

Figure 2 HALOPA Analytical Function Online

2.2.2 Dynamic risk assessment based on real-time data of safety control systems

According to the protection layer frequency analysis model, the failure probability of the protection layer lowering the frequency of a certain initial event can be regarded as θ , and the a priori risk probability distribution function is:

$$f_0(\theta) \propto \theta^a(1-\theta)^b \quad (3)$$

Its likelihood function $L(E|\theta)$ is

$$L(E|\theta) \propto \theta^m(1-\theta)^{n-m} \quad (4)$$

where

a, b=the prior distribution parameter of the failure probability of the protective layer;
m=The number of the posterior cumulative event occurred when the protective layer is valid;
n=The number of the posterior cumulative event occurred when the protective layer is failed.

With the posterior probability distribution function corrected by dynamic risk analysis as$f_1(\theta|E)$, the posterior probability distribution can be obtained by:

$$f_1(\theta|E) = \frac{f_0(\theta)L(E|\theta)}{\int f_0(\theta)L(E|\theta)d\theta} \propto \theta^a(1-\theta)^b\theta^m(1-\theta)^{n-m} \propto \theta^{a+m}(1-\theta)^{b+n-m} \quad (5)$$

Since m and n are parameters of the age change and the posterior probability distribution function seems to be continuous, the dynamic failure frequency of the protective layer at a certain moment can be obtained by the following formula[4-6].

$$P_i(\theta) = \int_{i-1}^{i} f(\theta)d\theta = (1 - F_i(\theta|E)) - (1 - F_{i-1}(\theta|E)) = F_{i-1}(\theta|E) - F_i(\theta|E) \quad (6)$$

In where $F_i(\theta|E)$ is $f_i(\theta|E)$ the cumulative failure frequency function as

$$F_i(\theta|E) = \int_0^i f(\theta|E)d\theta \quad (7)$$

And then bring in the following formula to get the dynamic risk frequency of the device or process unit:

$$f_I^C = f_I \times P_{enable} \times \prod_{j=1}^{J} P(\theta)_{Ij} \quad (8)$$

and

$$f_I^C = f_{IPL\_first} \times P_{enable} \times \prod_{j=2}^{J} P(\theta)_{Ij} \quad (9)$$

Where

$P(\theta)_{Ij}$=Posterior failure probability value of the jth independent protection layer

Based on this, the PHAMS system realizes the real-time update of the protection layer risk reduction frequency, and intelligently analysis the overall risk value and grading of the unit to plant.

3 SIL Intelligence Identification Online System

3.1 SIL Identification Modelling

The analysis data of the SIS protection layer of the HALOPA evaluation system in PHAM is updated and interconnected by means of the safety reliability database. After the HALOPA dangerous event using the same safety instrument loop is automatically brought into the SIL identification, the PHAMS system takes the same kind of consequences. C, to calculate without the cumulative frequency F^C before the SIF

$$F^C = \sum_{i=1}^{n} f_i^C \qquad (10)$$

where

$$f_I^C = f_I \times P_{enable} \times \beta \times \prod_{j=1}^{J} PFD_{IJ} \times \prod P_{condition} \qquad (11)$$

$$\beta = \frac{2^N - 1}{N} \qquad (12)$$

where

f_I^C= The occurrence frequency of the dangerous event I with consequence C

β= the common factors caused by several SIFs

N=the number of SIFs for the dangerous event I

Based on the modelling, the PHAMS system can intelligently group common risk events and assess the risk reduction factor RRF of SIF in each group [7] :

$$\text{RRF} \geq \frac{F^C}{TMEL} \qquad (13)$$

and

$$1 \times R_{\text{Factor}} < RRF_{SIL1} \leq 10 \times R_{\text{Factor}} \qquad (14)$$

$$10 \times R_{\text{Factor}} < RRF_{SIL2} \leq 100 \times R_{\text{Factor}} \qquad (15)$$

$$100 \times R_{\text{Factor}} < RRF_{SIL3} \leq 1000 \times R_{\text{Factor}} \qquad (16)$$

where

R_{Factor}= risk reduction factor between1 to 10.

Based on the SIL identification results of each group, the optimal solution is intelligently selected as the RRF of this safety instrument loop.

Figure 3 SIL Intelligence Identification Online

3.2 The SIL Identification Based on SINOPEC Risk Matrix in PHAMS

With the aid of data warehouse technology, PHAMS system can compare HALOPA assessment data which involved with SIS protection layers of the initial situation, the consequences classes and non SIS protective layer of data, and automatically take the residual risk of the events into the risk matrix. Then based on SINOPEC risk matrix, the SIL identification module can intelligently couple and assess the residual risk and RRF.

according to Sinopec's safety management focusing on high severity impact, PHAMS sets an independent TMEL requirements (as shown in the following table), and can automatically selects the most serious impact event as the final SIS identification data of this event. .

Table 1 TMEL of SIL Intelligence Identification[8-10]

Severity Classes	*Health & Safety (r/y)*	*Property Damage (r/y)*	*Social Impact (r/y)*
A	$\leq 1\times10^{-1}$	≤ 1	≤ 1
B	$\leq 1\times10^{-2}$	$\leq 1\times10^{-1}$	$\leq 1\times10^{-1}$
C	$\leq 1\times10^{-3}$	$\leq 1\times10^{-2}$	$\leq 1\times10^{-2}$
D	$\leq 1\times10^{-5}$	$\leq 1\times10^{-4}$	$\leq 1\times10^{-4}$
E	$\leq 1\times10^{-6}$	$\leq 1\times10^{-5}$	$\leq 1\times10^{-5}$
F	$\leq 1\times10^{-7}$	$\leq 1\times10^{-6}$	$\leq 1\times10^{-6}$
G	$\leq 1\times10^{-7}$	$\leq 1\times10^{-6}$	$\leq 1\times10^{-6}$

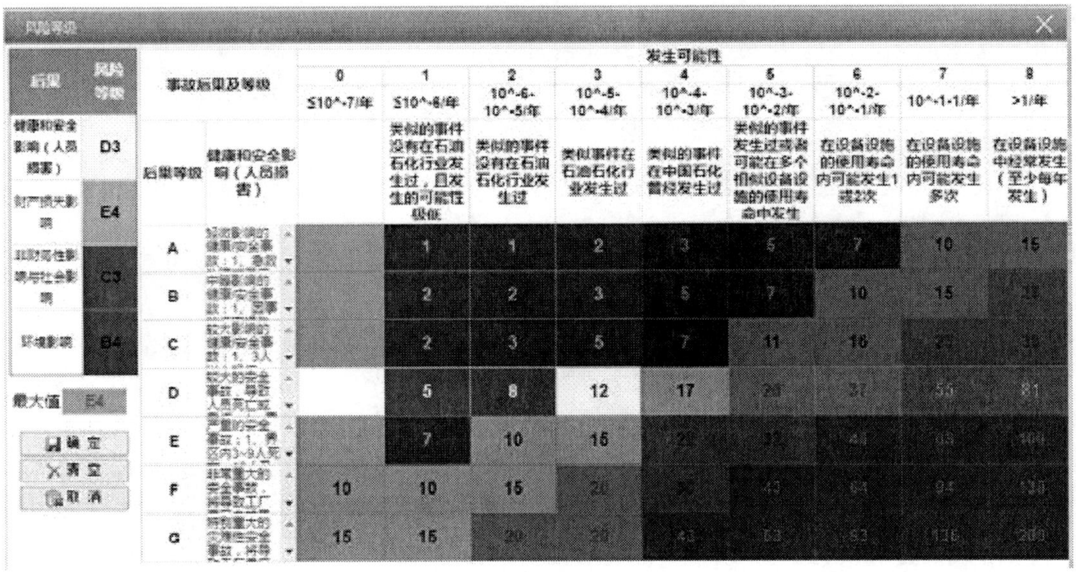

Figure 4 SIL Intelligence Identification on SINOPEC Risk Matrix

On the basis of SIL identification, the PHAMS system can also implement SIL verification based on multi-factor model, and evaluate the structural constraints, PFDavg and MTTF respectively. The verification result data is consistent with the accuracy requirements of IEC61508 and IEC61511.

4 Risk Management and Decision Support System

With the powerful safety and reliability database and expert experience database of PHAMS system, as well as the data mining of the result data of HALOPA and SIL system, the Risk Management and Decision Support System can implement the cumulative risk frequency calculation of independent devices and plants in construction or in service, to assess the overall risk based on different impact severities through the linear comparison [11-12].

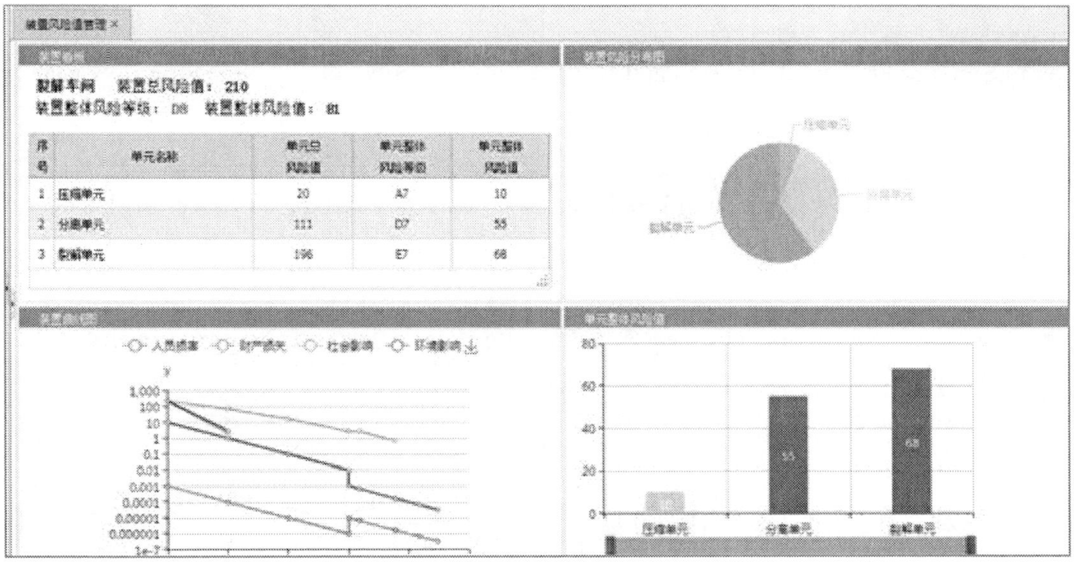

Figure 5 Overall Risk Rating and Risk Event Distribution Analysis of Ethylene Cracking Facility

5 Conclusion

1. PHAMS system sets up 9 categories of safety reliability database, expert experience data and Sinopec risk matrix, etc, which can support unified SINOPEC risk assessment and management standards, risk analysis template, process data storage and data mining for PHAMS system , as well as the interconnection and real-time update of data of each system.

2. HALOPA system implements plant overall risk analysis based on protection layer data, and performs data coupling evaluation from static analysis based on equipment cycle and dynamic risk correction to realize real-time evaluation of risk assessment technology and device operating conditions. The dynamic risk analysis research will be further strengthened to realize the system risk prediction function.

3. The SIL identificaiton and verification system can realize the evaluation of safety instrument function of in service and in construction. In the long run, it helps to improve the safety management level of the plant's overall safety instrument function

4. The risk management and decision support system can realize the overall risk assessment and risk distribution based on various evaluation results, through in-depth mining and statistical analysis, the PHAMS can realize the big data analysis management of Sinopec branches' risk management.

6 References

[1] The Bureau of Safe Production Supervision and Administration of SINOPEC. Guide for Safety Risk Identification and Assessment of SINOPEC[S].2018

[2] AQ/T 3054-2015. Guide for Application of Protective Layer Analysis Methods. Chapter 8： Scenery Frequency Calculation [S].

[3] Khan,Seyed Javad Hashemi,Nicola Paltrinieri,Paul Amyotte,Valerio Cozzani,Genserik Reniers.

Dynamic risk management: a contemporary approach to process safety management[J]. Faisal Current Opinion in Chemical Engineering.

[4] CCPS. Risk-based Process Safety [M]. 2013.

[5] Maryam Kalantarnia,Faisal Khan,Kelly Hawboldt.Dynamic risk assessment using failure assessment and Bayesian theory[J].Journal of Loss Prevention in the Process Industries . 2009 (5)

[6] Rakesh Ranjan,Sonam Singh,Satyanshu K. Upadhyay. A Bayes analysis of a competing risk model based on gamma and exponential failures[J] . Reliability Engineering and System Safety . 2015

[7]Zugravu-Soilita N. How does Foreign Direct Investment Affect Pollution? Toward a Better Understanding of the Direct and Conditional Effects[J]. Environmental & Resource Economics, 2017:1-46.

[8] Khakzad N, Yu H, Paltrinieri N, et al. Chapter 5 – Reactive Approaches of Probability Update Based on Bayesian Methods[J]. Dynamic Risk Analysis in the Chemical & Petroleum Industry, 2016:51-61.

[9] Villa V, Paltrinieri N, Khan F, et al. Towards dynamic risk analysis: A review of the risk assessment approach and its limitations in the chemical process industry[J]. Safety Science, 2016, 89:77-93.

[10] Mark Adrian van Staalduinen,Faisal Khan,Veeresh Gadag. SVAPP methodology: A predictive security vulnerability assessment modeling method[J]. Journal of Loss Prevention in the Process Industries.

[11] Podofillini L, Dang V N. Conventional and dynamic safety analysis: Comparison on a chemical batch reactor[J]. Reliability Engineering & System Safety, 2012, 106(4):146-159.

[12] R A F. Problems with risk analysis in the chemical industry. A detailed examination of the theoretical and practical problems faced by the risk analyst in the study of a chemical plant[J]. Plant/operations Progress, 1983, 2(3):185–190.

549550 Process Hazard Analysis

Process Safety Management System starts with managing the process related risks and from the first step of establishing a new process unit that is needed to evaluate the risks and define the necessary measurements. Process Hazard Analysis methodologies are the systematic tools which help to evaluate the risks in a process. For different needs there are several methodologies like DOW FETI, HAZOP, What if, HAZID, Fault Tree and Event Tree Analysis, Scenario Modelling, Facility Siting etc. Some of these methodologies are semi quantitative some are quantitative methodologies. In order to define and prevent major industrial accidents in our refinery we use DOW FETI, HAZOP, FTA and ETA and Scenario modelling. Moreover for some kind of projects we may also apply What if and HAZID analysis instead of HAZOP depending on the complexity. Facility Siting is another tool for risk related studies which is used to evaluate the risks for buildings in order to protect people inside the buildings. The workflow for a full unit PHA study starts with DOW FETI calculation which is used for risk ranking of al static equipment inside the unit. DOW FETI calculation is done in accordance with the equipment temperature, pressure, hold-up amounts, components and also reaction types if there is any. Then risk assessment process continues with HAZOP studies which is for scenario identification. In HAZOP, first step is chopping the unit into small parts to be evaluated which is called nodes. After that all nodes are evaluated using P&IDs and PFDs by keywords like temperature, pressure, flow, level etc. HAZOP helps to understand what kind of scenarios are possible inside the unit and these critical scenarios are taken through FTA and ETA studies. HAZOP is a semi quantitative risk assessment method where FTA and ETA are quantitative methods. FTA is applied for critical scenarios in order to find out the probability of top event then ETA is then applied in order to find out the probability of scenarios like fire, explosion, toxic dispersion etc. caused by this top event. FTA and ETA analysis all use the statistical PFD and failure rate values of all initiating events and barriers defined in HAZOP for the related scenario. Scenario Modelling is the following step which is used for understanding the effect of the probable scenario. The most probable scenarios coming from ETA is modelled to see the effect zones. These effect zones are used to plan Emergency Response Studies for all these potential major accidents.

548517 Site-Wide Asset Integrity Analytics & Benchmarking

All too often, reliability engineers focus on the equipment that is deemed as most critical or covered by some regulatory reference, while sometimes forgetting about those areas where something as simple as water can cause major issues. Non-destructive testing technicians and API inspectors alike can help facilities gather and compile data to perform predictive analytics and even benchmark sites with similar assets. This data can be used to identify assets where others may be seeing issues. This also allows facilities to compare reliability engineering effectiveness. This presentation will share some sample results of non-destructive testing and inspections of both regulatory required and non-required assets in the fuel ethanol production sector as well as an example of how that data can be assembled and shared across any sector to educate and learn from each other. Afterall, there are no secrets in safety!

548330 Reliability, Availability, and Maintainability (RAM). What Does It Provide?What Should You be Aware of before You Conduct a RAM Study?

Reliability, Availability and Maintainability (RAM) studies can significantly add value to your Process Safety Management (PSM) activities, especially with regards to Mechanical Integrity (MI). RAM studies are an excellent way to determine which equipment limits you from achieving optimum availability and throughput. Using industry failure rate data or even site-specific data can help you identify areas for improvement and determine how best to maintain your equipment. But RAM studies can go beyond purely theoretical studies by using the experience and knowledge gained from operating and maintaining similar facilities to develop a pragmatic approach to solving mechanical integrity-related issues. This paper will cover the pros and cons of site versus industry data and will present a case example of utilizing experience to better define the proposed problems and solutions. If done correctly, RAM studies can be more than statistics; they can be roadmaps to process reliability improvements.

546707 Driving Asset Integrity and Process Safety Consistency across a Multinational Organization

Mexichem is in the process of implementing world-class process safety management (PSM) and Asset Integrity Management (AIM) programs globally for their facilities. This is a major undertaking as Mexichem, to a large extent, has grown through acquisitions. Therefore, each plant has a safety culture and management system unique to that asset. As we all know, it's difficult to drive global consistency when dealing with different types of operations in various regions throughout the world under differing regulatory regimes. This is especially difficult when there is no common understanding of process safety. One of the first steps Mexichem has taken is to develop a "fit-for-purpose" management system with an emphasis on developing a common understanding of process safety and ensuring the culture is robust enough to support the system when it is implemented.

To aid Mexichem in this overwhelming task, Risktec Solutions were engaged to develop and deliver a bespoke set of interactive training workshops on a variety of PSM / AIM topics. The first of which was related to the identification and management of safety critical equipment (SCE). This presentation will outline the plan and methodology Mexichem and Risktec have used to transfer such knowledge and discuss some of the successes and challenges along with way.

548216 Alarm Floods and Their Connection to Plant Incidents and Near Misses

Investigations of many catastrophic plant events have shown alarm flooding to have been significant distractions for the operators just prior to the incident. Alarm floods have also been determined as a cause or contributing factor to smaller incidents and near misses.

In fact, many of the more recent findings by U.S. Chemical Safety and Hazard Investigative Board (CSB) include alarm management as one of the key issues in the investigative reports.

Even after several years of trying, many plants still struggle with controlling alarm floods. Many also have problems generating consistent and accurate alarm metric reports for the entire plant. Now that the ISA 18.2 Alarm Performance Metrics are considered "generally accepted good engineering practices," we should make haste to demonstrate where each of our plants fall (particularly the PSM plants) versus these metrics before the regulating authority moves in this area. If we are not able to prove we meet the metrics (particularly for alarm flooding) under all operating conditions, we should remediate.

Static rationalization can reduce your average number of alarms, but without controlling the alarm floods, there is no help for the operator when he needs it the most. For most, we must begin the process by recognizing that alarm floods occur most of all during a change of process state.

The means of controlling alarm floods has been available to us for many years. Unfortunately, only a select few have implemented the solution and are achieving the desired results today. These plants have achieved equal to or better than ISA 18.2 since before it was published.

This session will cover the justification for true alarm management from the safety and economic perspective with several practical steps that everyone should take if they are not currently meeting the ISA 18.2 specifications during all operating conditions.

548318 Do Incident Data & Lessons Learned Really Facilitate Process Improvement?

Regulations are a set of guidelines for the industry for protecting the people and environment, when it comes to process safety systems. While some companies tend to find that complying with regulatory requirements are always a challenge, on the other hand some tend to go overboard in their bid to stay compliant. In spite of the route taken, there are incidents – major, minor and near misses.

Other than the accumulation of this data, the investigations also result in lessons learned. Over the past four decades, lot of incident data and lessons learned has accumulated, been assimilated, and also analyzed. Complementing that are the audits, system reviews, hazard identification exercises, changes to regulations, and knowledge sharing forums.

It is a very efficient and effective mechanism to leverage these vast sources of current information developed, and implement its learnings, in the operations of existing processes. While the core concepts of safety are universal across the various industrial sectors, its application may differ. But once again, the lessons learned databases are effective sources of making this leveraging process more efficient.

This paper will discuss the some case studies of how the concept of lessons learned and the knowledge of the incident data has been leveraged by the authors in various manners ensuring success from one industry to another – example (1) Learnings from an incident due to gaps in the Emergency Response Plan, bringing about success at another chemical manufacturing site; (2) Incident data from a steel plant, when analyzed and applied to a chemical plant resulted in improved of effectiveness.

546673 Structured & Scrutinized Approach Towards HSE Excellence; A Success Story

Author: Usman Ali, Process Safety Engineer

Co-Author: M Ahsan Sarfraz, Unit Manager HSE

This paper studies the practices & programs that support reporting of Near misses, minor unsafe conditions, incidents and quantitative risk assessment strategies in our organization for analyzing and learning from past events of operational failures, injuries, safety and environmental incidents.

A near miss is a leading indicator to an incident that, if scrutinized appropriately, can prevent injuries and accidents. Some persons may be tempted to write off near misses as "no harm, no damage" situations, but organizations who track near misses, determine how and why they occurred, and take corrective action can prevent similar or more serious incidents from occurring in the future. Fatima Fertilizers Limited believes that, "We don't look at what happened; we look at what could have happened".

Keeping in mind the importance of near miss reporting, FFL initiates drive and implement systems for

1. Capturing and increasing near miss reporting at whole site.
2. Logging data for analysis and using this valuable data for identifying gaps in existing process safety system.
3. Implementing sustainable improvement plans and safety campaigns (Safe hands, Safe Routines, Job Cycle Checks, Safe Drive etc) in our process safety systems for making our site a safer workplace.

These actions has improved the site performance which is reflected as Site has now over 45 Safe Million Man Hours, Total Record-able Injury Rate (TRIR) is below 0.2 since last 03 years and No Loss time injury since last 06 years.

548162 Events Report Database Analysis for SHE/PSM Programs Improvements

Incident databases contain records of events that have occurred in a petrochemical facility during recent years. The most of chemical industries report databases are filled with accident valuable data (frequency and severity). Harnessing data mining to explore incident databases, it was used data mining to analyze the incidents database of the studied plant. Inferent approaches are ineffective for low frequency, high consequence events because of their rarity. Given this information limitation, this work uses database analyses to forecast incident frequencies, their relevant causes, equipment involved, and their possible consequences, in specific petrochemical plants.

Databases information analysis, also help to avoid future accidents, thereby reducing the risk. More specifically, this study presents dynamic analyses of incidents in a petrochemical plant ubicated in Mamonal industrial park, in Cartagena- Colombia. Differences in the chemical-specific predictions with the overall predictions are demonstrated. Furthermore, value theory is used for consequence of rare events by formulating distributions for events over a threshold value. Finally, the accidents number is used to estimate the capital at risk within an industry utilizing the frequency and loss-severity distributions.

In conclusion, it was recommended to update de maintenance program (predictive, preventive and corrective) in cases of abnormal changes in production rates, comparing the current production levels, with the analyzed plant production forecast. Also, it was suggested to improve the report culture, due to gaps founded regarding to events classification.

549483 The Process Safety Analysis and Management System in Sinopec

The big data technique has been in trend for the process safety assessment. In china, it's the very need to transfer the qualitative safety management to quantitative safety management, which should rely on the professional risk analysis technique, as well as the standard safety reliability database. Based on the above, it's very important to integrate several professional risk analysis technique, as HAZOP, LOPA, SIL,BOWTIE etc. to realize the integral risk identification and evaluation, grading and management plan making, supporting by the proprietary SINOPEC risk metrics and grading standards, as well as safety equipment reliability database,which is developed as the Process Safety Analysis and Management System for SINOPEC, to support all SINOPEC branches to make dynamic risk analysis and management online

The key data for HAZOP, SIL and Management module in PHAMS is the dependent protection barriers, which we do focusing on the failure probability, at the same time, we do use of the Bayesian Correction Modelling to verify the failure probability of safety equipent and protection barriers on site, which can validly save the grading cost on time and personnel, all of this is towards to the deeper research on dynamic risk analysis for PHAMS.

Thanks to the PHAMS, we has set up and implemented several standards on HAZOP, LOPA and SIL for SINOPEC.

552170 Analysis and Improvement of Safety Integrity Level Considering Human Factors

Safety Instrumented System (SIS) is an irreplaceable active protection layer that prevents the propagation of process accidents and plays an important role in reducing the risk of accidents. In addition to the sensor response triggering the SIS, a human emergency button is often provided in practice to ensure that the SIS is triggered by personnel in the event of the sensor part failure. Due to human factors added, the emergency button will improve the reliability of the SIS. However, there is no research to study the human reliability of the SIS upon sensor part failure in a continuous operation chemical plant, and how human factors can improve and affect the overall reliability of the SIS still remain undiscovered. In this paper, a verification methodology was proposed which integrated with Cognitive Reliability and Error Analysis Method (CREAM) and Reliability Block Diagram (RBD) to emphasize human reliability analysis of the SIS verification. The operator behavior is subdivided into several steps and the Human Error Possibility (HEP) is calculated by a modified CREAM. Then the RBD method is used to verify the SIS reliability considering human factors. An SIS system of one reacting furnace was taken as a case study using this method to show how human factors can improve the reliability of the SIS. The proposed methodology allows SIS verification considering human factors to improve safety integrity level in real practice thus help stakeholders saving costs and making more decisions on SIS implementation.

551560 Ethics for Registered Engineers, a 1PDH Professional Ethics Presentation

LAPELS staff will discuss the Board, enforcement activities and investigative procedures, recent cases, the duties of a firm's supervising professional, ethical issues and other common questions.

Humans Make Mistakes…Which Cause Incidents

Michelle Brown, CSP
FMC
2929 Walnut St.
Philadelphia, PA 19104
Michelle.Brown@fmc.com

Prepared for Presentation at
American Institute of Chemical Engineers
2019 Spring Meeting and 15th Global Congress on Process Safety
New Orleans, LA
March 31 – April 3, 2019

Humans Make Mistakes…Which Cause Incidents

Michelle Brown, CSP
FMC
2929 Walnut St.
Philadelphia, PA 19104
Michelle.Brown@fmc.com

Keywords: human error, investigation

Abstract

When reviewing investigations, it is not uncommon to find human error identified as a cause. However, if you take a closer look at these human errors, most of them are being made by a single group of humans – operators. Are these really the only humans making errors which contribute to or cause events? This paper seeks to answer this question by reviewing all the identified causes of a set of events within a company to determine if errors made by other groups of humans were identified.

1 Introduction

The role of human error in incident causation has been recognized for decades. Many companies have expended significant effort to improve their ability to identify human error in the incident investigation process. A simplified investigation process is shown in Figure 1 below.

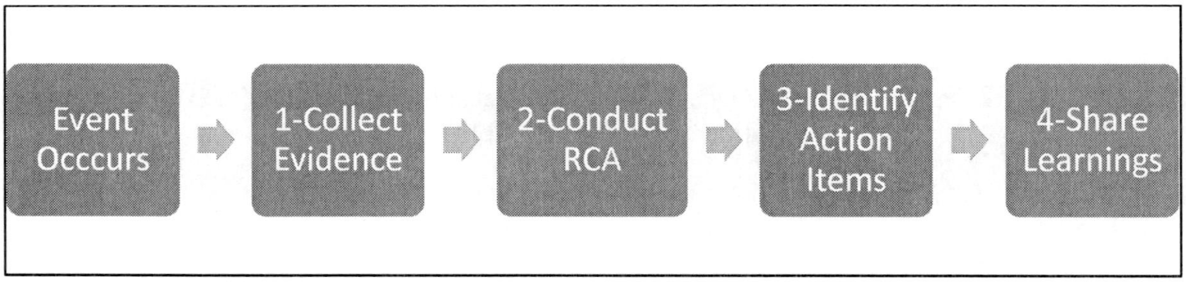

Figure 1. Simplified Investigation Process

The RCA is where various causes are identified. Often, causes are further sorted into different categories. For example: Behaviors, Conditions, and Management Systems. The Behavior category is where human errors associated with the event would most likely be identified.

To evaluate how effectively the investigation process identified human error, I reviewed OSHA Recordable injury event records for my company between 2015 and 2018. Of the 75 injury events which occurred during this time – 19 (25%) identified at least one human error. However, reading through the event descriptions, I would have expected a higher number to have identified one or more human errors within the investigation. What were we missing?

Looking more closely at the 19 investigations which had identified one or more human errors, I found in all 19 cases the human error identified was attributed to the injured party – which in most cases was an operator. This puzzled me – could it be true operator error was the only human error associated with these injury events? Perhaps, what we were missing was the identification of errors by humans other than operators.

Humans are involved in all facets of chemical production. Humans design the equipment used to make chemicals. Humans develop the operating procedures used to make chemicals. Humans supervise crews or shifts of other humans who make chemicals. Humans develop and implement the safe work practices and permits used to make chemicals. Humans make the business decisions needed to make chemicals. Humans maintain the equipment and instrumentation used to make chemicals. Humans program the software used to make chemicals. In fact, without humans we cannot make chemicals.

With this expanded consideration of human involvement in all facets of chemical production, I selected three injury events to review and determine whether the investigation had missed human errors made by humans who were NOT operators.

2 Case Review

Case 1:

Incident Description: An operator needed to transfer a viscous material from a drum to a nearby vessel. Following the transfer, the procedure called for a flush of the transfer pipe and hose with hot water to prevent blockages. A source of hot water was not available in the operator's immediate work area. Therefore, in order to have hot water available to complete the required flush, the operator (wearing the correct PPE) filled a 5-gallon bucket with hot water from a nearby condensate return line (~90 °C) and placed the bucket near the drum. The operator then removed the extra PPE he was wearing to protect himself from the hot water and proceeded to complete the transfer. As he manevered the flexible transfer hose, it hit the bucket, overturning it onto his leg.

Figure 2 below shows the work area just after the event occurred. You can see the flexible hose in the upper right corner of the picture and the overturned bucket that contained hot water in the bottom right.

Figure 2: Work area following event

The investigation revealed the following root causes:

1. Operator did not recognize the hot water hazard once the bucket was filled and placed in work area and removed his PPE once he had filled and moved the bucket of hot water
2. Procedure required flushing of line with 10L of hot water but did not specify additional PPE to protect operator from the hot water hazard
3. Established site practice existed to use hot condensate (~ 80°C) as the flush material and transport it to the work area in a bucket

The first cause was clearly an error on the part of the operator who did not recognize the hot water hazard was still in his work area after filling the bucket. However, the second cause involved the operating procedure, which did not include any information or warnings regarding the hot water required to complete the flushing step. The procedure was written by a human – however, this was considered a management system issue. The third cause revealed a site practice had developed over time to use available hot condensate as the flush water because there were no other sources of hot water available in the immediate area. In other words, the human who wrote the procedure did not provide any information to the operator on where and how they were supposed to safely get the hot water needed to complete the flush. This left it up to the operator and perhaps their supervisor to figure out where to get hot water. All three of the causes identified contained a human error; however, only the first one was identified in the investigation.

Case 2:

Incident Description: An operator was washing out a centrifuge with hot water (~40 °C) wearing the PPE identified in the procedure. The centrifuge had a steam jacketed chute which discharged to another piece of equipment located on the floor below where the operator was working. The chute became completely blocked as the operator was washing – wash water accumulated above the blockage. The operator stopped washing the centrifuge and moved to the floor below and began spraying water up into the blocked chute through a sample door located at the bottom of the chute. Suddenly the blockage released, discharging very hot material (~100 °C) onto the operator. The sample door the operator used to spray into and attempt to unblock the steam jacketed chute can be seen in Figure 3 below.

Figure 3: Sample port on lower portion of steam jacketed chute from centrifuge

The investigation revealed the following root causes:

1. Procedure did not provide information on how to wash the centrifuge
2. Operator was trained; but there was no verification of understanding
3. PPE used did not provide enough protection
4. Centrifuge was not dewatered properly which increased the wash frequency from 2 to 8 times per day. However, the process continued to be operated without taking corrective actions

The first cause was related to the procedure being inadequate. Procedures are often written by engineers and experienced operations personnel. However, in this case the procedure writer did not provide enough information on how the operator was to safely perform this task. The second cause was related to training. The accountability for effective training is a shared one between the operator being trained and the person conducting the training. It is not enough to ensure all the necessary information is included in the training – we must also ensure the person being trained understands the information. The third cause was related to the PPE being inadequate for the hazard of very hot water. Operators do not specify PPE – they use it. PPE specification usually comes from an assessment of the job task which are often done by EHS professionals working with operations personnel. The final cause was related to the operation of the centrifuge itself. In the weeks prior to the event, the need to wash the centrifuge had dramatically increased; however, management elected to continue running the process rather than shutting it down to determine why the centrifuge was not dewatering properly. All four of the causes identified contained a human error; however, only the second, one related to the operator, was identified.

Case 3:

Incident Description: While making rounds, an operator noticed a holding pot was overflowing due to a tripped pump. To stop the overflow, the operator quickly connected a flexible hose to a drain valve on the suction side of the pump and routed the flexible hose to a nearby collection pit. As he opened the valve on the drain line, the flexible hose suddenly disconnected from the drain line releasing hot water into the operator's shoe.

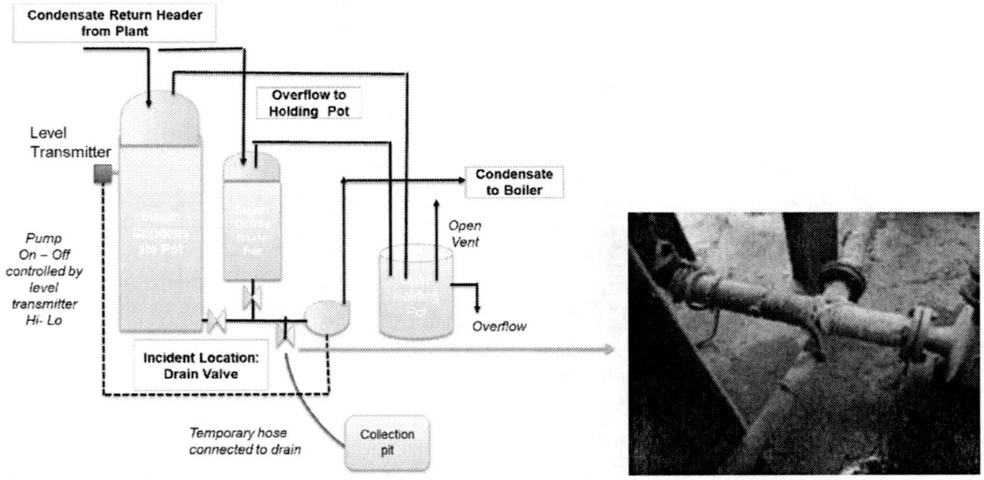

Figure 4: Diagram of Condensate System and picture of drain connection used in event

The investigation revealed the following root causes:

1. Risk assessment for task was not completed correctly
2. Flexible hose was available without guidance regarding pressure rating, service, or connection methods
3. Condensate pot overflow was not piped to a safe location
4. Inadequate vapor venting system allowed excess vapor to accumulate in the condensate causing a vapor lock in the line and tripping the pump
5. Procedure did not include actions to take if pump tripped

The first cause was related to the risk assessment of the task performed by the operator. Although, the task was not part of the procedure, the operator did not complete a safe work permit or other similar risk assessment before acting. This was clearly an error on the part of the operator. The second cause was related to the site not having a flexible hose management process in place. Without an established process, operators were not trained on which hoses and couplings were appropriate for a task based on the operating conditions. Operators do not establish flexible hose management programs – this is done by site leadership working with technical and EHS resources. Once the program is established, operators are trained on the requirements and appropriate hose materials and couplings would be made available. The third and fourth causes related to the design of the process itself. The operator did not design the process, this is done by engineers. The fifth cause was related to the procedure which did not include any information on what actions an operator should take if the pump tripped. Again, procedures are usually not written by an operator – so the lack of this information in the procedure would be the error of the procedure writer. All five of the causes identified contained a human error; however, only the one related to the operator was identified.

As can be seen from the three cases reviewed, most of the human errors (which contributed to the events) were not identified. When investigations identify a condition or management system issue which contributed to an event; it should probe deeper to uncover any potential human errors hidden within them. In this way actions can be taken to address all of the human errors contributing to the event minimizing even further the potential for a similar event to occur in the future.

19 AIChE Spring Meeting
15TH GLOBAL CONGRESS ON PROCESS SAFETY
NEW ORLEANS • 2019

The Influence of Cognitive and Motivational Biases While Establishing Consensus During a HAZOP

Kathryn Grant
Icarus-ORM
931 38 Ave. SW, Calgary AB, Canada
kathryn@icarus-orm.com

Joshua van Lindenberg, Monika Kang, Erik Sherwin, Starr Tze
Icarus-ORM
#240 70 Shawville Blvd SE, Calgary, AB, Canada, T2Y 2Z3
joshua@icarus-orm.com, monika@icarus-orm.com, erik@icarus-orm.com, starr@icarus-orm.com

Prepared for Presentation at
American Institute of Chemical Engineers
2019 Spring Meeting and 15th Global Congress on Process Safety
New Orleans, LA
March 31 – April 3, 2019

The Influence of Cognitive and Motivational Biases While Establishing Consensus During a HAZOP

Kathryn Grant
Icarus-ORM
#240 70 Shawville Blvd SE, Calgary, AB, Canada, T2Y 2Z3
kathryn@icarus-orm.com

Joshua van Lindenberg, Monika Kang, Erik Sherwin, Starr Tze
Icarus-ORM

Keywords: HAZOP, human factors, consensus, biases

Abstract

Hazard and operability studies (HAZOPs) are common process hazard analysis (PHA) tools which have been shown to be an effective method of identifying risks within chemical processing facilities. However, one of the biggest challenges in HAZOPs is assigning the risk ranking of a scenario based on the severity of the consequence and the likelihood of the initiating event. Biases introduced at this stage of the process, including motivational biases and cognitive biases, can lead to inconsistencies in results and inaccurate assessment of the risk. One method of establishing consensus is voting on the severity and likelihood. Voting is first performed prior to detailed discussion of the scenario. Next, the scenario and voting results are discussed to elevate the collective knowledge of the HAZOP participants, and then a second, final vote is performed. In this study, a series of PHAs were performed on a variety of different petrochemical processes (upstream, midstream, and downstream) for multiple different companies (149 HAZOP participants). Participant polling data across different consequence categories (health and safety, environment, financial) was evaluated to determine if there were any trends in voting patterns based on the participant role (engineer vs. operator vs. project manager vs. facilitator) or the number of years of experience of the participants. The two primary factors which were evaluated were risk acceptance (how severe an individual deemed the consequence to be compared to the group consensus) and persuadability (how likely an individual was to change their mind during the second vote after group discussion). This information is useful for determining biases that arise while assessing the severity and likelihood for HAZOP scenarios and determining which key people should participate in a HAZOP. The information from this study can be used to help develop strategies that can be used to reduce these biases, with the ultimate goal of improving facilitator techniques and HAZOP results.

1 Introduction

Hazard and operability studies (HAZOPs) are common process hazard analysis (PHA) tools which have been shown to be an effective method of identifying risks within chemical processing facilities. However, one of the biggest challenges in HAZOPs is determining the inherent risk ranking of a scenario, which is based on the severity of the consequence and the likelihood of the initiating event. Biases introduced at this stage of the process, including motivational biases and cognitive biases, can lead to inconsistencies in results and inaccurate assessment of the risk. This in turn can lead to either elevated risk to the facility if the risk is underestimated, or a requirement to add additional safeguards, which may not enhance the safety of the facility, if the risk is overestimated.

In order to overcome this challenge, Icarus-ORM has used a polling technique to survey the HAZOP participants and determine the appropriate risk ranking. The polling results from HAZOP sessions in the petrochemical industry were analyzed to determine trends in voting patterns and assess biases present when determining the inherent risk ranking. Polling results from 17 different HAZOPs were analyzed. The facilities assessed were all upstream, midstream, and downstream petrochemical processes. The study evaluated projects for five different companies and a total of 149 HAZOP participants.

2 Methods

2.1 Polling Methodology

When assessing the inherent risk for a HAZOP scenario, polling was used to determine the risk ranking. Each participant was given a chance to vote for the severity that they thought was representative of the worst credible consequence for that category. Severity categories assessed include health and safety (H&S), environment (ENV), and financial (F). The facilitator then initiated a discussion and elicited explanations for the various votes, thus allowing all participants to express their views. A second vote was then conducted on the same question, and results were reviewed again. The same methodology was followed to determine the likelihood of the initiating event (L).

2.2 Analysis of Results

2.2.1 Conservatism

The Conservatism Score assigned to participants was based on the deviation between an individual's vote and the average vote, during the first vote (prior to group discussion).

Example:

Participant	Severity for health and safety (SV)	Conservatism Score (C)
Project Manager	3	-1
Engineer	4	0
Operator	4	0
Facilitator	5	1
Average	4	

$C_{Facilitator} = SV_{facilitator} - SV_{ave} = 5 - 4 = 1$

A higher severity score indicates a more serious consequence and a higher likelihood score indicates a higher likelihood for the initiating event (Note: The exact scale and severity definitions for each company varied slightly). A positive conservatism score indicates the individual voted more conservatively than the average of the group (i.e. the individual voted that the consequence would be more severe). A negative conservatism score indicates the individual voted less conservatively than the average of the group (i.e. the individual voted that the consequence would be less severe).

2.2.1 Persuadability

The persuadability score assigned to participants was based on the difference between an individual's second vote (after group discussion) and their first vote (prior to group discussion).

Example:

Participant	Severity for health and safety on first vote (SV1)	Severity for health and safety on second vote (SV2)	Persuadability Score (P)
Process Safety	3	3	0
Engineer	4	3	-1
Operator	4	4	0
Facilitator	5	3	-2

$P_{Engineer} = SV2_{Engineer} - SV1_{Engineer} = 3 - 2 = -1$

A value of zero for the persuadability score indicates that the participant did not change their mind for the second vote. A positive persuadability score indicates the participant voted more conservatively in the second vote than the first vote (i.e. the individual voted for a higher severity consequence), and a negative persuadability score indicates the participant voted less conservatively in the second vote than the first vote (i.e. the individual voted for a lower severity consequence).

2.2.1 Categories

Results were split into categories based on participants' role in the project and their self-reported experience level. Roles which were assessed included engineers, operators, project managers, process safety, and facilitators. Experience levels assessed include <5 years, 6-10 years, 11-15 years, 16-20 years, and >20 years.

3 Results

3.1 Conservatism

Overall it was observed that engineers had the smallest conservatism score and were most likely to align with the average severity or likelihood ranking. People in a facilitator or process safety role were more conservative than average for all categories, while project managers were less conservative on health and safety and likelihood, but more conservative on environmental and financial consequences. Operators were very close to the average value overall, with less conservative rankings in the health and safety category and more conservative rankings in the environmental and likelihood categories. There were no obvious trends observed for years of experience. There was higher variability in the 5-10 years of experience and 11-15 years of experience categories than other experience levels.

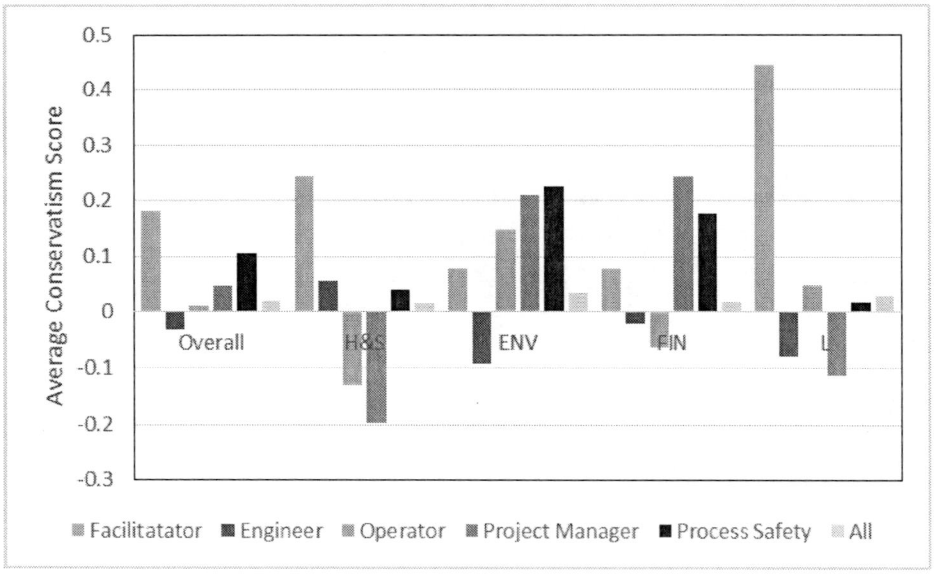

Figure 1. Average conservatism score by category for different roles

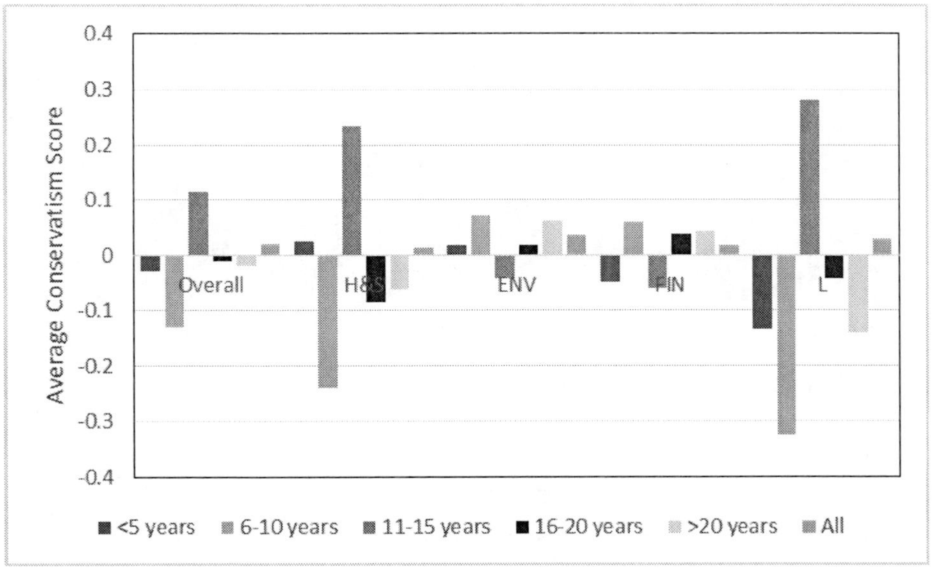

Figure 2. Average conservatism score by category for different years of experience

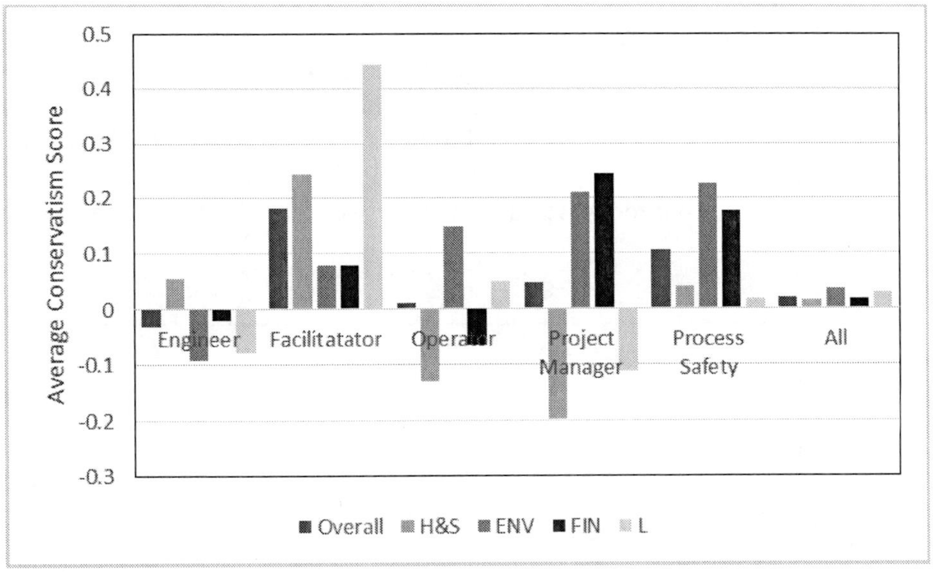

Figure 3. Average conservatism score by role for different categories

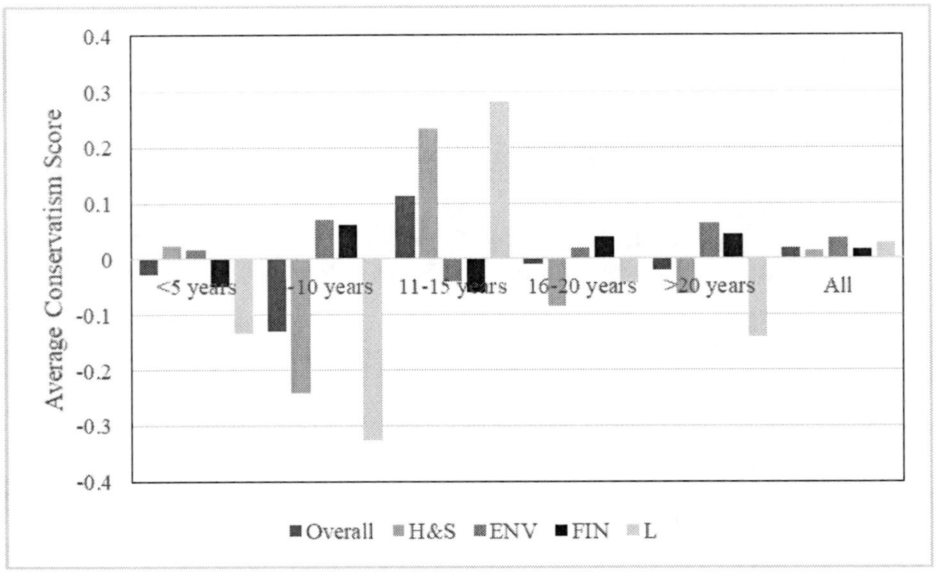

Figure 4. Average conservatism score by years of experience for different categories

3.2 Persuadability

Overall, the entire HAZOP group appeared to vote more conservatively on the second vote than the first vote. The likelihood category had the largest difference compared to the other consequence categories. Out of all the roles, facilitators had the smallest change between their first and second votes. Engineers voted more conservatively on their second vote in all categories, while operators voted more conservatively on their second vote in all categories except the environmental severity. There were no obvious trends observed for years of experience.

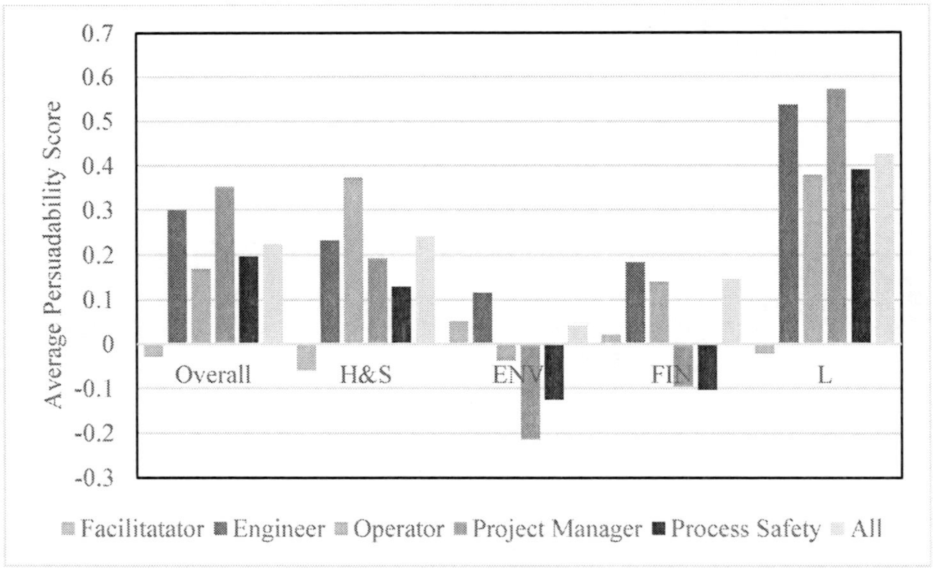

Figure 5. Average persuadability score by category for different roles

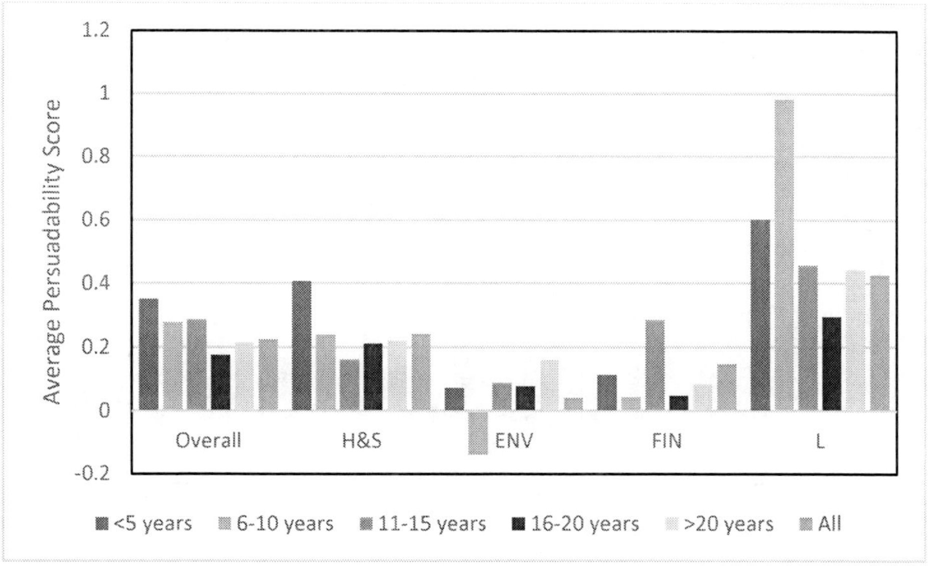

Figure 6. Average persuadability score by category for different years of experience

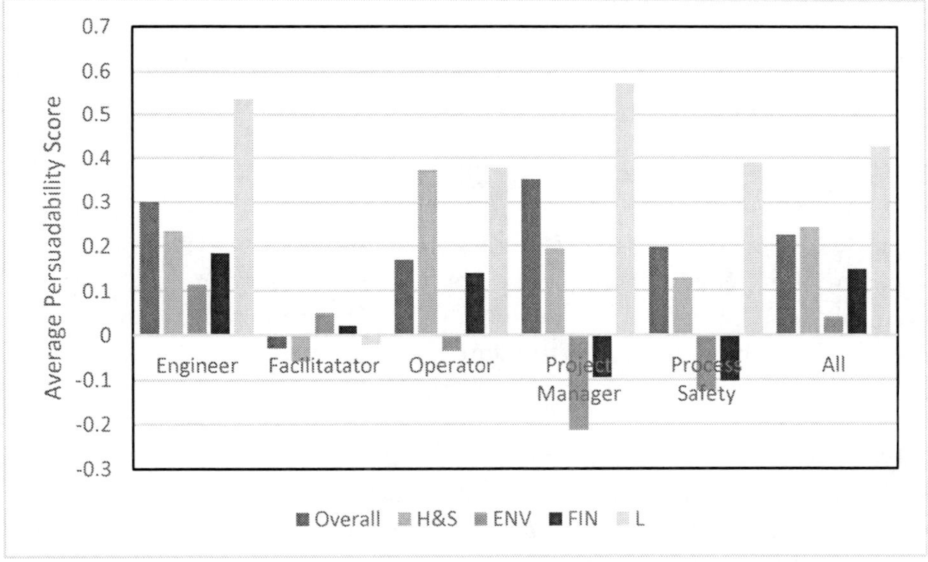

Figure 7. Average persuadability score by role for different categories

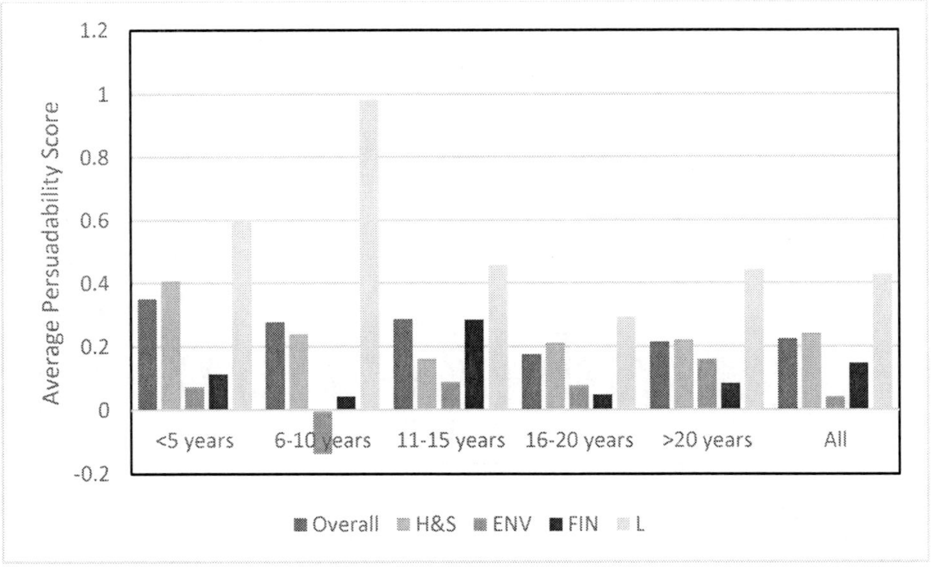

Figure 8. Average persuadability score by years of experience for different categories

4 Discussion and Conclusions

It was observed that individuals in facilitator or process safety roles were more conservative than average in all categories. Both these roles involve identifying potential hazards and preventing incidents on a regular basis, as well as being familiar with historic events. It is also interesting to note that facilitators had the smallest change between their first and second votes. This indicates that they were less likely to be influenced by others in the group. Project managers were found to be more conservative than average on financial and environmental consequences, but less conservative on health and safety and likelihood than the overall group. The overall HAZOP group voted more conservatively on the second vote than on the first vote. This shift may be the result of a strong argument made by those voting more conservatively in the first round.

The variety of opinions between different roles highlights that there may be conflicting objectives between those who are responsible for identifying the hazards and those who are responsible for implementing changes and budgeting for additional safeguards. This demonstrates the importance of having multiple perspectives in the HAZOP, including someone such as a facilitator or process safety expert who may have more exposure to historical incidents. Overcoming motivational biases caused by factors such as project budget is important to find the best long-term solution for the project. A facilitator who can ask the appropriate questions to take advantage of the expertise of all members of the group can help bring value to the group by ensuring all opinions are considered.

The largest difference in votes before and after the group discussion was observed in the likelihood category. This is interesting, because we have observed that during the group discussion this is one of the areas in which there appears to be the most uncertainty among the group. This emphasizes the importance of good data either from operating experience or databases to support decisions surrounding the likelihood of failures. This study highlights this area as one of the areas with the most uncertainty and where a facilitator or PHA expert can help provide added value to the HAZOP process by providing the group with resources and engaging the appropriate people in the discussion.

There were no obvious trends observed based on the number of years of experience of participants in either persuadability score or conservatism score. There was greater variability in the 6-10 years category and the 11-15 years category, but this can likely be attributed to the higher number of participants in these categories. Based on these observations, it appears that experience does not have an impact of either the conservatism or persuadability of HAZOP participants.

5 Future Studies

While this study highlights several trends observed regarding the conservatism and persuadability of HAZOP participants when establishing consensus around consequence and likelihood rankings, it is important to note there are several limitations to this study. Since data was taken from real HAZOP studies and different companies had slightly different risk matrices, the observations in this study represent trends, and were not analyzed for statistical relevance.

In future studies it would be interesting to compare data across different companies to see if there was a difference based on the safety culture of a given company. In this case, only relative risk rankings were assessed, because the scenarios evaluated varied for each HAZOP. Other plans for future studies include expanding the data set to get more relevant data, including data from other industries.

6 References

[1] S. Tze. "Process Hazard Analysis Handbook: Advanced Edition," Icarus-ORM Academy, 2018.

[2] P. Baybutt. "A Framework for Critical Thinking in Process Safety Management," Process Safety Progress, pp. 337-340, December 2016.

[3] Z. Tsiga, M. Emes, and A. Smith. "Attitudes to risk in petroleum projects," Procedia Computer Science, pp. 305-312, October 2016.

[4] I. Sutton. "Quantifying Process Safety With Expert Judgement," Society of Petroleum Engineers, March 2009.

[5] P. Baybutt. "Process safety incidents, cognitive biases and critical thinking," Hydrocarbon Processing, pp. 81-82, April 2017.

[6] H. Duhon and I. Sutton. "Why We Don't Learn What We Should From HAZOPs," SPE Projects, Facilities & Construction, pp. 104-109, June 2010.

Best Practices for Addressing Human Factors during PHAs/HAZOPs
… Especially during PHA of Non-normal Modes of Operation (Startup, Shutdown, & Online Maintenance)

By: William Bridges
Process Improvement Institute, Inc. (PII)
1321 Waterside Lane, Knoxville, TN 37922 USA
wbridges@piii.com

PROCESS IMPROVEMENT INSTITUTE

By: Rashed Al-Zahrani
UNITED, a SABIC affiliate
ZahraniRA2@united.sabic.com

Prepared for Presentation at
American Institute of Chemical Engineers
2019 Spring Meeting and 15th Global Congress on Process Safety
New Orleans, LA
March 31 – April 3, 2019

Best Practices for Addressing Human Factors during PHAs/HAZOPs … Especially during PHA of Non-normal Modes of Operation (Startup, Shutdown, & Online Maintenance)

By: William Bridges
Process Improvement Institute, Inc. (PII)

By: Rashed Al-Zahrani
UNITED, a SABIC affiliate

Keywords: PHA, Human Factors, Process, Hazard Evaluation, HAZOP, Safety, Factors, Human Error, Operations, Maintenance, Hazards, Process Safety, Review, HRA, PSM, Inspection, Programs, Reliability, Ammonia, Emergency, Recommendations, Checklist, FMEA, Instrumentation, Process Hazard Analysis, Risk Management, Initial, LOPA, Materials, Relief, P&ID, Risk Assessment, Investigatory, Layer of Protection Analysis, Risk-Based Process Safety, RBPS

Abstract

Hazard evaluations, also called process hazard analysis (PHAs) have been performed formally in gradually improving fashion for more than five decades. Methods such as HAZOP and What-If analysis have been developed and honed during this time. Some weaknesses identified 30 years ago still exist in the majority of PHAs performed around the world. Critically, most PHAs do not thoroughly analyze the errors that can occur during startup, shutdown, and other non-routine (non-normal) modes of operations; sadly the commonly used approaches for PHA of continuous mode of operation only find about 5 - 10% of the accident scenarios that may occur during startup, shutdown, and online maintenance. This is true even though about 70% of major accidents occur during non-routine operations. Instead of focusing on the most hazardous modes of operation, most PHAs focus on normal operations (e.g., HAZOP of equipment nodes). In a majority (perhaps more than 80%) of both older operations and new plants/projects, the non-routine modes of operations are not analyzed at all. This means that perhaps 70% of the accident scenarios during non-routine operations are being missed by those PHAs. **If the hazard evaluation does not find the scenarios that can likely occur during these non-routine operations, the organization will not know what safeguards are needed against these scenarios.**

Chapter 9 of "Guidelines for Hazard Evaluation" (3rd Ed, CCPS)[1], requires hazard evaluations of all hazards of the process during all modes of operation and it describes the Best Practices for this analysis. The US OSHA PSM regulation requires PHA of all hazards during all modes of operation as well, and several key citations since 1990 have focused on PHA of non-normal modes. US CSB and other agencies have also recognized this weakness in PHAs.

Another key weakness is that most PHAs fail to truly evaluate human factors when brainstorming and analyzing potential accident scenarios. Most PHA leaders are not trained in human factors and most do not know the limits of control we all have on human errors.

This paper describes an approach for integrating human factors considerations into hazard evaluations of process designs, operating procedures, and management systems. In the description of our approach, we cite OSHA's and EPA's definitions for consideration of human factors during PHAs. Critical issues related to human factors can be identified and addressed in different phases of a hazard evaluation. Case studies illustrate the effectiveness of this strategy. This paper also explains the business case for doing PHAs of procedure steps for non-routine modes of operation, while also describing the growing regulatory pressure from US OSHA and others. The paper recaps the practical ways to efficiently and thoroughly analyze the step-by-step procedures that are used to control non-routine operating modes, as well as those for batch and between batch operations.

INTRODUCTION

Human error in research, design, construction, installation, operation, maintenance, manufacturing, inspection, management, etc., can be considered the cause of almost all industrial accidents. (Experts typically quote that about 85% of accidents are caused by human error, though some say that except for natural disasters this figure is 100 %.) However, simply attributing these incidents to "human error" without evaluating the root cause implies that the errors are inevitable, unforeseeable, and uncontrollable. Nothing could be further from the truth.

Human errors are sometimes mistakenly called procedural errors. This is not true anymore than saying all equipment errors are due to design errors. People make mistakes for many reasons, but PII estimates that only about 10% of accidents due to human errors in the workplace occur because of *personal* influences, such as emotional state, health, or carelessness; most human error is due to weaknesses in the control of human factors. Over the past five decades of industry research and observation in the workplace on human error, we have come to know that human error probability depends on many factors.

These factors (described in more detail in *Human Factors Missing from PSM* [2]), include those shown below (note that the percentages shown below were developed by PII after analysis of more than 15,000 process safety, safety, and operational incidents):

- Procedure accuracy and clarity (the number one, most cited root cause of accidents):
    - A procedure typically needs to be 95% or better accuracy to help reduce human error; humans tend to compensate for the remaining 5% inaccuracies in a written procedure.
    - A procedure must clearly convey the information (there are about 25 rules for structuring procedures to accomplish this) and the procedure should be convenient to use.
    - Checklist features – These should be used and enforced either in the procedure or in a supplemental document.

- Training, knowledge, and skills
    - Employees should be selected with the necessary skills before being hired or assigned to a department.
    - Initial Training – There must be effective training. The initial training should be demonstration-based training on each proactive task and each reactive (e.g., response to alarm) task.
    - Ongoing validation of human action is needed and usually should be repeated (in either actual performance or in drills/practice) at least once per year (as discussed later in this paper). For human IPLs or safeguards, the action should be demonstrated to be "fast enough" as well.
    - Documentation – the human performance should be documented and retained to demonstrate the error rates chosen are valid.

- Fitness for Duty – Includes control of many sub-factors such as fatigue (a factor in a great many accidents), stress, illness and medications, and substance abuse.

- Workload management – Too little workload and the employee becomes bored (reducing alertness, increasing distractions), while too much overwhelms the employee (increasing stress, decreasing time per task); both cases can increase human error.

- Communication – Miscommunication (of an instruction or set of instructions or of the status of a process) is the second or third most common cause of human error in the workplace. There are proven management systems for controlling communication errors (such as repeat back, use of common jargon).

- Work environment – Factors to optimize include lighting, noise, temperature, humidity, ventilation, and distractions.

- Human System Interface – Factors to control include layout of equipment, displays, controls and their integration to displays, alarm nature and control of alarm overload, labeling, color-coding, fool-proofing measures, etc.

- Task complexity – Complexity of a task or job is proportional to the (1) number of choices available for making a wrong selection of similar items (such as number of similar switches, number of similar valves, number of similar size and shaped cans), (2) number of parallel tasks that may distract the worker from the task at hand (leading to either an initiating event or failure of a protection layer), (3) number of individuals involved in the task, and (4) judgment or calculation/interpolation, if required. For most chemical process environments, task complexity is typically low (one action per step), but for response actions (human IPLs) there are almost always other tasks underway when the out-of-bounds reading occurs or the alarm is activated.

In addition to the human factors listed, other considerations for use of a human as an IPL include (1) time available to perform the action and (2) physical capability to perform the action safely.

Other papers provide much more detail on each human factor and the relative weighting of each.[2, 3, 4]

These human-error causes (human factors), which in turn result from other human errors, are all directly within management's control. When using human error data for controlling initiating events (IEs) and independent protection layers (IPLs), the site should ensure that the factors above are consistently controlled over the long-term and that they are controlled to the same degree during the mode of operation that the PHA, HAZOP, What-if, FMEA, or LOPA covers. For instance, if workers are fatigued following many extra hours of work in a two week period leading up to restart of a process, then the human error rates can increase by a factor of 10 times or 20 times during startup.[5]

Although this paper focuses on the requirements of a PHA (a redo or new hazard evaluation of an entire process), the approach is equally effective for other hazard evaluations such as preliminary and detailed design reviews (for new/revised processes) and large management of change hazard reviews.

INCREASING REGULATORY PRESSURE

Industry has taken some initiatives on resolving this problem. One initiative was to improve the focus on PHA of non-routine procedures as part of the update to "Guidelines for Hazard Evaluation"[1]. A new Chapter 9, Section 1 was added that necessitates hazard evaluations of all hazards of the process during all modes of operation. This textbook also explained why, when, and how to perform such analysis of step-by-step procedures.

Many companies have taken the initiatives to do the same, including about 20% of the largest chemical, petrochemical, and refining companies. But, the vast majority of companies who should be analyzing step-by-step deviations are not; and the major accidents continue to occur partly because of this. As a result, US regulators are beginning to increase pressure on regulated companies to perform PHA's of All modes of operation.

US OSHA Regulation and Enforcement:

The US OSHA PSM regulation requires PHA of all hazards during all modes of operation as well, and several key citations since 1990 have focused on PHA of non-normal modes.

- **Before there was a PSM regulation from US OSHA**, the agency published CPL 2-2.45 (Systems Safety Evaluation of Operations with Catastrophic Potential)[6]. In this guidance document, OSHA stated that a human error analysis should address:
  - *Consequences of failure to perform a task.*
  - *Consequences of incorrect performance of a task.*
  - *Procedures and controls to minimize errors.*[6]

  This approach is still the fundamental analysis method for PHA of non-normal modes of operation.

- **Phillips 66 "PHA" Citation —A citation with 566 instances was issued to Phillips 66 in Pasadena, TX, following their 1989 disaster** that killed 23 workers.[7] **The citation was related to a violation of the General Duty Clause (Section 5(a)(1) of OSH Act of 1970).** US OSHA cited Phillips against the General Duty Clause, since the PSM standard (29 CFR 1910.119) had not yet been issued. OSHA cited Phillips for not protecting its workers from hazards of fire/explosion by, among others, not performing a PHA that should have included an evaluation of the effect of design modifications on operator performance, and the identification of the source of observed human error and the identification of human factors that could result in incident event sequences. The citation stated, "**This review should result in a systematic listing of the (1) types of errors likely to be encountered during normal or emergency operation, (2) factors contributing to such errors, and (3) proposed system modifications to reduce the likelihood of such errors".**

  The settlement agreement[8] between US OSHA and Phillips included the following requirements for process hazard analyses (PHAs) of the rebuilt and surviving units:
  - "Phillips will analyze each process…and will include human factors analysis … [and] will be …led by an independent consultant."
    - William Bridges (of JBF Associates at the time, now with PII) led these PHAs. Before these PHAs began, OSHA, Phillips, and Mr. Bridges decided that the best approach for finding all human error scenarios was to perform a HAZOP of deviations of the steps for the procedures governing

activities for startup, shutdown, and particularly online maintenance.
- "Phillips will provide OSHA an independent consultant's evaluation of the adequacy of its settling leg maintenance procedures performed while the polyethylene reactors are in operation..."
    - As part of the settlement to meet this requirement, it was decided by JBFA, Phillips and OSHA to perform a Human Reliability Analysis (HRA) of the Setting Leg online maintenance procedure, to ensure that the statistical risk of the accident recurring is less than the background risk of driving to work.

The PHA and HRA resulting from the Phillips settlement agreement is presented as a Case Study later in this paper for sake of clarity.

- **Paragraph (e) of the US OSHA regulation on PSM, 29 CFR 1910.119,[9] and similar requirements in US EPA's rule for risk management programs (RMP), 40 CFR 68.24,[10]** specifically require that PHAs consider and address hazards of the process, i.e., all hazards regardless of the mode of operation (routine or non-routine).
    - 29 CFR 1910.119(e)(1) states that the PHA, "shall identify, evaluate, and control **the hazards** involved in the process"
    - 29 CFR 1910.119(e)(3)(i) states that the process hazard analysis shall address "**The hazards of the process**".
    - 29 CFR 1910.119(e)(3)(vi) states that the process hazard analysis shall address human factors.
    - Appendix C to the OSHA PSM standard states that both routine and non-routine activities need to be addressed by the PHA of the covered process.

There is no qualifier that limits the OSHA PHA requirement to only routine modes of operation. PSM requires that **all hazards** related to the process be addressed, regardless of the mode of operation or activity (routine or non-routine).

- **OSHA Inspection No. 103490306 (Nov 2, 1992).[11]** In the first major PSM inspection in 1992 using 29 CFR 1910.119, OSHA assessed a serious violation when the PHAs did not address "human factors such as board operator error, line breaking mistakes, and improper lockout and isolation of process equipment," all of which are errors originating from failure to either perform tasks or perform them correctly.

- **US OSHA published an internal document on Program Quality Verification of Process Hazard Analysis in 1993 (by Henry Woodcock, of OSHA).[12]** This document states that a PHA should include analysis of the "procedures for the *operation* and *support* functions" and goes on to define a "procedure analysis" as evaluating the risk of "skipping steps and performing steps wrong." The authors concur and PII has found the same true in PHAs that we have performed using various methods; a 2 Guideword HAZOP approach is normally optimal for PHA of procedures.

- **OSHA Inspection No. 123807828 (Nov 18, 1993)**[13] – Ashland Oil, Catlettsburg, KY. Several operators were preparing to ignite a 2-B-3 crude heater after a two week turnaround. The lead operator had two very inexperienced workers helping him light the heater. A large quantity of fuel gas entered the heater before the pilot light was ignited. The resulting explosion killed one employee, who received fatal injuries to the back of his head. The operators bypassed safety shutdown features; poor engineering allowed this to occur and should have been discovered in the PHA. In addition, they did not check the firebox to ensure that it was gas-free before lighting the heater.

  The Kentucky OSHA citation read: *The PHA did not address all hazards of the #2 Crude unit;.... The PHA did not address the hazards associated with the startup of the crude unit after a turnaround, ...emergency shutdown..., emergency operations and normal shutdown of the unit. The process hazard analysis that was completed by the PHA team for the #2 Crude unit only evaluated the hazards associated with normal mode of operation of the #2 Crude unit.*

  Settlement: All procedures were re-written and all PHAs were redone to include a PHA of deviations from procedural steps for all non-continuous modes of operation.

- **Recent US OSHA PSM National Emphasis Programs** for Chemical Processes[14] and also for Refineries[15] underscore the need for companies to identify potential accident scenarios during non-routine modes, and to reduce the frequency and consequences of such errors as part of an overall process safety management (PSM) program.

OSHA recognizes that CCPS/AIChE has added as Chapter 9.1 in the 3rd edition of *Guidelines for Hazard Evaluation*[1] to further emphasize the need for a PHA to include hazard evaluations of all modes of operation and that this chapter has added best-practice detail on the approach for doing the hazard evaluation of startup, shutdown, and online maintenance modes of operation. Despite the specific OSHA standard that requires PHAs of covered processes must address all hazards, many PHAs still do not address hazards during all modes of operation. Further, many of the regulated community have stated "Well, OSHA did not tell us to perform a PHA of procedures for non-routine modes of operation." On the other-hand, OSHA did not state to do only a hazard evaluation of normal mode of operation and stop there.

To highlight the importance that PHAs address hazards during all modes of operation and activities (routine and non-routine), OSHA is considering issuing a Hazard Alert that would incorporate the concepts in Chapter 9.1 of *Guidelines for Hazard Evaluation Procedures, 2008, CCPS/AIChE.*[1] Also, as stated above, OSHA has an enforcement initiative, CHEM NEP, that utilizes a list of dynamic questions that OSHA compliance officers use to evaluate compliance at facilities covered by the program. **It is possible that future dynamic list questions could address PHAs of all modes of operation, and is further possible that this CHEM NEP update is drafted and waiting for release.**

Pressure from the US Chemical Safety and Hazard Investigation Board (US CSB)

The CSB has commented on the need for PHAs to address all hazards of the process during all modes of operation. Their clearest statement was in the report 2008-08-I-WV-R1[16] on the Bayer CropScience accident in Institute, WV, 2008. In that report, CSB asks Bayer to:

- Revise the corporate PHA policies and procedures to require:
    a. Validation of all PHA assumptions to ensure that risk analysis of each PHA scenario specifically examines the risk(s) of intentional bypassing or other nullifications of safeguards,
    b. **Addressing all phases of operation and special topics including those cited in chapter 9 of "Guidelines for Hazard Evaluation Procedures" (CCPS, 2008),**
    c. Training all PHA facilitators on the revised policies and procedures prior to assigning the facilitator to a PHA team, and
    d. Ensure all PHAs are updated to conform to the revised procedures.

US EPA's RMP Regulation

In the Risk Management Program rule (40 CFR 68.24)[10] EPA also recognizes the importance of procedural analysis, by defining the purpose of a PHA to **"examine, in a systematic, step-by-step way, the equipment, systems, and *procedures* (emphasis added) for handling regulated substances."**

> *A well-done PHA should identify all failure scenarios that could lead to significant exposure of workers, the public, or the environment…….For toxics under PSM, however, you may plan to address a loss of containment by venting toxic vapors to the outside air. In each circumstance, a PHA should define how the loss of containment could occur. However, for EPA, the PHA team should reassess venting as an appropriate mitigation measure. (From EPA RMP Guidance, Chapter 7, pgs 7-6 & 7-7; General Risk Management Program Guidance.[17])*

Example of Local Regulation: Contra Costa County Hazard Materials Program

The counties in California are the implementation and enforcement agencies for the US EPA RMP regulation, which in California is termed, California Accidental Release Prevention (CalARP) regulations. One premier implementer is Contra Costa County (CCC). In addition to the standard requirements found in EPA's RMP regulation (which has requirements essentially identical to OSHA PSM), CCHMP has also added their own initiatives to improve how the 10 major facilities in CCC address human factors and PHAs of all modes of operation. The Industrial Safety Ordinance (ISO)[18] specifically requires that each site perform a PHA of procedures, just to be certain PHAs of all modes of operation are performed. One question in the county's auditing protocol is:

- Did the Stationary Source perform Procedural PHAs to evaluate potential active failures or unsafe acts in the procedure such as missed or out of sequence steps and including raising questions regarding the availability of personnel to perform a task as specified in the procedure? [Section B: Chapter 4.3 of the CCHMP Safety Program Guidance Document]

Conclusions on Regulatory Pressure

Clearly, the regulatory pressure is increasing for industry to perform a PHA that thoroughly addresses hazards during all modes of operation, including deviations from steps in startup, shutdown, and online maintenance procedures.

By the way, a similar focus is underway by the same government entities listed above to improve the coverage of all damage mechanism (corrosion, erosion, external impact, etc.) within a PHA. (such as CalOSHA's proposed rule for refineries). The 2008 update to the book *Guidelines for Hazard Evaluation Procedures*[1] was also to address weaknesses observed (across the industry) by US CSB (and others) in coverage of damage mechanisms within PHAs; US CSB requested these changes from CCPS.

Approach for Fully Addressing Human Factors in Hazard Evaluations - 4 Steps

To fully address human factors during PHA/HAZOP, a four-step approach is suggested.

- Step 1: Ensure education and experience of PHA/HAZOP Leaders in human factors and human error prevention. Also ensure the PHA/HAZOP Leaders are competent in PHA of procedures (which is very different than PHA of equipment nodes)
- Step 2: Ensure the brainstorming of scenarios includes consideration of human error and even multiple human error as causes; be specific as possible on the human error.
- Step 3: Have the PHA team perform a hazard review of procedure steps using a HAZOP or What-If analysis to uncover potential human errors associated with modes operations such as startup, shutdown, and online maintenance.
- Step 4: Supplement the PHA/HAZOP of the individual scenarios with a checklist analysis of general human factor issues, to ensure all major categories were addressed.

STEP 1 - Ensuring PHA/HAZOP Leaders are Competent in Human Factors and PHA of Procedures

The PHA Leader's understanding of human factors and human error prevention (HEP) will be the most significant indicator in whether the PHA will be successful in identifying the hazards caused by human error (the team will not be equipped to do this on their own, especially for scenarios that have not yet occurred). This understanding required of the PHA Leader is not a given trait that all leaders share (though it should at least be covered partially as part of PHA Leadership Training), it is a unique skill set that must be developed deliberately by the Leader, specifically by completing focused courses in human factors/human error prevention, participating in coaching by mentor/senior PHA Leaders, and studying the relevant statistics regarding human factors and human errors in industry. This training, along with several years in operation or safety roles to provide experience, will give the leader the ability to quickly recognize potential human factors (human error modifiers such as high fatigue or miscommunication) in the modes of operation being analyzed, and will provide the necessary tools to systematically brainstorm with the team in a way that maximizes hazard identification and meeting effectiveness. A large portion of these human error related hazards can only be identified with analysis of procedure steps (PHA of procedures, as part of the non-normal/non-continuous modes of operation), so the leader must be adept in techniques required for such analysis, including methods such as What If or 2 or 7 Guideword brainstorming of procedure step deviations, and the rules for analyzing and documenting procedure deviations and safeguards listed, discussed below in *STEP 3*.

STEP 2 - Ensuring PHA/HAZOP Leaders Look for and Find Human Error Causes during Routine and Non-routine modes of Operation

The training discussed above in human factors and human error prevention is mostly wasted if it is not applied to non-routine modes of operation (normal modes of operation generally are less dependent on human error). Many companies have not learned the basic truth that accidents that can occur during startup, shutdown, and online maintenance have little to do with accidents that occur during continuous more of operation. In fact, safeguards that adequately protect against accidents in normal mode, are of little or use for some scenarios that are unique to these non-routine modes of operation. Finding the unique safeguards necessary for the each mode of operation is critical. In addition to standard brainstorming of step deviations, the leaders and scribes most look for multiple simultaneous causes of an accident scenario (double jeopardy). This may seem like a conflict with traditional HAZOP rules, but in fact, there has never been an official rule against consideration of double jeopardy, though there has been recognition that none of the qualitative PHA methods are thorough in finding ALL double-jeopardy scenarios. But during PHA of steps of procedures, it is fairly common for a double jeopardy to occur.

Therefore, the **company MUST require PHA of all modes of operation, and the analysis must be done the right way (as described in this paper).** This must be in a standard that is enforced by the company. At a minimum, the standard must:

1. Ensure the PHA Leaders are trained (Step 1) in the methods that best suit those modes of operation (like those in Steps 3 and 4), providing all procedures and related materials they will require (and ensuring that these materials are accurate and up to date), and that

the PHA team includes the necessary expertise (experienced operators, senior operators/trainers as needed, etc.).

2. Ensure that the right methods are in fact used for leading and documenting the PHA of non-routine modes:
   a. This includes auditing the PHA meeting, as it is in progress, to ensure PHA is complying with Steps 3 and 4.
   b. Auditing the PHA report to ensure completion of the analysis per Step 3 and 4. Was there a PHA of steps? Were the right methods used? Is there evidence of unique safeguards being found for unique scenarios for each mode of operation? Is the evidence of discussion of double-jeopardy scenarios?

STEP 3: PHA of Non-Continuous Modes of Operation

During the period 1970 to 1989, 60% to 75% of major accidents in continuous processes occurred during non-routine modes of operation (principally startup and online maintenance modes).[19] This trend has continued unabated for most of the process industry to the present day. A compilation of 47 major process safety accidents from 1987 to 2010 was provided in an earlier paper on this topic (Bridges, et al) [20]; of these, 69% occurred during non-routine operations. In addition, a poll of over 50 clients indicates that 70% of their moderate and major accidents occurred during non-routine modes of operation. This data is particularly disturbing when factoring in the time at risk, since most continuous processes are typically shut down 5% or less per year. Therefore, for many continuous processes the workers and other stakeholders are 30 to 50 times more likely to have a major accident during the time frame of startup, shutdown, or on-line maintenance modes of operation.

One reason for processes being at higher risk during these operating modes is many of the safeguards (independent protection layers; IPLs) are bypassed or may not be fully capable in these modes. A hazard evaluation is necessary to help a company identify the layers of protection necessary to lower the risk to acceptable levels. To fulfill this need, a company operating a continuous process should **fully** evaluate the hazards during **all** modes of operation. Unfortunately, in the first four decades of hazard evaluation use (beginning after the Flixborough disaster in the UK in 1974 – an accident that occurred during startup in a temporary, poorly engineered configuration), many companies have done a poor job of identifying and evaluating accident scenarios during startup, shutdown, and online maintenance modes of operation, while usually doing a good job of evaluating hazards of normal modes (continuous or normal batch modes) of operation.

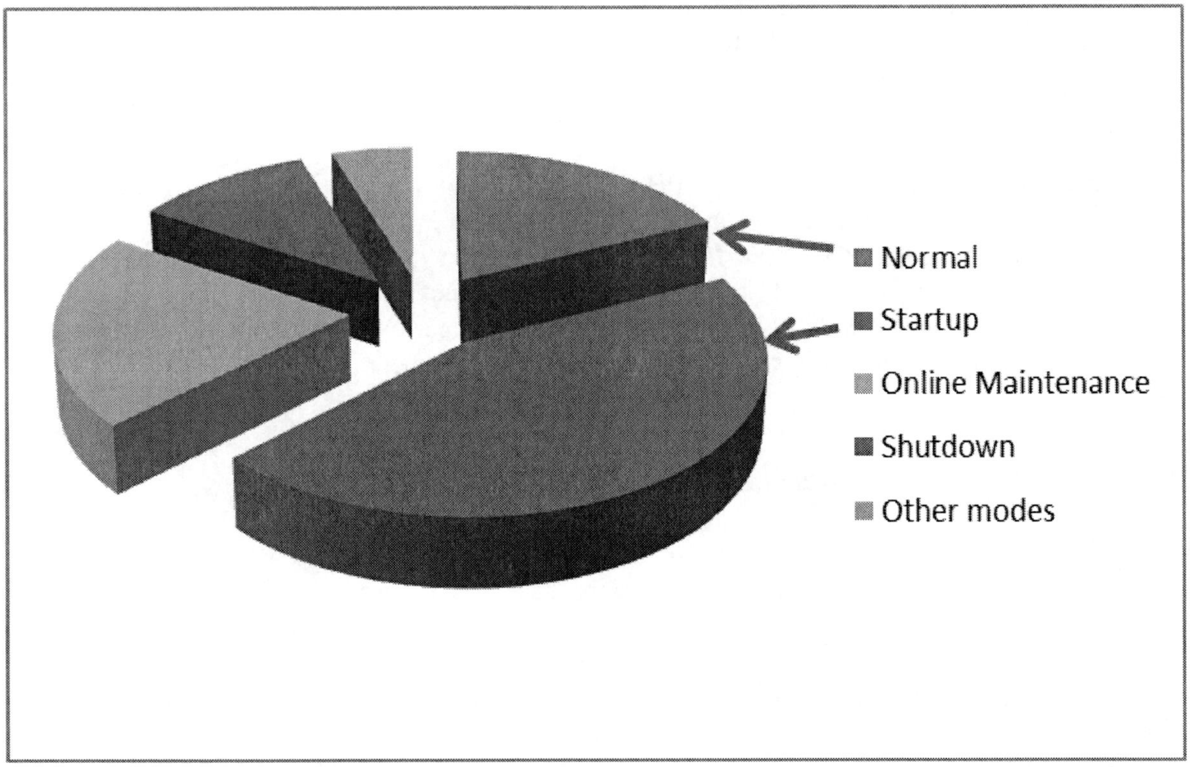

Figure 1: Distribution by Operating Mode of the 47 Largest Process Safety Accidents between 1987 and 2010[20]

How does someone responsible for coordinating or performing hazard evaluations (including PHAs) uncover potentially important accident scenarios during all modes of operation without consuming too many resources? To correctly answer this question, we must (1) understand the root causes of human error and (2) develop a strategy for systematically finding the scenarios that are caused by human error, during all modes of operation. The strategy must be thorough, yet provide for a practical allocation of resources. This paper provides a strategy that uses widely accepted hazard evaluation techniques (such as those referenced by OSHA and EPA for PHAs, which include what-if analysis and hazard and operability [HAZOP] analysis). This strategy has proven effective for hundreds of facilities over the past two decades since it was first published.[21] In addition to identifying accident scenarios during non-routine modes; this approach helps to more fully address human factors, which is a specific requirement of OSHA's PSM regulations and EPA's RMP rule.

Human factor deficiencies can make operations during non-routine modes extremely hazardous – since operators generally have less operating experience for non-routine modes, and these types of operations rely heavily on operator decision-making and tasks. In addition, there are usually less layers of protection in effect during non-routine operations. Analyzing procedure steps can identify steps where the operator is most likely to make mistakes and suggest ways to reduce risk of an accident scenario, ranging from adding hardware to improving management systems.

The approach outlined in this work applies equally to any hazard evaluation where the steps for a non-routine mode of operation are well defined (i.e., written), including PHAs of existing units, hazard evaluations during preliminary and detailed design phases of projects (for new/revised processes), and large or small management-of-change hazard reviews.

Overview of Methodology for Hazard Evaluation of Non-Routine Modes of Operation

The hazard evaluation of non-routine modes of operation involves reviewing procedures using a HAZOP, simplified HAZOP, or What-if analysis to uncover potential accident scenarios associated with non-routine operations, for continuous or batch operations. As mentioned earlier, human error is more likely and more critical during non-routine operations. By analyzing procedural steps where human error is more likely, and where human error or component failure could lead to a consequence of interest, risk can be reduced. The objective for the hazard evaluation team is to evaluate the risk associated with skipping steps and performing steps incorrectly.

FMEA cannot be applied to procedure-based deviations, unless you create a "human" component, in which case you have simply merged HAZOP deviations for "steps" into FMEA. Pre-Hazard Analysis (P_rHA) and other hazard evaluation methods are not applicable for accomplishing a detailed hazard evaluation of non-routine modes of operations.

Checklist of human factors issues (see an earlier paper[21] and also Guidelines for Hazard Evaluation Procedures[1]) can be very useful after the detailed hazard evaluation of deviations of steps. Such analysis can indicate where generic weaknesses exist that can make errors during any mode of operation more likely, or that can make errors during maintenance more likely. Such human factors checklists are normally used at the end of the analysis, they can be done piecemeal during an analysis (on breaks from the meetings) by individuals on the team, and then the results of each individual review can be discussed as a team at the end.

As with scenarios uncovered during continuous modes of operation, the company may need to perform analysis (including semi-quantitative analysis such as LOPA or HRA) to more fully address any unresolved or complex issues raised in the hazard evaluation of non-routine modes of operation.

Case studies presented later in this paper illustrate the analysis approach and the usefulness of this strategy.

Purpose of Hazard Evaluation of Procedures-Based Modes of Operation

Although incorporating human factors considerations into hazard evaluation studies of continuous operation is straightforward by asking why the human might make a mistake that leads to a parametric deviation, this approach only addresses a small fraction of the potential human errors that can affect process safety. Many analysts have tried to find accident scenarios in non-routine modes of operations by adding generic guide words such as "deviations during startup" and "deviations during maintenance/sampling" to the hazard evaluation of equipment nodes/sections. Unfortunately, this only catches a fraction of the accident scenarios that can occur in non-routine modes since a hazard evaluation team is focused on "continuous" mode of operation during HAZOP or What-if of equipment sections/nodes.

From an informal survey of more than 100 companies, most do not currently perform process hazard evaluations of procedures, although many do perform some type of job safety analysis (JSA). The JSA is an excellent starting point for an evaluation of procedures because a JSA identifies the tasks that workers perform and the equipment required to protect workers from typical industrial hazards (slips, falls, cuts, burns, fumes, etc.). Unfortunately, a typical JSA will not usually identify process safety issues or related human factors concerns. For example, from a JSA perspective, it may be perfectly safe for an operator to open a steam valve before opening a feed valve; however, from a process safety perspective, the operator may need to open the feed valve before the steam valve to avoid the potential for overheating the reactor and initiating an exothermic decomposition; a PHA of procedural step deviations would likely find this deficiency, but a JSA will miss this process safety issue. The primary purpose of a JSA and other traditional methods for reviewing procedures has been to ensure that the procedures are accurate and complete (which is required of employers in 29 CFR 1910.119(f)(3)).[9]

By contrast, the purpose of a hazard evaluation is ***not*** to ensure the procedures are accurate and acceptable, but instead, to ***evaluate the accident scenarios if the procedures are not followed***. Even the best procedure may not be followed for any number of reasons, and these failures to follow the prescribed instructions can and do result in incidents. In fact, in the chemical industry and most other process industries the chance of an operator or other worker making a mistake in following a procedure is greater than 1/100, and in some cases much greater. When considering common human factor deficiencies that accompany non-routine operations, such as fatigue, lack of practice, the rush to restart and return to full production, etc., the probability of errors can climb to 1/10 chances per task (a task being about 1 to 10 detailed steps).[5]

The purpose of a hazard evaluation of non-routine modes of operation (governed by written procedures) is to make sure an organization has enough safeguards for the inevitable instance when a step is either performed wrong or skipped (inadvertently or due to shortcutting or other reasons)

Industry has found that a HAZOP or what-if analysis, structured to address procedures, can be used effectively for finding the great majority of accident scenarios that can occur during non-routine modes of operation.[1, 21, 22, 23] Experience shows that reviews of non-routine procedures have revealed many more hazards than merely trying to address these modes of operation during the P&ID driven hazard evaluations.

> ***Example:*** *For the BP Texas City, Texas refinery, pre-2005, if the isomerization unit had a hazard analysis of the startup mode (using What-If and 2-Guide Word analysis [explained later in this work]), the team would have likely identified that the high-high level switch in the column was a critical safety device. They also may have recommended moving the switch to a location higher in the column and then interlocking the high-high level switch to shutdown feed to the column. However, the site performed a parametric deviation HAZOP of the equipment nodes which focusing on continuous mode of operation, and so the team decided that the high-high level switch was not critical since devices in upstream and downstream process units (during continuous operation) would indicate possible level problems in the column – and besides, the operator would certainly notice the high-level condition on the sight glass during the rounds twice per shift. Unfortunately, these are not necessarily safeguards during startup of the column (1) since the routine practice was to overfill the bottoms (raise the level above the upper tap of the level controllers transmitter and above the nozzle for the high-high level switch and (2) since swings in upstream and downstream units are expected (and so likely such swings would not have led to intervention by the operators of the other units).*

To reinforce the need for and to explain the method for analysis of deviations of steps in a procedure, Section 9.1 was included in the 3rd Edition of *Guidelines for Hazard Evaluation Procedures,* 2008[1]; this was one of the major changes to the hazard evaluation procedures.

HAZOP Method for Analyzing Deviations of Procedural Steps

The Hazard and Operability (HAZOP) method has two major variations; one for the continuous mode of operations (where the team brainstorms what would happen if there were deviations of parameters) and procedure-based (where the team brainstorms what would happen when the steps of a procedure are not followed correctly). The procedure-based variation of HAZOP is the oldest form of HAZOP (from ICI in 1960s)[24]. It was an expansion of a Hazard Evaluation method based strictly on asking:

- **What happens if the step is skipped?**
- **What happens if the step is performed wrong?**

In turn, the "pre-HAZOP" method for brainstorming accident scenarios from not following procedures (including because the procedure is itself wrong) is based on the understanding that human errors occur by someone not doing a step (errors of omission) or by doing a step incorrectly (errors of commission). So, simply asking what would happen if the operator omitted a step or performed a step wrong is one way to structure a hazard evaluation of a step-by-step procedure. (We will discuss the usefulness of this simple approach to hazard evaluation of steps later.)

Seven (7) Guide Word Method

In an effort to be more thorough, the inventors of HAZOP (at ICI) broke these two types of errors into subparts and agreed on using the following 7 Guide Words:

Omission:	**Skip (or Step Missing)** **Part Of**
Commission:	**More** **Less** **Out of Sequence** **As Well As** **Other Than** **Reverse**

In the early 1990s, the guide word Skip was augmented by adding the option of discussing "are there any steps **missing** from the procedure."[21]

To apply HAZOP to procedural steps for startup, shutdown, online maintenance, and other modes of operation, the facilitator (or team) first divides the procedure into individual actions. This is already done if there is only one action per step. Then, the set of guide words or questions is systematically applied to each action of the procedure resulting in procedural deviations or what-if questions. The guide words (or procedural deviation phrases) shown in Table 1 were derived from HAZOP guide words commonly used for analysis of batch processes. The definition of each guide word is carefully chosen to allow universal and thorough application to both routine batch and non-routine continuous and batch procedures. The actual review team structure and meeting progression are nearly identical to that of a process equipment HAZOP or what-if analysis, except that active participation of one or more operators is even more important and usually requires two operators for a thorough review; a senior operator and a junior operator.

For each deviation from the intention of the process step (denoted by these guide words applied to the process step or action), the team needs to dig beyond the obvious cause, "operator error," to identify root causes associated with human error such as "inadequate emphasis on this step during training," "responsible for performing two tasks simultaneously," "inadequate labeling of valves," or "instrument display confusing or not readable." The guide word *missing* elicits causes such as "no written procedural step or formal training to obtain a hot work permit before this step," or "no written procedural step or formal training to open the discharge valve before starting the pump."

Table 1: Definitions of 7-8 Guide Words for HAZOP of Procedure-Based Operations

Guide Word	Meaning When Applied to a Step
Missing (optional guide word)	A step or precaution is missing from the written procedure prior to this step (similar to "Out of Sequence", except the missing step is not written)
Skip (No, Not, Don't)	The specified intent of this step is not performed
Part-of	A portion of the full intent is not performed. Usually only applies to a task that involves two or more nearly simultaneous actions ("Open valves A, B, and C".)
More	Too much of the specified intent is done (does not apply to simple on/off; open/close functions); or it is performed too fast
Less	Too little of the intent is done, or it is performed too slowly
Out of sequence	This step is performed too early in the sequence
As well as	Something happens, or the user does another action, in addition to the specified step being done correctly (could be a short cut)
Other than (or Reverse)	The wrong device is operated, selected, read, etc., or operated in a way other than intended. Or the wrong material is selected or added. "Other than" errors always imply a "Skip" as well.

Two (2) Guide Word Method for Analyzing Deviations of Procedural Steps

A more streamlined guide word approach has also proven very useful for (1) procedures related to less hazardous operations and tasks and/or (2) when the leader has extensive experience in the use of the guide words mentioned previously and can therefore compensate for the weaknesses of a more streamlined approach. The two guide words for this approach (as defined in Table 2 below) encompass the basic human error categories: errors of omission and commission. These guide words are used in an identical way to the guide words introduced earlier. Essentially "omit" includes the errors of omission related to the guide words "skip," "part of," and "missing" mentioned earlier. The guide word "incorrect" incorporates the errors of commission related to the guide words "more," "less," "out of sequence," "as well as," and "other than" mentioned earlier. Note that these two guide words (Table 2) fill the basic requirements for a human error analysis as outlined in OSHA's CPL 2-2.45.[6]

Table 2: 2 Guide Word (Guide Phrases) for Modified-HAZOP of Procedure-Based Operation

Guide Phrase	Meaning When Applied to a Step
Step not performed	The step is not done or part of the step is not done. Some possible reasons include the employee forgot to do the step, did not understand the importance of the step, or the procedures did not include this vital step
Step performed wrong	The employee's intent was to perform the step (not omit the step), however, the step is not performed as intended. Some possible reasons include the employee does too much or too little of stated task, the employee manipulates the wrong process component, or the employee reverses the order of the steps.

Table 3: Example of 2 Guide Word HAZOP of a Critical Step in a Procedure

Drawing or Procedure: SOP-03-002; ***Cooling Water Failure***	**Unit:** HF Alkylation	**Method:** 2 Guide Word Analysis	**Documentation Type:** Cause-by-Cause

Node: 23	**Description:** STEP 2: Block in olefin feed to each of the 2 reactors by blocking in feed at flow control valves				
Item	**Deviation**	**Causes**	**Consequences**	**Safeguards**	**Recommendation**
23.1	Step not performed	Operator failing to block in one of the reactors, such as due to miscommunication between control room operator and field operator; or control valve sticking open or leaking through	High pressure due to possible runaway reaction (because cooling is already lost), because of continued feeding of olefin (link to 11.7 - High Rxn Rate; HF Alky Reactor #1/#2) High pressure due to high level in the reactor, because of continued feeding olefin (link to 11.1 - High Level; HF Alky Reactor #1/#2)	High temperature alarm on reactor High pressure alarm on reactor Field operator may notice sound of fluid flow across valve Flow indication (in olefin charge line to reactor that is inadvertently NOT shutdown) Level indicator, high level alarm, and independent high-high level switch/alarm	
		Operator failing to make sure bypass valve is also closed, since this precaution is not listed in the written procedure; or the bypass valve leaks through	High pressure due to possible runaway reaction (because cooling is already lost), because of continued feeding of olefin (link to 11.7 - High Rxn Rate; HF Alky Reactor #1/#2) High pressure due to high level in the reactor, because of continued feeding olefin (link to 11.1 - High Level; HF Alky Reactor #1/#2)	High temperature alarm on reactor High pressure alarm on reactor Operator skill-training requires checking bypasses are closed, when blocking control valves Field operator may notice sound of fluid flow across valve Flow indication in olefin charge line (but likely not sensitive enough for small flows) Level indicator, high level alarm, and independent high-high level switch/alarm	
		Operator failing to close low control valve manually from the DCS because the phrase "block in" is used instead of the word "close"	Valve possibly opens full at restart, allowing too much flow to reactor at restart, resulting in poor quality at startup and/or possibly resulting in runaway reaction and high pressure	Control room skill training requires always manually commanding automatic valves closed before telling field operator to block in control valve	37. Implement best-practice rules for procedure writing, which includes using common terms.
23.2	Step performed wrong	Operator closing the olefin charge flow control valves before shutting down the charge pump, primarily because the steps are written out of the proper sequence	Deadheading of charge pump, leading to possible pump seal damage/failure and/or other leak, resulting in a fire hazard affecting a small area (link to 5.12 - Loss of Containment; Olefin Charge Line/Pump)	Step 3 of procedure that says to shutdown charge pump The step to shut down the charge pump (Step 3) is typically accomplished before Step 2 (in practice)	41. Move Step 3 ahead of Step 2.
		Field operator closing both upstream and downstream block valves	Possible trapping of liquid between block valve and control valve, leading to possible valve damage (due to thermal expansion)	Field operator skill training stresses that only one block valve should be closed	

What-if Method for Analyzing Deviations of Procedural Steps

The What-if method for analyzing procedure-based modes of operations is free brainstorming without the aid (or constraints) of guide words. This method is described in detail the *Guideline for Hazard Evaluation Procedures (CCPS)*[1]. The hazard evaluation team using this method would read the procedure and then answer the question: "What mistakes will lead to our consequences of interest?" The team would list these mistakes and then brainstorm the full consequences, causes, and existing safeguards – the same analysis approach described for the guide word approaches mentioned earlier in this section. What-if brainstorming **is not** applied to each step of the procedure, but rather covers the entire task (procedure) at one time.

Choosing the Right Method for Analysis of Non-Routine Modes of Operation

Obviously the What-if approach takes far less time than the 2-Guide Word method, and the 2 Guide Word method takes much less time than the 7-8 Guide Word method of HAZOP of procedures. Experience has shown that hazard evaluation facilitators, newly trained in the three techniques above, tend to overwork an analysis of non-routine procedures, so a tiered approach is best. In this tiered approach, the first step in choosing the right method of analysis in the hazard evaluation of procedures is to screen the procedures and select only those procedures with extreme hazards. These procedures should be subjected to a detailed HAZOP analysis (7-8 guide word set) presented above. The 2-Guide Word set is efficiently used for less complex tasks or where the consequences are lower. The What-if method is applicable to low hazard, low complexity, or very well understood tasks/hazards.

Experience of the leader or the team plays a major part in selecting the procedures to be analyzed, and then in deciding when to use each guide word set.

Figure 2 shows the typical usage of the three methods described above for a typical set of operations procedures within a complex chemical plant or refinery or other process/ operation. Most of the procedures are simple enough, or have low severity hazards to warrant using the What-if method. Currently, the 7-8 Guide Word approach is used infrequently, since most tasks do not require that level of scrutiny to find the accident scenarios during non-routine modes of operations.

Figure 2. Relative Usage of Techniques for Analysis of Procedure-Based Modes of Operation

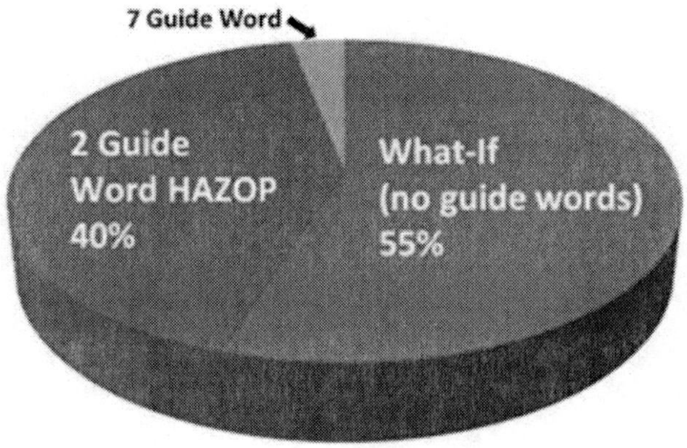

The experience of the leader or the team plays a major part in selecting the method to use for each task/procedures to be analyzed. However the first decision will always be "Are these procedures ready to be evaluated to determine risk?" If the procedures are up-to-date, complete, clear, and used by operators, then the best approach for completing a complete hazard evaluation of All modes of operation, including routine modes of operation, is shown in Figures 3A and 3 B below:

Figure 3A: PHA of a ALL Modes of Operation for a CONTINUOUS Process

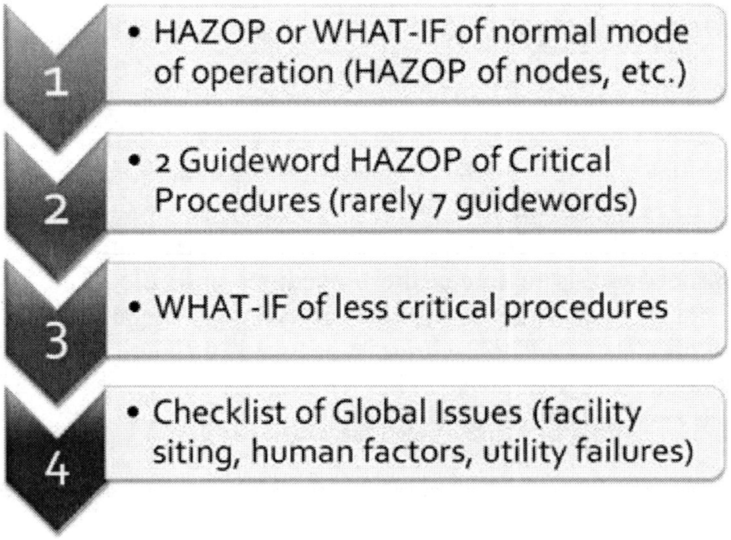

Figure 3B: PHA of a ALL Modes of Operation for a BATCH Process

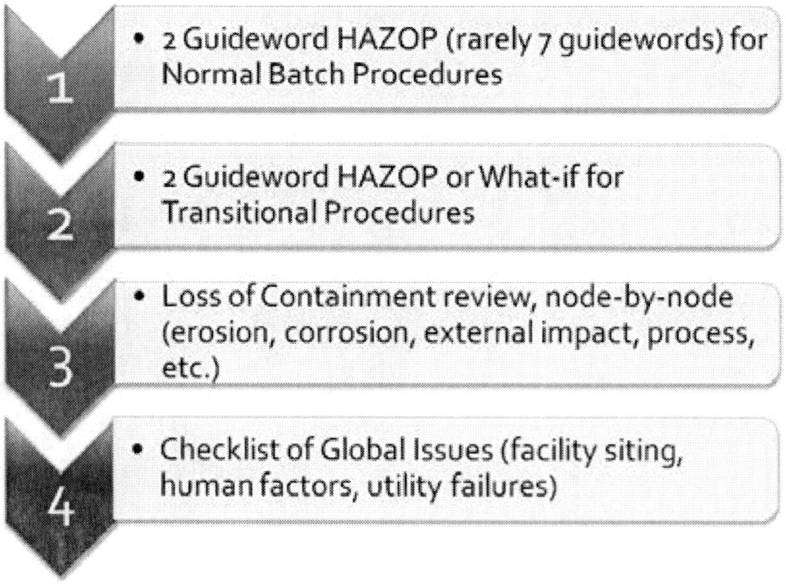

If procedures are not at least 90% accurate (with 95% accuracy being the target), then the best approach is to develop accurate and up-to-date procedures as quickly as possible and afterwards do a PHA of the newly issued procedures.

Any procedure (even a computer program) can be analyzed using these techniques. Reviews of routine procedures are important, but reviews of non-routine procedures are even more important. As mentioned earlier, the nature of non-routine procedures means that operators have much less experience performing them, and many organizations do not regularly update these procedures [though this should change as companies comply with 29 CFR 1910.119(f)][9]. Also, during non-routine operations, many of the standard equipment safeguards or interlocks are off or bypassed.

Using the approaches above, a company doing a complete hazard evaluation of an existing unit will invest about 65% of their time to evaluate normal (e.g., continuous mode) operation and 35% of their time for evaluating the risks of non-routine modes of operation.

Many companies do **not** perform a thorough analysis of the risk for startup, shutdown, and on-line maintenance modes of operation; the reason normally given is that the analysis of these modes of operation takes "too long." Yet, the hazard evaluation of the normal mode is taking too long and so the organization feels it has no time left for the analysis of procedures for startup and shutdown modes of operation. But, if these hazard evaluations for the normal mode of operation are **optimized** (such as using rules presented elsewhere[25]), the organization will have time for thoroughly analyzing the non-routine modes (typically discontinuous modes) of operation and the organization will still have a net savings overall! This point is critical since 60-75% of catastrophic accidents occur during non-routine modes of operation. Figure 4 illustrates (for a continuous process unit) the typical split of meeting time for analysis of routine mode of operation versus non-routine modes of operation.

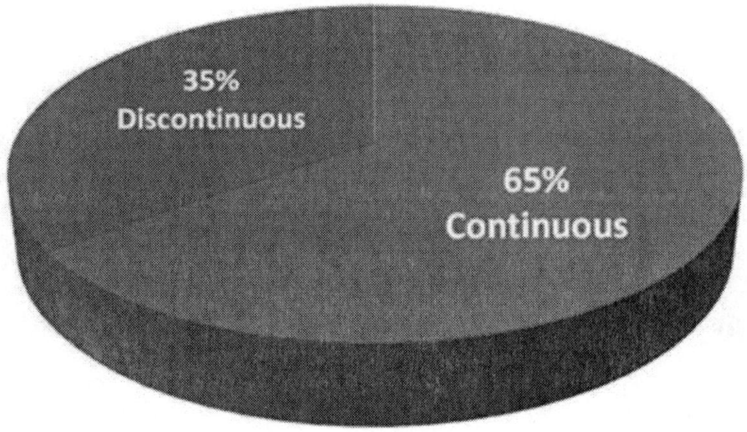

Figure 4. Relative Amount of Meeting Time Spent for Analysis of Routine and Non-routine Modes of Operation for a Continuous Process

General Guidelines for Analyzing Non-routine Modes of Operation or Batch (Step-by-step) Processes

- Define the assumptions about the system's initial status. "What is assumed to be the starting conditions when the user of the procedure begins with Step 1?"
- Define the complete design intention for each step. "Is the step actually 3 or 5 actions instead of one action? If so, what are the individual actions to accomplish this task?"
- Don't analyze safeguard steps that start with ensure, check, verify, inspect, etc., or where the consequence of skip is "loss of one level of safeguard/protection against ….." There is no reason to analyze these steps since they will show up as safeguards of deviations of other steps. This approach is similar to not analyzing a PSV during a HAZOP of continuous mode (parametric deviation analysis); instead the PSV is shown as a safeguard against loss of containment.
- Together with an operator before the meeting, identify the sections of the procedures that warrant use of:
    - 7-8 Guide Words (extremely large consequences can happen if deviations occur)
    - 2 Guide Words (the system is complex, mistakes are costly, or several consequences could occur)
    - On others, use What-If (no guide words or guide phrases; for use on simpler or lower hazard systems)
- Decompose each written step into a sequence of actions (verbs)
- Apply guide words directly to the intentions of each action

The Following Preparation Steps May Also Be Needed:

- Walk through procedure in the plant with one or more operators to see the work situation and verify the accuracy of the written procedure. This is optional and should have also been performed as part of validation of the procedure after it was originally drafted.
- Determine if the procedure follows the best practices for "presentation" of the content; the best practices will limit the probability of human error.
- Discuss generic issues related to operating procedures, such as:
    - staffing (normal and temporary)
    - human-machine interface
    - worker training, certification, etc.
    - management of change
    - policy enforcement
- Review other related procedures such as lock out/tag out and hot work.
- **IF the procedures are NOT >90% accurate, then re-write the procedures first!**

STEP 4 - Using Checklist analysis to Help Ensure Thorough Coverage of All Human Factors

A general, template based Human Factors checklist questionnaire should be included in the scope of any large, unit-sized PHA to help ensure generic human factors are thoroughly considered. These checklists are available in the default templates of some PHA documentation software, such as LEADER ™ (from ABS Consulting). Even if not specifically called out in the PHA scope, it's worth going through template checklists during the last few hours of a meeting. If possible, the more efficient strategy is to send the checklists to team members a month or so before the meeting, so the team can go over the questionnaire, walk down the items on the checklists in the field (with an operator or other PHA team member, if possible), and then submit their responses to the PHA Leader prior to the meeting. Then, on the last day of the meetings, then Leader/Scribe reviews the consolidated list of issues found by the team members. This saves meeting time and allows for more in-depth consideration of the listed issues. (Note that generic checklists for Human Factors and Facility Siting are often completed together; and though the items listed on the siting checklist are not categorized directly as human factors, they include design considerations that may have been overlooked during the project/initial PHAs that can in turn lead to more human error.). Adding the Human Factors checklists to the analysis scope typically identifies more minor to moderate consequences (compared to those found in HAZOP of procedure steps or in HAZOP of normal mode of operation), and will generally yield far fewer recommendations (typically only 1-5% of the total recommendations come from use of the checklists). However, given the almost negligible additional costs to include the checklists in the analyses scope, the benefit from catching the last few hazards that might have slipped through the cracks otherwise will always justify such costs.

For an example, see the human factors checklist listed in ***Exhibit A.***

CASE STUDIES

PII and others have performed tens of thousands of PHAs, including many thousands of PHAs of all modes of operation. Everyone who applies the lessons above find hundreds to thousands of scenarios that are unique to startup, shutdown, and online maintenance. For example:

> *Results from thousands of PHAs of non-routine modes of operation indicates that about 3% to 6% of the relief valves across the industry have been discovered to be too small and so had to be resized to account for accident scenarios that were only found during analysis of non-routine modes of operation.*

The following case studies illustrate the usefulness of the process outlined in this paper.

Case Study 1: Phillips Polyethylene Plant 6, Pasadena, TX

In 1991-1992, a PHA was performed for the first of the rebuilt polyethylene plants at the Phillips 66 plant in Pasadena, TX. The accident there two years prior claimed 24 lives, injured hundreds of others, destroyed all three polyethylene plants, and cost Phillips an estimated $1.4 billion. Following the investigation of the accident, one of the requirements of the settlement agreement between Phillips and the US government was to ensure the PHA of the rebuilt units addressed hazards during ***All modes of operation***.

The PHA team varied in size, but always included at least two operators. The team leader was a process engineer with 15 years of experience, who was also trained in human factors. The PHA first covered the continuous mode of operation for the approximately 200 nodes of equipment (from feed stock through pellet handling) using the "parametric deviation" form of HAZOP (and some What-If). Then, to complete the analysis of all modes of operation, the PHA team performed a step-by-step analysis of all steps of all startup and shutdown and online maintenance procedures (about 700 steps changed the state of the system and each of these steps were analyzed) using the 7 Guide Word HAZOP method (2 Guide Word analysis was not known to the team at this time). For deviations such as "operator skips a step," the causes identified by the team included "the operator doing this step miscommunicates with the operator who performed steps earlier in the day and went to the wrong reset panel/switch in the field". In this example, an "other-than" error led to the "skip" error; so two errors occurred at once: the wrong switch was flipped and the correct switch was not flipped. Other causes included: "label not distinct enough" or "thinking/believing the previous operator completed this step." The additional safeguards suggested by the PHA team sometimes lower the likelihood of the error by addressing a human factors weakness. But in many cases, the solution was a change to the hardware or instrumentation, including adding new interlocks (these would be called Safety Instrumented Functions today) and adding mechanical interlocks and installing larger relief valves. In a couple of cases, isolated sections of the process were redesigned to lower the inherent risk, such as adding error-proofing (Poke Yoke) features.

Figure 5: Typical Settling Leg Assembly for Phillips Polyethylene

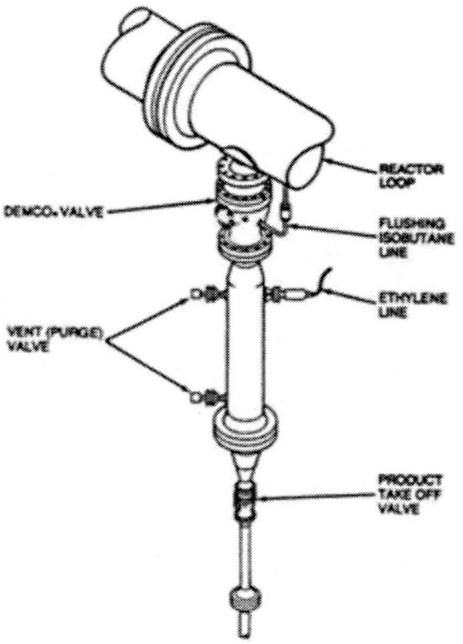

The 7 Guide Word HAZOP of non-routine modes of operation took 2.5 weeks of meetings, 40 hours a week. This was in addition to the 3.5 weeks of meetings to complete the parametric deviation analysis HAZOP of the continuous (normal) mode of operation (as mentioned before, 200 nodes of equipment); some all this the Normal PHA or Traditional PHA, but that is a misnomer. *Note that if the team had known of and been trained in 2 Guide Word HAZOP for procedure steps, they likely would have chosen that for many of the tasks and it is estimated that the meeting time for analysis of non-routine procedures would have been reduced to less than 2 weeks, with little or no loss of thoroughness.* The completed PHA report was submitted to US OSHA for review and was approved almost immediately; OSHA particularly reviewed the analysis of all modes of operation and coverage of human factors.

After the PHA was completed, the settlement agreement also required a quantified human reliability analysis (HRA) of the online maintenance task of clearing a plugged settling leg – mistakes during this task led to the accident in 1989. The HRA was performed similar to those performed for nuclear reactor risk assessments, using a human reliability event tree to model the error probability for the task. The HRA results indicated that if the three IPLs in the new design are maintained, the probability of a similar event occurring when using this procedure was less than the risk of fatality when driving to work. This report was also approved by OSHA for settlement purposes.

The HRA shed light on new aspects of making errors and recovering from the errors during this task, but the HRA results did not result in changes to the process steps (at least not much), or the training program, or the human factors engineering, or the hardware/IPLs.

In this instance, the HRA validated the results of the 7 Guide Word HAZOP, but did not make new recommendations. The HAZOP of the procedures had already found the major accident scenarios and had already identified well enough the changes needed to reach tolerable risk.

Case Study 2: HUNTSMAN

Qualitative analyses of non-routine operating procedures is an extremely powerful tool for uncovering deficiencies that can lead to human errors and for uncovering accident scenarios during all modes of operation. Huntsman recognizes that regulators have required similar approaches for decades and that the regulators continue to note lack of analysis of the risk of non-routine operations and lack of risk review of changes to procedures.

The Huntsman Geismar site implementation of PHA of non-routine modes of operation in conjunction with the traditional PHA creates a complete process hazard analysis of all modes of operation. Using PHA of non-routine modes of operation, Site teams have already identified hazardous scenarios that exist only during startup that were not identified by the traditional PHA approach. Teams have also identified existing safeguards and potential new layers of protection to prevent the newly identified hazardous scenarios from occurring. The site has already installed inherently safer designs identified using the new methodology that significantly reduce the risk.

The Geismar site believes our implementation of PHA of non-routine modes of operation is a world class initiative that is necessary to identify all potential hazardous scenarios at our facility.

Case Study 3: UNITED (a SABIC affiliate)

المتحدة
united

In January 2019, PII completed a 'Re-do' PHA of the UNITED Ethylene Plant (Jubail, Saudi Arabia; a SABIC affiliate), which is to serve as that plant's new baseline PHA. This new hazard analysis included a PHA of Procedures, in compliance with SHEM 02.01, Rev 8, specifically section 5.12.2, which requires the PHA to consider all modes of operation, and section 5.12.2.9, that the PHA cover control system failures, including user interfaces and human factors.[26]

The meeting time was set at 19 days, with 14 days allocated for HAZOP of continuous/normal mode of operation and 5 days dedicated to PHA of Procedures (Step 3 of this paper) which was used to cover non-normal modes of operation: shutdown, startup, and online maintenance; and 3 hours for checklists reviews (such as Step 4 of this paper). For the continuous/normal mode of operation the plant was sectioned and analyzed in the typical HAZOP style, deviating each node's parameters as such as high and low deviations of level, flow, pressure, temperature, etc. as suitable for each node. The PHA of Procedures was done in the last 5 days, so the team was well aware of the major hazards and safeguards (at least for normal modes) relating to the equipment listed in each procedure. The procedure list was reviewed with the team on the first day to decide which procedures presented major process hazards (consequences of interest, in this case non-occupational hazards/serious injury or fatality consequences), so that more time could be focused on highest risks containing procedures. These procedures were typically more complex and usually longer in length. For these identified with significant process safety potential impact, the 2 Guideword Method was used and for those with less hazards, the What-If method was used. As usual, the goal was identifying specific scenarios of interest for those steps, capturing safeguards and safeguard steps in the documentation process. The few hours of the meetings included analysis using checklists to cover any hazards that weren't otherwise identified, and the team was given additional time outside of meeting to respond to the individual questions in both the Human Factors and Facility Siting Checklists, yielding an additional 3 recommendations deemed safety critical.

The PHA team identified many hazard during the meeting in both the normal mode HAZOP and the PHA of Procedures, listing 115 Safety Critical Recommendations (as defined by UNITED) and 7 Operability Recommendations (not safety critical, but have some impact to communication or effective operation). Of these Recommendations, 42 (or 36% of total) were identified during the PHA of Procedures; and while many of these were simple fixes needed to step order or wording,***14 were in response to critical (high risk) consequences identified in the procedure analysis, requiring new or upgraded independent protection layers*** to bring the risk to acceptable levels.

> *EXAMPLE:* During the procedure analysis for the acetylene reactors, it was discovered that there were no adequate safeguards against run-away reaction during start up, meaning the reactor shell could reasonably be expected to fail at some point due to human error during startup, which would likely cause a large explosion with the potential for multiple fatalities. The startup process required an extremely slow ramp-up in temperature (1°C per 5 minutes), and was controlled entirely by control room operators (by manually changing set points as they monitored for temperature spikes). In this case, the team recommended new logic and safety instrumented functions to protect against the catastrophic reactor failure and explosion.

These hazards, many of which had NO safegaurds in place, would have been missed if the PHA scope failed to include procedure/non-normal modes of operation, which worth noting as statistically it is these types of errors/failures that lead to the majority of catastrophic consequences (non-normal modes of operation are when 80% of major accidents occur, as discussed previously). UNITED sees that the PHA of Procedures added extreme value to their current high value PHAs; and fully complies with corporate standards. As mentioned elsewhere, accurate procedures are needed to support a good PHA of procedures.

In summary, the PHA of procedures yeilded great benefit to UNITED by finding dozens of dangerous scenarios that were not found during PHA of normal mode of operation.

Another perhaps equally important aspect of human factors that was looked at carefully during the PHA was the fact that certain scenarios required more robust human intervention, esepcially when responding to some process alarms. These alarms are named "Critical Alarms" and directly trigger a Human Response IPL. These critical alarms required more attentive approach, where failure to respond adequately could results in severe consequences. Therefore, and in order to address the concerns of human errors that were specifically outlined in the work of Tew and Bridges (2010)[16], a common recommendation was made to UNITED to look at all alarms that are titled "critical" during the PHA study, and ensure that Training, Knowledge, and Skills are properly managed for these. For such Human Response IPLs, the pertinent operators must be provided a written trouble-shooting guide and each operator must be specifically trained for how to respond in time to each individual critical alarm. Moreover, response to these critical alarms must be tested and evaluated to ensure effectiveness of the Human Response IPL, and the time to respond (TTR) must be less than the safe span of time (within the Process Safety Time [PST]), based on simulated scenarios.

Another key human factor aspect covered in our PHA related to reducing specific human errors in the inspection and maintenance of Safety Instrumented Systems (SIS), which is a relatively new concept, largely introduced and proved by the work of Bridges and Thomas (2012)[28], and as it becomes increasingly important to maintain robust and effective high Safety Integrity Levels (high SILs). The PHA team closely looked at the concern of a possible increase in the likelihood of human error affecting the SIF's Probability of Failure on Demand (PFD). As a result, a recommendation was written for all of UNITED to consider maintenance staggering concept as a method to minimize the systematic probability of human error during proof testing and inspection. In essence, if a SIF has multiple channels (such as multiple sensors) that are voted 1oo2 or 1oo3, etc., then if the same person performs the calibration check of each channel, then the person should have 3 days separation between Channel 1 calibation and Channel 2, and similarly for Channel 3; otherwise the probability of repeating a failure on the 2nd and 3rd channel is about 90%. This is indeed an important consideration that needs to be always looked at when discussing human factors.

Case Study 4: SINOPEC-SABIC Tianjin Petrochemical Company (SSTPC)

The process plants at SS-TPC currently are:

- Ethylene (ET)
- MTBE & Butadiene (BD/MTBE)
- Phenol/Acetone (PHAC)
- High Density Polyethylene (HDPE)
- Linear Low Density Polyethylene (LLDPE)
- Polypropylene (PP)
- Pyrolysis Gasoline (DPG)

- Ethylene Oxide & Ethylene Glycol (EO/EG)
- Tank farm and Storage
- Utilities

PII led the PHA of the units, for all modes of operation. Besides a HAZOP or What-if of continuous modes of operation, the PHA team also used the Two Guideword or What-if approach to complete a PHA of startup, shutdown, and online maintenance modes of operation. The PHA of the non-routine modes of operation took about 20% of the total meeting time and was done at the end of the unit node-by-node analysis for continuous mode of operation.

Hundreds of scenarios were found during analysis of procedure-based modes of operation, resulting in many recommendations that had not been found during PHA of normal mode of operation. These resulted in implementation of new instrumentation and permissives and interlocks to reduce the accident scenario likelihoods.

Conclusions

Qualitative analysis of non-routine operating procedures is an extremely powerful tool for uncovering deficiencies that can lead to human errors and for uncovering accident scenarios during all modes of operation. This approach of step-by-step HAZOP and/or What-If analysis is not new to industry, and regulators have required similar approaches for decades. And regulators continue to note lack of analysis of the risk of non-routine operations and lack of risk review of changes to procedures.

> ***From the Wall Street Journal[87] referencing the presidential commission investigating the Deepwater Horizon accident of April 2010:*** *BP had rules in place governing procedural changes, but its workers didn't consistently follow them, according to BP's September [2010] internal report on the disaster and the report released earlier this month [January 2011] by the presidential commission on the accident. "Such decisions appear to have been made by the BP Macondo team in ad hoc fashion without any formal risk analysis or internal expert review," the commission's report said. "This appears to have been a key causal factor of the blowout."[27]*

> ***From CSB Report on August 2008 Bayer CropScience Explosion:[16]*** *"The accident occurred during the startup of the methomyl unit, following a lengthy period of maintenance ... CSB investigators also found the company failed to perform a thorough Process Hazard Analysis, or PHA, as required by regulation...In particular, for operational tasks that depend heavily on task performance and operator decisions, the team should analyze the procedures step-by-step to identify potential incident scenarios and their consequences, and to determine if the protections in place are sufficient."*

More regulatory pressure is sure to follow, since major accidents continue to occur during non-routine modes of operation.

Regardless of what hazard evaluation technique is employed, it is imperative for PHA teams to ask, "Why would someone make this mistake?" whenever a human error is identified as a cause of a potential accident. "To err is human" may be a true statement, but the frequency and consequences of such errors can be effectively reduced with a well-designed strategy for analyzing risk of non-routine operating modes.

Most critically, such analyses make it easier to provide a thorough consideration of human factors.

References

1. "Guidelines for Hazard Evaluation Procedures, 3rd Edition, with Worked Examples," Center for Chemical Process Safety (CCPS), AIChE, New York, 2008.

2. Tew, R. and Bridges, W., "Human Factors Missing from PSM," *Loss Prevention Symposium (part of the Global Congress on Process Safety [GCPS]),* AIChE, March, 2010.

3. "The SPAR-H Human Reliability Analysis Method," NUREG/CR-6883, U.S. Nuclear Regulatory Commission, prepared by Gertman, D.; Blackman, H.; Marble, J.; Byers, J. and Smith, C., of the Office of Nuclear Regulatory Research, Washington, DC, August 2005.

4. "Human Event Repository and Analysis (HERA)," NUREG/CR-6903, U.S. Nuclear Regulatory Commission, prepared by B. Hallbert, A. Whaley, R. Boring, P. McCabe, and Y. Chang, 2007.

5. Bridges, William G. and Collazo-Ramos, Dr. Ginette. "Human Factors and Their Optimization." *8th Global Congress for Process Safety Proceeding (GCPS).* American Institute of Chemical Engineers (AIChE), 2012.

6. U.S. Department of Labor: Systems Safety Evaluation of Operations with Catastrophic Potential. Occupational Safety and Health Administration Instruction CPL 2-2.45, Directorate of Compliance Programs, September 6, 1988.

7. OSHA Inspection Number 106612443 - Phillips 66 Company, Houston Chemical Complex, Citations 1-1 through 1-566, Issued 4/19/1990.

8. *Stipulation and Settlement Agreement,* Phillips 66 Company ("Phillips") and Lynn Martin, Secretary of Labor, United States Department of Labor, 8/22/1991.

9. "Process Safety Management of Highly Hazardous Chemicals; Explosives and Blasting Agents," US OSHA Final Rule, 29 CFR 1910.119, February 24, 1992.

10. "Risk Management Programs for Chemical Accidental Release Prevention," US EPA Final Rule, 40 CFR 68, 1994-2000.

11. OSHA Inspection Number 103490306, Issued November 2, 1992.

12. Woodcock, Henry C., “Program Quality Verification of Process Hazard Analyses (for instructional purposes only),” US OSHA, 1993.

13. OSHA Inspection Number 123807828; Issued November 18, 1993.

14. U.S. Department of Labor: PSM Covered Chemical Facilities National Emphasis Program. Occupational Safety and Health Administration CPL 03-00-014, Directorate of Enforcement Programs, November 29, 2011.

15. U.S. Department of Labor: Petroleum Refinery Process Safety Management National Emphasis Program. Occupational Safety and Health Administration CPL 03-00-010, Directorate of Enforcement Programs, August 18, 2009.

16. “Investigation Report: Pesticide Chemical Runaway Reaction Pressure Vessel Explosion, at Bayer CropScience, LP, Institute, WV, on August 28, 2008”, US Chemical Safety Board, Report No. 2008-08-I-WV, January 2011.

17. *EPA RMP Guidance, Chapter 7, pgs 7-6 & 7-7; General Risk Management Program Guidance.*

18. Contra Costa County Hazardous Materials Programs, *Contra Costa County Industrial Safety Ordinance (ISO)*, by CCHMP. June 15, 2011.

19. Rasmussen, B. "Chemical Process Hazard Identification," *Reliability Engineering and System Safety*, Vol. 24, Elsevier Science Publishers Ltd., Great Britain, 1989.

20. Bridges, W. and Clark, T., “How to Efficiently Perform the Hazard Evaluation (PHA) Required for Non-Routine Modes of Operation (Startup, Shutdown, Online Maintenance),” *7th Global Congress on Process Safety [GCPS]),* AIChE, March, 2011.

21. Bridges, W.G., et. al., “Addressing Human Error During Process Hazard Analyses,” *Chemical Engineering Progress*, May 1994.

22. Hammer, W., “Occupational Safety Management and Engineering, 3rd Ed.,” Prentice Hall, 1985.

23. Kongso, H.E., “Application of a Guide to Analysis of Occupational Hazards in the Danish Iron and Chemical Industry,” *International Conference on Hazard Identification and Risk Analysis, Human Factors and Human Reliability in Process Safety,* Center for Chemical Process Safety, AIChE, New York, 1992.

24. S. Mannan, ed., *Lees' Loss Prevention in the Process Industries, 4th Ed.,* Elsevier Butterworth-Heinemann, ISBN 978-0-12-397189-0, Oxford, UK, 2012.

25. Tew, R., et.al., "Optimizing Qualitative Hazard Evaluations (or How to Complete A Qualitative Hazard Evaluation Meeting in One-Third the Time Currently Required)," *5th Global Congress on Process Safety,* AIChE, April 2009.

26. *SHEM 02.01 Process Safety Risk Assessment, EHS-PRC-SM-002.01-07,* Rev 08, April, 2018, by SABIC

27. W. Bridges and H. Thomas, "Accounting for Human Error Probablity in SIL Verification Calculations" 8th *Global Congress on Process Safety*, AIChE, Houston, 2012

28. *Wall Street Journal*, January 29, 2011.

Exhibit A - The latest Human Factors Checklist

Company:	**Plant:**	**Site:**	**Unit:**	**System:**
Method: Checklist	**Type:** Human Factors Checklist	**Team Members:**		

Item	Topic	Questions/Issues	Responses	Recommendations
1.1	Housekeeping and General Work Environment	Are adequate signs posted near maintenance, cleanup, or staging areas to warn workers of special or unique hazards associated with the areas? Are adequate barriers erected to limit access to maintenance, cleanup, or staging areas? Are working areas generally clean? Are provisions in place to limit the time a worker spends in an extremely hot or cold area? Is noise maintained at a tolerable level? Are alarms audible above background noise both inside the control room and in the process area? Are normal and emergency lighting sufficient for all area operations? Is there adequate backup power for emergency lighting? Is the general environment conducive to safe job performance?		
1.2	Accessibility/ Availability of Controls and Equipment	Are adequate supplies of protective gear readily available for routine and emergency use? Are workers able to perform both routine and emergency tasks safely while wearing protective equipment? Is emergency equipment accessible without presenting further hazards to personnel? Is communications equipment adequate and easily accessible? Would others quickly know if a worker is incapacitated in a process area? Are the right tools (including special tools) available and used when needed? Is the workplace arranged so that workers can maintain a good working posture while performing necessary movements to conduct routine tasks? Is access to all controls adequate? Can operators/maintenance workers safely perform all required routine/emergency actions, considering the physical arrangement of equipment (e.g., access to equipment, or proximity of tasks to rotating equipment, hot surfaces, and hazardous discharge points)? Are valves that require urgent manual adjustments (e.g., emergency shutdown) easily identifiable and readily accessible?		
1.3	Labeling	Is all important equipment (vessels, pipes, valves, instruments, controls, etc.) legibly, accurately, and unambiguously labeled? Does the labeling program include components (e.g., small valves) that are mentioned in		

Item	Topic	Questions/Issues	Responses	Recommendations
		the procedures even if they are not assigned an equipment number? Has responsibility for maintaining and updating labels been assigned? Are emergency exit and response signs (including wind socks) adequately visible and easily understood? Are signs that warn workers of hazardous materials or conditions adequately visible and easily understood?		
1.4	Feedback/ Displays	Is adequate information about normal and upset process conditions clearly displayed in the control room? Are the controls and displays arranged logically to match operators' expectations? Are the displays adequately visible from all relevant working positions? Do separate displays present similar information in a consistent manner? Are automatic safety features provided when a process upset requires rapid response? Are automatic safety features provided when a process upset may be difficult to diagnose due to complicated processing of various information? Are the alarms displayed by priority? Are critical safety alarms easily distinguishable from control alarms? Is an alarm summary permanently on display? Are nuisance alarms corrected and redundant alarms eliminated as soon as practical to help prevent complacency toward alarms? Have charts, tables, or graphs been provided (or programmed into the computer) to reduce the need for operators to perform calculations as part of the operation? If operators are required to perform calculations, are critical calculations independently checked? Does the computer check that values entered by operators are within a valid range? Do the displays provide an adequate view of the entire process as well as essential details of individual systems? Do the displays give adequate feedback for all operational actions? Are instruments, displays, and controls promptly repaired after a malfunction? Do administrative features exist that govern when instruments, displays, or controls are deliberately disabled or bypassed and that govern their return to normal service at the appropriate time? Does a formal mechanism exist for correcting human factors deficiencies identified by the operators (e.g., modifications to the displays, controls, or equipment to better meet operators' needs)?		
1.5	Controls	Is the layout of the consoles logical, consistent, and effective? Are the controls distinguishable, accessible, and easy to use? Do all controls meet standard expectations (color, direction of movement, etc.)?		

Item	Topic	Questions/Issues	Responses	Recommendations
		Do the control panel layouts reflect the functional aspects of the process or equipment? Does the control arrangement logically follow the normal sequence of operation? Can operators safely intervene in computer-controlled processes? Can process variables be adequately controlled with the existing equipment? Do operators believe that the control logic and interlocks are adequate? Does a dedicated emergency shutdown panel exist? If so, is it in an appropriate location?		
1.6	Workload and Stress Factors	Is the control room always occupied (i.e., assigned duties do not require the control room operator to be absent from the control room)? Are the number and frequency of manual adjustments required during normal and emergency operations limited so that operators can make the adjustments without a significant chance of mistakes as a result of overwork or stress? Is the number of manual adjustments during normal operations sufficient to avoid mistakes as a result of boredom? Have the effects of shift duration and rotation been considered in establishing workloads? Is the number of extra hours an operator must work if his or her relief fails to show up sufficiently limited so that worker safety is not adversely affected? Is the number of hours an operator or maintenance worker must work during startup or turnarounds sufficiently limited so that worker safety is not adversely affected? Can additional operators (e.g., from other areas or from off site) be called in quickly to help during an emergency? Is the staffing level appropriate for all modes of operation (normal, emergency, etc.)?		
1.7	Procedures	Do written procedures exist for all operating phases (i.e., normal operations, temporary operations, emergency shutdown, emergency operation, normal shutdown, and startup following a turnaround or after an emergency shutdown)? Are safe operating limits documented, providing consequences of deviating from limits and actions to take when deviations occur? Are procedures current (i.e., are they revised when changes occur)? Do operators believe that the procedure format and language are easy to follow and understand? Are the procedures accurate (i.e., do they reflect the way in which the work is actually performed)? Is responsibility assigned for updating the procedures, distributing revisions of the procedures, and ensuring that workers are using current revisions of the procedures? Are temporary notes or instructions incorporated into revisions of written operating procedures as soon as practical? Do procedures address the personal protective equipment required when performing routine and/or nonroutine tasks?		

Item	Topic	Questions/Issues	Responses	Recommendations
1.8	Training (Employees and Contractors)	Are new employees trained in the hazards of the processes? Do operators and maintenance workers receive adequate training in safely performing their assigned tasks before they are allowed to work without direct supervision? Does operator and maintenance worker training include training in appropriate emergency response? Do operators practice emergency response while wearing emergency protective equipment? Do operators practice emergency response during extreme environmental conditions (e.g., at night or when it is very cold)? Are periodic emergency drills conducted? Are emergency drills witnessed by observers and critiqued? Does a periodic refresher training program exist? Is special or refresher training provided in preparation for an infrequently performed operation? When changes are made, are workers trained in the new operation, including an explanation of why the change was made and how worker safety can be affected by the change? Are operators and maintenance workers trained to request assistance when they believe they need it to safely perform a task? Are operators and maintenance workers trained to report near misses as part of the incident investigation program? Are operators trained to shut down the process when in doubt about whether it can continue to operate safely?		

551629 Ethics As an Underpinning to Safety

Ms. Grubbe will demonstrate and define ethics in support of good safety, and will show how ethics is eroding in academia, government and in industry. She will also suggest a way for academics and university industrial board members to consider inserting ethics into every engineering curricula.

551630 Case Study Reviews in Ethical Failures Contributing to Process Safety Incidents

Ms. Grubbe will use her knowledge and experience of several serious process incidents, and will tie the messages of the ethical basis of safe operations to key instances where ethical failures occurred. She will discuss the hard dynamic that occurs when an engineer chooses to remain silent vs. speak out about a noted issue.

551893 Ethical Issues in Safety- Optimizing Safety Culture and Leadership to Rise Above the Challenge!

In today's fast paced and demanding business environment engineers and operators are faced with competing priorities and pressures. During these complex times, everyday decisions must be made without compromising the Ethics of Safety and allowing safety priorities to rise to the top . During this session, the importance of a strong safety culture and leadership behaviors will be discussed as key to successfully navigating the Ethics of Safety. Practical short and long term solutions will be shared that can be applied across a broad spectrum of industries.

19 AIChE Spring Meeting
15TH GLOBAL CONGRESS ON PROCESS SAFETY
NEW ORLEANS • 2019

Set Your Operators Up for Success:

Practical Guidelines for Addressing Human Factors in PHA Causes and Protection Layers

Richard Carter, P. Eng., F.S. Eng. (TÜV Rheinland)
ACM Facility Safety Inc.
#300, 926 5th Ave SW, Calgary, AB, Canada
rcarter@acm.ca

Prepared for Presentation at
American Institute of Chemical Engineers
2019 Spring Meeting and 15th Global Congress on Process Safety
New Orleans, LA
March 31 – April 3, 2019

Set Your Operators Up for Success:
Practical Guidelines for Addressing Human Factors in PHA Causes and Protection Layers

Richard Carter P. Eng., F.S. Eng. (TÜV Rheinland)
ACM Risk Sciences and Development, Inc.
#300, 926 5th Ave SW, Calgary, AB, Canada
rcarter@acm.ca

Keywords: Human Factors, Process Hazards Analysis, PHA, HAZOP, LOPA, What-If, causes, safeguards, protection layers

Disclaimer:

Abstract

Everyone makes mistakes. Therefore, every Process Hazards Analysis should address the potential for error when humans interact with a processing system. It is unreasonable, however, to ask a PHA team to address every error that could be made by a person, so guidance must be given on the types of human factors to address.

Slips, mistakes and violations are all types of human factors that could result in a hazardous scenario, or prevent a protection layer from working correctly during a critical hazardous scenario. A PHA must consider human error in the identification of causes and in assigning safeguards or protection layers, but this must be done in a way that keeps the quality and efficiency of the PHA session at a high level.

To effectively incorporate human factors into PHAs it is essential to understand why people take these incorrect actions. Once this is understood it is possible to address them effectively, improving the identification of potential causes of hazardous scenarios and the PHA team's ability to validate safeguards that require a human response.

This paper will investigate the types of errors and violations people make, and discuss the reasons that people act in this way. It will then explain how to effectively incorporate this knowledge into a plan for addressing human factors in PHAs when identifying potential causes of hazardous scenarios and in deciding which protection layers to rely on. To achieve this, it will reference existing literature and draw on the author's experience of facilitating over 200 PHAs for a wide variety of petrochemicals and hazardous materials processing companies throughout North America.

1 Introduction

In 1998, there was an explosion at the Esso gas plant at Longford, Victoria, Australia. There are many lessons regarding human factors to take away from the incident, but one of the most telling is an operator's response to a question about responding to a certain alarm. The process involved in the incident was often operated for extended periods of time with a vessel operating above the maximum condensate level that could be measured by instrumentation. An alarm set at 85% of the measurable level was routinely ignored. When an operator was asked why the control room operator did not maintain the level below 100%, or act to bring the level below 100% as soon as possible, the answer was:

> *I would accept in theory that would be desirable but in practicality it's a very difficult thing to do with that particular control system.*[1]

This quote highlights the importance of addressing human factors in Process Hazards Analysis (PHA): It is important to consider the realities of the process and how it is operated in practice, not just in theory, when analyzing the potential failures that can occur, as well as the protection layers that are considered to protect against the resultant hazardous consequences. If such considerations are ignored, the PHA team is setting operations up to fail by imposing unrealistic expectations of perfection and ability on them.

"PHA" is used in this paper as an umbrella term for many different methodologies used to identify and address potentially hazardous scenarios, such as a HAZOP or What-If analysis. A PHA that fails to adequately consider the realities of human factors is likely to expect operators to perform their jobs perfectly all the time, and sometimes demands the impossible from them when responding to alarms and identifying issues during rounds. The PHA team must acknowledge that operators are human, and therefore will occasionally make mistakes; the PHA team's job is to ensure that a minor mistake will not lead to a catastrophic event. Similarly, when considering the safeguards intended to prevent hazardous events, the team must consider whether an operator is reasonably capable of identifying the problem and performing the corrective task that is being required of them, or if they are asking for an impossible task or a superhumanly fast response. It is essential to include experienced operations representation in the PHA team to ensure that all of these concerns are addressed adequately.

This paper will discuss the types of errors that all people make and how they relate to PHAs, and then discuss how this knowledge should be used when identifying the potential errors that could lead to a hazardous scenario. It will then discuss the factors to consider when determining valid safeguards to prevent hazardous scenarios from occurring.

This paper addresses human factors in PHAs relating to the operation of a facility, and does not include a discussion on human factors during construction or commissioning activities. These are usually the focus of their own specialized study, although many of the same human factors issues apply. The management systems of a company and how they affect the process and personnel actions are considered for the identification of potential causes and the validity of safeguards. Effective change to these factors, however, is usually outside the scope of a PHA[2] and therefore is not discussed in this paper.

By following the guidelines in this paper, the analysis team can set realistic expectations of human factors in the causes and safeguards of their PHA, and help to set their operators up for success in managing the process.

1 Factors that Affect Human Performance

There are many different factors that might affect how an individual will act at a given facility. Although not all of them need to be explicitly analyzed in a PHA, they affect how the facility is designed, operated and maintained. Therefore, it is essential for the team to be aware of them when determining the likelihood of human error, or the ability of an operator to take a certain action.

The following is a summary of important items to bear in mind[3]:

- Economic and political environment: How do the legislative, regulatory and economic conditions affect how workers perform? For example, the stability of an employee's job may affect how likely they are to commit a violation to maintain facility uptime.

- Corporate Policy and Management Systems: What are the company's culture, policies, and management attitude? For example, does management "walk the walk" or do they prioritize production and revenue at the expense of safety? Are adequate personnel and resources allocated for the task in question?

- Process environment: How inherently hazardous are the process and materials being handled? How complex is the process, and how frequently and in what ways do humans interact with it? Does the operator need to invent shortcuts and workarounds as a result of inadequate or out of date procedures, which may introduce hazards[4]?

- Physical work environment: Is there adequate lighting, is the level of noise reasonable, and at what temperature is the environment where the work is conducted?

- Work control systems: How well-trained are the workers? How often do they perform the task in question? How effective are the standard and emergency procedures, and are they enforced?

- Working conditions: How do the shift rotations and working hours affect personnel performance? How is the work distributed between workers? How do contractors interact with the process and with workers?

- Operator attributes: How well-trained and experienced are the workers generally? Do they have adequate skill and expertise for the task in question? Do they have adequately high morale and motivation to attempt the correct course of action, and how are these affected by the above factors?

- Risk Perception: Do the workers know what the potential consequences are if a hazardous event occurs? Do they believe that it could happen at this particular facility?

When considering human factors, the PHA team must consider both active factors (those that directly cause or have an immediate impact on a hazardous situation) and latent factors (those that do not have an immediate impact, and the consequences may only occur after a period of time).

Furthermore, these factors also affect the PHA team's own attitudes and thought processes during the PHA session, and therefore they must strive to identify and compensate for any potential negative impacts of these factors on the results of the analysis.

1.1.1. Case Study: Tunnel Vision at Deepwater Horizon

The causes of the 2010 Macondo well failure in the Gulf of Mexico (at the Deepwater Horizon drilling rig) are numerous and complex. However, there is a key piece of the puzzle that can act as an important lesson for PHA teams: tunnel vision. It is easy to grow complacent with a potential hazard, especially for workers at a facility who are at risk of that hazard every day. During the PHA session, therefore, it is common to initially think of the process as "safe" until it is proven to be otherwise, which can result in the PHA team understating or ignoring potential hazards that can later cause a catastrophic incident.

Prior to the Macondo incident, the drilling team performed a series of tests to verify that the well cement job was effective and that the well was adequately sealed prior to the drilling rig disconnecting from the well head. The cement job had already been labeled a success when it was completed, and therefore those performing the tests subconsciously "knew" that the cement job was good[5]. This led to the team discounting the indicators that the well had failed the test, developing inaccurate explanations for the anomalous results, and repeating the test in different ways until they achieved the positive result they were expecting. Unfortunately, that result was based on incorrect analysis of the data, and this was a major factor that contributed to the blowout.

In this same way it is possible for a PHA team member to enter an analysis session subconsciously believing that the facility is already safe, often based on the fact that it has been operating for some time without an incident. In a HAZOP, for example, the PHA team will assign a likelihood and severity ranking to a scenario that they believe to be correct. Frequently, however, if the risk ranking is returned as higher than they expected, their first thought will be to reduce the severity or likelihood ranking they had given the scenario because of their subconscious assumption that the scenario "can't be that dangerous" or "can't occur that often".

2 Types of Human Factors

There are two main categories of human factors that need to be considered during a PHA: errors and violations[3]. An error is performed by someone who believes they are taking the correct action, whereas a violation is performed by someone who knows they are taking a different action from the one prescribed.

2.1 *Errors*

An error is an unintentional incorrect action or a lack of correct action[6]. An incorrect action is known as an error of commission, and failing to take a correct action is known as an error of omission. Errors take the form of either **slips** or **mistakes**, both of which can be an error of commission or an error of omission.

One type of error is a **slip**, during which a person who is competent, knows what they are doing and has the correct intention, either performs an incorrect action or fails to perform a correct action. The person most likely knows what the consequences are of the incorrect action or failure to perform this action, but on this occasion they are rushing, distracted or otherwise compromised and the slip occurs.

Examples of a slip include an operator shutting down boiler #1 when they mean to shut down boiler #2, unloading a truck of caustic solution into a sulphuric acid tank, or missing a step in a procedure.

The other type of error is a **mistake**. These differ from a slip in that during a mistake, the individual is taking an action that they believe to be correct, but which turns out to be incorrect. They may be misinformed about the consequences of their actions, have determined their actions based on flawed reasoning or incorrect logic, or be so focused on getting one part of their task correct that they fail to adequately focus on the rest. A mistake also applies to a person failing to take a required action, incorrectly believing that it is not necessary at the time.

2.1.1. Case Study: Mistake Causing Incorrect Chemical Unloaded Into Storage Tank

As described in a CCPS Process Safety Beacon[7], a toxic gas release was caused by mixing a material named Chemfos 700 with sodium nitrite. The delivery truck driver had informed the plant pipefitter assigned to help him that he was delivering Chemfos 700, for which there was a loading connection with the label "Chemfos 700". The pipefitter, however, inadvertently connected the truck unloading hose to an adjacent pipe labeled "Chemfos Liq. Add". The pipefitter was unaware that there were two pipes with very similar names for drastically different chemicals and the resultant chemical reaction generated a cloud of toxic gases, requiring the evacuation of 2,400 people and treatment of 6 people for injuries from inhaling the toxic gases.

2.2 *Violations*

Violations differ from both slips and errors in that the individual is aware that they are performing a "wrong" action, or failing to perform a required one. They will usually understand the consequences or increased risks that their action carries, but have decided that it is the best thing they can do at that time. There are many different reasons why they may act in this way, but it is often due to either an external pressure such as the requirement to maintain facility uptime, or a belief that the "correct" action is unnecessary or inefficient.

Examples of violations include bypassing a safety system to prevent a "nuisance" shutdown, purposely deviating from the prescribed steps in a procedure, or knowingly driving faster than the legal speed limit.

Most times that a person performs a violation, there is some additional risk that the person is accepting which they have determined is allowable due to the benefits they receive. In processing facilities, this usually means that they are also accepting this increased risk on behalf of everyone around them; if someone bypasses a safety-critical equipment they may be placing everyone in that facility or unit in danger without their knowledge.

2.2.1. Case Study: Violation of Safety-Critical Device During Startup

This case study illustrates that sometimes operators are required to perform violations of procedures or standards to even start a process.

During a HAZOP facilitated by the author for a major oil and gas company in Alberta, Canada, a striking example was brought up of necessary violation of a safety critical device. The HAZOP team was discussing the startup of an emulsion treater, a piece of equipment that uses heat and residence time to separate the components of an emulsion of oil, water and gas. The treater has an interlock to shut down the fired heater portion of the treater if the liquid level in the vessel is too low, to prevent the heating tubes from being exposed and therefore overheating and failing.

Figure 1. A photograph of an example emulsion treater.[8]

This interlock also closed automated valves on the inlet and outlet connections of the vessel. Prior to startup, the treater would be empty and therefore there would be no liquid level in the vessel, which would keep the low-level interlock in the active state and prevent the inlet valves from being opened. This prevented the treater from being started up at all unless the safety interlock was bypassed. The operators had discovered that by wedging a coin underneath the external portion of the float, it would keep the level element from activating the low-level shutdown switch, which would allow the treater to be started up. When the level rose higher than the interlock activation point, the internal float would lift the element off the coin and the coin would fall out onto the floor, allowing the low-level interlock to function as normal.

The operators were not pleased to have to use this workaround, but at the time had not been able to convince management to implement a process modification to mitigate the concern. This system has since been modified so that the violation is no longer necessary.

3 Addressing Human Factors in PHA: Causes

When performing a cause-based PHA such as HAZOP, the first task is to brainstorm possible events that could occur which have the potential to lead to hazardous scenarios. Suggested re-write: It is easier to determine the potential failure modes when the events are due to electronic or mechanical systems failure than when human factors were the cause. Therefore, the PHA team needs to consider what type of failures will and will not be discussed. The details and nuances of how human factors are addressed will differ between facilities, so it is important the approach is appropriate for the specific facility in question.

3.1 Personnel Competency

When determining the potential for human error as causes during a PHA, some assumptions must be made regarding the competency of the personnel working with the equipment. A discussion of the potential for a piece of equipment to be built, repaired or maintained incorrectly is usually outside the scope of the PHA due to time constraints, so this discussion usually focuses around potential actions that an operator may take or fail to take. In most cases, it can be assumed that any operator on-site has a certain level of training and competency. Therefore, discussions of human factors can be limited primarily to errors that are deemed reasonable for an operator of that level of competency to make.

3.1.1. Case Study: High Likelihood of Human Error at Technical College

When considering personnel competency, it is important to consider the realities of the process, administrative controls, working conditions, and other factors that may influence the likelihood of an operator to perform a slip, mistake or violation. For example, a HAZOP that the author facilitated for a power engineering laboratory at a technical college required consideration of almost every conceivable human factor. The laboratory included two fully-operating steam generation boilers and numerous other processes and pieces of equipment that were required to support these and to practice other aspects of power engineering.

The laboratory lead instructor requested that we consider all conceivable human errors, as these systems were operated by power engineering students who were not yet qualified to run these processes without supervision. Furthermore, the instructor was keen that the students have the opportunity to experiment with the systems in order to gain a better understanding for how the process would respond, without keeping their interaction with the process so restricted that they would not be able to learn how it worked. The types of errors that had to be considered included closing valves in main process lines, entering control setpoints one or two orders of magnitude too high (e.g. entering a pressure control setpoint of 5000 kPag instead of 50 kPag), and isolating safety-critical instrumentation when the process is in operation. The instructor cited situations where he had been talking with a student and the student had absent-mindedly closed the isolation valve on a nearby pressure transmitter without paying attention to what they were doing.

Despite the seemingly vast number of potential errors that one of the students could make, the HAZOP successfully addressed these scenarios by grouping them together. Entering the incorrect setpoint into a controller, for example, was included in the likelihood of a control valve malfunctioning open or closed, and the potential for isolated safety devices was addressed by not accepting these devices as valid safeguards unless steps were taken to ensure they could not be taken out of service by a student.

3.2 *External Pressures*

It is important during a PHA to consider the realities of the process and working environment when investigating potential causes of hazardous scenarios. For example, if management prioritizes facility uptime over safety, it may increase the likelihood of a violation. If there is insufficient staffing for the work required, there may be an increased likelihood of a slip as people rush to complete tasks or need to work extended shifts. If there is a high rate of employee turnover, there is a greater likelihood of a mistake due to the reduced familiarity of the new employees with the process. Refer to the section 'Factors that Affect Human Performance' for a more in-depth discussion of these factors.

3.3 *Procedures*

When addressing the potential for error or violation when performing a task that is governed by an existing procedure, consider whether there is a good culture for developing well-written procedures, following them, and maintaining them. Investigate any statements made by a member of the team that imply that a given procedure might not be followed, and consider the likelihood of erroneous interaction with the process accordingly.

Although all PHAs must consider these points when determining the validity of a procedure as a safeguard, it is time-intensive to perform an in-depth analysis of procedures. Therefore, procedures can be the subject of their own Procedure PHA where the team reviews the procedure line-by-line to determine whether the action is correct and clearly written, the consequences of failing to perform each step or taking a different action than that prescribed have been identified, and whether the operator is aware of those consequences.

3.3.1. Case Study: Violation of Procedures Required at BP Texas City Refinery

Although there are many valuable learnings from this tragedy and the subsequent investigation, this case study is concerned with the reasons for the competent, well-meaning operators to violate operating and startup procedures.

Prior to the 2005 explosion at BP's refinery in Texas City, TX, the operators would routinely deviate from prescribed procedures out of necessity to protect equipment and operate the process effectively[9]. Procedures were not regularly updated, even though the process had changed over time, and therefore the procedures were outdated documents to be treated as guidelines and not to be strictly followed. The operators developed informal work practices based on knowledge of past experiences, which were relayed from the more experienced operators to the less experienced operators out of necessity, as the written procedures did not reflect the actual configuration or operating conditions of the facility. This was a significant contributing factor in the explosion event.

The operations representative on the PHA team will be able to provide the rest of the team with knowledge of how up-to-date and closely-followed operating procedures are, which will help the team determine the likelihood of a deviation from a procedure on purpose or by accident.

3.4 Manual Valves

Manual valves being left in an incorrect position, or incorrectly being manipulated during operation, may be addressed if it is reasonable to do so. It is usually more effective, however, to create a set of assumptions to focus on those valves that have a reasonable likelihood of being in the incorrect position, or the consequences of a mistake are drastic. The following are some examples of rules that may be used to determine which manual valves do and do not require analysis.

- Car Seals and Locks: As long as the facility's car seal or locking process is robust and reliable, these controls can be considered adequate to prevent the valve from being mis-operated, as it provides a physical protection from the valve being operated incorrectly. Some companies, however, choose to evaluate the consequences of incorrect operation of these valves in order to provide a record of the reason for the car seal or lock, reducing the likelihood that it is removed in the future without due process.
- Valves that are difficult to operate: If a manual valve is located high in a piperack with no access, or a similarly hard-to-reach location, it can be assumed that the valve will not be inadvertently operated as the amount of effort required to reach these valves is disproportionately high. Very large valves that require a long time and many turns to open or close may be less likely to be operated incorrectly, making a slip unlikely.
- Valves that can be reasonably expected to not be mis-operated: Most processes contain many drain and vent valves as part of the normal piping configuration. However, a discussion of each of these being inadvertently open when the process is running is not usually valuable as it can be reasonably expected that the operator will not open a drain or vent valve at such a time. This is not applicable to all facilities, however, and depends on the piping configuration and operating philosophies.
- The mis-operation is included in another cause: In the case of manual isolation valves around an automatically actuated valve, the consequence of one of the manual valves being inadvertently closed is usually the same as if the actuated valve malfunctions in the close position Therefore, as long as the likelihood of either of the manual valves being inadvertently closed is less than the likelihood of the actuated valve malfunctioning closed, it does not add value to the PHA discussion to review them as separate causes, and instead it is more efficient to include them in the likelihood of the actuated valve malfunctioning.

On the other hand, manual valves that are considered "critical" by the PHA team should be analyzed for mis-operation. A "critical" manual valve is one that is either likely to be inadvertently opened or closed, or one that has potential severe consequences as a result of inadvertent operation

- High severity of consequences: If the potential hazardous consequence of mis-operation of a manual valve is severe, and the valve is not removed from consideration by one of the previous points, the scenario should be analyzed. Even if it is unlikely that the valve is mis-operated, it should still be analyzed to determine if there is a requirement to protect against the hazardous consequence. If the PHA team is unsure whether the potential consequence is severe, it is better to document the valve mis-operation during the 'cause brainstorming' portion of the analysis, and then revisit it when determining consequences.

- High likelihood of mis-operation: If there is reason to believe that a reasonably trained and competent operator may inadvertently open or close a valve, it should be analyzed. This increased likelihood may be due to a requirement to operate the valve frequently, thereby increasing the likelihood of a slip, or due to a confusing valve arrangement, thereby increasing the likelihood of a mistake. The below photograph shows an extreme example of a process where the likelihood of operating the wrong valve is very high, and would certainly require analysis of valves being incorrectly manipulated.

Figure 2. 1918 photograph of the control room of German Submarine *U.B. 110*.[10]

4 Addressing Human Factors in PHA: Safeguards

When identifying safeguards requiring personnel response in a PHA, the team must consider the same potential for human error as discussed above, and must also consider the ability for a person to respond in the way requested by the safeguard. If the PHA team is relying on the operator to take a certain action to prevent the propagation of a hazardous event, the operator should be aware of what is required of them, able to perform the action even in a high-pressure emergency situation, and must not be expected to perform multiple tasks simultaneously unless it is part of their training and expertise to do so.

4.1 Alarms

Whenever a process alarm is used as a safeguard in a PHA, the team must ensure that it is possible to respond to the alarm and successfully bring the system to a safe state through some action of the operator. There must be a corrective action or actions available to the operator from the control room or the field, and they must have adequate time to take this action before the hazardous scenario occurs. This action must not put the operator in harm's way. If there is no action that the operator can safely take, the alarm is not a valid safeguard. Furthermore, it must be reasonable that the alarm will be acted upon rather than being ignored or bypassed. The example in this paper's introduction of the operator's answer to a question about responding to an alarm at the Esso Longford gas plant illustrates that due to a facility's operating realities, it is not always possible to take action to correct the process deviation that the alarm indicates.

The safeguard wording should include an indication of the action the operator should take to control the situation, such as "LAH-1001 will alarm with operator action to divert flow from T-101 Storage Tank to T-102 Storage Tank through the DCS." For some alarms there may be many different actions the operator could take depending on the nuances of the exact situation; in such cases, it may be appropriate to document the fact that the operator would be able to determine the correct action at the time that a response to the alarm is required.

In order to successfully bring the process to a safe state after receiving an alarm, the operator needs sufficient process safety time to acknowledge the alarm, identify the issue, determine the correct course of action, and then follow it, all before the hazardous scenario occurs. For example, overpressure scenarios often occur within minutes or seconds, and the PHA team cannot expect the operator to respond in a matter of seconds every time the alarm sounds. If there is insufficient time for all of the response steps, the alarm is not a valid safeguard. When determining the required process safety time for an alarm, consider the following points:[11]

- Is the alarm presented in a clear and easily understood manner to the operator?
- How many other alarms would the operator get at the same time in this scenario? An excessive number of alarms will make it difficult for the operator to successfully identify the correct problem and necessary course of action.
- Is there an action that the operator can take to bring the system to a safe state, without putting the operator in harm's way?
- Can the operator perform that action from the control room or is someone required to go into the field?
- Would an operator be available to undertake that action or could all operators potentially be busy with other urgent activities?
- Is the response action a documented, clear and well-known procedure or do they need to troubleshoot "on the spot"?
- Are there enough personnel in the control room and in the field to adequately respond to urgent situations?

4.2 Procedures

Suggested re-write (confusing paragraph): In most cases, existing procedures are not a valid safeguard, because the PHA team will have made an assumption at the start of the PHA that standard procedures are in place and are being followed. Therefore, the cause likelihood will already include the possibility that the operator has not followed the relevant procedure.

Some facilities will allow a procedure as a safeguard only if it has been designated a "critical procedure", and has been confirmed and signed off by a supervisor to be adequate, up-to-date and to be separate from the cause of the hazardous scenario.

If the procedure is considered a valid safeguard based on these factors, but it has not been written or the facility has not yet been constructed, an action item must be made to ensure that the procedure is developed and that the key required action is included.

4.2.1. Case Study: Inadequate Wording of Procedures

Many procedures will detail how to perform a task but fail to inform the operator of the potential consequences of deviating from the steps listed, which should be a routine aspect of all procedures[12]. ACM Facility Safety Inc. performed a study to review the validity of safeguards used in a HAZOP for a gas processing facility; the HAZOP was conducted internally by the operating company. There were several procedures listed as safeguards for high criticality items such as startup of equipment. One of these procedures involved starting up a steam turbine, and instructed the operator to open the low-pressure steam return line before opening the high-pressure steam supply line. This order is important because if the high-pressure line is opened first it can overpressure the steam return line resulting in a loss of containment of high-pressure steam near the operator. The overpressure scenario was protected by a pressure safety valve (PSV), designed to relieve the steam pressure to a safe location if this occurred. The wording of the procedure, however, was that the operator must open the valves in the prescribed order "to prevent the PSV from lifting", and did not adequately describe the potential consequences of failing to follow this step exactly. The PSV should have been considered a last line of defense in case all else failed.

A different procedure for another steam turbine at the same facility simply stated, "Open the low and high pressure steam valves," without clearly defining which order to open them in or the consequences of failing to do so.

This is similar to a scenario described in a CCPS Process Safety Beacon[13] in which a steam locomotive operator considered the boiler pressure adequate during startup when the pressure relief valve opened. He was not aware that the relief valve was a last line of defense against the boiler exploding, and through his method of operating the boiler, he was defeating a layer of protection that was part of the design.

4.3 Operator Rounds

Operators can often identify developing scenarios before they become hazardous during rounds, by noticing strange sounds, smells, vibrations or an instrument reading. In order for operator rounds to be an effective safeguard, however, it is important that there is something that the operator can reasonably be expected to notice, and that adequate time is provided for rounds to occur before the hazardous event happens. When deciding whether to use an operator rounds observation as a safeguard, consider the following points:

- Whether the indication of the developing scenario, such as an instrument reading, would be checked by the operator. It should be an item on the operator's rounds sheet before it can be considered a valid safeguard.
- Whether there is adequate time between the indication first becoming identifiable to the operator and the hazardous event occurring. If the process safety time is less than the time between rounds it is not a valid safeguard. Including the potential for the indication being missed or even entire rounds being postponed, however, it is better to allow for at least two sets of rounds to increase the likelihood that the issue is identified.

- Whether there is any action the operator can take after identifying the issue to reduce the impact of the hazardous scenario. For example, in the case of a spill to the ground, operator rounds are not an effective safeguard because by the time the operator has noticed the spill the damage has been done. An operator identifying secondary containment filling up, however, may be a valid safeguard against overflow of the containment area.

4.4 Violation of Safeguards

Regardless of the type of safeguard being considered, always consider the potential impact of human factors on the reliability of that safeguard. Safeguards may be bypassed, removed, or ignored. A common example is an operator bypassing a process alarm because it is known to activate frequently and unnecessarily. If the alarm cannot be bypassed but the setpoint can be changed, the operator may set the alarm value outside the range that the process would ever see, which has the same effect as bypassing it because the alarm will never be activated.

Efforts to prevent this may not be effective. In a HAZOP facilitated by the author, it was suggested that a certain alarm be secured by requiring a password to change it; the operator's response was "then everyone knows the password and it's irrelevant." An operator in a different HAZOP commented that when they have a momentary loss of power, which happens several times each year, they receive hundreds of alarms at the same time, almost all of which are nuisance alarms from instrumentation losing power. They commented that due to the countless hours it would take to carefully consider each alarm, the only practical course of action was to clear whole pages of alarms at a time, acknowledging that it meant they might miss a key alarm that indicates an issue they had overlooked. This practice had been a contributing factor in an incident that resulted in loss of containment of high-pressure steam.

Most safeguards can be defeated in one way or another, and therefore it is important to understand the practices at the facility in question to know when to rely on them. If there is knowledge that a safeguard is, or is likely to be, bypassed or otherwise defeated then it is not a valid safeguard. The exception would be if that defeat is part of a stringent "critical defeat" process to ensure it is monitored and returned to service in a timely manner, and that contingency plans are enacted where necessary to ensure safe operation in the meantime.

4.4.1. Case Study: Violation of Signage

In most cases, signage is not an effective safeguard. It may be ignored, obscured by equipment, dust or snow, or may be misleading due to poor wording. In the below example seen at an office by the author, someone has put up a sign instructing people to put their dishes in the dishwasher and not leave them in the sink. Clearly this has not worked in the past, which has led to that person putting up two more signs with the exact same wording, and with different highlighting around the instruction not to leave dishes in the sink. However, people are still leaving cups in the sink.

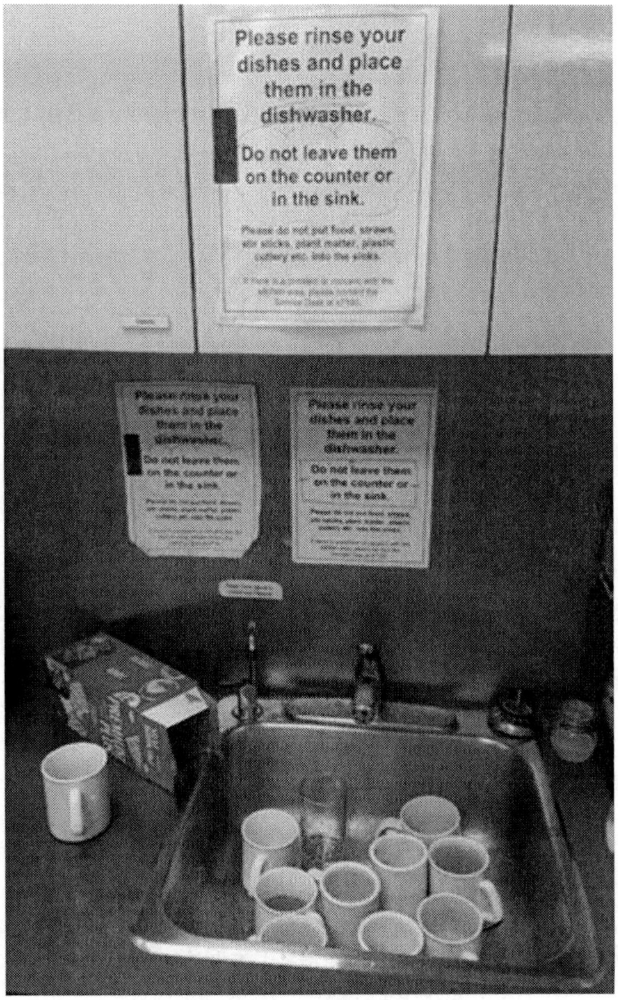

Figure 3. Example of violation of signage. Photo credit: Richard Carter

Why did all these people ignore the signs and put their cups in the sink? It could be that they felt that they didn't have time to put their cup away, or maybe the dishwasher was full and they didn't know what else to do with it, or perhaps one person violated the sign and then subsequent people saw their cup in the sink and assumed the dishwasher was full without checking. Strictly interpreting the sign, however, reveals that it is instructing people on what to do with dishes and says nothing about what to do with mugs and glasses! It is doubtful that the violation of the signs occurred due to a literal interpretation of the sign in this case, however, but it could be an essential distinction in a high potential hazard scenario.

Although this is a frivolous example compared to the hazards involved in processing facilities, it illustrates the issue with relying on signage. A more alarming example is an acquaintance of the author who mistakenly drank oily water, even though it had a sign next to it saying, "Oily Water Don't Drink".

5 Identifying Human Factors Concerns

In some cases, the concerns addressed in this paper may not be apparent. Facility personnel may be used to bypassing alarms, deviating from procedures, or working with minimal staffing levels, with the attitude of "that's how we do things here". They may also choose not to bring up concerns because they don't feel like any change will happen if they do, or even because their manager is in the room and they are worried about complaining. In order to identify concerns regarding human factors it may be necessary to incorporate further analysis into the PHA process, such as one of the following studies.

5.1 Human Factors Checklist Analysis

A checklist analysis can be used to assist in the identification of many of the concerns discussed in this paper. The checklist needs to be built by an experienced team knowledgeable in how human factors can affect PHAs. It can then be used in future PHAs to determine the validity of causes and safeguards, and identify potential areas of improvement. A generic checklist can be made for an operating company, but it is important that this template is then reviewed and adapted to be made specific for each facility, as some concerns will be unique, and others will not be applicable to all facilities. The team will feel like their time is being wasted if they respond to many questions with "not applicable" while at the same time their existing concerns are not being addressed.

It is also important that the results of this analysis be reviewed by an appropriate member of the management team of the facility's operating company and that concerns are followed up on, otherwise the PHA team will not see it as a useful exercise. One company the author facilitated for has a mandatory human factors checklist analysis as part of every PHA. The team answered the questions regarding adequate training by saying the training process was adequate for control room operators but not for field operators. This was not followed up on by management; these concerns had already been considered by the team during the PHA analysis, but the checklist analysis did not add value as no actions came out of it.

5.2 Staffing Requirement Study

In addition to the staffing requirements for normal operation, it is important that facilities maintain a sufficient number of personnel, and in the correct roles, to respond to emergency scenarios that may occur. These personnel may be on-site or on-call, as required by the potential hazardous scenarios that may occur.

In order to determine the required staffing at their facilities, one company commissioned a series of Staffing Requirement Studies from ACM Facility Safety Inc. These used the What-If methodology to identify potentially hazardous scenarios of a certain severity or greater, and then determined the number of personnel required to respond to each scenario, as well as identifying which specific roles were required. From this, it was possible to determine the minimum number of workers for each role that were required to be on-site or on-call at any given time.

When considering staffing requirements, it is also important to consider start-up and planned shut-down situations, as the likelihood of a hazardous event to occur is much greater during these activities.

6 Conclusion

When incidents occur at a processing facility, blame is often initially assigned to the operators for taking a particular action or failing to respond adequately to the situation. As we have seen, however, there are many other factors that can affect the actions of the operators, and sometimes they are expected to perform impossible feats.

When brainstorming potential causes of hazardous events, consider the factors surrounding the operator's day-to-day tasks and investigate cases where they may make reasonable slips or mistakes. The PHA team should be attempting to guard against any understandable error leading to a catastrophic event. Also consider any factors that might force an operator to perform a violation of any procedure, practice or standard, and investigate any solutions for the root cause of this violation.

When determining the validity of a safeguard that requires operations response, consider whether the action is possible and reasonable considering the potentially high-stress situation, and ensure that they have sufficient time to identify the potential issue and perform the correct response.

It is imperative that experienced operations representation is included in the PHA team, as they will have unique experience and insight into the factors discussed. In the case of a greenfield facility, the operations representative must have experience operating a similar process, which may require hiring external consultants.

If all of the above is considered, then the PHA team is setting their operators up for success by affording them a robust process that tolerates normal human errors, and does not force them to develop deviations and shortcuts. This will not only increase the safety of the facility and employees, but it will also generate a PHA that is more closely representative of the process and therefore improve any decision-making processes based on the PHA results.

References

[1] A. Hopkins, *Lessons from Longford*, CCH Australia Limited, Sydney, NSW, Australia, 2000

[2] Center for Chemical Process Safety of the American Institute of Chemical Engineers, *Layer of Protection Analysis: Simplified Process Risk Assessment*, John Wiley & Sons, Inc., Hoboken, New Jersey, 2001

[3] ACM Facility Safety Inc., *PHA / HAZOP Facilitation Workshop*, ACM Facility Safety Inc., Calgary, AB, Canada, 2005-2018

[4] Center for Chemical Process Safety of the American Institute of Chemical Engineers, *Guidelines for Technical Management of Chemical Process Safety*, John Wiley & Sons, Inc., Hoboken, New Jersey, 1989

[5] A. Hopkins, *Disastrous Decisions: The Human and Organisational Causes of the Gulf of Mexico Blowout*, CCH Australia Limited, Sydney, NSW, Australia, 2012

[6] I. Sutton, *Process Risk and Reliability Management: Operational Integrity Management*, Elsevier Inc., Burlington, MA, 2010

[7] Center for Center for Chemical Process Safety of the American Institute of Chemical Engineers, *What if you unload the wrong material into a tank?,* April 2012

[8] Petro-Techna International, *Heater Treater*, Available at http://www.petro-techna.com/heater-treater/, Accessed January 16, 2018.

[9] U.S. Chemical Safety and Hazard Investigation Board, *Investigation Report: Refinery Explosion and Fire, BP, Texas City, Texas*, Available at https://www.csb.gov/bp-america-refinery-explosion/, 2007

[10] Tyne & Wear Archives & Museums, *U-Boat 110, The Control Room Looking Aft, Starboard Side*, Available at https://www.flickr.com/photos/twm_news/8770687732/in/album-72157633570788096/, Accessed on January 14, 2019.

[11] Center for Chemical Process Safety of the American Institute of Chemical Engineers, *Guidelines for Initiating Events and Independent Protection Layers in Layer of Protection Analysis*, John Wiley & Sons, Inc., Hoboken, New Jersey, 2015

[12] Center for Chemical Process Safety of the American Institute of Chemical Engineers, *Guidelines for Risk Based Process Safety*, John Wiley & Sons, Inc., Hoboken, New Jersey, 2007

[13] Center for Center for Chemical Process Safety of the American Institute of Chemical Engineers, *Process Safety Beacon, Safety Device or Control Device?,* March 2016

About the Author

Mr. Carter is a Professional Engineer, and is qualified as a Functional Safety Engineer (F.S. Eng.) in Process Hazards and Risk Analysis through the TÜV Rheinland Functional Safety Program.

He is an experienced facilitator of Process Hazards Analysis (PHA) studies, such as Hazard and Operability (HAZOP), Layer of Protection Analysis (LOPA), Hazard Identification (HAZID) and What-If Analysis. He has facilitated more than 200 PHA studies for a wide variety of processes and for some of the largest operating and engineering organizations in Canada and the United States, in addition to field experience in oilsands and petrochemicals facilities.

Mr. Carter is a training instructor, and teaches the *P&ID/Engineering Drawings Interpretation*, *Introduction to Process Hazards Analysis*, and *PHA/HAZOP Facilitation* courses at ACM Facility Safety Inc. He has also managed ACM's education and training program for public and private offerings, and develops customized PHA procedures and education materials to meet the specific needs of clients.

19 AIChE Spring Meeting
15TH GLOBAL CONGRESS ON PROCESS SAFETY
NEW ORLEANS • 2019

Accounting for Emergent Failure Paths in LOPA

Dave Grattan, P.E., CFSE
aeSolutions
10375 Richmond Ave. Suite 800
Houston, TX 77042
dave.grattan@aesolns.com

Prepared for Presentation at
American Institute of Chemical Engineers
2019 Spring Meeting and 15th Global Congress on Process Safety
New Orleans, LA
March 31 – April 3, 2019

Accounting for Emergent Failure Paths in LOPA

Dave Grattan, P.E., CFSE
aeSolutions
10375 Richmond Ave. Suite 800
Houston, TX 77042
dave.grattan@aesolns.com

Keywords: Systems thinking, Emergence, Meta-Uncertainty, Human Reliability, Human Factors

Abstract

One of the fundamental assumptions made when using standard LOPA (Layer of Protection Analysis) is that the barriers selected for a common threat path are independent. In most cases the analysis made by the LOPA team is adequate to judge the degree of independence between barriers. However, this may not always be the case, especially when the desired LOPA target is less than 1e-4 per year. In these cases, LOPA is more susceptible to unaccounted for system effects, than to independent random failures (what LOPA assumes). Another way to say this is that whenever a model (for example, LOPA) predicts that a failure will occur with a negligible chance, the probability that the model can fail becomes important.

Potential failure paths can emerge between barriers in a common threat path due to what is known as "system effects." That is, to the interaction between otherwise independent barriers due to common support systems (for example, Maintenance), or other Operational or Management impacts. Emergence is a system effect that cannot be identified through other methods, such as IPL (Independent Protection Layer) validation. However, Human Factors methods exist that provide a framework for discovering emergent failures between barriers due to system effects.

This paper will discuss the application of one such system technique known as "NET-HARMS" (Networked Hazard Analysis and Risk Management System). The NET-HARMS technique is a combination of two well-established Human Factors methods, the first being HTA (Hierarchical Task Analysis) and secondly, a modified SHERPA (Systematic Human Error Reduction and Prediction Approach) as the taxonomy used to classify system failures. Both methods are easy to use and can be learned quickly with a little practice. The author has several years' worth of experience applying these methods to difficult LOPA problems involving administrative controls, and will show how this analysis can be extended to include hardware barriers as well.

1 Introduction

The principles of Systems Theory (or "systems thinking") originated in the 1950's as a way to holistically analyze problems [16]. This is opposite to what is called Reductionism, which is when a system is broken down, and the parts analyzed individually. Reductionism is how most engineering activities occur. Systems thinking holds, that when the system is broken down for analysis, some properties that belong to the system will be lost. The properties that are lost, are called emergent properties, i.e., they are properties of the system, not the parts of the system. Emergence has been called the single most fundamental systems idea [16]. Mechanical systems provide the easiest way to visualize emergence. The parts of a car have different properties than the car as a whole (system). In this paper, we will utilize the Systems Theory concept of Emergence as applied to a Barrier System (see **Fig. 1**). Emergence involves the following concepts: (1) expand the barrier boundary to include people, (2) include focus on the interactions between people, and between people and hardware, (3) don't only focus on hardware reliability, and (4) determine how to define and measure Safety.

Many catastrophic accidents involve failure of multiple discrete barriers (safeguards and Independent Protection layers). It is unlikely (from a probabilistic standpoint) that multiple barriers would all be failed randomly (by chance) as is assumed in typical hardware centric probabilistic calculations. Instead, systemic (human based) conditions creating dependencies are more likely the cause. It makes sense then to study why and how multiple and otherwise independent barriers can fail by systems issues. Of course, the frustrating aspect of these incidents is, if just one barrier had worked, the accident would have been prevented. Therefore, understanding human behavior related to a specific barrier will be discussed as well.

Why do we focus on Barriers, in lieu of say, causes or enabling conditions? It is because the pathway to a potential hazard must go through a set of Barriers, no matter how the scenario starts or is enabled (this assumes we have properly identified the hazard and its Barriers). **Figure 2** shows the People in a Barrier system shown generically in **Fig. 1**.

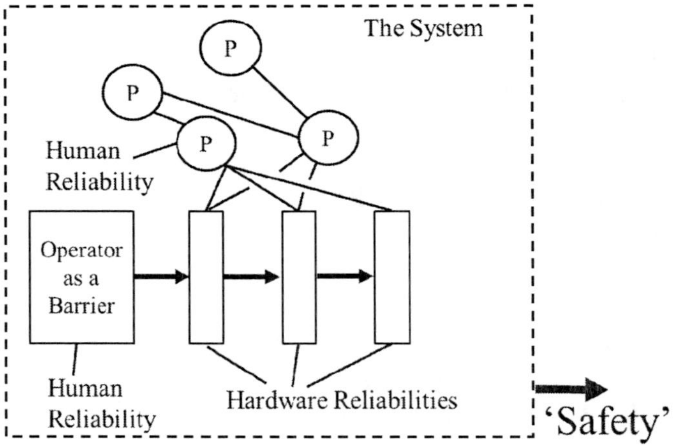

Figure 1. A Barrier System. 'Safety' is a system property that emerges from the interactions between the parts of the system. Component properties (for example, Hardware) are different than system properties. 'P' are people (or groups or departments) that are part of the system.

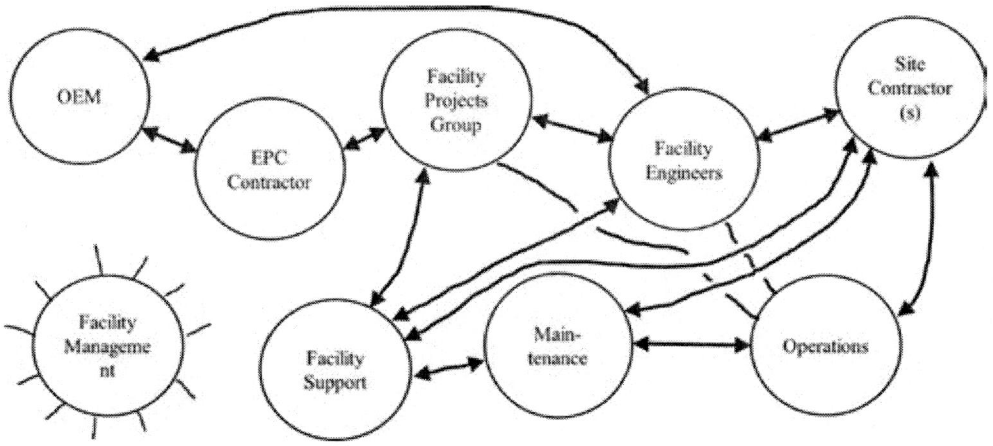

Figure 2. The People in a Barrier System [12]. The physical Barrier is not shown, to recognize the fact that even hardware barriers are fundamentally human. The links between the groups represent task dependencies such as information and expectations that must be communicated. Gaps in the links represent latent failure conditions that could affect one or multiple barriers in the same threat path. The Safety Integrity of the Barrier "emerges" not from its hardware reliability but from the System as a whole.

2 Thinking in Systems

Systems thinking is embedded in the new paradigm known as "Safety Differently" or "Safety 2.0" [1]. New accident models such as STAMP (Systems-Theoretic Accident Model and Process) [2] and FRAM (Functional Resonance Accident Model) [1] utilize systems thinking. Systems thinking is coming to PSM (Process Safety Management), specifically to the way potential hazards are identified, quantified, and controlled. The following is a short list of elements that make up systems thinking [15]. Each element is discussed individually below this list.

- High component reliability is not enough to prevent accidents. Instead, it is necessary to recognize that Reliability and Safety are different properties. Reliability resides at a hardware component level. Safety resides at a system level (it is an emergent property of the system, i.e., it is more than the sum of its parts, it is the product of the *interactions* between its parts). More generally, this is referred to as reductionism vs. holism.
- Safety is not defined by the absence of negative events (e.g., freedom from harm). Instead, Safety is defined by the presence of something (e.g., barriers, reliability, resilience, effort, etc.).
- Accidents are not a linear chain of events. Instead, real accidents are complex and non-linear, often involving drift or "creeping change."
- Probabilistic calculations based on coincidental (i.e., random) failures creates over-confidence (e.g., LOPA). Instead, there are better ways to communicate risk, for example, using qualitative and quantitative Bow-ties.
- Operator error is not the cause of most accidents. Instead, operator error is a consequence of the system (i.e., poor design, production pressure, etc.). At most, normal

operator (human) variance is a local trigger for latent conditions (not necessarily failures) to line-up causing an accident.

- Hindsight bias distorts our view of past events and leads us to believe we can predict future events better than we are able to. Instead, our focus should be on predictive tools, not relying solely on lessons learned from past events to make us safer.

2.1 Reliability versus Safety

It is a common belief that improving hardware reliability will improve safety. Who could argue against this? The reality is however, even if equipment *never failed*, accidents would still happen, simply due to the human element. We need both, hardware reliability and human reliability. Furthermore, Reliability is a component property, while Safety is a system property that "emerges" (i.e., does not reside at the component level) from the interaction between the parts of a system. In this paper, the "system" is the identified barriers (e.g., hardware) used to protect against a potential accident, and includes the people that directly interface with said barriers (Operations, Maintenance, Technical staff, Purchasing, Stores, Contractors, etc.) as well as more remote people such as Management (see **Fig. 2**).

2.2 How to define Safety

The goal of Process Safety (to prevent catastrophic accidents) suffers from what is called the Problem of Induction [3]. Inductive statements can be verified by, for example, experience or statistics. The issue is, attempting to show that a process is safe (e.g., by predicting a rare event frequency meets some acceptably low target), may not be possible to any significant degree of confidence. It is similar to trying to prove a negative (i.e., show that something, for example accidents, won't or can't happen). Negatives cannot be demonstrated using statistics because there is not enough data (e.g., on rare events) *for my plant*. Justifying a process is safe by experience (i.e., "we've operated for 30 years without that happening") is a weak argument because it takes just one accident to disprove the process was safe, and by then it's too late. The issue is encapsulated by the following saying "Absence of Evidence, is not the same as Evidence of Absence." It is not possible to gather "evidence of absence." What can we do?

The best we may be able to do to validate a process is "safe" is to use a combination of feedback performance (i.e., field data) of our Barriers, along with a short (several years) predictive evaluation of the quality of Barrier management. Bayes theorem could be used.

A Bayesian Prior could be developed from the following conceptual equation:

Prior = Barrier Hardware Reliability + Barrier Human Reliability + Barrier qualitative assessments.

The qualitative assessment can be converted into a quantitative probability distribution because Bayes theorem is an epistemological statement (i.e., based on knowledge) versus a statement of frequencies and proportions. We can use all of our qualitative knowledge of how Barriers are being managed as an input to the Bayes Prior.

Later, as we gather new evidence (i.e., the field data on our Barriers) we can update our Prior to calculate a Posterior. This cycle repeats itself every several years as new data is gathered (i.e., the Posterior becomes the new Prior). See **Fig. 3**.

Figure 3. Both Bayes Theorem and Gaussian style statistics are need to demonstrate the confidence in a predicted rare event frequency.

2.3 Accident Models

A good accident model will have the ability to incorporate system components or layers into its boundary. Reason's [4] Swiss Cheese Model (SCM) is one such model. The SCM is superior to all other system accident models because of its ability to communicate in a simple way the nature of complex accidents through the graphical image of Swiss Cheese layers. The SCM is sometimes mistaken for a linear model, because the failure trajectory is shown as being linear through the holes in the Swiss Cheese. This is a misperception. It is very difficult to draw a non-linear accident model in any meaningful way (i.e., that is useful to practitioners).

LOPA (Layer of Protection Analysis) does not represent a practical implementation of the SCM, although they look similar (i.e., the Onion layers concept). LOPA stops at human error (i.e., does not look at systemic factors), and does not even treat active (front-line) errors effectively (i.e., from a human factors perspective). To be clear, the SCM although popular and well-known, is not being used as an accident model in many Process Safety Management programs.

How do we implement the SCM in a practical way? The answer: HFACS (Human Factors Analysis and Classification System) (see **Table 2**). HFACS was created to make Reason's Swiss Cheese accident model useful to the practitioner. That is, to define the holes in the layers of protection (i.e., the Swiss Cheese slices). The ability of HFACS to describe both active (i.e., front-line) errors as well as latent conditions (systemic weaknesses in the organization that manifest themselves at the worst possible time) makes HFACS particularly appealing to use. The HFACS method can be used to identify, classify, and correct these system factors *before* an accident occurs.

2.4 Meta-Uncertainty

When a LOPA calculation shows an event to have a predicted likelihood of occurrence of 1e-4 per year (or less), the result is subject to more than random (classical) uncertainty. At these low numbers, meta-uncertainty (the uncertainty of uncertainty) can begin to dominate [3]. These are the "unknown unknowns" (i.e., what we don't know we don't know) and even worse are the "unknown knowns" (i.e., what we should know (for example, excessive bypassing is occurring), or what we know but refuse to acknowledge is a problem (for example, my LOPA calculation based on coincidental failures is a true measure of my risk)). The issue is referred to as "unsampled-randomness" that no amount of generic data or advanced Markov or Monte-Carlo methods can account for. To be clear, confidence intervals based on classical (frequentist based) statistics are "knowns" and do not represent system uncertainty (i.e., meta-uncertainty).

The Bow-tie graphical presentation which can include safeguards, IPLs (Independent Protection Layers), administrative controls, Degradation factors, Degradation factor controls, audit results, incident results, with the ability to quantify many of these events, makes Bow-tie a far superior tool than LOPA to communicate Risk (see **Fig. 4**). In addition, a Bow-tie diagram can be created for a single barrier or a set of Barriers in a common threat path, showing all the ways the Barriers can fail, including by common system effects, which can be very effective.

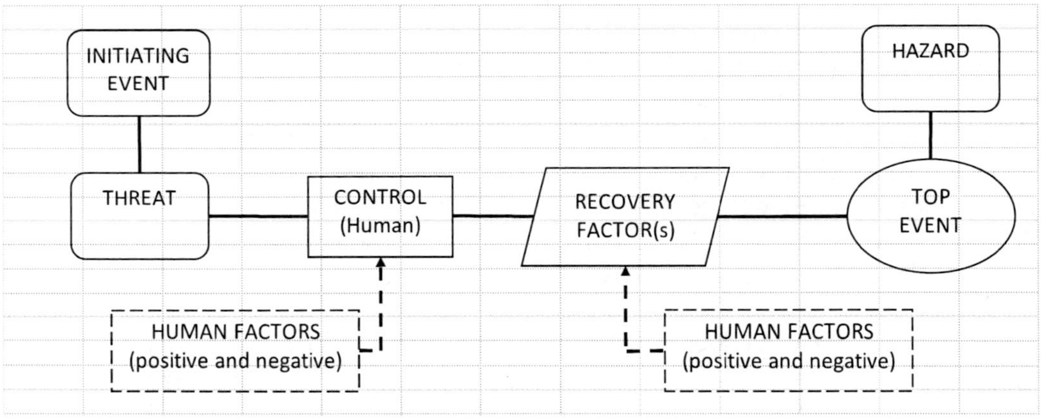

Figure 4. LHS (Left Hand Side) of a Bow-tie. This template can be used for documenting and quantifying a procedural control associated with a catastrophic risk. The human factors from **Section 4.2** can be coded into the spreadsheet. Use **Task Analysis** to complete the template.

2.5 Human Error

Humans (front-line workers) get blamed for 80 to 90 percent of industrial accidents, depending on the report you're reading. Don Norman, the famous Engineering Psychologist, puts the figure closer to 1 to 5 percent [5]. The difference exists because it's easy to stop at "human error" versus looking for (and correcting) the systemic issues that promote or make human error more likely. A different person in the same environment would have made the same error. It's not a front-line operator problem, it's a design problem.

2.6 Hindsight Bias

Hindsight bias is one of the more harmful cognitive biases [6]. It leads us to believe we understand the past more than we really do, and in turn gives us over-confidence in predicting the future (for example, the way potential accidents can happen). The past always appears more deterministic than it really is. When we connect-the-dots to trace back an accident path, we lose the randomness that existed at the time. Hindsight bias presents a serious challenge to (1) how much we can learn from past events, and (2) our ability to predict the occurrence of future events. In addition to learning from past accidents (albeit not as much as we think we do), predictive human factors tools such as Task Analysis paired with an appropriate Taxonomy, provide a systematic way to identify accidents before they occur. This is discussed further in **Section 4**.

3 Is it better to be Lucky or Good in Process Safety?

For a systems thinker, especially one who has practiced Human Reliability methods (see **Table 2**), studied meta-Uncertainty **(see 2.4)**, and the Problem of Induction **(see 2.2)**, probabilistic calculations based on coincidental (random) failure of barriers that purport numbers as low as 1e-4 per year or lower (anytime you have to "count the zeroes"), can look pretty silly. Does anyone actually believe these numbers? There have been many instances reported in the Literature of probabilistic risk assessments showing very low numbers, only to have the accident occur within one year of said calculation [2,7].

You can see how skepticism would creep in, challenging the value, even the necessity of spending resources to produce such unrealistic numbers (for example those produced by Safety Integrity Level calculations, or LOPA).

However, there is a trap in believing calculations based on random failures are useless (*no matter* how low the number is, and regardless of how bad the systemic factors are). The trap is called "ergodicity." Ergodicity is a rarely discussed probabilistic concept. It means "in the long run." And in the long run, rare events (events with low but non-zero probability) that occur only randomly, are *guaranteed* to happen (somewhere), given enough time or opportunity [14]. And in the Process Industries, there is ample opportunity for random accidents to occur. Ergodicity is also the reason to invest in both preventive and mitigative barriers (Left and Right-hand side of the Bow-tie, respectively). Ergodicity is the reason hardware reliability calculations based *only* on random failure, contributes positively to process safety.

So, is it better to be Lucky or Good? In Process safety *we need both.* (The word "Lucky" in this context does not mean haphazard. It means occurring by chance, but also following good reliability engineering principles for hardware Barriers. The "Good" refers to identify and fixing potential causal (deterministic) failure of hardware Barriers (or good human factors for Human Barriers), that can derange random based probabilistic calculations for said hardware or human).

4 Human Factors Tools vs. Traditional PSM Tools

PSM (Process Safety Management) utilizes a variety of tools (see **Table 1**) that have the potential to better incorporate human factors principles and methods. To be clear, PSM tools are not human factors methods (although many PSM tools do incorporate elements of human factors). Human factors and reliability methods such as Task Analysis are well established, see for example Stanton et al. [8] and Grattan [12], for a discussion of said tools. PSM does not need new human factors methods. What it needs however is to adopt some of the said methods into the practice of PSM, by integration or stand-alone use.

In general, PSM tools, while addressing many human factors elements, suffers from the following shortcomings when evaluating human factors:

1. PSM tools related to human factors fail to look at Behavior, instead focusing on States and Conditions **(see 4.1)**.
2. PSM tools related to human factors fail to leverage the predictive ability of human factors methods **(see 4.2)**.

Why do we look at behavior? The intent of observing behavior is *not* to change behavior; that is the goal of Behavior Based Safety (BBS) programs. Instead, the intent is to understand the "work-as-imagined" versus "work-as-done" principle [1]. Work-as-imagined by designers, managers, PHA teams, procedure writers, etc. is *never* the same (*can never* be the same) as the work-as-actually-performed by Operations and Maintenance staff. The differences can reveal latent conditions that will eventually bite.

Using checklists to "do" human factors is a good example of these deficiencies. Checklists presuppose that every human factor or error can be predefined and put into checklist form [2]. A checklist does not evaluate Behavior, and a checklist is certainly not predictive. Checklists are useful, especially for the "knowns," however beyond that they are inadequate. At the other end of the spectrum are Audits. Audits do look at behavior, and are a good candidate to incorporate some of the tools in **Table 2**.

Table 1. PSM Tools that represent an opportunity to Incorporate Human Factors Methods

PSM Tool
Functional Safety Assessments
Audits
Checklists
IPL Validation
Bow-Tie Analysis
Other specialized methods (e.g., Control System HAZOP, Failure

Modes and Effects Analysis, Procedural PHA, etc.) that provide an opportunity to investigate human factors.

Table 2. Some Human Factors Tools that can be used Stand-alone or Integrated with Table 1 to Improve the Practice of PSM related to Human Factors (see Stanton et al. [8])

Human Factors Tool	Application/ Purpose
Hierarchical Task Analysis	Understand the difference between "work-as-done" vs. "work-as-imagined"
Cognitive Task Analysis	Evaluate the conscious thinking a task involves (as opposed to the physical work)
Situation Awareness Analysis	Evaluate the "What?", "So what?" and "What-now?"
Human Error Taxonomy (e.g., HFACS, SHERPA)	Pair with Task Analysis or Human Reliability Analysis
System Error (e.g., NET-HARMS) [13]	Used to identify Emergent failure paths
Human Reliability Analysis (e.g., AEA, Petro-HRA)[12]	Used to quantify Human Error rates for a given Task

4.1 States and Conditions vs. Behavior

One of the goals of human factors analysis to understand how work is actually performed (i.e., behavior). **Table 3** gives examples of the behavior associated with a state or condition of equipment, design, or procedures, etc.

Table 3. Any State or Condition has a corresponding Behavior associated with it

State/ Condition	Behavior
Is equipment properly labelled?	Will the operator actually **read** the label?
Are the proper Tools available?	Does the operator or maintenance worker **use** the tools?
Are procedures available?	Does the operator or maintenance worker **use** the procedures?

What-if this step in the procedure is omitted?	How **likely** is the operator to skip that step and **why**?
Is the alarm configured as priority?	How will the console operator **interpret** the alarm?
Is the alarm audible above background noise in the control room?	Will the console operator **turn down** the alarm volume because it is annoying?
Is adequate information about normal and upset process conditions clearly displayed in the control room?	How likely is the console operator to **proactively monitor** said information?

4.2 Error Promoting and Error Likely Situations

Human Factors methods have the ability to be predictive, i.e., to be able to predict that human error will occur (or be more likely to occur) when a given condition(s) exists.

Two references that are useful for helping to identify bad human factors that can be corrected are found in Norman's [5] discussion of design induced error, and Taylor's [9] discussion of human error syndromes. In addition, Miller's [10] "Basic Human Needs to be Fulfilled in Design," which form the basis for many industrial accidents is an excellent reference. All of these can be applied in an existing operating facility as part of a Task Analysis.

5 Human Factors Tools to Support LOPA Barrier Analysis

The tools listed in **Table 2** can be used stand-alone or in conjunction with the Tools in **Table 1** to analyze any single Barrier identified in LOPA. The Barrier can be hardware or human based (administrative controls, etc.). Even hardware Barriers contain critical human influence (e.g., Operations, Maintenance, Contractors, etc.) that should be recognized (see **Fig. 2**). A human barrier can also be the "Cause" of a catastrophic accident (see **Figs. 5 and 6**). Typically, resources do not exist to evaluate every barrier from a human factors perspective, therefore some prioritization will have to be made.

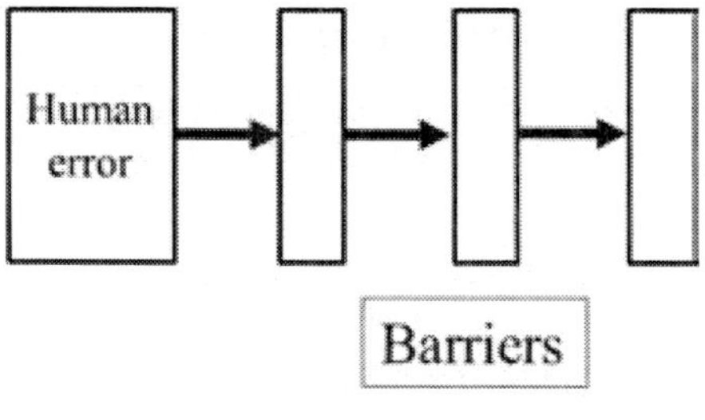

Figure 5. PHA-LOPA viewpoint of Human Error as a Cause (this is a fallacy) [12]

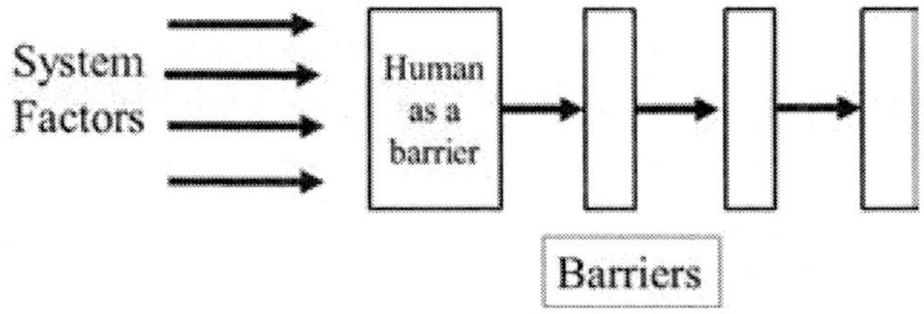

Figure 6. When human error is viewed as a consequence (not a cause), the human becomes an important Barrier in the sequence of events [11]. Any human error which results in a catastrophic consequence should become a candidate for a Human Factors analysis [12].

6 Emergent Failure Paths in LOPA

Fig. 1 gives a generic representation of a barrier system that can create emergent failure, that is, between the links (interactions) of the parts. The Barrier(s) can be hardware or human based.

Networked Hazard Analysis and Risk Management System (NET-HARMS) is unique because it is not a new systems model of accident causation. Instead, it utilizes existing Human Factors tools found in **Table 2** to identify (predict) emergent errors and risk due to interaction between the parts of the system (in this case a system related to Barrier integrity). The same tools can be used to evaluate barrier integrity inside the circles of **Fig. 2** as well.

Let's sharpen the pencil and look in detail at some of the latent conditions that can exist between the parts of **Fig. 2** (the links). **Fig. 7** shows that expectations related to a barrier must be communicated to (in this case) Facility Support, so that, for example, adequate maintenance, failure tracking, and management of change for the Barrier can occur.

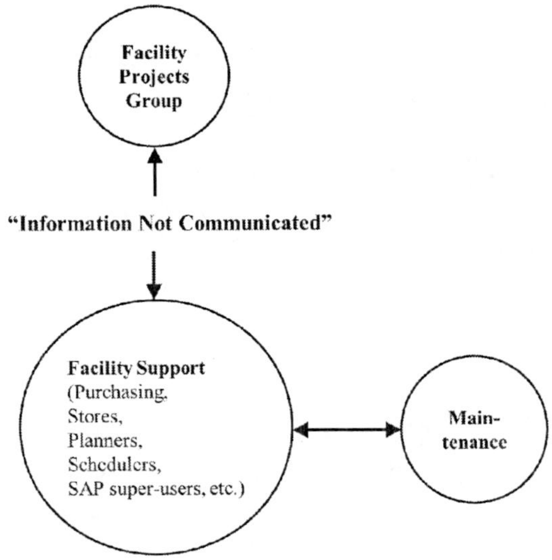

Figure 7. Facility Support personnel are a common link in all barriers. Not communicating expectations to Facility Support personnel will create Latent failure conditions that can eventually fail multiple barriers in the same threat path.

Table 4 list the taxonomy published for NET-HARMS [13] that can be used to identify latent failures between parts of the Barrier system. This taxonomy should be used along with a Task Analysis for those parts of the system where latent weaknesses are suspected.

Table 4. NET-HARMS Taxonomy to Apply to the System Links [13]

Code	Description
T1	Task mistimed
T2	Task omitted
T3	Task completed inadequately
T4	Inadequate task object
T5	Inappropriate task
C1	Information not communicated
C2	Wrong information communicated
C3	Inadequate information communicated
C4	Communication mistimed

7 Conclusion

A Barrier system is more just hardware reliability. It is a collection of people (system elements) that are linked together in such a way that the Barrier integrity emerges through the interactions between the people, and between people and hardware. This is a systems perspective.

Generally, PSM Tools are not Human Factors methods, in that PSM tools cannot predict the likelihood of human error, nor they do not focus on said behavior. However, Human Factors methods can be used in select applications, for example, Barrier Analysis, to provide an improved evaluation of the integrity of the Barrier(s). In this way, both hardware reliability requirements and potential systemic failures, are addressed.

NET-HARMS was presented as a unique Systems tool that can be used to identify emergent failure in a single Barrier or System of Barriers protecting a common threat path. NET-HARMS uses Task Analysis paired with a modified SHERPA Taxonomy to identify potential latent failure conditions between parts of a system.

Experienced PSM practitioners should begin to use Human Factors methods, and with practice will improve Process Safety related to Barrier integrity.

8 References

[1] Hollnagel, E., 2014. Safety-I and Safety-II The Past and Future of Safety Management, CRC Press, Boca Raton.

[2] Leveson, N.G., 2012. Engineering a Safer World: Systems Thinking Applied to Safety. MIT Press, Cambridge.

[3] Taleb, N.N., 2004. Fooled by Randomness – The Hidden Role of Chance in Life and in the Markets. Random House, New York.

[4] Reason, J., 1990. Human Error. Cambridge University Press, Cambridge.

[5] Norman, D., 2013. The Design of Everyday Things, Revised and Expanded ed, Basic Books, New York.

[6] Kahneman, D, 2011. Thinking Fast and Slow. Farrar, Straus & Giroux, New York.

[7] Perrow, C., 1984. Normal Accidents - Living with High Risk Technologies. Princeton University Press, New Jersey.

[8] Stanton, N.A., Salmon, P.M., Rafferty, L.A., Walker, G.H., Baber, C., Jenkins, D.P., 2013. Human Factors Methods – A Practical Guide for Engineering and Design, second ed. Ashgate, England.

[9] Taylor, J.R., 2016. Human Error in Process Plant Design and Operations – A Practitioner's Guide. CRC Press, Boca Raton.

[10] Miller, Gerry, 2017. Presented at: Forum on Human Factors to Support Safer and Effective Offshore Energy Operations, OESI, TAMU MKO, 2017.

[11] CIEHF (Chartered Institute of Ergonomics & Human Factors), 2016. White Paper: Human Factors in Barrier Management. United Kingdom.

[12] Grattan, David, 2018. Improving Barrier Effectiveness Using Human Factors Methods, Journal of Loss Prevention in the Process Industries, 55 (2018) 400–410.

[13] Dallat, C., Salmon, P.M., Goode, N., Identifying risks and emergent risks across sociotechnical systems: The NET-worked Hazard Analysis and Risk Management System (NET-HARMS).

[14] Bennett, Deborah J., 1998. Randomness. Harvard University Press.

[15] Meadows, Donella H., 2008. Thinking in Systems, A Primer. Chelsea Green Publishing, Vermont.

[16] Chapman, Jake, 2004. System Failure, second ed., Demos, London.

19 AIChE Spring Meeting

15TH GLOBAL CONGRESS ON PROCESS SAFETY

NEW ORLEANS • 2019

HUMAN FACTORS: WHAT ROLE SHOULD THEY PLAY IN RISK ASSESSMENTS AND HOW SHOULD THEY BE HANDLED

David Rowe
Global Risk Consultants
Conroe, Texas
David.rowe@tuvsud.com

Prepared for Presentation at
American Institute of Chemical Engineers
2019 Spring Meeting and 15th Global Congress on Process Safety
New Orleans, LA
March 31 – April 3, 2019

HUMAN FACTORS: WHAT ROLE SHOULD THEY PLAY IN RISK ASSESSMENTS AND HOW SHOULD THEY BE HANDLED

David Rowe
Global Risk Consultants
Conroe, Texas
David.rowe@tuvsud.com

Keywords: human factors, risk assessments, hazard analysis, PSM

Abstract

"Human factors" relate to the effects that operators, management, etc exert on a process or system. While it is, typically, easy enough to determine the mechanical or chemical hazards of a process or system, deciphering the human element can be more difficult. There can include task / workplace design, leadership / culture paradigms or even letting shortcuts become the norm. When the assessment of risk of a process or system is conducted, the affect of the employees, contractors and management must be taken into account or critical hazards may be missed.

1 Introduction

Accidents, other than natural disasters, have a human factor influence. Decisions are made that affect the systems whether in the design, the installation or the operation of a process or system. Humans design processes and install them. Humans develop the procedures to operate them and choose the safeties to protect them. And lastly humans, for the most part, operate them. Humans influence other humans in the process – not only do humans cause accidents (unintentionally) by making errors directly related to the process itself, but they also cause errors by creating deficiencies in the design and the implementation of management systems. Ultimately management systems are created in an effort to manage these risks by directly or indirectly influencing the process.

2 What are Human Factors?

What does "Human Factors" mean? According to the World Health Organization (WHO) – "human factors refer to environmental, organizational and job factors, and human and individual characteristics which influence behavior at work in a way that can affect health and safety."

Human factors can be divided into categories. 1) Tasks, workload and work patterns 2) Working environment and workplace design 3) Workplace culture and communication 4) Leadership and resources 5) Policies, programs and procedures 6) Worker competency and skill and 7) Employee personality, attitude and risk acceptance. These are not stand-alone categories and their interactions can be critical.

2.1 Tasks, workload and work patterns.

If tasks or workflows do not make logical sense, then operators may make adjustments to make the flow simpler. Also, if operators are asked to go all over the facility to complete certain tasks, a work flow may be developed to best accommodate the employee and not necessarily the best flow for the process.

2.2 Working environment and workplace design.

The design and layout of the workplace can affect the way that employees interact with the process. If the working environment is not safe, then tasks may be adjusted or avoided because no one wants to get hurt. If the workplace is not designed well, the poor design can make maintenance tasks and even operations more difficult thus promoting work arounds or short cuts.

The system's controls and warnings should be consistent and understood by all. For example, does red mean de-energized and thus safe or does it mean danger? Either approach is acceptable as long as it is consistent throughout a process. Along this same line are the Human Machine Interfaces (HMI's) understandable to those that must use them or only to the engineers that designed them? Operators need to be able to understand the layout and use of the HMIs in case of an event. They should not have to toggle between dozens of screens to make a situation safe.

Controls should be accessible. Controls include HMIs, as mentioned above, as well as any manual controls that must be accessed for normal operations or in case of an event. Labeling should be complete and referenced in procedures and other instructions. For example, in an ammonia system, a manual King or Queen Valve Control accessibility defines whether an incident is a minor leak or will become a catastrophic release. Labeling, including flow direction, valve tagging, and proper signage should be complete and consistently referenced in procedures and other instructions. Failure to have labeling and valve tagging in place, and the failure to have services and valve tag numbers accurately depicted on P&IDs, increases response time for newer operators or facility first responders, and can increase the potential for incorrect actions taken during maintenance procedures.

2.3 Workplace culture and communication.

It cannot be stressed enough that the culture of the facility affects how the workers will respond. If the culture emphasizes "get the job at all costs", then that is what will happen. Short cuts will be taken to ensure that production is on time, even though safety as well as quality may be sacrificed. Obviously, the point of a business is to produce a product and if it is not making money you will not stay in business. However, a safe environment can be a productive one as long as it's built into the "culture". If an organization routinely pushes its employees to take double shifts or work extra hours then they are putting stress on the employees that could, and usually do, negatively affect safety, efficiency and quality.

An organization's ability to communicate effectively has an impact on safety as well on the operation of the process. I have heard many times – "well that's not what I understood", or "I didn't hear that", or "I was never told to do it that way". Most of us struggle with effective communications at some point in the process and have room for improvement. Many people need to have instruction on how to communicate effectively with those superior to them as well as those that work for them.

2.4 Leadership and resources.

Leadership is as critical a human factor as many of these other points. If leadership (supervisors up to plant management) does not believe in a safe operation, then it will not happen. Stating that Safety is #1 at the tool box talk or at the weekly safety meeting is not promoting safety. Ensuring good, accurate JSAs are in place or ensuring the adherence to procedures are examples of promoting safety. If an employee comes to their supervisor with a question or concern, it is important that they are listened to and not dismissed as wasting time.

Leadership style can make a difference on the operation of an organization. There was once a company that would have a morning meeting with line supervisors, and it was well known that someone would be berated during that meeting. This added stress to the supervisor who then passed it to his line employees. Either way – the company was creating stress and distractions that could and did directly affect quality as well as safety. After the morning meeting the some of the line supervisors would take their frustration from the meeting out on the line workers. Yelling at them, "taking them off their game", providing mental distractions that would lead to mistake.

OSHA, especially in the PSM standard, discusses the importance of employee involvement. Employees should have access to resources that make their job safe and efficient. Safety Data Sheets, procedures, action items from audits and hazard reviews are a few examples of resources that employees should have at their disposal and know how to find.

Other resource related human factors include the use of appropriate equipment to perform the job correctly. There should be sufficient resources available, be it at the site or the tool room, to effectively perform the job. If resources have to be replenished, it should be done before needed. It is this point that communication is important. It may be the responsibility of the worker to inform supervision of a need.

Along the line of resources is manpower. Enough people to safely complete a job with minimal extra time is critical. Most of us at one time or another have taken a job because it had overtime, but eventually that catches up with a person and affects their performance. At what point should rest become required?

One of the most dangerous things an organization can allow is let shortcuts become the norm. Once an action has become the normal way of operating, it can be very difficult to return to the proper method of operation. For example, I saw a company that had spurious alarms all the time and it was "normal" to simply silence them. Watching the operator, an alarm would go off and without looking up they would silence it. While this may work for a seasoned operator, this was taught to the new operators as opposed to having them investigate and ensure that the alarm wasn't something more critical. We have all heard "but I have done it that way 100 times and not gotten hurt. To me that says, "I was allowed to do it wrong and got lucky 100 times".

2.5 Policies, programs and procedures.

Accurate procedures are critical. Not only do they help with the operation of the process, they ensure that new employees are trained the same way every time. If a procedure is found to be wrong or maybe there is a better way of performing that procedure, there should be a formal process for modifying the procedure not just adopting the new method. The process for procedure modification should also include training all affected employees on the revised procedure to ensure the change is clearly communicated to everyone to reduce, if not eliminate, the potential for errors which could lead to equipment damage or an incident.

In addition to formal procedure modifications, there should be an acceptable method for reporting hazards or unsafe acts and conditions. If employees are aware that they can bring these things forward and not be chastised for them, then more will be learned about the facility and processes.

2.6 Worker competency and skill.

A facility's hiring practices should be reviewed to ensure that not only are competent people being hired, but they are also being effectively trained. I watched a construction contractor hire workers to fill roles without verifying knowledge. These positions included Iron Workers, Pipe Fitters and even Riggers.

An organization's hiring department should work closely with the facility to ensure that a crew, shift, or department is not left shorthanded and ensure employees are not asked to do more than: 1) they were trained to do, 2) they are capable of doing, or 3) are not put under unnecessary stress.

While hiring competent and well-trained personnel is not always possible, it is possible to ensure they are trained to the point that they are competent. Employees should be aware of not only how to perform a job or task but, they should also be aware of the hazards associated with that job or task. An experienced welder knows how to weld. But does he know what happens if you weld on the wrong line in a live plant?

2.7 Employee attitude, personality and risk tolerance.

An employee's attitude towards a job or task is greatly influenced by the attitude of his supervisor and peers. Thus, proper leadership is critical to ensure a safe operating environment. Some other factors that influence an employee's attitude toward a job or task may include:

Ownership – the employee should feel that his actions are serving a purpose and not just earning a check. Ensure that employees have a voice and that management addresses the concerns of the employees.

Distractions – Are cell phones in the work environment? What other distractions are allowed that might distract the employee from completing the task at hand? Distractions will affect the actions of the employees.

Creating the right team, shift or crew can be an important and daunting task as well. Some people just don't work well with certain people, and while both could be competent or exemplary at their position, they may need to be on different crews. Supervision should be aware of how people interact and work to ensure smooth operations.

3 How to address Human Factors in Assessing Hazard

3.1 Process Safety Management (PSM)

The effects of Human Factors can be seen throughout Process Safety. It is more often addressed during a Process Hazard Analysis but is a part of most of the elements. Simply limiting the number of people in an area or minimizing the time that someone has to be in a "hazardous situation" a facility can dramatically change the risk to employees by reducing the potential human influence to a process.

Human Factors need to not only be reviewed during a Process Hazard Analysis (PHA) but need to be factored into procedures, training, MOC, incident investigations as well. A few of these are discussed below.

3.1.1 Process Hazard Analysis (PHA)

The effect of the PHA as with the compliance audit, related to human factors, will be to ensure that they are taken into account within all aspects of a process. The PHA should be effective in identifying gaps as Human Factors can act as safeguards. Within a PHA, the team will need to review the process to ensure that the human affect is minimized as much as possible. The structure of most PHAs does not emphasize the effect of the human interaction in a process. HAZOP for example is very detail and process oriented. It is easy to work through a HAZOP and forget about the human influence.

Factors such as pipe and valve labeling can result in critical errors if incorrect or insufficient. HMIs should be organized to best assist the operators. An operator should not have to scroll through multiple pages to make a piece of equipment safe should an event occur. Color schemes should be consistent through-out the HMIs. If Red means de-energized and thereby safe for one HMI it should also represent that on another. Inconsistencies lead to mistakes that can be critical to the operation of a process. HMIs and control panels need to be developed to aid the operator in understanding the process and controls and not designed for the engineer.

It is important to address human factors in all modes of operation. Many organizations do not fully analyze for errors during all modes of operation, as such accidents that start with a human error may not be predicted. Normal operating conditions are what procedures are usually designed around. However, typically more human interaction is required for start-up and shut down and reaction to emergencies than during normal operation. These modes of operation should be reviewed, steps analyzed for where human involvement could affect the safety of the process, and those steps addressed in procedures, work instructions, or engineered out.

3.1.2 Procedures / SOPs / Training

It can be stated that procedures are not really risk assessments but provide the instructions for completing a task. However, the thoroughness of a procedure can minimize the human effect on a process. The procedure should not only explain how to run the system, but it should point out critical safeties that can protect a process and explain consequences of deviation.

Many times, operating instructions address human factors with, what to do should an upset occur, but what about maintenance instructions. Maintenance personnel should understand what level of influence they may have over the safety and operability of a process by the tasks that they perform. If they are working on gauges and meters then an error in their work or calibrations could be critical to the system. If the maintenance department does not follow proper permitting structure then critical equipment could be left with incomplete repairs and still undergo start-up. It is important that when considering procedures or work instructions that the maintenance ones be considered as well.

Training needs to reinforce the procedure. While training is often referenced in recommendations it needs to be determined if the training or the learning was insufficient. If the learning was ineffective, then retraining may prove effective but if the training itself does not effectively address the hazards and procedures then the training needs to be readdressed. Training should include not only normal operations but upset and emergency situations as well.

3.1.3 Incident Investigations

If a facility or organization has already factored human factors into their operation and design, then any investigation team should have no trouble understanding where the system and / or the employee failed. However, with many organizations it is still a knee jerk reaction to quickly blame an employee for a mistake when other human factors and the organization's ability to address them was the larger issue. Many companies lack an awareness of where human factors fit into an accident sequence.

When an employee gets hurt doing a task, there should be a few areas explored. Training should be sufficient to enable an employee to deal with unexpected events. As with everything, there should be proof of competency, to ensure that the training was effective. Awareness of the hazards should be sufficient that employees understand the potential effects of mistakes and incidents. Along with training and understanding, adequate procedures should be developed to address all aspects of operation. Finally, management and leadership should be trained and understand the hazards so as to not allow shortcuts to become acceptable.

Too often deviations or short cuts becoming the norm is observed. An employee gets a job done using a shortcut. His supervisor is aware of it, but nothing happened as long as he got the job done. On one occasion the employee gets hurt and was faulted for not following the rules when in fact, by doing nothing, the supervisor had made what the worker was doing acceptable.

We recognize that mistakes are made and sometimes people do not make the best decisions, but with good management systems, these theoretically could be eliminated. We should strive to that theoretical point – even knowing it is not truly attainable.

3.2 JSA / JHA

(Job Safety and Job Hazard Analysis) are designed to assess the hazards of a job immediately prior to performing a job or task. It is important that during the development of the JSA / JHA (with the workers) that the human interaction with the equipment is taken into consideration. The effects of live vs purged piping and equipment need to be addressed. The state of the employees needs to be addressed as well. It must be known and understood by supervision the effects of multiple or long shifts without breaks.

The culture should actively respect workers for coming forward with legitimate concerns. This will minimize employees hiding issues and potentially getting hurt. It is easy to pencil whip a hazard analysis because "I have done this before and these were the hazards, and no one got hurt". It is easy to allow the supervisor to write the JSA/JHA and the employees to sign but not review it, but this is not how they were designed to be utilized. Management needs to stress JSAs, adequate work instructions, or complete SOPs.

If situations change, then the JSA / JHA or even PHA should be reassessed. Management and supervision should understand the potential hazard of the change or at least that there may be associated hazards. These should be addressed in a new or revised JSA/JHA or PHA.

3.3 Engineering Reviews

When designing a system, facility, or equipment, human factors need to be considered. The equipment should be laid out not only to fit a footprint but to allow for safe operation and maintenance.

Example – a construction job was to install a bank of 6 pumps on the first floor of a new structure. There was piping overhead and structure all around. Once the pump bases were laid out, it became clear that maintenance would be a nightmare with very little clearance, and no ability to get lifting equipment to the pumps. As an after-thought a trolley system was installed and some changes were made to the piping to allow for easier access.

Design engineers, in the past and even today, tend to focus more on the cost and efficiency of the equipment, not the ease of use, and definitely not the ease of maintenance. If operations and maintenance tasks are not considered in the design phase, then the resulting situation will be highly conducive to shortcuts and work arounds. The worker will determine what can be done to make a task easier. Examples include: Not installing all the bolts or not tightening them all to the proper torque with the thought that there are a couple that are difficult to get to, they are on the bottom, there are plenty of others and no one will notice.

When engineering a new product or system, sometimes it takes a little extra time and questions to make designers consider that not only does this have to be built, but it also has to be operated and maintained.

4 Conclusion

Human factors influence the operation of any organization. It is not just the sum of the pieces of equipment that determine the hazards. The effects of human interaction can greatly impact the safety, as well as the efficiency, of an organization. Everyone from the top down has a role to play, and if each role is not played then, an organization will suffer.

Hazard assessments must include Human Factors to get a thorough understanding of the potential hazards and to minimize the rate at which human factors can affect a process. However, human factors should also be applied to engineering, operating procedures, as well as Incident Investigations.

5 References

[1] William Bridges & Revonda Tew. "Human Factors Elements Missing from Process Safety Management (PSM)" March 2010

[2] Guidelines for Hazard Evaluation Procedures, 3rd Edition, CCPS/AIChE, 2008

[3] Les Chambers "A Hazard Analysis of Human Factors in Safety Critical Systems Engineering" http://crpit.com/confpapers/CRPITV55Chambers.pdf

[4] Chartered Institute of Ergonomics and Human Factors " Manufacturing Increasing Safety and Security"

[5] International Association of Oil and Gas Producers "Human Factors in QRA" Report No. 434-5 March 2010

19 AIChE Spring Meeting
15TH GLOBAL CONGRESS ON PROCESS SAFETY
NEW ORLEANS • 2019

Why Choose a Race Car When Economy Will Do – Competency with Safety Instrumented Systems

James E. Torres, FSP
W. R. Grace & Co.
1201 Gulf States Utilities Road
Baton Rouge, LA 70805
james.torres@grace.com

Nick Ashley, Ph.D., P.E.
W. R. Grace & Co.
1201 Gulf States Utilities Road
Baton Rouge, LA 70805
nick.ashley@grace.com

Prepared for Presentation at
American Institute of Chemical Engineers
2019 Spring Meeting and 15th Global Congress on Process Safety
New Orleans, LA
March 31 – April 3, 2019

Why Choose a Race Car When Economy Will Do – Competency with Safety Instrumented Systems

James E. Torres, FSP
W. R. Grace & Co.
1201 Gulf States Utilities Road
Baton Rouge, LA 70805
james.torres@grace.com

Nick Ashley, Ph.D., P.E.
W. R. Grace & Co.
1201 Gulf States Utilities Road
Baton Rouge, LA 70805
nick.ashley@grace.com

Keywords: safety instrumented systems, competency, process controls, layers of protection

Abstract

There are qualitative and quantitative approaches to hazard assessment that justify the need for instrumentation with reliable performance and rigorous maintenance to ensure mitigation of high-risk events. Safety Instrumented Systems (SIS) are logical solutions to obtain increased protection from hazardous process deviations but can increase capital expenditure with additional hardware and maintenance coverage requirements. System cost significantly increases with increased performance capability. SIS performance is measured by defined Safety Integrity Levels (SIL) specified by the international SIS standard (IEC-61511, "Safety Instrumented Systems for the Process Safety Industry Sector"). The IEC standard contains strong language for personnel competence and accountability requirements to ensure functional safety objectives are met when implementing these systems. Smaller companies may struggle with internal personnel proficiency to properly design, implement, and maintain a SIS throughout the safety life-cycle. This discussion will offer practical suggestions for making choices between simple SIS design and more complex systems, accounting for factors such as actual process risk reduction requirements, competency and availability of site resources for ongoing maintenance and proof testing, project economics, and contesting myths and misinformation from other sources.

1 Introduction

The use of Safety Instrumented Systems (SIS) as protection or mitigation against undesirable events has become increasingly common across the chemical process industries (CPI). With the widespread adoption of computer-based tools for design, data collection, testing, and maintenance of these systems, operating companies are turning to SIS as a key element of the risk mitigation strategy for their facilities.

At the same time, many misconceptions have arisen as to the true requirements and application of the IEC 61511 standard. Computer-based tools which make SIS design more efficient become misused and create confusion amongst site-level staff who are charged with the design, operation, and maintenance of the installed systems. Statements such as, "You can't have a SIS unless it's SIL-3" unfortunately serve to further perpetuate the lack of understanding and often originate from well-meaning, but misinformed, individuals within an organization.

Those who often struggle the most are individuals representing smaller to mid-size companies, who lack the large staffs of corporate control engineers or dedicated SIS specialists found in larger enterprises [1]. This paper is primarily aimed at those who find themselves drowning in unexplained jargon, acronyms, and conflicting information, and looking for a roadmap to sort it all out. The site personnel responsible for identifying the need and basic design of a SIS must understand the foundational requirements of the IEC 61511 standard and basic SIS design and selection criteria in order to avoid under- or over-designing the system [2].

The purpose of this paper is to help site-level practitioners understand how to determine what is required for their specific application, and select the right tool for the job. Many companies who market SIS hardware, especially large, high-I/O capacity logic solvers, offer a very large, very expensive box, and the "brains" of the system occupy multiple cabinets which can fill up one-quarter to one-half of an I/O room. Some companies offer limited capacity (5-15 I/O) safety PLC logic solvers. Still other companies' offerings include a smaller box, about the size of a sandwich, which can be installed directly in the operating unit. Is bigger always better? How does one know when to choose the large, very expensive, high-I/O capacity option versus the small, cheaper, but still highly reliable option? Those are the questions we aim to answer in this paper.

We refer to the large, expensive box as the racecar, and the smaller, homely looking, more economical box as the economy car. Which option is needed for a given application depends on what one needs the car to do. If the intent is to compete in a highly competitive race, such as the Indianapolis 500®, the Daytona 500®, or the Monaco Grand Prix™, a high-performance racecar is required. If the need is for a good, reliable vehicle to travel between home and work, go to the grocery store, and bring the kids to soccer practice, a good economy car will do just fine. In this paper, we intend to provide practical advice on how to match the right tool with the right application.

2 Understanding SIS Performance Requirements

The first step in selecting the right SIS for a particular application is to clearly understand its performance requirements. To do this, one must understand the hazardous consequence that is being protected against, and the required level of risk reduction necessary to achieve a tolerable risk, along with any other independent protection layers (IPLs) that serve to reduce the likelihood or severity of the scenario under consideration.

In this paper, we will only address SIS operating in low demand mode, when the expected challenge to the safety instrumented function (SIF) occurs once per year or less frequently, as is typical of most specialty chemical process applications. High demand (SIF challenged more than once per year) and continuous mode (SIF continuously maintains the process in a safe state) are beyond the scope of the present discussion.

Before we delve into understanding performance requirements, the first option when assessing IPLs should always be to *avoid* installing a SIS whenever practicable. In the hierarchy of controls, the first option is always hazard elimination. If the system can be re-designed to eliminate the hazard, or if a cost-effective inherently safer alternative can be found, this by default is the preferred choice. While safety instrumented systems are highly reliable and serve a critical purpose, they also come with numerous requirements for specification, documentation, and maintenance over their entire lifecycle. Proceed with evaluating a SIS only when hazard elimination and inherently safer approaches have been exhausted.

2.1 Identify Scenarios for Additional Risk Reduction

The process hazards analysis (PHA) must identify scenarios with significant consequences. These scenarios often involve potential for one or more fatalities, adverse community impacts, environmental damage, loss of critical assets, significant business interruption, and/or intense media coverage. The PHA team plays a critical role, as the team must correctly and thoroughly identify potential process hazards, and determine the credible, worst-case outcomes. In this regard, the PHA is foundational not only for identifying scenarios which may need to be mitigated by an instrumented system, but for all process safety activities at the site.

Once scenarios with potentially highly hazardous outcomes have been identified, the remaining gap to reach a tolerable risk can be determined, based on each company's internal risk criteria. This can be evaluated using one of several methods.

2.1.1 Qualitative PHA

A purely qualitative PHA may be used to determine the level of risk reduction required to reach a tolerable risk. Using a company's internal risk matrix, it is a simple exercise to determine the gap between the risk ranking from the PHA, and the additional safeguards required to reduce the likelihood to tolerable risk.

The advantage of this method is that it is simple to execute and does not require any specialized training in quantitative or semi-quantitative risk analysis. The drawback of the purely qualitative approach is that the best determination can only be an order-of-magnitude estimate, where each "box" on the risk matrix is assumed to represent one power of 10. The gap between the current risk assessed by the PHA team and the tolerable risk could be one, two, three, or more orders-of-magnitude, corresponding to a risk reduction factor (RRF) of 10, 100, 1000, or more. The qualitative approach, while simple to employ, will often result in conservative overdesign with the costliest SIS implementation due to the order-of-magnitude uncertainty with which the risk reduction factor can be known.

For example, consider a PHA scenario which has identified a hazardous event of the greatest severity. By qualitative methods, the likelihood of this event has been selected to be "2" on the general risk matrix shown below in Figure 1.

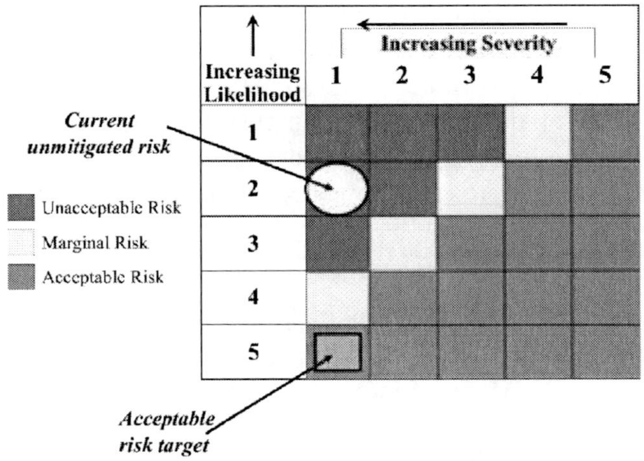

Figure 1. Qualitative PHA identification of scenario with unacceptable risk

Using the order-of-magnitude approach described to find the required risk reduction factor using qualitative methods, the acceptable risk target is located three levels below the current unmitigated risk, or requires a reduction in likelihood of the event from "2" to "5." Assuming each row on the risk matrix represents one order of magnitude, a three-row risk reduction from likelihood "2" to "5" would result in a risk reduction factor of 1000 as the target for the SIS performance requirement. This is shown graphically in Figure 2.

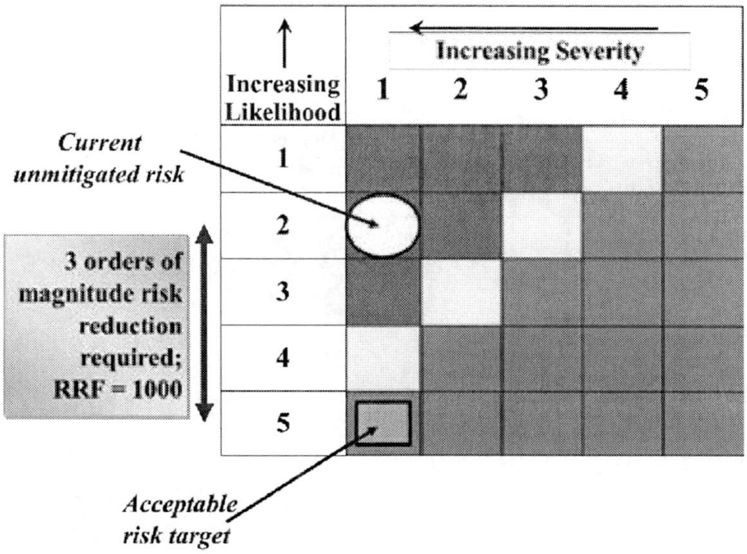

Figure 2. Risk Reduction Factor (RRF) determination by qualitative order-of-magnitude approach.

2.1.2 Semi-Quantitative Layer of Protection Analysis

Among the most common ways of determining the risk reduction required by a SIS is the semi-quantitative layer of protection analysis (LOPA) [3, 4]. When practicing LOPA, the severity of a scenario is determined qualitatively by the PHA team's analysis, often using specific corporate guidance. The likelihood is determined by the frequency of the initiating event, sometimes modified by enabling conditions or conditional modifiers [5, 6]. The combination of severity and likelihood gives a semi-quantitative risk rank, which can be compared to the company's quantitative criteria for acceptable risk. Any non-SIS independent protection layers (IPLs) are applied before assessing the residual risk reduction factor to be mitigated by a SIS. An IPL is a safeguard that mitigates the potential consequences of an undesired event by interrupting the hazardous event sequence; an IPL is independent of both the initiating event and any other applied safeguards [6]. The required risk reduction factor is calculated by dividing the scenario mitigated event frequency by the acceptable risk criteria for the given severity. A sample semi-quantitative risk matrix is given below in Figure 3. An example application of this method is provided in section 2.2.1.

The advantages of using LOPA to determine the SIS required risk reduction are that the method is generally straightforward to apply, and gives a more quantitative measure of the risk reduction than the purely qualitative PHA removing unnecessary conservatism (and resultant overdesign). The drawbacks to the LOPA method are that it requires a competent LOPA practitioner, corporate quantitative risk criteria, and a sufficiently large database of initiating event frequencies and IPL credits so that the current risk and required risk reduction can be determined with a high degree of certainty. In addition, LOPA focuses only on single cause-consequence pair events and is not designed to evaluate complex event scenarios.

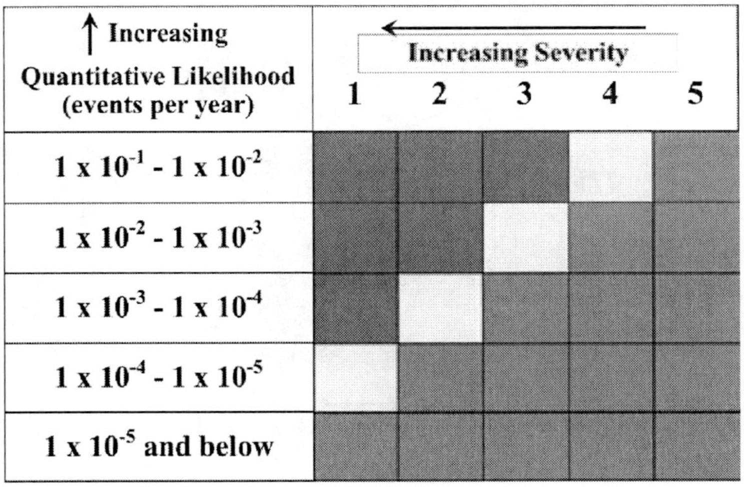

Figure 3. Example semi-quantitative risk matrix with quantitative risk criteria.

2.1.3 Fault Tree Analysis

Fault Tree Analysis (FTA) is a more advanced method that is useful for assessing the failure frequency associated with mechanical and electromechanical components. Using combinations of "AND" and "OR" logic gates, more complex scenarios involving compound initiating events, dependencies, or sequences of events can be assessed using this method [7]. The resulting overall scenario likelihood is compared with the corporate quantitative risk criteria to determine the required risk reduction to be implemented by the SIS.

Advantages of this method are that fault tree analysis can analyze more complex scenarios, and is not limited to a single initiating cause. As such, some scenarios that cannot properly be assessed using PHA and LOPA techniques can be understood using FTA. The disadvantages are that it requires a practitioner highly trained and skilled in the methodology, tools being used, corporate quantitative risk criteria, and a large database of individual component failure rates.

2.1.4 Quantitative Risk Analysis

The final methodology for determining required SIS performance is fully quantitative risk analysis (QRA). The QRA re-assesses the scenario originally identified by the PHA team and evaluates both the severity, and the likelihood quantitatively. The severity is determined using probit functions or modeling results which assess blast, fire, or toxic exposure impacts to a defined population [8, 9]. Likelihood is again determined by using industry failure rate data for given components or equipment. QRA can also be used to evaluate larger scenarios, including domino effects, not typically addressed during a PHA [8, 9].

The advantages of using QRA are that a wide variety of scenarios can be analyzed, including highly complex scenarios requiring consequence modeling to determine the extent of effects from blast, fire, or toxic releases. QRA, being a fully quantitative method, yields the most detailed calculation to determine the required risk reduction factor to be provided by a SIS. The primary disadvantages of QRA is that it is both highly complex and highly specialized, and therefore also expensive and time intensive. Few small or mid-size companies have an internal resource fully competent at performing QRAs on a regular basis. Outside consultants must routinely be utilized.

2.1.5 Recommendations

When selecting an analysis methodology, the level of risk to be managed will determine the degree of analytical rigor required. For most applications, a semi-quantitative LOPA is sufficient to determine the required risk reduction factor to be achieved by the SIS with a high degree of certainty. LOPA provides an optimal balance between technical rigor, required resources, facilitator competency, and complexity involved in performing the analysis. For more complex applications involving compound initiating events, complex hazard scenarios, or very high levels of risk, more quantitative methods such as Fault Tree Analysis or QRA may be required.

For any of the quantitative or semi-quantitative methods (LOPA, FTA, QRA), quantitative corporate risk tolerance criteria are required. When this is not available, the only choice is qualitative, order-of-magnitude PHA.

2.2 Risk Reduction Factor and SIL Performance Requirements

Once the potentially high-consequence scenarios have been identified and screened through a more rigorous methodology, such as those discussed above, the safety integrity level (SIL) performance requirements can be identified. The calculated risk reduction factor (RRF) sets the performance target of the safety instrumented function (SIF), and gives the design criteria for the SIL calculations. Application of the methodology will be demonstrated by way of example.

2.2.1 Example determination of RRF and SIL performance requirement

A batch, multi-products plant operates under a minor source environmental permit, which allows flexibility of operation for research & development campaigns of potential new products. Overflow of a solvent tank, with a resulting spill that leaves the site in the plant effluent, could potentially jeopardize the site's license to operate and result in revocation of the minor source permit.

The PHA team has identified a potential overfill scenario as shown in Figure 4, which requires further analysis. LOPA methodology using a semi-quantitative risk matrix has been selected as the preferred method by the company.

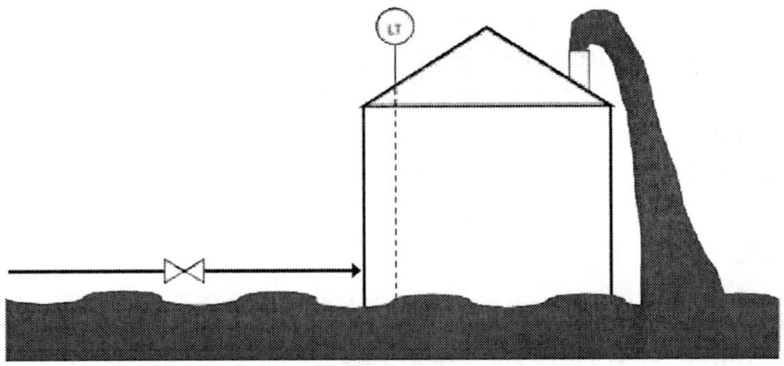

Figure 4. Solvent tank overfill.

Based on company criteria, an environmental consequence that could jeopardize a site's operating permit is defined to be a very high severity event. The PHA team identified the initiating event for this scenario as failure of the single level transmitter on the tank. The transmitter is not associated with any BPCS control function. The company database of initiating event frequencies for this component gives a frequency of $1.1x10^{-3}$ events per year. The resulting risk ranking is shown on the risk matrix in Figure 5.

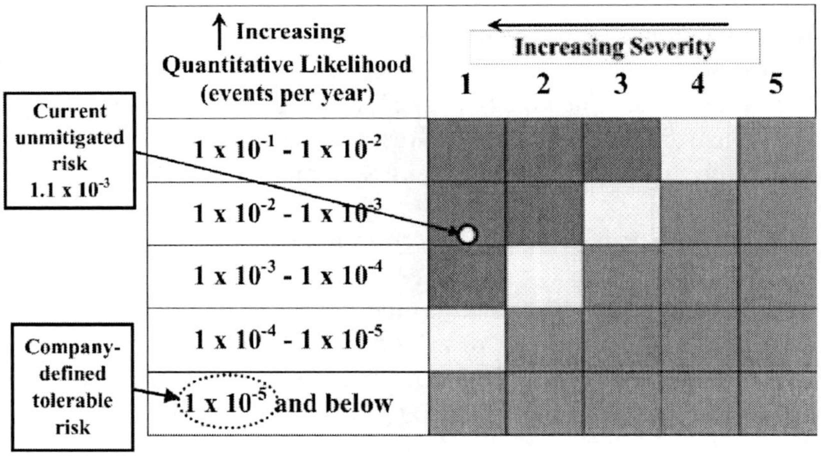

Figure 5. Risk ranking and risk reduction factor determination for tank overfill by LOPA.

Assuming no other safeguards can be applied, the current unmitigated event frequency is $1.1x10^{-3}$ events per year. From the risk matrix, the company has selected a tolerable risk criterion of $1x10^{-5}$ events per year. Dividing the unmitigated event frequency by the tolerable risk criterion gives a risk reduction factor of 110.

Comparing this RRF to Table 1, we see that the required performance results in the need for a SIL-2 SIS. Since RRF is the inverse of probability of failure on demand (PFD_{avg}), the design target PFD_{avg} for the SIS is 0.0091.

By comparison with Figure 2, the strictly qualitative approach resulted in the more conservative requirement for a SIL-3 SIS. Application of the semi-quantitative LOPA methodology enabled the unmitigated event frequency to be determined with greater certainty, resulting in the revised performance requirement for a very minimal SIL-2 SIS. When possible, application of a semi-quantitative or quantitative methodology to determine the SIS performance requirement is encouraged.

Table 1. Risk reduction factor (RRF) and probability of failure on demand average (PFD_{avg}) for each safety integrity level (SIL) [10]

SIL Rating	Risk Reduction Factor (RRF)*	Probability of Failure on Demand Average (PFD_{avg})*
SIL - 1	10 - 100	0.1 - 0.01
SIL - 2	100 - 1000	0.01 - 0.001
SIL - 3	1000 - 10,000	0.001 - 0.0001
SIL - 4	10,000 - 100,000	0.0001 - 0.00001

*Low-Demand Mode only

NOTE: SIL - 4 is extremely difficult to achieve, and is not commonly encountered in the process industries

A potential solution to the overfill scenario is shown in Figure 6. The solution involves a very simple SIS containing a SIL compliant level detection device, logic solver, actuator, and ball valve to shut off the source of material inflow upon high level detection. The system is easy to install, proof test and maintain. Because the required risk reduction factor for this scenario was a reasonably low SIL-2 system (RRF = 110), a simple, highly reliable solution will suffice. A high-I/O capacity logic solver and redundant field device architecture solution is not warranted to implement reliable overfill protection on a single vessel to prevent undesirable environmental consequences.

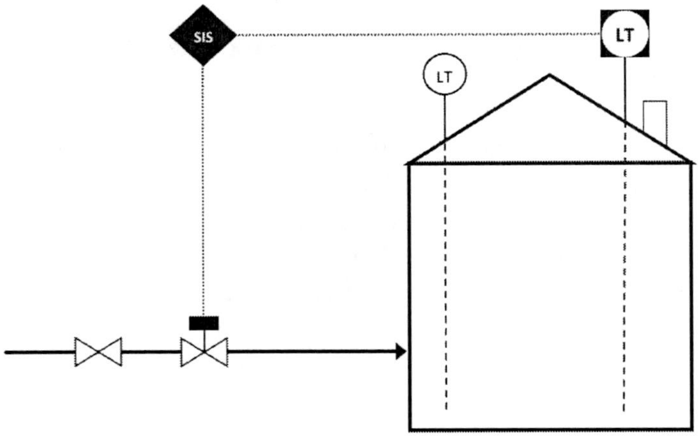

Figure 6. Potential simple, highly reliable SIS for tank overfill protection.

2.3 Establishing Minimum Competency in Performance Requirements

Understanding the performance requirements of the SIS is critical, before launching into detailed SIS design or engaging contractors or consultants to perform these activities. Just as one would not buy a car without first understanding the minimum requirements of the vehicle they are looking for (sports car, sedan, minivan), any attempt to begin detailed design of the SIS without first understanding the hazard scenario and required risk reduction factor would result in significant frustration and wasted effort.

There are two basic parts to understanding the SIS performance requirements: *What* the SIS needs to do, and *How* it needs to do it. The *What* is determining the risk reduction factor previously described. The *How* is understanding the control action needed by the SIS to interrupt the chain of events in the hazard sequence. To develop minimum competency in understanding these two parts of the SIS performance requirements, we suggest a two-pronged approach.

2.3.1 *What* the SIS needs to do: Detailed understanding of the hazard analysis technique

Larger companies often have entire departments at the site or corporate level which specialize in hazard analysis application and SIS performance requirement determination. Smaller to mid-size companies may not have this luxury. At many smaller companies, the sole process safety site resource may be tasked with performing all process safety functions, including SIS specification, and corporate assistance may or may not be available.

For those in smaller to mid-size companies who find themselves in this position, we make the following recommendations:

1. Training in hazard analysis techniques is essential. There are many high-quality training opportunities available at affordable cost. Many LOPA training classes are ~ 2 days in length, and will also include discussion of SIS specification and determining risk reduction factor. For those situations where travel to a training course is prohibitive, online offerings

are also beginning to emerge. Make a strong case to your plant manager or supervisor that this training is critical to managing the risk at your facility, and to further developing your competency as a valued process safety resource.

2. Network, both within your company and externally, to find trusted colleagues who can help answer questions and guide you in the right direction. Your site is likely not the only one within your company that is facing this challenge – see if another facility has similar hazards and how they have addressed them, or if a more experienced professional at another site would be willing to mentor you. Once internal options have been exhausted, external networks can open a wealth of knowledge and experience. Both the American Chemistry Council (ACC) and the American Fuel and Petrochemical Manufacturers (AFPM) have regional process safety networks that are open to member companies. The regional networks are powerful and engaging, and provide the opportunity to learn from peers at other companies.
3. Publications, such as those by CCPS, provide excellent references and examples for understanding hazard evaluation techniques and their application to SIS performance requirements [5-7, 11].
4. Become the expert. If your company does not have internal expertise in this area, why not seek to develop that competence yourself? The combination of the above efforts, along with practical experience at your facility and continuous learning provide a strong foundation for long-term success. Your span of influence would extend outward from your own site to the entire company, helping to manage risk across the entire enterprise.

2.3.2 *How* the SIS needs to do it: Understanding basic control actions and solutions

In many cases, the action required by the SIS is to *stop* something from occurring – stop flow from a source, or stop the temperature or pressure from increasing. For batch, multi-product plants, the SIS action can also act as a permissive, giving the BPCS "permission" to take an action or proceed with batch operations if certain conditions are met [12-14]. A key example may be to ensure that an agitator is running before charging reactants and catalysts to a reactor.

The key to understanding the *How* action of the SIS is this: the process dynamics determine the required action to interrupt the hazardous chain of events from reaching the undesired consequence. We suggest a two-step analysis:

1. Determine the key variable(s) that must be measured and controlled to prevent a hazardous event from occurring.
2. Evaluate the dynamics of the process at the time the hazardous event can occur.

2.3.2.1 Determine key process variables

The hazard event sequence, analyzed in the PHA and further described in the semi-quantitative LOPA or other quantitative analysis provides the details needed to perform this assessment. It is also important, when performing this analysis, to identify what can be measured and what can be controlled.

Consider the simple example of overflow protection from Figure 6 again. This is a very straightforward system in which the parameter to be measured is level, and the parameter to be controlled is flow into the tank. The action of the SIS then would be to measure the level in the tank and to stop the inlet flow once the level has reached a pre-determined point, to prevent overfilling the tank.

Sometimes, selecting the right parameters to measure and control can be more complex. Consider the process shown in Figure 7 in which two reactants combine to generate a gaseous product in a semi-batch reactor. At first glance, it may seem as though pressure is the easiest variable to measure. However, the reaction may be operating under high pressure as a means of controlling the reaction equilibrium and production rate. Thus, fluctuations in the pressure indicative of an uncontrolled reaction may not be easily detected by the pressure transmitters. Other options for measurement include the temperature of the reaction. However, this may be highly sensitive to the point in the reactor where the temperature is measured, and a complex temperature measurement scheme would likely not be an efficient way to run the process. If the reaction kinetics are well-understood, the feed rates of the reactants may be an excellent choice to measure and control the reaction. Flow transmitters monitored by the SIS, in addition to those monitored by the basic process control system (BPCS), can provide a reliable measure of reactant feed rates, and thus the product vapor generation rate. Shutting off the flow of reactants stops the reaction, and prevents the overpressure associated with an uncontrolled reaction.

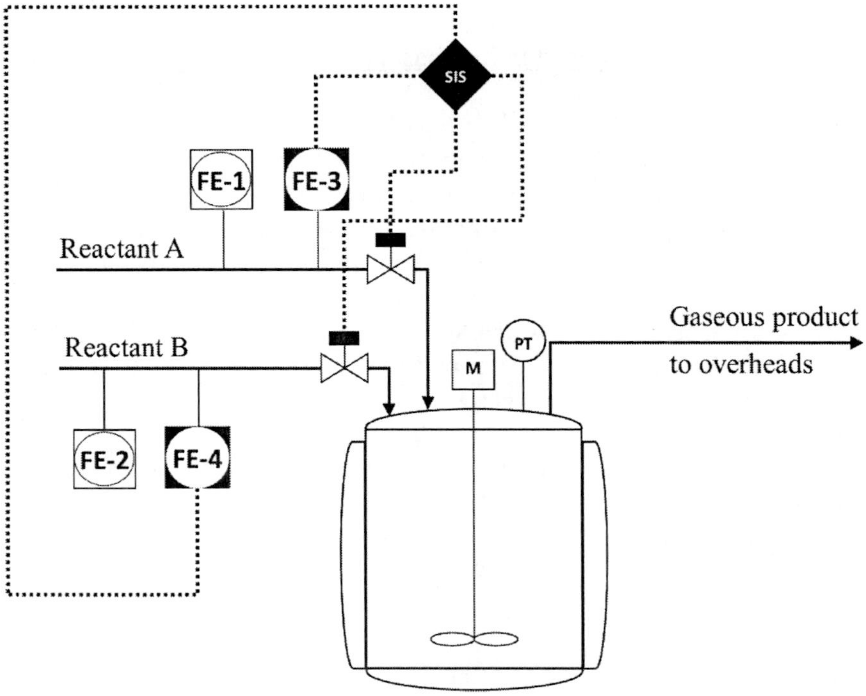

Figure 7. Potential SIS design for semi-batch reaction which generates gaseous product.

2.3.2.2 Evaluate process dynamics at the time of the hazardous event

Batch, specialty chemical processes often involve complex, multi-step processes with multiple reaction steps. Consider the process shown in Figure 8 which has the potential for a runaway reaction to occur, generating large amounts of heat and pressure, which could overpressure and rupture the vessel. In this case, understanding the reaction kinetics, vessel heat transfer, and the process dynamics are essential to selecting the right parameters to measure and control.

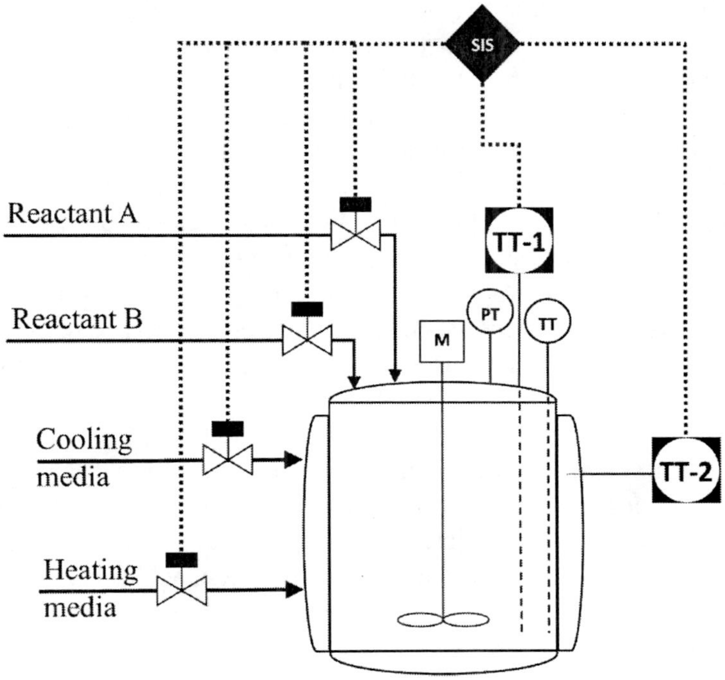

Figure 8. Potential SIS arrangement to mitigate a runaway reaction in a batch reactor.

If the reactive hazards of this process have been properly evaluated and understood, a maximum temperature limit (MTL) or 24-hour adiabatic induction temperature (TAIT-24) should be available. These safe operating limits would set the temperature below which the reaction must be controlled, and above which the reaction must be terminated. Clearly, temperature is a good choice for the measured variable.

However, the process dynamics must be evaluated to determine the proper setpoints and control actions to mitigate the potential runaway reaction. Heat transfer in large, jacketed vessels is often somewhat inefficient, and reactor contents can be slow to respond to changes in jacket temperature, and can be compounded by the potentially significant time delay associated with the time constant of the thermowell/sensor combination [15, 16]. Vessel material of construction may also further limit temperature rate-of-change to avoid vessel or jacket damage, as in glass-lined reactors. For evaluating these types of hazards, a transient energy balance which accounts for the vessel heat transfer rate and any limitations on jacket temperature adjustment must be generated. Due to heat transfer dynamics, the internal vessel temperature of an exothermic reaction may continue to rise after additional cooling has been applied to a vessel jacket, until the heat removal rate exceeds the heat generation rate. The chosen setpoint for initiating the maximum reactor cooling is critical, as the process dynamics must be considered such that the reactor contents do not reach the point of no return and runaway before adequate cooling can be applied. In extreme cases, direct injection of cooling media to the reactor itself has been required.

The setpoint temperature at which the SIS activates must sometimes be set sufficiently below the maximum temperature limit to account for the process dynamics. This is the concept known as the *process safety time* (PST), which is the time interval between a process deviation or failure occurring and the occurrence of an undesired consequence resulting from the failure. Time-to-trip (TTT) is the time interval between the process deviation and the SIS setpoint. SIF response time (SRT) is the time interval between the SIS activating at its setpoint and the final elements reaching their response position. SRT is the total time for sensors to measure a process variable, the logic solver to receive the process measurement and command a response action, and the final elements moving to their response position to interrupt the sequence of hazard progression. Process response time (PRT) is the time lag between the final element response action and the process reaching a safe-state condition. A generous safety cushion, or safety margin time (SMT), must be applied to ensure the system does not overshoot and reach the undesired consequence while the process responds to the final element action. As a simplified approach, industry best practice is that the SIF response time should be no greater than 50% of the process safety time determined from evaluation of the process dynamics [17]. Competency in understanding process dynamics is gained through a combination of complete understanding of process hazards, reactive hazard analysis, and fundamental chemical engineering tools such as heat transfer models and the dynamic energy balance.

2.4 Common Misconceptions

In this section, we address a few common misconceptions regarding SIS performance requirements:

Misconception # 1: Safety Instrumented Systems are only installed for personnel safety

This is a statement commonly heard by the authors. There are many good reasons why a company may choose a SIS as part of its risk mitigation strategy. While personnel safety is a common reason for installing a SIS, other valid reasons include environmental protection, as well as the protection of unique or difficult-to-replace assets [18]. All are explicitly referenced in IEC 61511-1:2016 (clause 1, parts j and k) [10]. In short, a SIS may be used to protect anything upon which a company places high value, including human life, environmental compliance, and its operating assets.

Misconception # 2: SIL-3 is the only "real" Safety Instrumented System

The IEC 61511 and ISA 84 standards have recognized for many years the range of safety integrity levels and corresponding risk reduction factors associated with SIL-1, SIL-2, and SIL-3 safety instrumented systems [10, 19]. The key is understanding the SIS performance requirements and the necessary risk reduction for the specific hazardous scenario to be mitigated. Not every SIS needs to be SIL-3; sometimes a SIL-1 or SIL-2 system are all that is required. The goal should be to effectively match SIS performance ability with risk reduction requirements at minimal cost and complexity. This is determined by an appropriate performance requirements analysis, described in this section.

Misconception # 3: Utilizing multiple, independent SIFs to reach the required Risk Reduction Factor is not allowed

This misconception is often employed when the calculated risk reduction factor (RRF) to mitigate a hazardous event results in a value greater than 1000, or the requirement for a SIL-3 system. The design, documentation, operational and maintenance rigor increases significantly as the safety integrity level of a SIS increases. For numerous reasons, many companies do not wish to install a SIL-3 SIS because of the onerous burden it places on site maintenance and I&E resources. For operational and maintenance reasons, it is often much easier to install two independent systems, one SIL-1 and one SIL-2, to reach the required risk reduction determined in the performance analysis. This is also explicitly addressed in IEC 61511-2:2016 (Annex A, clause A.9.2.4) [20].

3 Safety Requirement Specification (SRS)

The Safety Requirement Specification (SRS) is developed immediately after the Hazard & Risk Analysis (PHA) effort and is the start of the SIS realization phase. The SRS is one of the most critical elements to implementing a SIS. This document at its most basic level specifies a functional description and SIL requirement for each SIF comprising the safety instrumented system. At an expanded level this document provides insight to all parameter requirements to implement the SIF and is the central hub to which all safety life-cycle activities must be verified and validated against. This document must be easily understood, verifiable, maintainable, feasible, and provide clear insight to all requirements necessary for the SIF to effectively mitigate the identified hazard. Periodic changes to the protected process or upgrades in instrumentation may be required and as such the SRS becomes a living document which must be updated accordingly to reflect the current process, installed equipment, and risk reduction performance. This document is typically handed off to an outside consultant, engineering firm, or equipment supplier to definitively design the SIS. As stated by Paul Gruhn in *How Safe Are Your Systems?*: "Ultimately the SIS is only effective as the people who are designing it are informed," [2] thus it is critically important that the individual preparing the SRS provides sufficient information to accurately describe SIS function and required performance.

The current IEC 61511-1 standard (clause-10.3.2), specifies 29 items that must be addressed in the SRS [10]. Smaller companies may not have internal personnel proficient to routinely produce quality SRS documents that include all the required elements. However, developing this competency is strongly encouraged. Site process safety resources may not need to perform detailed SIF design, but should be competent to develop a quality SRS so that outside resources can effectively design the SIF to meet the required risk reduction. Unfortunately, since the IEC 61511 standard is performance based and not prescriptive, there are no guidance example SRS documents provided. The end user must develop their own structure and format when developing the SRS. The lack of formal guidance or examples in the standard places a disproportionate burden on personnel from small to mid-size companies who lack the support staff or corporate resources to generate the SRS that are often found in larger enterprises. This is a potential opportunity for process safety personnel to seek trusted guidance from an outside consultant to aid in the preparation of the safety requirement specification. Such expert guidance will aid in developing personal competency. Competence in this area will allow effective communication when SIF designers present options to mitigate the hazard. Lack of competence in either understanding SIS performance requirements or development of the SRS can result in over-, under-, or mis-design of the SIF [1].

4 Component Selection

Once performance requirements have been established for the SIF, the designer can now turn to component selection to identify the specific hardware required to meet the target risk reduction factor. At its most basic level, a SIS is comprised of a combination of these basic components: sensor, logic solver, and final element, as shown in Figure 9. The specific elements identified, along with their arrangement and configuration (architecture), determines the risk reduction factor that can be achieved.

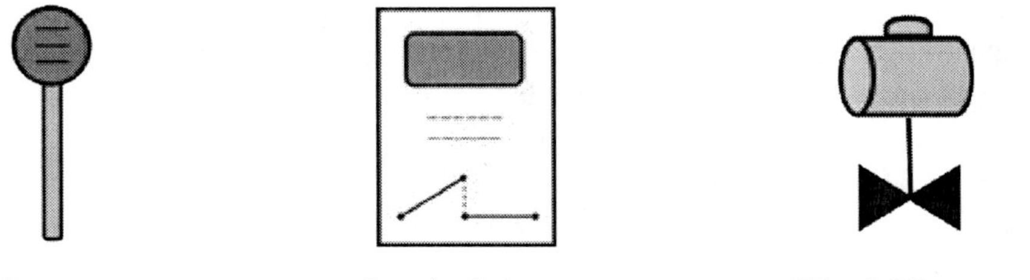

Figure 9. Basic SIS components.

Component selection and design is another area in which personnel from smaller companies may struggle due to lack of resource availability. This is an area where consultants can be particularly helpful, once a performance requirement has been determined and minimum design standards have been detailed in the safety requirements specification (SRS). However, a basic understanding of component selection and limitations is useful before reaching out to a consultant to perform the detailed design calculations.

For purposes of this paper, we will focus only on developing the basic competency around the function and selection of sensors, logic solvers, and final elements. Detailed analysis of SIS design, including architecture (redundancy and voting configuration), spurious trip rate, safe failure fraction, diagnostic coverage, proof test interval, and other components, as well as execution of the detailed SIF calculations, are beyond the scope of this paper, but are available in the IEC 61511 standard, and elsewhere.

It is also helpful to note here that while consultants and equipment vendors can be very useful in performing SIS design, this is not to imply that the methodology or calculations are so complex that they cannot be performed by a competent process safety engineer. Most of the math, particularly for simple systems, can be done with basic algebra, provided that the user understands the methodology behind SIS design [21]. Recommendations on developing this competency are provided in Section 7.

The key to appropriate component selection lies in understanding that reliability is inversely proportional to failure rate. A higher failure rate generally equates to reduced reliability and higher probability of failure on demand. Components certified for functional safety application generally have significantly lower specific failure rates as compared to generic components. Therefore, increased reliability results from using functional safety certified components across the entire system. Simply by choosing certified components, highly reliable, minimal SIL-1 or SIL-2 systems can typically be installed for simple architectures and reasonable test intervals. Each of the major components will be briefly reviewed below, then the solvent tank overfill example will again be used to illustrate component selection for a simple SIL-2 design.

4.1 Sensors

Sensors are the measurement devices used to detect key process variables, such as pressure, temperature, level, and flow. Sensors are generally comprised of two types: analog and discrete. Analog sensors include transmitters which measure the process variable over a range, such as temperature or pressure transmitters. Discrete devices measure the process variable and activate at only a single point using an "on/off" distinction. Discrete sensors include devices such as level and flow switches.

Selection of which component is appropriate for the SIS will depend upon the characteristics and dynamics of the process, type of technology already being employed by the basic process control system (BPCS), and the kinds of failures characteristic to each type of device.

4.1.1 Process Characteristics & Dynamics

As previously described in Section 2, the process dynamics must be understood to select the right variables for both measurement and control. The dynamics are also relevant to component selection, as the scan rate or detection ability of the sensor must be able to detect changes in the process variable fast enough to allow the logic solver and final element action to move the system to a safe-state within the process safety time constraint. For this reason, transmitters with high scan rates or discrete switches may be chosen for specific applications where the process safety time is very small. Furthermore, the process characteristics are also key in selecting the proper sensor. Services which are corrosive, fouling, or can otherwise degrade or interfere with a sensor's performance must be considered when selecting a measurement device. The user's own experience with the process fluids and operational characteristics is critical to this evaluation. Physical arrangements, such as nozzle size and availability, and space available for transmitters and wiring, can also affect the sensor selection.

4.1.2 Sensor Technology Employed by the BPCS

Another consideration in sensor selection is to evaluate the technologies already in use on the process. Sometimes there are advantages in selecting a different sensor technology for the SIS than what is already in use by the BPCS, to reduce the risk of both devices failing due to a common cause failure. For example, if a level transmitter is already in use by the BPCS, a level switch may provide an option for an alternative technology. Or if a thermocouple is in use as a temperature-sensing element, an RTD device may be an alternative to consider.

In other cases, due to process characteristics and user experience, a similar technology to what is currently being used is the right choice. In those cases, using a functional safety certified device for the SIS, or employing a similar technology supplied by a different manufacturer, can help to reduce potential for common cause failures.

4.1.3 Sensor Failure Characteristics

The types of failures common to analog and discrete technologies should also be considered, to determine what is the best selection for the SIF to be installed. For analog devices, such as transmitters, a common failure mode results in an improper measurement at either end of the variable range. For example, a temperature transmitter may fail "low," providing a false low temperature indication when the actual temperature inside the vessel is greater. The opposite, "high" failure can also occur, providing a false high temperature indication compared to the actual process temperature. For most processes, one of these directions is a dangerous failure with regards to process safety, and the other is often only an operability or product quality result. Most modern "Smart" sensors operate with a 4-20 mA analog current output range and utilize internal diagnostics to detect under-current or over-current signal fault failures as well as other internal fault conditions. The devices can be configured to produce a fault alarm annunciation and/or artificially drive the output current to an appropriately defined high or low safe-state value allowing a downstream logic solver to direct action to move the system to a safe-state.

For discrete devices, a common failure mode is the failure to trip at the required setpoint. For example, a level switch failing to trip at the desired high-level inside a vessel. For many years, discrete devices were thought to be inadequate for functional safety, due to their single-point failure potential, difficulty of calibration, and other issues. However, modern instruments with internal diagnostics have proven to be significantly more reliable than their predecessors, and some can even transmit a continuous signal indicating their current health status to a site's maintenance or instrumentation group. As a result, functional safety certified devices now exist for both analog and discrete sensors, and the user can select from among a wider variety of devices and technologies than ever before.

4.2 Logic Solvers

The logic solver is the component of the SIF that receives input from the sensor, evaluates the input against some pre-determined criteria, and when required, commands the final element to take an action. There are three major types of logic solvers: electro-mechanical, electronic, and programmable electronic [10]. Examples of each type include relays, solid-state devices, and programmable logic controllers (PLCs), respectively.

Electro-mechanical relays, solid-state devices, and PLCs are commonly used in designing safety instrumented systems, and all have benefits and drawbacks. The choice of logic solver can greatly impact the reliability, complexity, and cost of the SIS.

4.2.1 Electro-mechanical Relays

Relays have been used in process control and safety applications for several decades. Their failure modes are well-understood and predictable [21, 22]. Quality, functional safety certified relays have very high reliability. The most common failure mode is to fail-safe with the relay contacts open and the fail-dangerous mode (contacts welded closed) can be adequately mitigated with current-limiting fuse protection.[21].

Relays are very simple devices and find practical application for any SIF in which both the input from the sensor(s) action to be taken by the final element(s) are in the form of a discrete signal. For example, tripping a high level switch to take the air off a solenoid to close a valve actuator; or a temperature switch tripping at the desired setpoint to remove the power from an electric heater to stop heat input. The main limitations of relay-based systems are that they are capable of handling a relatively small number of inputs and outputs, are difficult to modify, and typically have no diagnostic capabilities. For this reason, SIFs which contain more than a few sensors and/or final elements, and any that require redundancy, will usually require an electronic logic solver.

4.2.2 Programmable Logic Controllers

Programmable logic controllers (PLCs) are also in widespread use in process control and safety functions. They can often accept multiple inputs, with a single or redundant PLC configuration serving as the logic solver for tens to hundreds of SIFs, and can be programmed to perform a wide variety of functions. Unlike relays, the failure modes of PLCs are not predictable, and have the added failure complexities associated with software failures, as well as hardware failures [21-23]. The hardware components of a PLC can fail-safe or fail-dangerous, via open-circuit or short-circuit, respectively. Because of these additional complexities, PLCs often require greater internal redundancy and diagnostic detection than relays, and require specialized programming software and experienced programmers. Verification and validation of the software components of PLCs, as well as the documentation required by the standard, is greater than for relay-based systems [10, 22]. For very large, complex PLCs, site resources may not have the right tools and level of competency to program and maintain these systems, and vendor contracts may be required.

Functional safety certified PLCs are used all over the world as logic solvers in safety instrumented functions, due to the versatility of tasks that can be performed. Any function that can be programmed can be executed by the PLC. Functional safety certified PLCs range from small, field-mounted devices, to modular expandable devices which can be added over time to support new processes or units, to very large, high-I/O capacity PLCs which can simultaneously process hundreds of SIFs and may occupy most of an I/O room. Cost of these units tends to scale with size and I/O capacity, therefore the project and process economics are very important when considering PLC logic solvers for safety instrumented systems.

4.2.3 Solid-State Devices

Solid-state devices came into common use in the 1970s in process control applications. They can accept more sensor inputs than relays, and use basic logic such as "AND" and "OR" configurations to determine when to execute action by the final element. These devices contain no software. As such, solid-state logic solvers are *configurable*, but not programmable. Many solids state systems contain internal diagnostics and can provide detailed fault information to site maintenance or instrumentation personnel.

Because solid-state devices are electronic systems, they have the same limitations as PLCs with regards to failure modes being non-predictable, and nearly evenly split between safe and dangerous failures [22]. However, since solid-state logic solvers don't contain software, they are usually less resource-intensive and can often be configured and maintained by qualified site I&E personnel.

One example of the solid-state logic solver is the single-loop logic solver, which can provide a high reliability, low-cost solution for simple SIF's with only a few inputs. The system combines the configurable solid-state functionality with internal relays to cause final elements to take the desired action [24].

4.2.4 Recommendations

Ultimately, the choice of logic solver depends on the complexity of the SIF and its performance, data visibility, and diagnostic requirements. For very simple arrangements, such as overflow protection on a tank, a relay-based or solid-state logic solver may be appropriate. For more complex systems where multiple redundancy and multiple inputs are required, more robust PLCs may be the right choice. In all cases, the benefits and drawbacks, including the ongoing maintenance and documentation requirements of all logic solvers, should be considered.

4.3 Final Elements

The final element is the component of the SIS which takes action, in response to a measurement by the sensor and a command by the logic solver. Action by the final element should result in the process being moved to a safe state, and interrupting the chain of events in the hazard sequence.

A variety of items can serve as final elements in the SIF loop. The most common are solenoids, actuators, and valves, which can either close (to stop the flow of reactants or heating media), or open (to initiate flow of cooling media or provide venting). Other final elements include contactors, which open to de-energize some set of equipment, such as an electric heating element. The greatest reliability and risk reduction is achieved when using components with low certified failure rates. However, some components, such as ball valves may be difficult to find as functional safety certified devices. In those cases, decreased proof test intervals, partial stroke valve testing, and other tools may be used to improve reliability.

4.4 Example

Returning to the solvent tank overflow example of Figure 4, and the proposed simple SIS solution of Figure 6, we can now evaluate the various components of the design by selecting a sensor, logic solver, and final element, and assess the design against the performance requirements determined in Section 2. The performance requirements specified the need for a SIL-2 system, with a RRF of 110.

For the sensor, a functional safety certified non-contact radar level sensor/transmitter is selected, with a RRF of 2277. A solid-state, single-loop logic solver, certified for SIL-2 capability in a non-redundant architecture is chosen as the logic solver with RRF of 2037. For the final elements, a functional safety certified solenoid which de-energizes a functional safety certified actuator and ball valve with a composite RRF of 138 are chosen. After performing the detailed SIS calculations (beyond the scope of the current paper), a SIF with a risk reduction factor of 122 is obtained, which exceeds the minimum specified RRF performance requirements. This potential solution is shown in Figure 10. A system such as this is a very reliable and economical means of providing tank overflow protection, for which a complex, expensive, high-I/O capacity redundant architecture SIF solution would be significantly over-designed.

The SIF design, however, is not yet complete. There are two additional constraints which must be evaluated to determine if this design will in fact meet the full performance requirements and conditions for a SIL-2 system. These are systematic capability, and architectural constraints, which are addressed in the following section on SIL Verification.

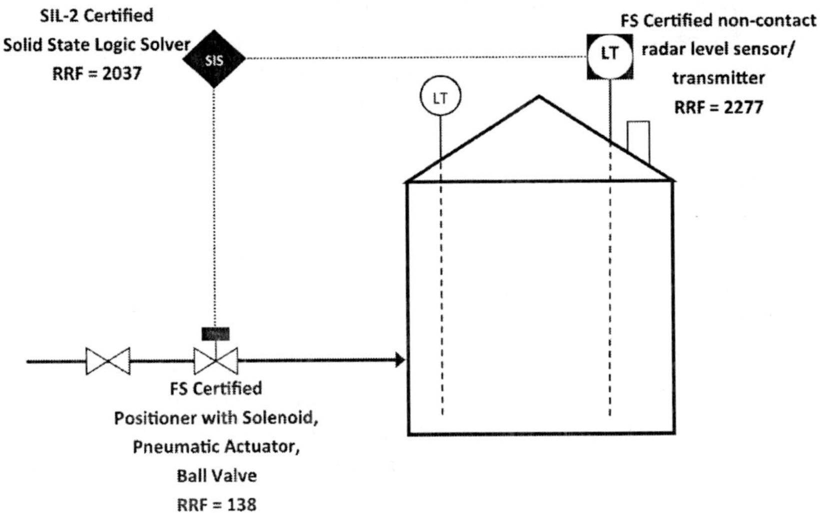

Figure 10. Potential SIF for overfill protection meeting minimum risk reduction factor requirements.

4.5 Misconceptions Related to SIS Components

As with SIS performance requirements described in Section 2, misinformation also exists regarding SIS components. We address three common misconceptions here.

Misconception # 1: A vendor is offering various sensors certified by a body I've never heard of and claims they are functional safety certified.

While there is no current list of approved Certification Bodies for functional safety, most functional safety certifications in accordance with IEC 61508 are performed by one of two major certifiers. This is not to say that the components offered may be not be properly certified, but rather that one should proceed with caution. If one does not feel they have the competency to review the safety certificate for the product themselves, a trusted consultant with experience in functional safety can be helpful. If the vendor cannot provide a certification document or the failure rate data and safety manuals for the devices, this is probably a red flag.

Misconception # 2: Simple relay devices can only be used in SIL-1 applications.

While relay-based logic solvers are simple and have been in existence for decades, they are also highly reliable and their failure modes and rates are well-known. A functional safety certified relay device can be used in any SIF up to its rated capability, including SIL-2 or SIL-3. The detailed SIF calculations which consider redundancy and system architecture, among other things, will partly determine whether the relay-based logic solver is the proper choice for the SIF being evaluated.

Misconception # 3: Selecting a SIL-3 rated logic solver means my entire SIF will be SIL-3.

As shown in Figure 9, a SIF loop is comprised of three main components: sensor, logic solver, and final element. While it is important to use functional safety certified components of the highest reliability required, having a SIL-3 certified logic solver does not make the entire SIF a SIL-3 system, any more than putting racing stripes on one's family car makes it a racecar. The performance of the SIF is determined by the entire system. The component which most often has the largest effect on the SIL calculation is the final element [25].

Misconception # 4: Redundancy is always required to have a SIL-2 system

Low risk reduction factor, minimal SIL-2 systems can be achieved using highly reliable, functional safety certified components with certified failure rate data, as shown in the example in this section. However, other factors including architectural constraints and systematic capability must be assessed to determine if the system will in fact qualify for SIL-2. These are described in the following section on SIL Verification. While it is not expected that a RRF of greater than about 150 can be achieved without the use of redundancy, it is physically possible by using good component selection to develop a minimal SIL-2 system without redundancy.

5 SIL Verification

Unlike many other standards, the IEC 61511 standard is performance based instead of prescriptive, i.e., specific guidance and instruction for equipment selection, configuration, architecture, and voting arrangement is not provided. It is up to the end user to design a SIF mitigation strategy and then quantitatively prove the selected configuration meets the requirements for the specified SIL target. While the major verification effort involves calculation of PFD_{avg} and corresponding risk reduction factor of the overall SIF loop, there are two additional requirements to be met to achieve overall SIL compliance. These are architectural constraints (AC) and systematic capability (SC).

The functional safety standards focus on two different failure categories: random hardware failure due to normal stress degradation and systemic failure due to flaws and weaknesses in design and manufacturing practice. Random failure modes are typically well understood and therefore predictable, whereas systemic failure can only be minimized or eliminated by modification of management systems which define the conduct of operations. PFD_{avg} and architectural constraint (hardware fault tolerance) are associated with random failure modes, whereas systematic capability is associated with systemic failures. All three criteria must be determined and accounted for when designing and verifying a SIF to meet a specified SIL target.

5.1 SIL Probability of Failure on Demand (SIL_{PFD})

Table-1 above, from IEC 61511 (clause-9.2.4), indicates the PFD_{avg} and RRF ranges that must be achieved by a proposed SIF to meet the risk reduction requirements specified by a target SIL. The majority of the SIL verification effort is conducted to determine average probability of failure on demand over a proof test interval (PFD_{avg}) for the complete SIF loop. According to Goble and Cheddie, "The standards allow any internationally accepted methodology for performing the probabilistic calculations, and the most common methods are through simplified equations, fault trees, and Markov models." [26] While discussion and guidance for the methodologies is beyond the scope of this paper, personnel who are not competent with the design and verification aspects should at a minimum be aware of the requirement to verify the proposed SIF design meets the PFD_{avg} and RRF requirements for the specified SIL target.

5.2 Architectural Constraint (SIL_{AC})

Architectural constraints are characterized by hardware fault tolerance (HFT), the ability of a system to continue to perform a required safety function in the presence of one or more dangerous hardware faults. Design redundancy is often employed to achieve an acceptable level of fault tolerance. For example: HFT=0 means there is no redundancy so loss of one component renders the complete system non-functional. HFT=1 means there is one level of redundancy so the system continues to function in the presence of dangerous fault conditions and loss of a single component does not render the system incapacitated. However, the system would be operating in a degraded capacity until it can be repaired.

Do all systems require redundancy? The answer depends on the required risk reduction and SIL target. IEC 61511 provides a revised table for allowable HFT requirements related to SIL requirements. The higher the risk and performance requirement, the higher the need for safety redundancy. Optimally matching true process risk reduction with SIL performance requirements can reduce HFT requirements and alleviate costly overdesigned systems.

Table 2. Minimum HFT Requirements According to SIL (IEC 61511, clause 11.4.5)

SIL Rating	Minimum Required Hardware Fault Tolerance (HFT)
SIL - 1 (any mode)	0
SIL - 2 (low-demand mode)	0
SIL - 2 (high-demand mode)	1
SIL - 3 (any mode)	1
SIL - 4 (any mode)	2

Hardware fault tolerance is a way to compensate for uncertainty in SIF design assumptions and component failure rate data. Performing failure probability calculations may not be sufficient if the failure rate data is of questionable accuracy or spans a large range, and can result in overly optimistic solutions.

The 2016 revision of IEC 61511-1 (clause 11.4) now allows different options for demonstrating hardware fault tolerance compliance. The end user can use the tables provided in the 61511 standard and perform their own analysis or may elect to use the hardware fault tolerance claim from the component's functional safety certificate determined in accordance with IEC 61508-2. Details of assessment methodology are beyond the scope of this paper. However, end users should be aware of the requirement to meet architectural constraints and should therefore review safety documentation to ensure minimum hardware fault tolerance for the SIL target is met.

5.3 Systematic Capability (SIL$_{SC}$)

Systemic failures are related to the underlying management and manufacturing systems, procedures, and practices that define a specific task or operation. For example, a programmer who inadvertently inserts a 'bug' into software code has introduced a systemic failure. This failure would be repeated in every instance of software usage. Due to these failures being related to human performance reliability and other management system processes, systemic failures are difficult to predict and quantify. Systematic capability (SC) is a qualitative measure (expressed on a scale of SC 1 to SC 4) of the confidence that the systematic safety integrity of an element meets the requirements of the specified SIL target (SIL-1 to SIL-4).

Systematic capability should be assessed and stated by the manufacturer when considering equipment selection [27, 28]. Manufacturing systemic failures may include manufacturing defects, component design flaws, software programming errors, assembly errors, etc. The end user should review safety documentation for systematic capability claims compared to the SIL target (i.e. components used in a SIL-2 SIF must have SC 2 or greater), within any stated usage restrictions for the application. Additionally, the end user has a further responsibility to minimize opportunities for systemic failure during the operation and maintenance portion of the safety life-cycle by assessing quality of implementation, usage, and maintenance procedures and practice.

It is no wonder with all the technical verbiage provided by the standards that few people develop necessary competency to understand the concepts and requirements of Safety Instrumented Systems. Personnel lacking competency will often understandably seek outside resources with in-depth knowledge and experience required to verify a proposed SIS design meets compliance with overall probability of failure, architectural constraints, and systematic capability. However, it is in the end users best interest to develop basic knowledge of the verification concepts and requirements so that meaningful communication occurs with outside resources to ensure that SIS requirements meet the user's needs and not be over or under designed.

6 Lifecycle Considerations – Suggestions for choosing between the Racecar and the Economy car

Many aspects of the process, facility, and company should be considered when evaluating potential SIS solutions.

6.1 New greenfield plant vs. retrofitting an existing facility

When constructing a new plant from the ground-up, one has the advantage of understanding and incorporating current design and risk criteria into the process design from its inception, which is the most economical phase of the project lifecycle to make changes. Knowing these criteria up front, the appropriate Safety Instrumented Functions (SIFs) can be identified and designed prior to any equipment being purchased or constructed. This is the point of greatest flexibility, and nearly all configurational options are on the table for selecting SIS architecture, components, and configuration as well as for incorporating inherently safer design options.

For projects where a large number of new SIF loops will be required, a greenfield plant represents an opportunity where choosing the large, modular safety PLC-based option may be more economical over the life of the process. It also reduces the overall complexity compared to several dozens of individual PLCs which must be properly proof-tested and maintained.

Retrofitting an existing facility to account for SIFs identified through recent hazard evaluations, updated corporate risk criteria, or new corporate design criteria, pose special challenges. First, the major fixed process equipment (reactors, drums, tanks, and other vessels) has likely been in operation for some time, with a fixed number of nozzles already in use. Additionally, the equipment spacing and layout, along with the piping arrangement, may not allow much room to install new instruments, valves, and actuators. Purchasing new, large process equipment may be prohibitively expensive, but a highly reliable instrumented solution is needed to meet the required risk reduction.

In the case of retrofitting an existing facility, the number of required changes could be limited to a small number of vessels, and may be on the order of less than a dozen total SIFs. In these instances, as well as due to space constraints in the process area, smaller, more economical systems may make the most sense, compared to a large system capable of handling many individual SIFs.

6.2 Process Complexity – Batch or Continuous

Continuous processes are the most commonly referenced in the IEC 61511 standard, as well as in most papers and guidance on its application [10, 21]. However, batch processes are extremely common in the specialty chemicals industry, and pose their own unique challenges: numerous manual operations and control, reactions which are highly sensitive to the timing or duration of reactant or catalyst addition, large volume heating and cooling requirements, and frequent sampling, among others [29]. Some literature has emerged regarding application of SIS to batch chemical processes [12-14].

While application to a continuous process is often more straightforward, application to batch processes is significantly more complex. These include: sharing components (instrumentation and/or final elements) between the basic process control system (BPCS) and the SIS; accounting for phase dependence within a batch (e.g., various operating steps within a batch may have different hazards, and therefore necessitate different SIFs to be active at different times); operator interaction with the SIS, including changing the active batch phase or SIS setpoint [12-14].

The complexity associated with a batch system often leads to the need for a flexible, economical, yet highly reliable instrumented solution. Systems which are difficult to reconfigure or reprogram may not be the most appropriate in this application. Individual, field-mounted, solid-state easily configurable devices can often provide this needed flexibility more readily than a large, high-I/O capacity safety PLC, which requires trained and skilled personnel to reprogram and validate.

6.3 Configurational Complexity – Single product or Multi-Product

Within batch processes, further configurational complexity exists depending on whether the process unit is setup for single-product or multi-product production. In a multi-product unit, recipe dependence is an additional complication, as various products may require completely different setpoints or SIF actions than another product, in addition to the phase dependence previously discussed [13].

In the authors' experience operating batch, multi-products specialty chemical plants, the greater the complexity of the process, the simpler the SIS design, operation, and maintenance needs to be. The SIS must be flexible enough to be re-set or easily reconfigured for various products and process operating conditions. However, the hazards of the process operated also necessitates a highly reliable protective system. A solution which is cumbersome and requires great effort to reprogram with each new recipe or product would result in great inefficiencies for a multi-product operation.

At this point, we must also say a few words about the criticality of management of change and administrative controls. The IEC 61511 standard requires these controls, which are a vital component of maintaining and safeguarding the SISs installed at the facility. For batch, multi-product plants, a robust management of change process, operating discipline, and management accountability is critical to ensuring that the flexibility required of a SIS on a multi-products unit is not compromised or abused. Site leadership must set expectations that a robust, disciplined management of change process will be followed for all process changes, including all SIS setpoint changes, without exception. The culture at the site must reinforce this expectation, and ensure that new personnel understand the reasons for this strict adherence to the standard. At facilities where this culture and discipline is lacking, means of protection other than SIS should be considered.

6.4 Facility Layout and Geography

The layout and location of a facility, especially with regards to the distance to the community, is another facet which must be considered in selecting an instrumented solution. The incident at the West Fertilizer Company in 2013, the Bhopal disaster in 1984, and others, demonstrate some of the undesirable community impacts when process hazards are not well-controlled. Operating companies must understand the worst-case consequences of their processes, and utilize inherently safer technologies and highly reliable engineering controls to prevent these consequences from occurring. For plants with potential community impacts, including those located immediately adjacent to the community or within highly populated areas, very robust protective systems (including SIS) may be required.

6.5 Site Maintenance and Instrumentation Resource Availability and Competency

Large companies or facilities with a small army of controls engineers, maintenance and instrumentation & electrical (I&E) technicians may well have the tools and training required to maintain a highly complex SIS. Smaller companies or facilities with limited maintenance and I&E resources may struggle to keep up with the maintenance and testing requirements that come with such a highly complex system. This is yet again another example of where simplicity in system design is better. The safety instrumented systems installed as part of the risk mitigation strategy at your facility should not be a "black box" that no one at the facility understands their purpose or function.

Returning to the racecar analogy, one can bring their family car to most any Automotive Service Excellence (ASE®)-certified mechanic, and expect them to have the right level of competence and training to perform basic maintenance on the car. The same cannot be said of a racecar. These machines require highly trained, highly specialized mechanics and tools to perform standard maintenance.

Understanding the level of competence and discipline of maintenance and I&E resources is critical to choosing the right solution. Saddling your valuable maintenance and I&E employees with a system they simply don't have the ability to work on will result in delayed, neglected, or incorrect maintenance and proof testing, which will only serve to weaken the reliability of the SIS.

A smaller, more manageable system which site maintenance and I&E resources have the tools and ability to test and maintain is often a better choice than a large, complex system which may require a maintenance contract with the manufacturer or an outside service provider to perform routine maintenance and proof testing. Scheduling flexibility is also greatly enhanced when existing on-site resources can be used, versus the planning required to bring in a vendor who may have to travel long distances and be scheduled months or longer in advance.

As addressed in one of the earlier misconceptions, local maintenance and I&E competency is one of the chief reasons for selecting multiple, independent lower SIL-rated SIS over one large complex SIL-3 system. If local resources are fully competent and have the proper tools to maintain SIL-1 and SIL-2 systems, but not a SIL-3 system, this should be accounted for when selecting the SIF design and analyzing performance requirements.

6.6 Cost and Project Economics

As with any project, the overall economics must be considered over the intended useful life of the SIS, or the defined payback period specified by corporate requirements. For new, greenfield plants, where the cost of the SIS can be depreciated over the life of the project, the high-performance racecar system may be more easily justified economically. For smaller retrofits to existing equipment, less expensive, yet still highly reliable systems may be more economical, especially when systems must be installed on expense budgets and are not capitalized.

6.7 Common Cause and Single Points of Failure

Good SIS deployment requires consideration of common cause failures and single points of failure of the SIS. When multiple safety instrumented functions (SIFs) are all routed to the same logic solver, the logic solver itself becomes a potential single point of failure which could render all SIFs inactive. Each facility and company will have to determine by their own analysis whether or not this is an acceptable risk. This should be evaluated from both a safety (reliability) standpoint as well as an uptime (availability) perspective.

Multiple, independent safety instrumented systems avoid this potential source of failure. Some of the more advanced safety PLCs are being offered with completely redundant PLC and I/O cards, to avoid precisely this type of single point of failure. However, not all manufacturers offer this feature, and the user should evaluate for themselves the details of the product offered before choosing a solution.

6.8 Similarity of Equipment and Processes

The similarity or dissimilarity of equipment and process operations is another key factor in determining the most appropriate SIS solution. This is best illustrated by way of example.

6.8.1 Storage Tank Overfill Protection

Consider again the storage tank with a requirement for overfill protection from Figure 6. Further consider a facility that has 10 of these storage tanks, at least some of which receive solvents or other materials from the same source. The similarity of operation of the tanks, along with the ability to shut off valves on solvent headers to prevent overflow of multiple tanks may best lend itself to the simplicity of a single logic solver performing multiple safety instrumented functions (SIFs) across the 10 storage tanks. This approach may simplify overall maintenance and testing of the system, as well as the number of final elements (actuators, valves, etc.) to be installed in the field.

6.8.2 Batch, Multi-Product Specialty Chemical Reactors

Consider now a batch, multi-product specialty chemical plant with 10 reactors of various sizes. Depending on the process being run in each of the reactors at any given time, different measured and controlled variables may be needed across different reactors, including temperature, pressure, flow, and level. Some of these parameters may change throughout the batch, and SIS setpoints may need to be adjusted for different phases of the batch. This situation lends itself more readily to the application of smaller, independent safety instrumented systems for each reactor, which provides greater flexibility for adjusting the measured and controlled variables and the SIS setpoints when necessary. This arrangement is also more flexible for maintenance and proof testing, as batch processes are inherently easier to schedule downtime than large, continuous processes which may run many months or years in between turnarounds.

6.9 Corporate Requirements

The final element guiding the selection of an instrumented system is the internal requirements specific to each company. Out of concerns for standardization across multiple facilities, there may be specific requirements for preferred vendors or standardized solutions. Cost and negotiated pricing can influence the types of elements to be procured. Operating experience with different types of sensors, logic solvers, or final elements in specific services should be consulted from corporate engineering standards.

Some practitioners with highly-specific corporate standards may have little-to-no flexibility to choose a solution for their facility or process unit. Others with greater flexibility should comply with their corporate standards, yet also design a solution which most economically achieves the required risk reduction at an appropriate level of design, operational, and maintenance complexity.

7 Competency with SIS

As has been described throughout this paper, there are many facets and nuances of SIS design which must be understood by both the end user and the SIS designer. Performance requirements, component selection, verification and validation, as well as maintenance and operation are all key portions of the safety lifecycle which must be well-understood to properly design, operate, and maintain an instrumented protective system [30]. Here we offer some suggestions for developing competency with the entire gamut of SIS design and the safety lifecycle.

7.1 Training

There is no substitute for good quality training by competent professionals. Many process safety consulting firms that offer training courses have offerings in safety instrumented systems, which range anywhere from 1-3 days at various levels of detail. For those with no experience whatsoever in designing or working with SIS, these trainings are often a good place to start, to help familiarize oneself with the language and basic components and functionality of SIS.

More advanced trainings are also available for those needing more detailed knowledge and starting to perform basic SIS calculations. Consulting firms or companies specializing in SIS design typically offer more advanced courses, which typically range from 3-4 days.

7.2 Printed & Online Resources

With the easy accessibility of information in the 21st century, much in the way of self-study can be done to gain competency, by consuming material from reputable sources. There are several options and various types of media here which can be helpful. First, the major SIS design firms offer blog posts, white papers, and webinars, most of which are free to attend. Blog posts and webinars are led by experts and cover a variety of topics from the introductory to the complex, and are a great, low-cost, self-paced way to begin to build competency in SIS. CCPS has published several books in this field which are helpful, including *Guidelines for Safe and Reliable Instrumented Protective Systems*, and *Guidelines for Safe Automation of Chemical Processes (2nd Ed)*. Proceedings from major U.S. and International conferences often feature papers and presentations by experts and experienced practitioners, and can provide a wealth of information at an affordable price. If conferences cannot be attended in-person, the papers can usually be purchased in a set of conference proceedings.

7.3 Certificate and Certification Programs

The highest level of recognition of SIS competency comes from earning a certificate or certification in SIS design. Various levels of these programs are currently offered by several organizations including ISA, exida, and TÜV Rheinland [2]. The programs vary among qualifications and requirements; most require the applicant to pass an examination. Some require specific training courses to be completed. Experience in SIS design is required, so the Certificate and Certification programs are not intended for beginners. However, with some training and self-study as described previously, one can develop the competency required to sit for a certificate or certification exam.

7.4 Networking

As with developing competency in hazard analysis and SIL determination methodologies, networking is also strongly recommended to promote competency development in safety instrumented systems. Internal communities of experts and SIS practitioners, as well as connections made through industry conferences, ISA's Safety and Security Division, ACC, or AFPM Regional Process Safety Networks, are invaluable in learning from others experience and expertise.

8 Conclusion

High reliability instrumented protective systems require competency to select, design, operate, and maintain, preserving their ability to perform the intended function over the useful life of the system. While vendors and consultants undoubtedly bring a great deal of value and assistance to the end-user, this does not eliminate the need for facility personnel to have at least minimum competency and understanding in system selection, operation, and maintenance. This paper has provided some practical tips to help facility personnel begin to develop the needed competency.

A key step in the selection of a safety instrumented system is to clearly understand the necessary performance requirements, and when a truly high-I/O capacity SIF with redundant architecture is or is not required. Just as a racecar is not needed for a routine trip to school or to the market, the family sedan falls far short of performance requirements on a racetrack. Understanding the difference between the two, and how to select the right tool for the task at hand, is critical to providing the right level of risk mitigation for the facility. This paper has also provided guidance on considerations for when each tool might be the most efficient for the hazard being addressed.

As batch, specialty chemical processes become more complex, the need for highly reliable instrumented protective systems will continue to grow. Process safety professionals with the competency to specify and select an appropriate SIS will be a valued resource for their facility and their company. Those already competent in this area are encouraged to mentor newer professionals, and pass along this critical knowledge to the next generation of process safety engineers.

9 Disclaimer

The views and opinions expressed in this publication are strictly those of the authors and do not necessarily reflect those of the authors' employer, organization, or any other group or individual. The information is presented without warranty or guarantee. Any action taken from the information provided herein (including recommendations) shall be at the user's sole risk.

10 References

1. General Factsheet: Safety Requirements Specification. Available at: https://library.e.abb.com/public/04cb0a3a45a9651ec1257b7a003680ad/FSM%20Safety%20Requirements%20Specification.pdf Accessed on January 5, 2019.
2. How Safe Are Your Systems? Available at: www.flowcontrolnetwork.com/how-safe-are-your-systems/ Accessed on January 5, 2019.
3. A.M. Dowell, "Layer of protection analysis for determining safety integrity level." ISA Transactions 37 (1998) p. 155-165.
4. R. Freeman, "Using layer of protection analysis to define safety integrity level requirements." Process Safety Progress 26 (2007) p. 185-194.
5. CCPS. *Guidelines for Enabling Conditions and Conditional Modifiers in Layer of Protection Analysis*. Center for Chemical Process Safety, American Institute of Chemical Engineers, New York, NY, 2014.
6. CCPS. *Guidelines for Initiating Events and Independent Protection Layers in Layer of Protection Analysis*. Center for Chemical Process Safety, American Institute of Chemical Engineers, New York, NY, 2015.
7. CCPS. *Guidelines for Hazard Evaluation Procedures, 3rd Edition*. Center for Chemical Process Safety, American Institute of Chemical Engineers, New York, NY, 2008.
8. CCPS. *Guidelines for Chemical Process Quantitative Risk Analysis, Second Edition*. Center for Chemical Process Safety, American Institute of Chemical Engineers, New York, NY, 2000.
9. "Guidelines for quantitative risk assessment: "Purple Book." The Hague: Directorate-General for Social Affairs and Employment, Netherlands, 1999.
10. IEC 61511-1 Ed 2.0, "Functional Safety: Safety instrumented systems for the process industry sector - Part 1: Framework, definitions, system, hardware and application programming requirements," IEC, Geneva, Switzerland, 2016.
11. CCPS. *Guidelines for Safe Automation of Chemical Processes, 2nd Edition*. Center for Chemical Process Safety, American Institute of Chemical Engineers, New York, NY, 2017.
12. I. van Beurden and R. Amkreutz, "Emergency batch landing." InTech Magazine, p. 30-32.
13. E.P. Steinhauser, "Addressing the challenges of implementing safety instrumented systems in multi-product batch processes." Journal of Loss Prevention in the Process Industries 57 (2019) p. 164-173.
14. NE 154: Functional Safety in Batch Processes, WIB/NAMUR, Interessengemeinschaft Automatisierungstechnik der Prozessindustrie e.V. , Leverkusen, Germany, 2016.

15. H.F. Rase. *Chemical Reactor Design for Process Plants - Volume 1: Principles and Techniques*. Wiley, New York, NY, 1977.
16. T. Vestal and J. Falin. "Applying Process Safety Time Concepts in Process Plant Applications," American Institute of Chemical Engineers 2018 Spring Meeting and 14th Global Congress on Process Safety. 2018. Orlando, FL.
17. D. Chastain-Knight. "Functional Safety Practices for Operations and Maintenance," American Institute of Chemical Engineers 2018 Spring Meeting and 14th Global Congress on Process Safety. 2018. Orlando, FL.
18. C.M. Pietersen. "The 'SIL Concept' in the process industry: International standards IEC 61508 / 61511," 11th Urea Symposium. 2008. Noordwijk, Netherlands.
19. ANSI/ISA-84.00.01-2004 Parts 1-3 (IEC 61511 Mod). "Functional Safety: Safety Instrumented Systems for the Process Industry Sector," ISA, Research Triangle Park, NC, 2004.
20. IEC 61511-2 Ed 2.0, "Functional Safety: Safety instrumented systems for the process industry sector - Part 2: Guidelines for the application of IEC 61511-1:2016," IEC, Geneva, Switzerland, 2016.
21. P. Gruhn, "The evaluation of safety instrumented systems - tools to peer past the hype." ISA Transactions 35 (1996) p. 25-32.
22. J.J. Sammarco, "Programmable electronic and hardwired emegency shutdown systems: A quantified safety analysis." IEEE Transactions on Industry Applications 43 (2007) p. 1061-1068.
23. J.T. Martel, "PLCs and safety PLCs: Lessons from pucker events." InTech Magazine, June 2008.
24. Safety instrumented systems: using single-loop logic solvers. Available at: https://www.processonline.com.au/content/safety/article/safety-instrumented-systems-using-single-loop-logic-solvers-630378803 Accessed on January 5, 2019.
25. M. Generowicz, *A New Approach to SIL Verification*. 2018, I&E Systems Pty Ltd.
26. W.M. Goble and H. Cheddie. *Safety Instrumented Systems Verification: Practical Probabilistic Calculations*. ISA, Research Triangle Park, NC, 2005.
27. IEC 61508-4 Ed. 2, "Functional Safety of Electrical/Electronic/Programable Electronic Safety-Related Systems - Part 4: Definitions and Abbreviations," IEC, Geneva, Switzerland, 2010.
28. IEC 61508-2 Ed. 2, "Functional Safety of Electrical/Electronic/Programable Electronic Safety-Related Systems - Part 2: Requirements for electrical/electronic/programmable electronic safety-related systems," IEC, Geneva, Switzerland, 2010.
29. J. Torres and N. Ashley, "A HAZOP Gap Query Approach for Hazard Assessment in Multiproduct Facilities." Process Safety Progress 37 (2018) p. 161-168.
30. I. van Beurden and K. Hildenbrandt. "Competency Requirements in Process Safety," American Institute of Chemical Engineers 2018 Spring Meeting and 14th Global Congress on Process Safety. 2018. Orlando, FL.

Perfecting Proof Test Effectiveness,

How more Extensive Testing can Improve Safety and Reduce Cost!

Iwan van Beurden
exida
80 N. Main Street, Sellersville, PA, 18960, USA
vanbeurden@exida.com

Kate Hildenbrandt
exida
80 N. Main Street, Sellersville, PA, 18960, USA
khildenbrandt@exida.com

Prepared for Presentation at
American Institute of Chemical Engineers
2019 Spring Meeting and 15th Global Congress on Process Safety
New Orleans, LA
March 31 – April 3, 2019

Perfecting Proof Test Effectiveness,

How more Extensive Testing can Improve Safety and Reduce Cost!

Iwan van Beurden
exida
80 N. Main Street, Sellersville, PA, 18960, USA
vanbeurden@exida.com

Kate Hildenbrandt
exida

Keywords: Safety Lifecycle, SIS, SIF, Cost, Proof Test, IEC 61511, SIL verification, Proof Testing.

Abstract

We all know that a periodic functional test of the components in Safety Instrumented Functions is necessary to ensure the appropriate safety integrity is maintained. These tests on some of the most reliable equipment in a plant are occasionally disruptive and intrusive in the process. However, in general the necessity of the tests is acknowledged. The second edition of the IEC 61511 functional safety standard for the process industry sector identifies the need to take the proof test effectiveness into consideration. This proof test effectiveness indicates the percentage of undetected failures revealed during a proof test. This parameter can have quite an impact on the overall achieved safety integrity. At the same time the parameter shows that more comprehensive tests will positively influence the achieved safety integrity. This paper will discuss the concept of proof test effectiveness and through a practical example, using a final element assembly including a trunnion ball valve, show how more extensive tests will improve safety and at the same time can reduce cost, e.g. maintenance cost due to lower proof test frequency requirements as well as equipment cost due to lower level of redundancy required.

1 Introduction

The international functional safety standard IEC 61511 [1] provides the safety lifecycle as a steadfast guideline to assess and mitigate risk for manufacturing processes including refineries, chemical, petrochemical, pulp and paper, and power plants. The Safety Life Cycle (SLC) is one of the fundamental concepts presented in the standard. The objective of the SLC is to optimize the Safety Instrumented System (SIS) design by matching the SIS design to the risk reduction requirements of the process.

The standard describes the various steps of the SLC in detail, but the overall objective of the Safety Life Cycle can also be summarized in three steps that are also illustrated in figure 1.

1. Analyze the problem
2. Design the solution
3. Verify that the solution solves the problem and track its performance

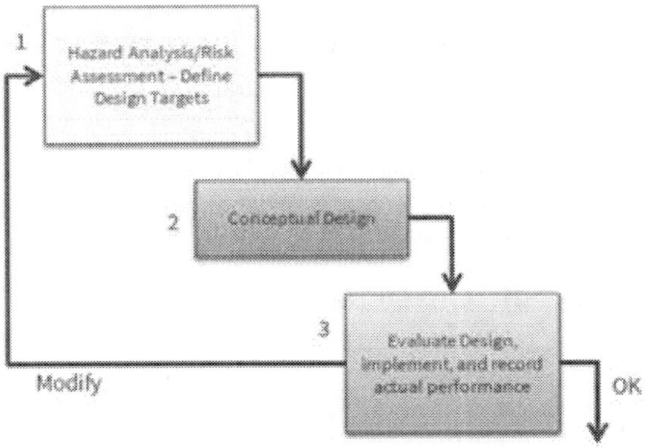

Figure 1. Basic Steps Safety Lifecycle

The analysis step involves the process hazard identification and risk analysis. Potential hazards with enough risk may warrant the design of Safety Instrumented Functions. For those hazards, a target Safety Integrity Level (SIL) is assigned according to risk reduction targets. In step 2 the individual Safety Instrumented Functions are designed to meet their specified SIL targets. This design process involves selecting the technology to be used, selecting equipment and configuring that equipment with enough redundancy to meet both the safety requirements and the process uptime (availability) requirements and subsequently implementing that design. The third and final step concerns the analysis that is performed to validate if the Safety Instrumented System as designed meets all specified requirements, operate the process, perform the required proof tests, and record any failures and process demands.

The typical Safety Instrumented Function (SIF) is used to bring the process to a safe state when a predetermined condition is violated, e.g. close a shutoff valve when the pressure in the main fuel line exceeds a certain value. This implies that if the predetermined condition is not violated, the SIF will do nothing, i.e. the valve in this example just sits there opened waiting for a request to act. The SIF as such acts as an insurance for the, sometimes very rare, occasion that the process violates the predetermined condition.

The static nature of a SIF leads to a set of failure modes not typically experienced in control/dynamic operation. Failure modes for SIFs are therefore classified into two main categories, i.e. safe and dangerous. A safe failure is typically defined as a failure that causes a false or spurious trip. Dangerous failures are typically defined as a failure that prevents a SIF from performing its intended protection function [2]. These two categories are further divided into detected and undetected categories, which account for the ability of any automatic detection mechanism to reveal the failure during normal operation. This leads to a total of 4 main failure mode categories:

- Safe Detected
- Safe Undetected
- Dangerous Detected
- Dangerous Undetected

Of these 4 categories, the Dangerous Undetected failures are representing failures that prevent the SIF from performing its intended function while the user/operator is completely unaware of this. These failures are the ones that are most dominant in determining the overall SIF's safety integrity. An example of a dangerous undetected failure in a SIF is a valve whose stem is stuck, preventing the valve from moving to the safe state in case of a violation of the predetermined condition.

2 Proof Testing

To reveal Dangerous Undetected failures, as well as any automatic diagnostic failure, users perform periodic manual tests. These manual tests are called proof tests. Hence a proof test is a manual test designed to reveal failures undetected by automatic device diagnostics. In case of the stuck stem example, a simple proof test could entail a full stroke of the valve to verify that it can still move and achieve the safe state when needed.

2.1 Proof Test Frequency

The frequency of the proof test is an important parameter in the design of the SIF, step 2 in figure 1. If more frequent proof testing of equipment is implemented, a higher calculated level of safety can be achieved. The simple proof test as described above implies that in order to perform the test, the process will be interrupted. Typically, there will be an overall requirement for the plant to avoid interruptions, so the SIF designer will need to balance process interruptions versus improving safety.

Figure 2 shows the Probability of Failure on Demand as a function of time for a complete SIF with a proof test interval of 1 year and 5 years respectively, for a 15-year mission time of the SIF. The black dotted line represents the average Probability of Failure on Demand (PFD_{AVG}) for the SIF. From the diagram, it can be easily concluded that the scenario with the more frequent proof tests has a lower PFD_{AVG}, and therefore achieved a higher safety integrity. Note the scenario assumes that you will reveal all dangerous undetected failures while testing. A simplified equation can be derived for it as shown in equation 1.

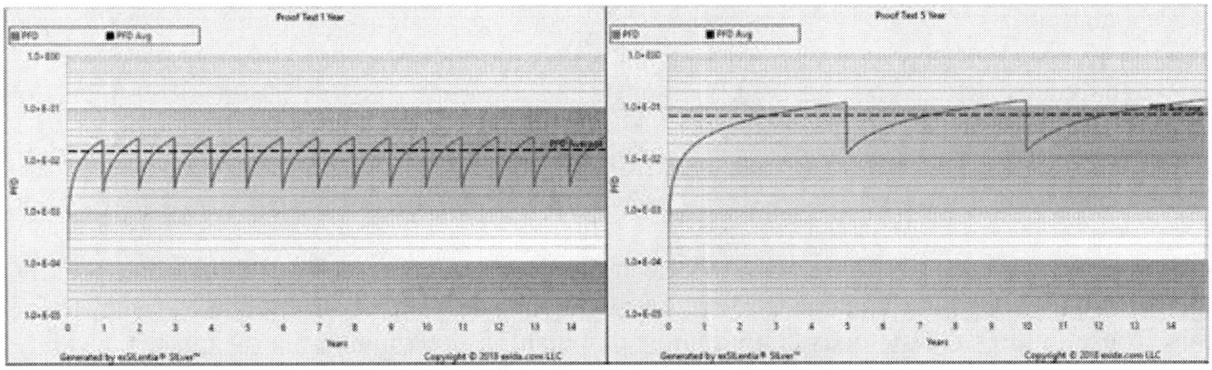

Figure 2. Impact of Proof Test Frequency

$$PFD_{AVG} = \frac{\lambda_{DU} \cdot PTI}{2}$$ **Equation 1**

Where: λ_{DU} is the Dangerous Undetected failure rate
PTI is the proof test interval

2.2 Proof Test Effectiveness

Not all proof tests are created equal and while assuming that the simplified test used in the example above would discover all undetected failures is very useful for illustration purposes, this is generally not the case with real world devices. The extensiveness of a test can have a significant impact on the number of dangerous undetected failures that are revealed as part of the test and therefore on the overall effectiveness of the proof test. Extending the valve proof test as described above by adding a leak test would potentially reveal dangerous undetected failures that are associated with the valve seat. This means that the extensiveness of the proof test must be taken into consideration in the PFD_{AVG} calculation. Equation 1 therefore needs to be expanded to include a portion representing failures that can be revealed during the test and a portion representing failures that cannot be revealed during the test. This leads to equation 2.

$$PFD_{AVG} = \frac{C_{PT} \cdot \lambda_{DU} \cdot PTI}{2} + \frac{(1 - C_{PT}) \cdot \lambda_{DU} \cdot MT}{2}$$ **Equation 1**

Where: C_{PT} is the proof test coverage factor
λ_{DU} is the Dangerous Undetected failure rate

PTI is the proof test interval
MT is the mission time of the SIF

Accurately accounting for the proof test effectiveness will have a negative effect on the achieved safety integrity, the calculated PFD_{AVG} will be higher, as the calculation now shows that certain failures are not revealed at all during the life of the system. This, however, provides a more realistic view of the expected performance of the installed system and correspondingly a more realistic view of the remaining risk in the process under evaluation.

Figure 3 shows the Probability of Failure on Demand as a function of time for a complete SIF with a proof test interval of 1 year assuming perfect testing and repair and a proof test interval of 1 year with realistic proof test coverage values for a 15-year mission time of the SIF. The diagram shows that certain failures will never be revealed and a gradual increase over time is the result.

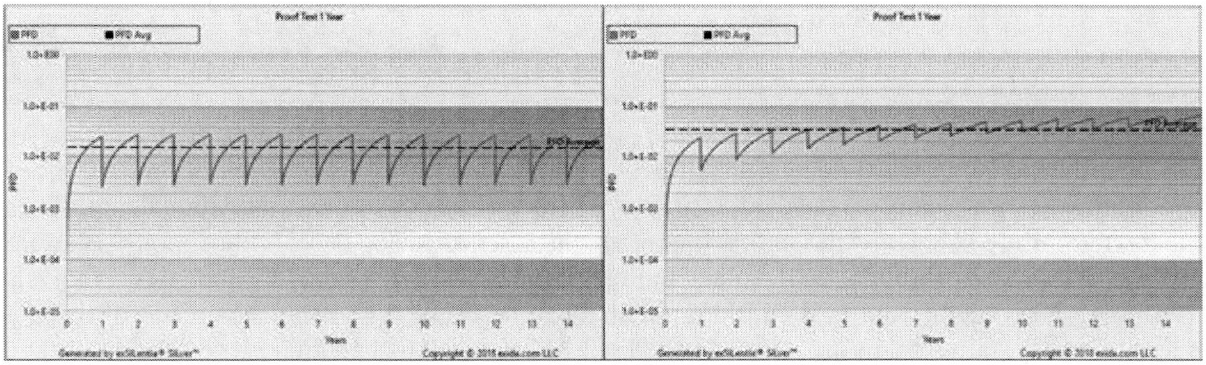

Figure 3. Impact of Proof Test Effectiveness

3 Determining Proof Testing Effectiveness

The previous section explained how proof test frequency and proof test effectiveness impact the overall PFD_{AVG} and therefore safety integrity of a Safety Instrumented Function. Though defining the test interval may be driven by the application and operating environment, establishing the correct proof test coverage to determine the overall effectiveness may not be. In order to determine the proof test coverage, it is important to understand what the particular device's Dangerous Undetected failures are and what failures can be revealed with that test. In general, the more extensive the test, the more dangerous undetected failures are revealed and the higher the proof test coverage. For final element assemblies five tests can typically be identified as shown in table 1 [3].

Table 1. Typical Final Element Assembly Tests [3]

Test Number	Test Name	Test Description
PT 1	Partial Valve Stroke Test	Move the valve 5-10% of its total stroke and record the duration
PT 2	Full Valve Stroke Test	Move the valve 100% of its full stroke, record the duration, and perform a physical inspection
PT 3	Full Valve Stroke Test at Operating Conditions	Perform test 2 at operating conditions.
PT 4	Full Valve Stroke Test and Leak Test	Perform test 2 with a leak test.
PT 5	Full Valve Stroke Test and Leak Test at Operating Conditions	Perform test 4 at operating conditions.

A detailed analysis of a particular device can establish the failure modes of that device, and with that identify the dangerous undetected failures that are intended to be revealed during the proof test. Table 2 shows components of a final element assembly and the associated dangerous undetected failure modes of the entire final element assembly. For each failure mode a failure rate is determined through a detailed Failure Modes, Effects, and Diagnostics Analysis (FMEDA). Next the portion of the failure rate that can be claimed revealed as a result of the test is identified [3] (failure rates are defined in FITs; 1 FIT is equal to 1 failure every $1 \cdot 10^9$ hours).

Table 2. Dangerous Failures Detected per Functional Failure Mode and Test [3]

Device	Final Element Assembly Functional Failure Mode	Dangerous Failure Rate [FIT]	Failures Detected by Test				
			PT 1	PT 2	PT 3	PT 4	PT 5
Solenoid Valve	Loss of Drive	230	207	207	228	228	228
Spring Return Scotch Yoke Actuator	Excessive Force	391	235	352	387	352	387
	Loss of Drive	149	0	89	89	148	148
	Leakage Past Seat	11	0	0	0	11	11
Trunnion Ball Valve	Excessive Force	355	213	320	351	320	351
	Loss of Drive	25	0	0	0	25	25

	Leakage Past Seat	120	0	0	0	119	119
Mounting and Coupling	Excessive Force	19	11	17	19	17	19
	Loss of Drive	7	0	0	0	7	7
	Total	**1307**	**666**	**985**	**1074**	**1225**	**1294**
	Test Coverage	**-**	**51%**	**75%**	**82%**	**94%**	**99%**

From the total dangerous undetected failure rate (1307 FITs) and the respective number of failures that are revealed as part of each of the 5 tests, a test coverage factor can be determined for the final element assembly for each test as shown in the bottom row of table 2.

4 Proof Test Effectiveness Impact

Given the test coverage factors derived in the previous section, the impact of the proof test effectiveness on a SIF's safety integrity can be determined. The SIF under consideration consists of 3 SIL 3 certified pressure transmitters voted 2oo3, a generic SIL 3 certified safety PLC, and a single trunnion ball valve, driven by a spring return scotch yoke actuator, and operated by a SIL 3 certified solenoid. The proof test intervals for the transmitters and logic solver are set to 5 years. For the final element assembly, various proof test intervals are chosen while considering each of the 5 tests identified in table 1 (in this example, even the partial valve stroke test is considered a proof test). The resulting calculated PFD_{AVG} values are documented in table 3.

Table 3. SIF PFD_{AVG} [RRF] for Various Final Element Assembly Tests and Test Intervals

Proof Test Interval	PT 1	PT 2	PT 3	PT 4	PT 5
1 year	5.30E-02 [18.9]	3.17E-02 [31.5]	2.54E-02 [39.4]	1.44E-02 [69.4]	9.84E-03 [101.6]
2 years	5.65E-02 [17.7]	3.70E-02 [27.0]	3.13E-02 [31.9]	2.13E-02 [46.9]	1.72E-02 [58.1]
5 years	6.71E-02 [14.9]	5.32E-02 [18.8]	4.91E-02 [20.4]	4.20E-02 [23.8]	3.91E-02 [25.6]

Table 3 shows a couple of interesting concepts. First of all, consider the 1-year proof test interval for PT 5, this shows an overall SIF PFD_{AVG} less then 9.84E-03 thereby meeting the requirements for SIL 2. In other words, a thorough test allows the SIF to consist of a single valve, thereby avoiding the need to use redundant valves and hence saving on initial procurement cost. The results in table 3 also show how the PFD_{AVG} almost doubles for PT 5 when extending the proof test interval to 2 years.

An interesting fact is that the calculated PFD_{AVG} for the various tests are slowly starting to converge as the test frequencies increase. This shows that the effectiveness of the test is starting to become less of an impact as the test intervals are increasing. In other words, if one decides to rarely test, the test interval becomes more dominant in the PFD_{AVG} calculation than the test coverage. Of note is that, at sufficiently long test intervals, the impact of proof test effectiveness becomes less pronounced.

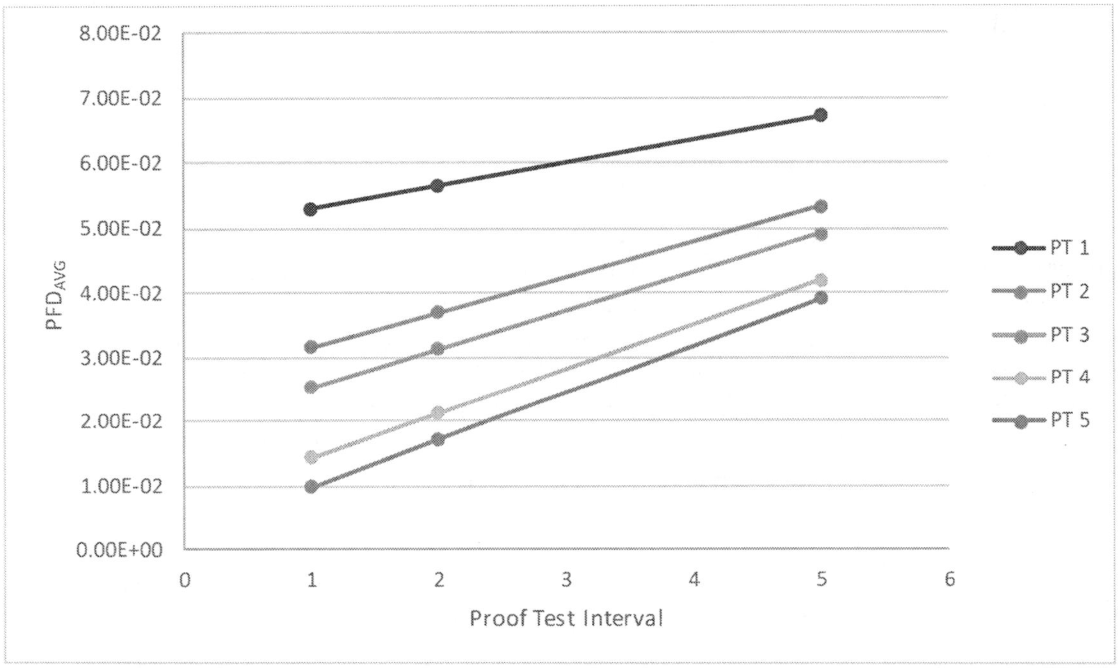

Figure 4. SIF PFD_{AVG} including Various Final Element Assembly Tests and Test Intervals

5 Conclusion

The proof test interval and proof test effectiveness are key inputs in the SIL verification that is part of the design and implementation phase of the IEC 61511 functional safety lifecycle. The specific application and operating environment are important factors in specifying the proof test interval. To determine the proof test effective, however, it is important to understand the specific test and what that test can reveal.

A detailed Failure Modes, Effects, and Diagnostics Analysis provides an overview of the failures that are not detected during normal operation and that should be tested for as part of the proof test. The specific test can then be compared to the design of a product to determine what failures are revealed during a test and what failures are not. This leads to a test specific proof test coverage.

For final element assemblies, 5 different tests are identified ranging from a partial valve stroke test to a full valve stroke test with leak test at operating conditions. The more extensive the test the higher the proof test coverage.

Depending on the properties of the Safety Instrumented Function that is being analyzed, various combinations of test interval and test coverage could be evaluated to determine the best match of the application at hand. In some cases, this may mean that a more detailed test could avoid the need to implement redundancy. Furthermore, at sufficiently long test intervals, the impact of proof test effectiveness becomes less pronounced.

6 References

[1] IEC 61511, *Functional Safety: Safety Instrumented Systems for the process industry sector*, 2nd edition, revision 1, International Electrotechnical Commission, Geneva, Switzerland, 2017.

[2] I.J.W.R.J. van Beurden and W.M. Goble, *Safety Instrumented System Design, Techniques and Design Verification*, International Society of Automation, Research Triangle Park, NC, USA, 2018.

[3] C.J. O'Brien, L.L. Stewart, L.D. Bredemeyer, Final Elements in Safety Instrumented Systems, IEC 61511 Compliant Systems and IEC 61508 Compliant Products, exida, Sellersville, PA, USA, 2018.

SIL-3, SIL-2, and Unicorns
(There Is a High Probability Your SIL 2 and SIL 3 SIFs Have No Better Performance Than SIL 1)

Presenter
A.M. (Art) Dowell, III, PE
Principal Engineer
Process Improvement Institute, Inc.
16430 Locke Haven Dr.,
Houston, TX 77059 USA
adowell@piii.com

Matias Massello
Engineer
Process Improvement Institute, Inc.
Argentina
mmassello@process-improvement-institute.com

William Bridges
President
Process Improvement Institute, Inc.
1321 Waterside Lane,
Knoxville, TN 37922 USA
wbridges@piii.com

Harold W. (Hal) Thomas, PE, CFSE
Specialist
exida.com LLC
80 North Main Street
Sellersville, PA USA
thomashw@exida.com

Prepared for Presentation at
American Institute of Chemical Engineers
2019 Spring Meeting and 15th Global Congress on Process Safety
New Orleans, LA
March 31 – April 3, 2019

SIL-3, SIL-2, and Unicorns (There Is a High Probability Your SIL 2 and SIL 3 SIFs Have No Better Performance Than SIL 1)

Presenter
A.M. (Art) Dowell, III, PE
Principal Engineer
Process Improvement Institute, Inc.
16430 Locke Haven Dr.,
Houston, TX 77059 USA
adowell@piii.com

William Bridges
President
Process Improvement Institute, Inc.
1321 Waterside Lane,
Knoxville, TN 37922 USA
wbridges@piii.com

Keywords: SIL, SIF, Human Factors, Systematic Error, Human Error

Abstract

Safety Instrumented System (SIS) standards improved the definition of interlocks and introduced requirements for improved management systems to enforce independence from other Independent Protection Layers (IPLs). SIS standards require verification that the performance of each Safety Instrumented Function (SIF) will be met during its lifetime; where the performance criterion is documented as the target Safety Integrity Level (SIL) or risk reduction factor for the SIF. The SIL is in turn tied to specific values of probability of failure on demand (PFD). The current SIS standards and the TR (Technical Reports, from ISA) that explain how to do SIL Verification calculations do not include accounting for specific human error probabilities -- this is a major deficiency as even the probability of a single human error can be much larger than the target PFD of 0.001 for a SIL 3 and oftentimes a little larger than the PFD of 0.01 for a SIL 2. The SIL Verification methods outlined in the standards and technical reports like ANSI/ISA TR84.00.02 facilitate consistency for the component-only failure rates. As user companies seek to obtain greater risk reduction from their SIS to satisfy their corporate risk criteria, failure to adequately address potential specific human failures can lead to overly optimistic results and a misallocation of resources intended to reduce risk

This paper shows that specific human error during testing, calibration maintenance, and restoration of a SIF is a significant contribution to the true PFD of the SIF for SIL 2 and dominates SIL 3 designs. Unless the human errors are accounted for and then compensated for, it is more likely to find a Unicorn than to actually get two or three orders of risk reduction from SIL 2 and SIL 3 SIFs.

Example methods for human error analysis related to a SIS are provided as well as some proven approaches for controlling human factors. The paper also discusses ways to prevent or else to detect and to recover from errors made in redundant channels (such as used in 1oo2, 1oo3, or 2oo3 voting).

1 Introduction

This paper is an update to a prior paper [1]. In the ensuing seven years, the international and USA global standards organizations have issued new guidance, but the issue raised in 2012 remains the same. Although technology and methods exist to evaluate specific human error probabilities, particularly during maintenance and operation of safety instrumented functions (SIFs), the standards continue to categorize all human error in the systematic failures category and ignore specific human errors in the verification calculation of safety integrity level (SIL).

The failure of SIFs can be due to a number of reasons. Common terminology in the industry characterizes these failures as either random hardware failures or systematic failures. This paper mainly focuses on the systematic aspects (which for convenience include specific human errors); however, all equations presented will also include the hardware failure contribution for completeness.

1.1 Systematic Failures versus Specific Errors

Systematic failures may manifest themselves via a number of failure mechanisms such as:

- Manufacturer design errors
- End user design errors
- Hardware installation errors
- Manufacturer software design errors
- End user programmable configuration errors
- Human error during operation and maintenance
- Management of change errors

As the list above shows, systematic error may be introduced by the manufacturer or the end user. This paper will focus on the end user, as equipment that has been properly reviewed and certified in accordance with IEC-61508 [2] undergoes a formal work process specifically looking to minimize systematic errors from the manufacturer. It is also likely that systematic errors that do occur will manifest themselves during the warranty period, allowing appropriate response to rectify the problem to a suitable level. In addition, the performance of each SIF is expected to be validated versus the safety requirements specification (SRS) during the factory acceptance testing and equipment certified in accordance with IEC 61508 is also expected to undergo proven in use validation by the manufacturer on a periodic basis to confirm the device SIL capability described in the certification. This validation of specific device SIL capability does not mean the performance of a SIF's safety integrity level is validated as that is an end user's responsibility beyond the control of the manufacturer."

Once under the control of the end user, the variability of application and control greatly increases, making control of specific human errors more difficult. This paper seeks to make the reader more aware of how specific human errors may occur and how they can impact the risk reduction of SIFs.

1.2 Human Error in SIF Failures

Human error during interventions with SIS can have a detrimental effect on the availability of one or more SIFs. There have been numerous cases where SIFs were left in bypass, etc., and an accident occurred. One notable event was at a facility in Institute, West Virginia, USA (in 2008). An SIF was bypassed to allow startup to proceed more smoothly. Reactant was allowed into a vessel without solvent and the temperature of the newly replaced residue treater ran away and it exploded, resulting in two fatalities and multiple injuries. In addition, it was also a near miss with respect to a potential large release of methyl isocyanate from a storage tank located just 80 feet from the explosion [3].

The focus of this paper will be on human interaction errors (usually called specific human errors in this paper). These errors include errors in the operation of the man-machine interface to the SIF (such as leaving an SIF in bypass during unit startup), errors during periodic testing of the SIF, and errors during the repair of failed modules in the SIF. This last type of human error includes the simple case of inadvertently leaving a root valve on an instrument closed.

The SIS standards of the mid-1990s through today recognized that human errors have a deleterious impact on the PFD of an SIF. This effect can be either errors that exist at Time Zero or specific human errors while operating (called systematic errors in some SIS literature and standards). The IEC standards qualitatively covered at length the need to control such errors [4], [5]. For example, Clause 3.2.69, Note 3 [5] says, "...However, safety integrity also depends on many systematic factors, which cannot be accurately quantified and are often considered qualitatively throughout the life-cycle..."

In the Technical Report (TR) from the International Society of Automation (ISA) for how to perform SIL verification calculations, ISA-TR84.00.02-2015 [6], Equation 8.1 (shown here as Equation 1) is given to calculate the SIF probability of failure on demand.

$$\boldsymbol{PFD_{SIF} = PFD_S + PFD_{LS} + PFD_{FE} + PFD_{SS}} \qquad \textbf{Equation 1}$$

where,

PFD_{SIF} is the PFD_{avg} for the SIF **Hardware**, actually $PFD_{hardware}$.

PFD_S represents the various sensors used to detect abnormal process conditions,

PFD_{LS} represents the logic solver used to make decisions based on the process conditions,

PFD_{FE} is the final element used to act on the process,

PFD_{SS} represents any required support systems, such as a power supply in ETT.

ISA-TR84.00.02-2002 [7] did include an additional term (shown here as Equation 2) for contribution of "dangerous system failures" in TR Equation 1a. This term was correctly omitted from the 2015 TR since true systematic failures are not random and base random human errors are not subject to improvement by decreased proof test intervals.

$$PFD_{SYS} = \left[\lambda_F^D x \frac{T_i}{2}\right] \qquad \textbf{Equation 2}$$

where,
PFD_{sys} = dangerous systematic failures
λ_F^D = dangerous failure rate, per year
T_i = test interval, year.

However, Equation 1 did NOT include a term for specific human errors related to interventions by the end user.

Also, the 2002 TR does not give any practical guidance as to which systematic and human errors are most significant. While the 2015 TR has a partial list of safe and dangerous systematic errors in Clause 9.1, it also does not provide guidance to determine which systematic and human errors are most significant.

Likewise, IEC 61511-2017 and ANSI/ISA 6511-1-2018 recommend a qualitative approach to control systematic errors, including human errors, by simply stating that the end user must control all human intervention error (easily said, difficult in practice).

In practice, most of the systematic error term results from specific human errors. These can include:

- Manufacturer contribution for certified equipment (believed to be negligible relative to end user errors)
- End user errors:
  - Design and installation errors
  - Introduction of undetected failures caused by proof testing
  - Bypass during operation

Of these end user errors, the dominating contribution is generally human errors that leave the protection failed at Time 0. These errors can occur during calibrations, proof tests, and re-commissioning of an SIF following routine maintenance interventions such as:

- Miscalibrations
- Leaving a root valve on an instrument closed
- Leaving an SIF in bypass, i.e.:
  - Bypassing the function due to a spurious trip and failing to remove the bypass.
  - Bypassing the function for startup because the system dynamics require it; however, the designers missed this need during startup mode of the process, resulting in an operational bypass that requires human intervention to remove the bypass rather than an automated design that removes the bypass.

- Bypassing the final element and failing to remove bypass when the test or repair is complete.

Therefore, a simple equation including the human error terms can replace the equations from ISA-TR84.00.02 (-2002 and -2012). The resulting improved equation is shown in Equation 3:

$$\boldsymbol{PFD_{SIF} = PFD_S + PFD_{LS} + PFD_{FE} + PFD_{SS} + PFD_{SYS}} \qquad \textbf{Equation 3}$$

We expand the PFD_{sys} using Equation 4.

$$\boldsymbol{PFD_{SYSi} = PFD_{SYS-PROCi} + P_{HUMi}} \qquad \textbf{Equation 4}$$

where,
PFD_{SYSi} = the systematic PFD of the ith part of the SIF,
$PFD_{SYS\text{-}PROCi}$ = the systematic errors and failures generated randomly by the process for the ith part of the SIF (such as plugging of instrument taps by process materials or contaminants),
P_{HUMi} = the probability of dangerous human error from the ith part of the SIF

(Note that $PFD_{SYS\text{-}PROCi}$ can be modeled as a hardware failure rate and a good inspection and testing program should be able to detect these failures.)

We expand the PFDs to show the sum of the individual device PFDs, (Equation 5).

$$\boldsymbol{PFD_{SIF} \approx \sum PFD_{Si} + \sum PFD_{LSi} + \sum PFD_{FEi} + \sum PFD_{SSi} + \sum PFD_{SYS-PROCi} + \sum P_{HUMi}}$$

Equation 5

Further, the overall human error term can be expanded and written as Equation 6.

$$\boldsymbol{P_{HUM} = P_{design\ error} + P_{installation} + P_{proof\ test\ error} + P_{bypassed}} \qquad \textbf{Equation 6}$$

Of the four terms in this equation, the first two can be detected and corrected during initial commissioning steps for the SIF. Experience has shown that the last two terms, $P_{proof\ test\ error}$ and $P_{bypassed}$ are likely to dominate the P_{HUM}, though more industry data is needed to support this observation. Assuming that P_{HUM} is dominated by $P_{proof\ test\ error}$ and $P_{bypassed}$, Equation 6 can be further simplified to Equation 7.

$$\boldsymbol{P_{HUM} = P_{proof\ test\ error} + P_{bypassed}} \qquad \textbf{Equation 7}$$

Experienced gained from many accident investigations and also from calculations, support the contention that for high SIL designs, the human errors during interventions, $P_{proof\ test\ error}$ and $P_{bypassed}$, dominate the calculated PFD_{SIF}. Unfortunately, most of the SIL verification calculations today use the truncated Equation 1 (Equation 8.1 in TR84.00.02-2015) instead of the more complete Equation 8 (below).

$$PFD_{SIF} = PFD_S + PFD_{LS} + PFD_{FE} + PFD_{SS} \qquad \textbf{Equation 1}$$

$$PFD_{SIF} = PFD_S + PFD_{LS} + PFD_{FE} + PFD_{SS} + PFD_{HUM} \qquad \textbf{Equation 8}$$

Remember that Equation 1 gives the $PFD_{hardware}$ (only) As a result, most SIL Verification calculations today effectively ignore specific end-user human errors when quantifying their risk reduction capability. This approach is equivalent to saying the system boundary for an SIF only includes the instrumented components (a subsystem). In LOPA (layer of protection analysis) and other quantitative risk assessments, the entire IPL system must be considered [8]. It is critical to include everything that can cause the SIF to fail in the IPL boundary [9].

1.2.1 IPL (SIF) Boundary Example

For example, for a low-level sensor to prevent high pressure gas blow-by (Figure 1), Equation 1 would evaluate only the devices in the narrow boundary:

- the level transmitter,
- the logic solver, and
- the solenoid valve and isolation valve to prevent gas blow-by.

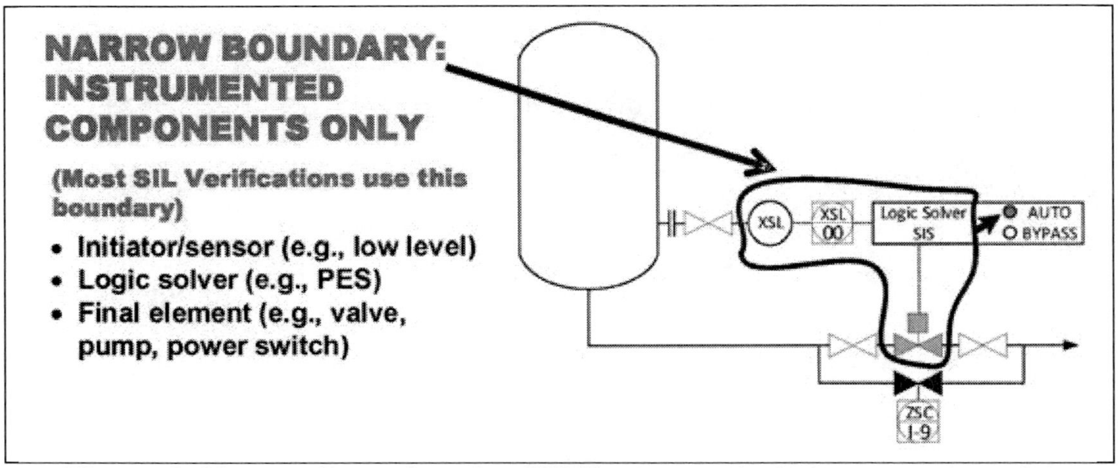

Figure 1: Narrow Boundary for SIF (instrumented components only)
Courtesy Process Improvement Institute, Inc., All Rights Reserved

Equation 8 would include the hardware in Equation 1 but would also include the human errors related to the SIF (Figure 2):

- the root valves and the vessel connection for the sensor(s)
- the auto/manual bypass switch for SIF
- the block valves around the isolation valve, and
- the bypass around the isolation valve.

The additional items in Equation 5 can contribute to the probability that the SIF is bypassed ($P_{bypassed}$) or that the SIF is rendered non-functional after the proof test ($P_{proof\ test\ error}$).

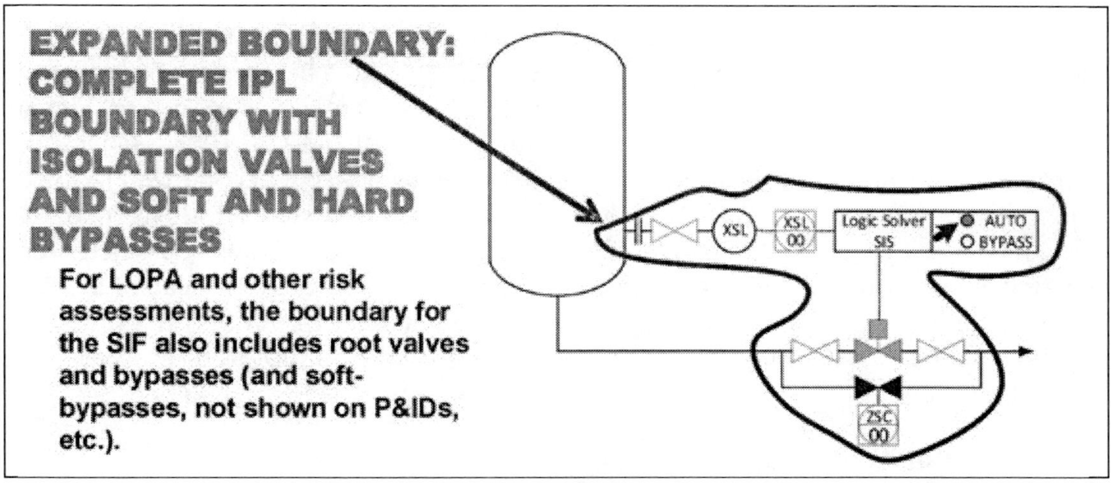

Figure 2: Expanded Boundary for SIF (instrumented components and isolation valves and soft and hard bypasses)

Courtesy Process Improvement Institute, Inc., All Rights Reserved

1.2.2 IPL (PSV) Boundary Example

In another example, if the IPL is a PSV, then the IPL system must include upstream and downstream features, such as isolation valves. Therefore, the probability of leaving an isolation valve closed should be included as a contribution to the overall PFD of the PSV IPL system.

For the remainder of this paper, we will concentrate on the effect of including the probability of human errors for calculation of PFD_{SYS}, and the effect this has on the resulting SIL. For convenience, this paper arbitrarily sets $\sum PFD_{SYS\text{-}PROC}$ = zero. This is being done so the reader can better focus on the human error aspect of systematic errors. $PFD_{SYS\text{-}PROC}$ is worthy of its own paper as its significance is generally dependent on the process fluid properties and/or ambient conditions.

The next two sections of this paper provide a basis for (1) the baseline error probability for human error during interventions and (2) the error probabilities given coupling of activities, such as occur with redundant systems. Following that, simple examples are provided to help show the relative impact of including human error terms $\sum PFD_{HUM}$ in the calculation of PFD_{SYS}.

2 Human Error Probability for a Single Execution of a Rule-Based Task

To calculate P_{HUMi}, the type of tasks must be defined and the baseline error probability for such a task needs to be established. Note that with excellent control of all of the human factors, a company can begin to approach the lower limits that have been observed for human error, but individual, specific human error probabilities may average about P_{HUMi} = 0.01 (which is a relatively large factor in SIL 2 and SIL 3 verification calculations). It is critical to provide detection and correction for specific human errors.

Excellent control of all human factors means a robust design and implementation of management systems for each human factor are achieved with a high level of operational discipline. The first well-researched publication detailing potential lower limits of human error probability was by Alan Swain and H Guttmann (NUREG-1278, 1983) [10] and by others. However, many times, the limits they referenced get used out of context. The lower limits in the NUREG-1278 assume excellent human factors, but such excellent control is rarely, if ever achieved. Additionally, some human errors listed by Swain and others were for a single error under highly controlled conditions, or on a "best day" instead of average error probability or rate over an average year of tasks. In general, Process Improvement Institute (PII) has found it best to use the average error probabilities as discussed in the following section.

2.1 Error Probability for Rule-Based Actions that are Not Time Dependent:

Actions that do not have to be accomplished in a specific time frame to be effective are not time dependent. It should be obvious then that these do not include response to alarms, or similar actions with time limits. Values listed below represent the lower limits for human error probabilities, assuming excellent control of human factors; these are expressed as the probability of making a mistake on any step:

- 1/100 - process industry; routine tasks performed 1/week to 1/day. This probability assumes excellent control of all human factors. Most places PII visits, the workers and managers and engineers believe this is achievable, but not yet achieved. {Actual data from the Savanah River Site indicated a Miscalibration error probability of 7.0E-3 and a Failure to Restore After Maintenance error probability of 5.1E-3 for an organization with excellent human factors control and data gathering (Table 3, [11]). It is noted that these values could be rounded to 1/100. Organizations are cautioned to determine the actual data for human error probabilities in their own management systems before using human probabilities lower than 1/100.}
- 1/200 - pilots in the airline industry; routine tasks performed multiple times a day with excellent control of human factors. This average has been measured by a few clients in the airline industry, but for obvious reasons they do not like to report this statistic.
- 1/1000 - for a reflex (hard-wired) action, such as either proactive or minor corrective actions while driving a car, or very selective actions each day where your job depends on getting it right each time and where there are error recovery paths (such as clear visual cues) to correct the mistake. *This is about the probability of running a stop sign or stop light, given no one is in front of you at the intersection; the trouble is measuring this error probability, since you would have to recognize (after the fact) that you made the mistake.*

See Bridges and Collazo (GCPS, 2012) [12] for more details on this topic.

2.2 Adjusting the lower limit probabilities to estimate a baseline probability at a site

As mentioned earlier, the lower limit probabilities assume excellent control of human factors in the industry mentioned. Note that airline pilots have a lower error probability than what PII has measured in the process industry. This is due, in part, to the much tighter control by the airlines and regulators on factors such as fitness-for-duty (control of fatigue, control of substance abuse, etc.). Excellent control of human factors is not achieved in many organizations; therefore, the human error probabilities will be higher than the lower limit, perhaps much as much as 20 times higher. Table 1 provides adjustment factors for each human factor. These factors can be used to adjust the lower limit of error probability upward or downward as applicable, but the factors should not be applied independently. For instance, even in the worst situations, we have not seen an error probability for an initiating event or initial maintenance error higher than 1/5, although subsequent steps, given an initial error, can have an error probability approaching 1 due to coupling or dependency.

- 1/5 - highest error probabilities with poor control of human factors; this high probability is typically due to high fatigue or some other physiological or psychological stress (or combination). This is the upper limit of error probabilities observed with poor human factors and within the process industry. *The error probabilities in the Isomerization Unit the day of the accident at BP Texas City Refinery [13] were about this probability. The operators, maintenance staff, and supervisors had been working about 30 days straight (no day off) on 12-hour shifts.*

For the examples provided later, this paper ***will use a baseline error probability of 0.02 (1/50) errors per step***, which is about average at the sites PII visited in the past 15 years. This value could be justified based on the fact that most chemical process sites do not control overtime during turnarounds and/or do not have a system for controlling verbal communication using radios and phones. In addition, for critical steps such as re-opening and car-sealing the block valves under a relief valve after the relief valve is returned from maintenance, the error probability is about 0.01 (1/100) to 0.04 (1/15) [14]; plus, the average probability of being in a "fail to function" state at time zero for a relief device is between 0.01 (1/100) and 0.02 (1/50) [15, 16, 17] (Bukowski, 2007-2009). Both of these tasks have multiple checks and have procedures (similar to what is done when servicing a SIF and when using bypasses for an SIF) and yet the observed human error probability remains between 0.01 and 0.02.

Table 1. SUMMARY TABLE of 10 HUMAN FACTOR CATEGORIES

Based in part on: Gertman, D.; et. al., *The SPAR-H Human Reliability Analysis Method*, NUREG/CR-6883, U.S. Nuclear Regulatory Commission, Office of Nuclear Regulatory Research, Washington, DC, August 2005 [18]. PII has modified the list slightly to account for general industry data and terminology and to incorporate PII internal data.

Human Factor Category	Human Factor Issue/Level	Multiplier for Cognitive & Diagnosis Errors
Available Time (includes staffing Issues) – ***for responses only***	Inadequate time	P(failure)=100%
	Barely adequate time (≈2/3 x nominal)	10
	Nominal time (1x what is expected)	1
	Extra time (at least 2x nominal and >20 min)	0.1
	Expansive time (> 4 x nominal and > 20 min)	0.01
Stress/Stressors	Extreme (threat stress)	5
(includes staffing issues)	High (time pressures such as during a maintenance outage; issues at home, etc.)	2
	Nominal	1
Complexity & Task Design	Highly complex	5
	Moderately complex (requires more than one staff)	2
	Nominal	1
	Obvious diagnosis	0.2
Experience/Training	Low	10
	Nominal	1
	High	0.5
Procedures	Not available in the field as a reference, but should be	20
	Incomplete; missing this task or these steps	8
	Available and >90% accurate, but does not follow format rules (normal value for process industry)	3
	Good, 95% accurate, follows >90% of format rules	1
	Diagnostic/symptom oriented	1
Human-Machine Interface (includes tools)	Missing/Misleading (violates populational stereotype; including round valve handle is facing away from worker)	20
	Poor or hard to find the right device; in the head calc	10
	Some unclear labels or displays	2
	Good	1
Fitness for Duty	Unfit (high fatigue level (>80 hr/wk or >20 hr/day, no day off in 7-day period; or illness, etc.)	20
	Highly degraded fitness (high fatigue such as >15 hr/day, illness, injury, etc.)	10
	Degraded Fitness (>12 hr day and >72 hr/wk)	5
	Slight fatigue (>8 hr per day; ***normal value for process*** industry)	2
	Nominal	1
Work Processes & Supervision	Poor	2
	Nominal	1
	Good	0.8
Work Environment	Extreme	5
	Good	1
Communication	No communication or system interference/damage	10
	No standard for verbal communication rules ***(normal*** value for process industry)	3
	Well implemented and practiced standard	1

3 Human Error Probability for Multiple Executions of a Rule-Based Task

Coupled (dependent) Error Probabilities: Coupling represents the probability of repeating an error (or repeating success) on a second identical task, given that an error was made on the first task. The increased probability of failure on subsequent tasks given that an error has already been made is known as dependence. The list below provides some starting point guidance on values to use:

- 1/20 to1/90 - if the same tasks are separated in time and if visual cues are not present to re-enforce the mistake path. This error probability assumes a baseline error probability of 1/100 with excellent human factors. If the baseline error is higher, then this probability will increase as well.
- 1/2 - if the same two tasks are performed back-to-back, and if a mistake is made on the first step of two. This error probability assumes a baseline error of 1/100 with excellent human factors. If the baseline error is higher, then this probability will increase as well.
- 8/10 to 10/10 - if the same three tasks are performed back-to-back and a strong visual cue is present (that is, the worker can clearly see the first devices he/she worked on), if a mistake is made on the first steps of the three.
- Two or more people become the same as one person (with respect to counting of errors from the group), if people are working together for more than three days; this effect is due to the trust that can rapidly build.

These factors are based on the relationships provided in NUREG-1278 10] and the related definitions of weak and strong coupling provided in the training course by Swain (1993) [19] on the same topic, as shown here in Table 2. The following relationship is for errors of omission, such as failing to reopen a root valve or failing to return an SIF to operation, after bypassing the SIF. The qualitative values in Table 2 are based jointly on Swain (1993) and Gertman [18] (SPAR-H, 2005 which is NUREG/CR-6883).

One can readily conclude that staggering of maintenance tasks for different channels of the same SIF or for related SIFs will greatly reduce the level of dependent errors. Unfortunately, most sites PII visits do not stagger the inspection, test, or calibration of redundant channels of the same SIF or of similar SIFs; the reason they cite is the cost of staggering the staff. While there is a perceived short-term higher cost, the answer may be different when lifecycle costs are analyzed.

<u>Simple Rule:</u> **Staggering of maintenance can prevent a significant number of human errors in redundant channels. In fact, the US Federal Aviation Administration (FAA) requires staggering of maintenance for aircraft with multiple engines or multiple control systems (i.e., hydraulics) (FAA Advisory Circular 120-42B, as part of ETOPS approval [20].** (ETOPS is Extended-range Twin-engine Operational Performance Standards, a rule which permits twin engine aircraft to fly routes which, at some point, are more than 60 minutes flying time away from the nearest airport suitable for emergency landing.)

Table 2: Guideline for Assessing Dependence for a within-SIF Set of Identical Tasks (based partially on SPAR-H, 2005 [18], and partially on field observations by PII)
Courtesy Process Improvement Institute, Inc., All Rights Reserved

Level of Dependence	Same Person	Actions Close in time	Same Visual Frame of Reference (can see end point of prior task)	Worker Required to Write Something for Each Component
Zero (ZD)	No; the similar tasks are performed by different person/group	Either yes or no	Either yes or no	Either yes or no
Zero (ZD)	Yes	No; separated by several days	Either yes or no	Either yes or no
Low (LD)	Yes	Low; the similar tasks are performed on sequential days	No	Yes
Moderate (MD)	Yes	Moderate; the similar tasks are performed more than 4 hours apart	No	No
High (HD)	Yes	Yes; the similar tasks are performed within 2 hours	No	No
Complete (CD)	Yes	Yes; the similar tasks are performed within 2 hours	Yes	Either yes or no

The level of dependency is determined from Table 2 by assessing whether the same person is doing the tasks, the proximity of the actions in time, the proximity of the actions in space (same visual frame of reference), and whether the work is required to make a record for each component:

1. Read down the "Same Person" column and find the applicable row(s), then
2. Read down the "Actions Close in Time" column and find the applicable row,
3. Then check the two columns on the right for that row.
4. The Level of Dependence is shown in the left-most column for the applicable row.

Table 2 has two rows for Zero Dependency (ZD) because ZD can be achieved two different ways:

- Tasks done by different persons or groups (staggered people), or
- Tasks done by same persons or groups, but tasks are done several days apart (staggered times).

Once the level of dependence is known, the probability of either repeat success or repeating errors on identical tasks can be estimated. For these probabilities, we use Table 3, which is a re-typing of Table 20-17 from NUREG-1278 [10] (and the similar table in SPAR-H [18] [Gertman, 2005]).

Table 3. Equations for Conditional Probabilities of Human Success or Failure on Task N, given probability of Success (x) or Failure (X) on Task N-1, for Different Levels of Dependence

Courtesy Process Improvement Institute, Inc., All Rights Reserved

Level of Dependence	Repeating Success Equations (but shown as error probability)	Repeating Failure Equations
Zero (ZD)	$P_{Success@N} = x$	$P_{Failures@N} = X$
Low (LD)	$P_{Success@N} = (1+19x)/20$	$P_{Failures@N} = (1+19X)/20$
Moderate (MD)	$P_{Success@N} = (1+6x)/7$	$P_{Failures@N} = (1+6X)/7$
High (HD)	$P_{Success@N} = (1+x)/2$	$P_{Failures@N} = (1+X)/2$
Complete (CD)	$P_{Success@N} = 1.0$	$P_{Failures@N} = 1.0$

4 Illustrative examples

To illustrate the impact (sensitivity) on PFD_{SIF}, we will look at two simple cases and will not provide the details on the calculation of the component aspects of PFD_{SIF}, but instead will provide the results of PFD_{COMP} to be the value obtained by using Equation 1, but without the systematic error terms (the same as using Equation 8.1 in TR84.00.02-2015). Then we will show a simple way to estimate the specific human error term (PFD_{HUM}) and show the resulting impact on PFD_{SIF}. Figures 3 and 4 show a candidate SIL 1 SIF and a candidate SIL 2 SIF, respectively.

4.1 Example 1 - Illustration of Estimate of PFD_{SIF}, for a SIL 1 SIF, with and without consideration of P_{HUM}

For the SIL 1 SIF in Figure 3, the component PFDs were estimated using standard, simplified equations for each, and using generic data available for the components. Based on this calculation, the PFD of the SIF without consideration of discrete systematic error yielded a $PFD_{COMP} = 0.039$. It is noted that the sensor/transmitter PFD contribution is 0.025; this value will be important in the second example included in Section 4.2.

For this example, the term $\sum P_{HUM}$ is next estimated by summing the

- Probability of leaving the root valve for the level switch (sensor/transmitter) closed
- Probability of leaving the entire SIF in BYPASS after maintenance or after some other human intervention (such as an inadvertent error or as a necessity during startup)
- Probability of miscalibration of the level transmitter/switch.

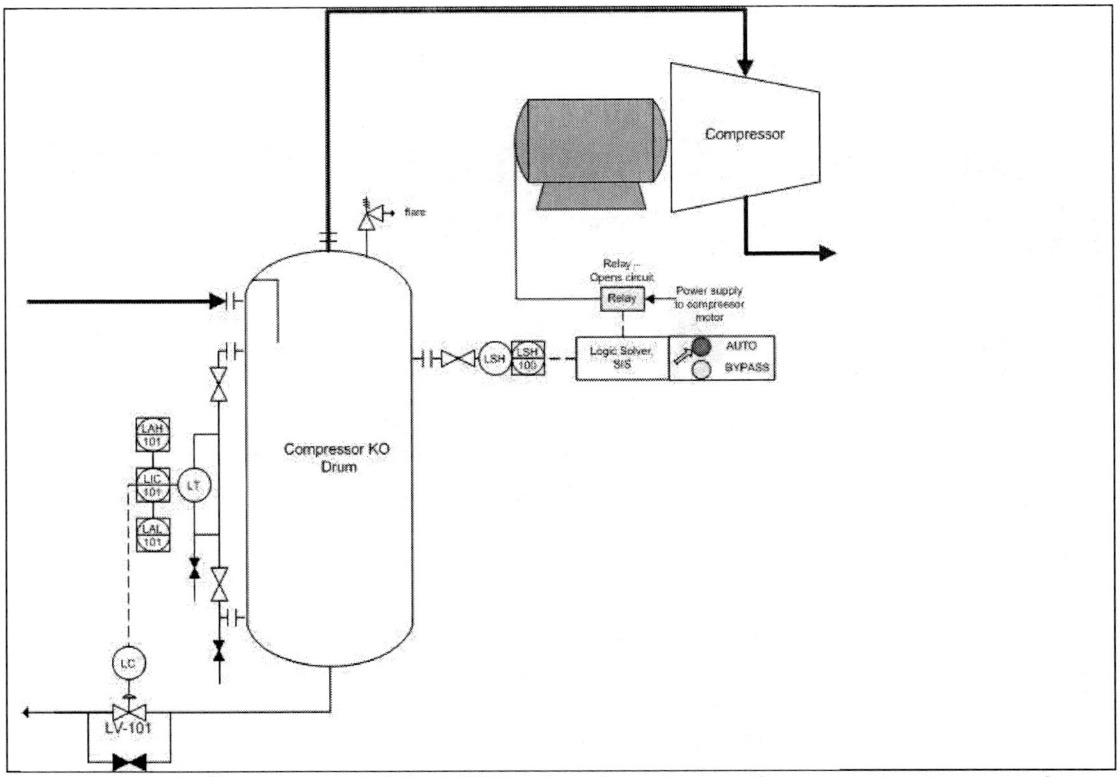

Figure 3. Example of SIL 1 SIF (high level trip of compressor motor)
Courtesy Process Improvement Institute, Inc., All Rights Reserved

Since these are all independent systematic errors, the error probability will simply be 0.02 (the base error probability provided) for each mistake, or:

$$\sum P_{HUM} = 0.02 + 0.02 + 0.02 = 0.06$$

This would then give an overall failure probability for the SIF of:

$$PFD_{SIF} = PFD_{COMP} + PFD_{HUM} = 0.039 + 0.06 = 0.099$$

Since the PFD_{SIF} is less than 0.1, the instrumented system for high level protection still qualifies as a SIL 1 SIF. But, suppose we wish to improve the reliability and independence of the instrumented system by using a smart sensor/transmitter for the high-level switch (LSH) which will detect null movement of the sensor reading (indicating the valve is closed or the tap is plugged) or suppose we put a limit switch (or captive key system) on the root valve. There is a probability that these safeguards against human error will also fail or be bypassed by the staff, but assuming the probability of that failure is the same as other human errors for this example, 0.02, then the overall system human error is reduced, because the probability of leaving the root valve closed is now ANDed with the probability of smart sensor/transmitter or limit switch failing:

$$\sum P_{HUM} = (0.02 * 0.02) + 0.02 + 0.02 = 0.04$$

Therefore, the revised PFD of the instrumented system becomes:

$$PFD_{SIF} = PFD_{COMP} + PFD_{HUM} = 0.039 + 0.04 = 0.079$$

Sensitivity to Baseline Human Error Probability: If the baseline human error probability increases to 0.04 due to fatigue or extra stress due to schedule constraints, then even with the extra instrumentation to detect valve closure, the PFD of the systematic human error will increase substantially:

$$\sum P_{HUM} = (0.04 * 0.04) + 0.04 + 0.04 = 0.082$$

The revised PFD of the instrument system becomes:

$$PFD_{SIF} = PFD_{COMP} + PFD_{HUM} = 0.039 + 0.082 = 0.121$$

In this modified case, which is applicable to about a third of the facilities PII has visited in the past 10 years (due primarily to fatigue), the instrumented system no longer qualifies as a SIL 1.

The human error for miscalibration is challenging to reduce, unless there are redundancy and voting of the level sensor/transmitters; then miscalibration errors can be essentially eliminated as an important contribution to human error. This case will be explored in Section 4.2 as part of Example 2. However, the redundancy and voting of transmitters cannot detect an error introduced by using the same calibrator (hardware) that has some error (drift).

The composite error of leaving the entire system in bypass is usually made up of

(1) the inadvertent error to return the system to AUTO after maintenance (a good design would generate a recurring alarm for an SIF in bypass, and a good management system would track and correct such alarms), and
(2) the probability that the staff will make the intentional decision to leave the SIF in bypass, for perhaps a reason not anticipated by the designers. Management of change (MOC) should address the latter case; a strong MOC system should have multiple personnel who would cross-check each other. A bigger issue is failing to recognize that an MOC is needed.

Therefore, this error probability potentially can be reduced by adding repeating alarms to alert the staff that the SIF is still bypassed. A strong management system is essential, else the staff may hear and acknowledge the alarms, but will leave the system in bypass intentionally. Thus, it is critical that the designers anticipate the need for a bypass (such as during startup) and that they provide an appropriate startup bypass that resets the SIF automatically.

4.2 Example 2 - Illustration of Estimate of PFD_{SIF}, for a SIL 2 SIF, with and without consideration of P_{HUM}

For the SIL 2 SIF described in Figure 4, the component PFDs were estimated using standard, simplified equations for each, and using available industry data for the components. For the case where the sensors are voted 2oo3, the PFD of the SIF without consideration of discrete systematic error yielded PFD_{COMP} = 0.008 (of which the 2oo3 voted sensor portion is 0.0025 and the 2oo3 voted logic solver is 0.003).

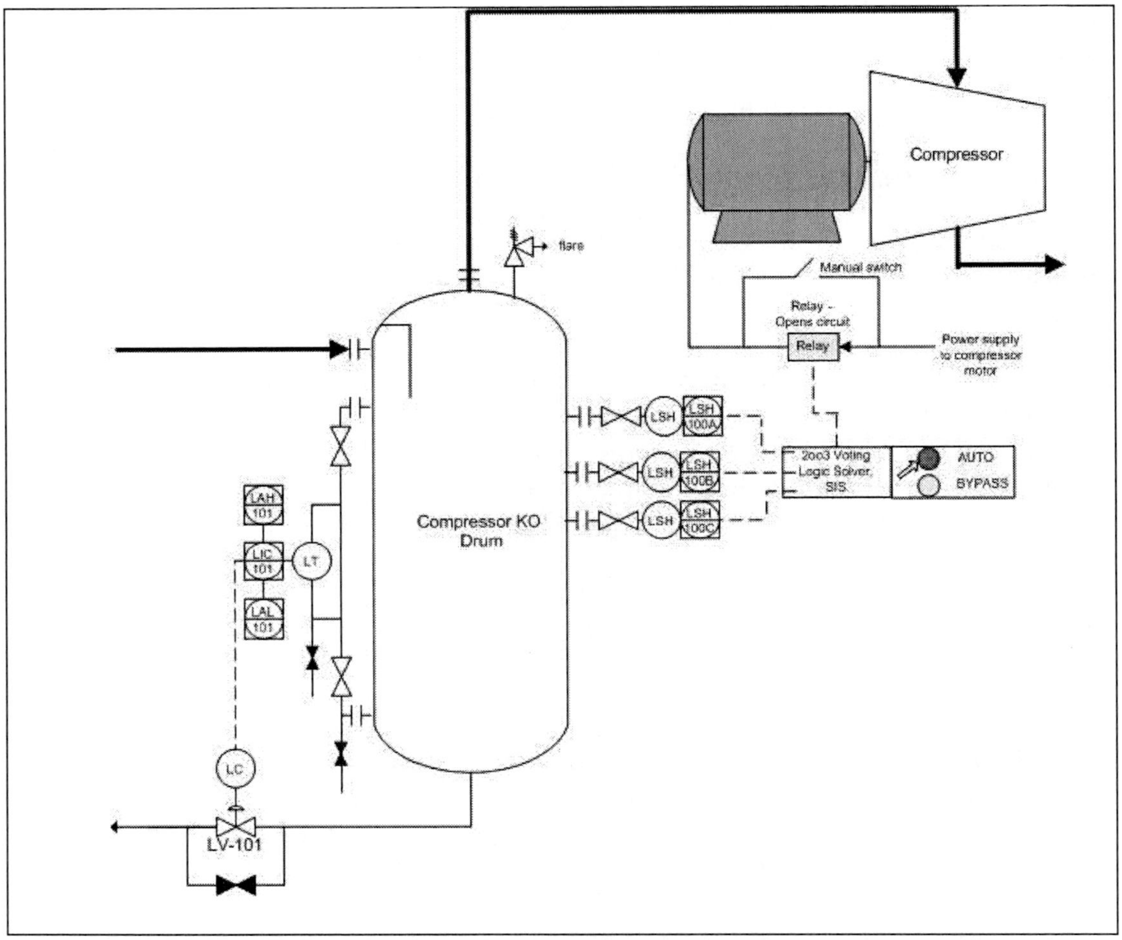

Figure 4. Example of SIL 2 SIF (high level trip of compressor motor)
Courtesy Process Improvement Institute, Inc., All Rights Reserved

For this example, the term $\sum P_{HUM}$ is next estimated by summing the

- Probability of leaving the level sensor/transmitters 2oo3 root valves closed, causing an unsafe failure. (This calculation is shown later).
- Probability of miscalibration of the level transmitter/switch. This calculation is shown later, but for this issue to be a significant probability, two of the three sensors/transmitters must be miscalibrated, unless there is comparison checking, then it would require miscalibration of all three transmitters.

- Probability of leaving the entire SIF in BYPASS after maintenance or after some other human intervention such as an inadvertent error or a necessity during startup; as before, we will use the base error probability of 0.02 as a starting point.
- Probability of leaving the relay bypass closed. As before, we will use the base error probability of 0.02 as a starting point.

$$\sum P_{HUM} = P_{root\ valve} + P_{miscal} + P_{SIF\ bypass} + P_{relay\ bypass}$$

4.2.1 Root valves closed

To aid in the calculation of the probability of leaving 2oo3 root valves closed, we use an event tree to show the conditional probabilities for leaving Valve B closed, given Valve A is open or closed, and similarly, the conditional probability of leaving Valve C closed, given Valve A or B are closed or both Valve A and B are closed. Figure 5 shows the results of this calculation. For the branch probabilities, the equations for high dependency (HD) of the human actions were used (see Table 3); this tree reflects the more prevalent case of redundant channels being maintained on the same day, by the same person, and that level valves are within the visual field of the worker. From Figure 5 the result for the probability of human error of leaving 2oo3 or 3oo3 of the root valves closed is 0.01765. But, the comparison checking between sensors/transmitters will alert the workers that a root valve is closed (assuming 100% detection and correction), so the only valid path is the 3oo3 path; the 3oo3 error case is the bottom row of the event tree in Figure 5. The probability of leaving all three root valves closed is 0.0077.

4.2.2 Sensors Miscalibrated

From the same figure, we can also extract the conditional probability of leaving 3oo3 sensors/transmitters bypassed; assuming comparison checking is in place to note deviations and correct the problem, only the case of 3oo3 errors is applicable. This represents a strong recovery path for the previous errors (100% detection and correction). The 3oo3 error case is the bottom row of the event tree in Figure 5. The probability of miscalibrating all three sensors/transmitters is 0.0077.

$$\sum P_{HUM} = P_{root\ valve} + P_{miscal} + P_{SIF\ bypass} + P_{relay\ bypass}$$
$$= 0.0077 + 0.0077 + 0.02 + 0.02 = 0.055$$

This contribution would then give an overall failure probability for the SIF of:

$$PFD_{SIF} = PFD_{COMP} + PFD_{HUM} = 0.008 + 0.055 = 0.063$$

Since the PFD_{SIF} is greater than 0.01, the instrumented system for high level protection in this example does not qualify as a SIL 2 SIF when accounting for human error probabilities related to interventions with the SIF.

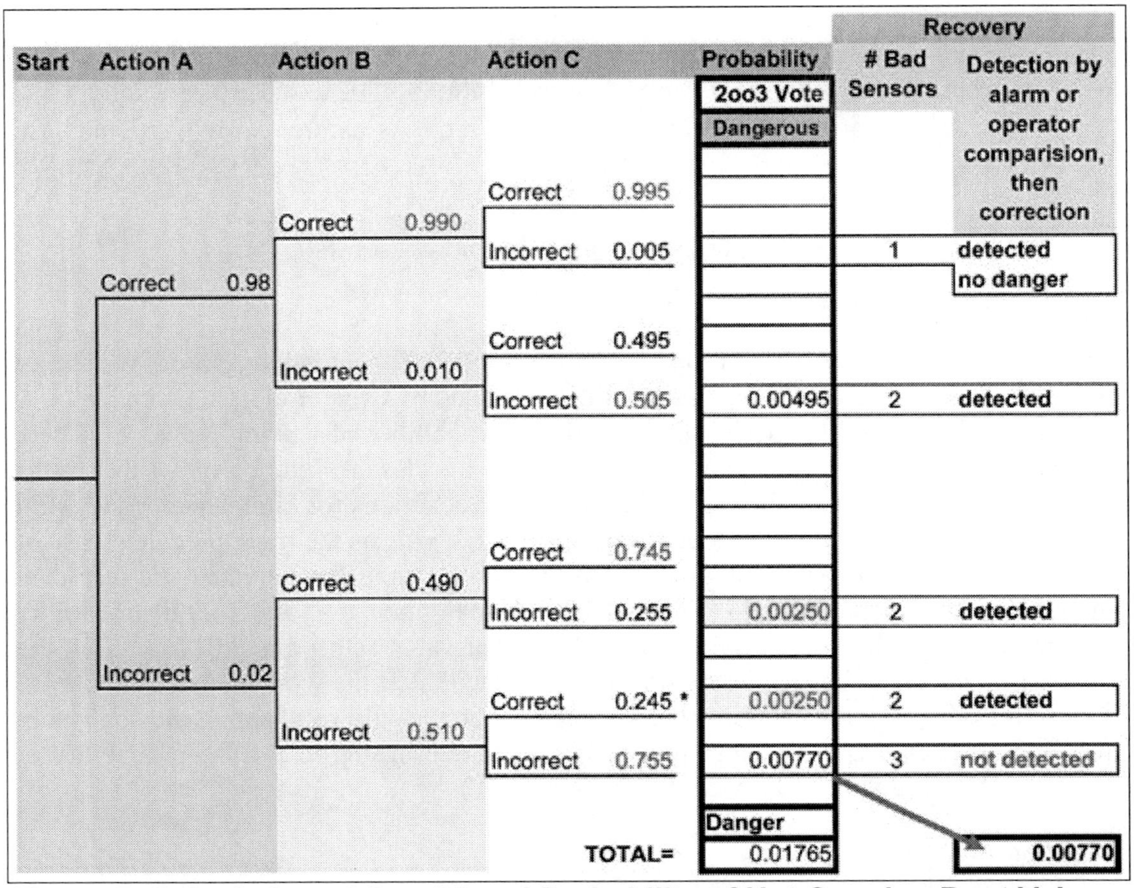

Figure 5. Calculation of Conditional Probability of Not Opening Root Valves; with the Last Column Showing the Probability of Leaving Two or Three Valves Closed (using High Dependence Equations)

Courtesy Process Improvement Institute, Inc., All Rights Reserved

Action Examples:

Action A: open Sensor A root valve after maintenance.
Action B: open Sensor B root valve after maintenance.
Action C: open Sensor C root valve after maintenance.

Action A: calibrate Sensor A.
Action B: calibrate Sensor B.
Action C: calibrate Sensor C.

Action A: remove bypass from device A.
Action B: remove bypass from device B.
Action C: remove bypass from device C

4.2.3 Detection and Correction of Sensor Errors

One means to improve the reliability and independence of the instrumented system is to use a smart sensor/transmitter for the LSH which will detect null movement of the sensor reading, indicating the valve is closed on the tap is plugged. Another possibility is to implement a limit switch (or captive key system) on the root valve. There is a probability that these safeguards against human error will also fail or be bypassed by the staff, but assuming the probability of that failure is the same as other human errors for this example, 0.02, then the systemic human error drops to about zero as the probability of leaving the root valve closed is now ANDed with the probability of smart sensor/transmitter or limit switch failing, as shown in Figure 6.

$$\sum P_{HUM} = P_{root\ valve} + P_{miscal} + P_{SIF\ bypass} + P_{relay\ bypass}$$

$$= 0.0000 + 0.0077 + 0.02 + 0.02 = 0.048$$

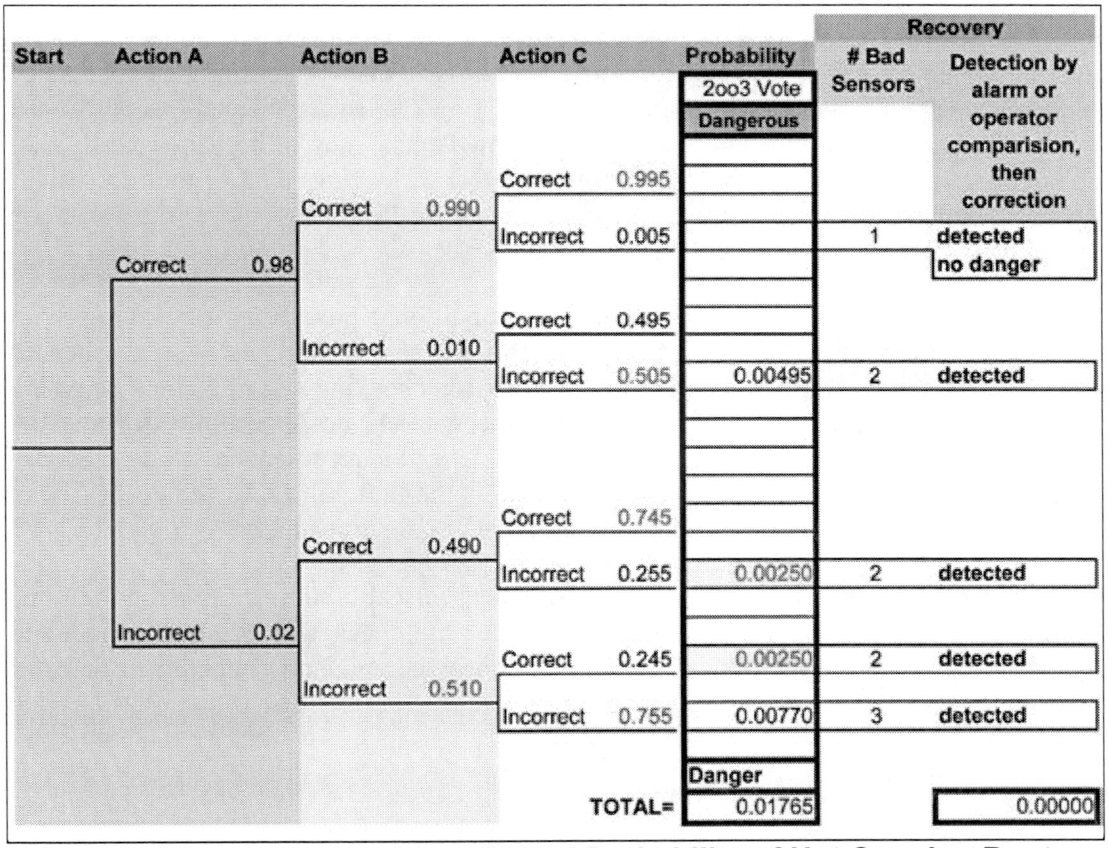

Figure 6. Calculation of Conditional Probability of Not Opening Root Valves; with the Last Column and Last Row Showing the Probability of Leaving Three Valves Closed, and with a Limit Switch or Smart Sensor to Check that Each Root Valve is Open.

Courtesy Process Improvement Institute, Inc., All Rights Reserved

Action Examples:

Action A: open Sensor A root valve after maintenance.
Action B: open Sensor B root valve after maintenance.
Action C: open Sensor C root valve after maintenance.

In this case the revised PFD of the instrument system becomes:

$$PFD_{SIF} = PFD_{COMP} + PFD_{HUM} = 0.008 + 0.048 = 0.056$$

Since the PFD_{SIF} is still greater than 0.01, the instrumented system for high level protection still does not qualify as a SIL 2 SIF when accounting for human error probabilities related to interventions with the SIF. But, we have reduced the errors related to dependent failures during checking of the sensors/transmitters as much as possible.

4.2.4 Staggered maintenance

As another alternative (instead of using smart sensors/transmitters or instead of installing limit switches on the root valves) we can reduce potential dependent human error by staggering maintenance activities across different shifts. This would drop the dependence to Low. The dependent error calculations using the Low Dependence equations of Table 3 is shown in Figure 7. From Figure 7, assuming low dependency of human error (LD on Table 3), the result for the probability of human error of leaving 3oo3 of the root valves closed is 0.00016 (assuming that comparison of sensor readings alerts the workers that one root valve is closed -- 100% detection and correction).

From the same figure, we can also extract the conditional probability of leaving 3oo3 sensors/transmitters miscalibrated. As before, only the case of 3oo3 errors is considered applicable, since it was assumed that sensor comparison checking was implemented where any transmitter not miscalibrated will provide the workers an opportunity to note the deviation and take corrective action to fix the problem; this represents a strong recovery path for the previous errors. The 3oo3 error case is the bottom row of the event tree in Figure 7. The probability of miscalibrating all three sensors/transmitters is 0.00016.

Note that sensor comparison checking will not detect miscalibration from using the same calibrator that has an error (drift). If the calibrators are checked frequently enough and corrected, staggering calibration would enable sensor comparison checking to detect a miscalibration.

$$\sum P_{HUM} = P_{root\ valve} + P_{miscal} + P_{SIF\ bypass} + P_{relay\ bypass}$$

$$= 0.00016 + 0.00016 + 0.02 + 0.02 = 0.040$$

This would then give an overall failure probability for the SIF of

$$PFD_{SIF} = PFD_{COMP} + PFD_{HUM} = 0.008 + 0.040 = 0.048$$

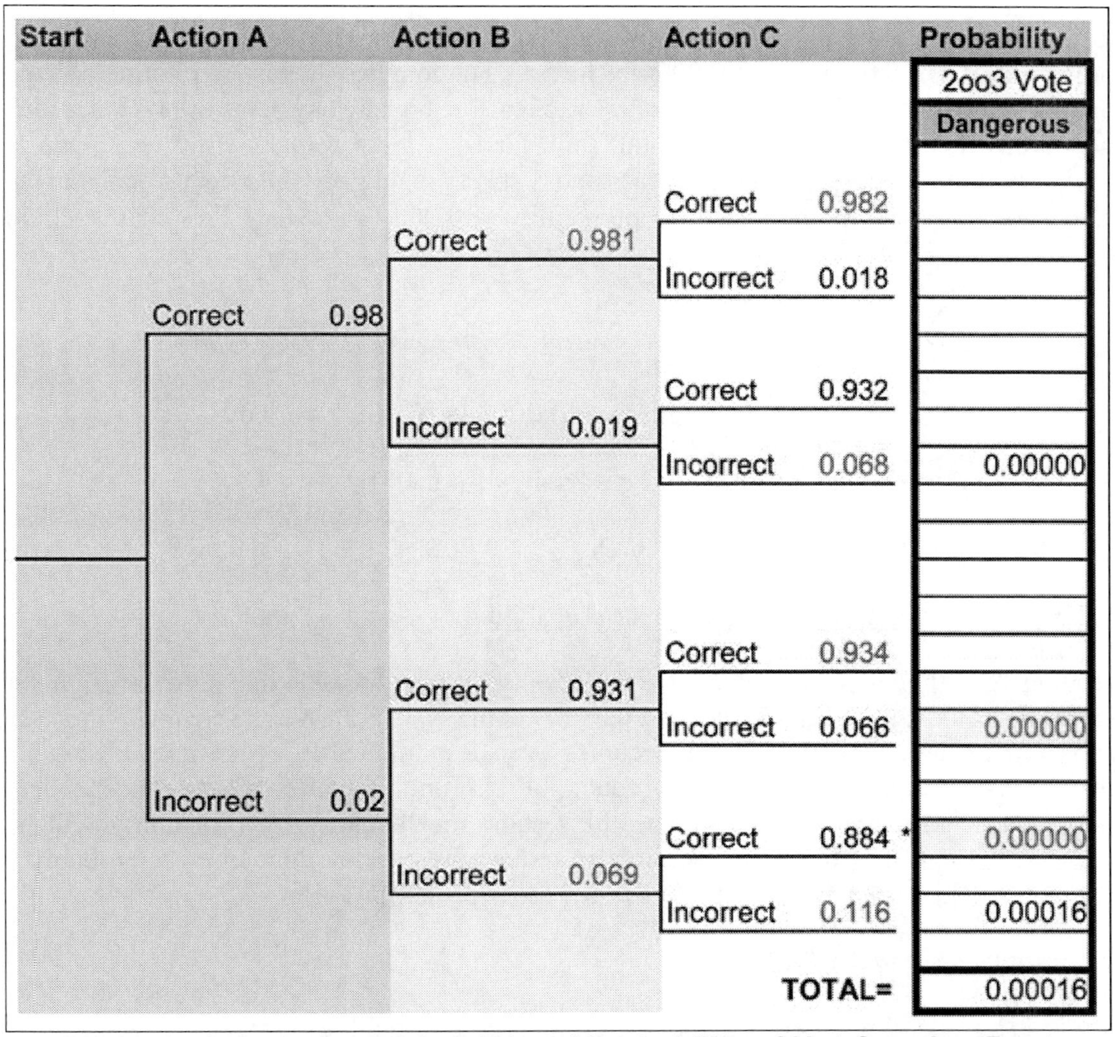

Figure 7. Calculation of Conditional Probability of Not Opening Root Valves; with the Last Column Showing the Probability of Leaving Two or Three Valves Closed (Using Low Dependence Equations)

Courtesy Process Improvement Institute, Inc., All Rights Reserved

Action Examples:
Action A: open Sensor A root valve after maintenance.
Action B: open Sensor B root valve after maintenance.
Action C: open Sensor C root valve after maintenance.

Since the PFD_{SIF} is greater than 0.01, the instrumented system for high level protection still does not qualify as a SIL 2 SIF when accounting for human error probabilities related to interventions with the SIF. The weak link in this design is again the human error probability of leaving either the relay bypass closed or the probability of leaving the entire SIF bypassed. This is a common concern on all SIF that have system bypasses. The most effective way to drop these error probabilities is to eliminate the capability for bypassing the relay and to eliminate the capability for bypassing the entire SIF. Or; we can install a parallel relay with a selector switch so that one relay (and only one) is aligned in the circuit to the motor of the compressor. This will likely drop the relay systemic human error probability from 0.02 down to 0.0004 or lower. The toughest bypass to eliminate is the one for the entire SIF. It is thought to be only feasible on batch systems or on continuous operations that can be shut down completely for each test interval. However, again, a design with a recurring alarm on bypass and a strong management system to correct the alarm could reduce the bypassed SIF probability to 0.0004 or lower. This change could achieve SIL 2 PFD.

$$\sum P_{HUM} = P_{root\ valve} + P_{miscal} + P_{SIF\ bypass} + P_{relay\ bypass}$$

$$= 0.00016 + 0.00016 + 0.0004 + 0.0004 = 0.001$$

This would then give an overall failure probability for the SIF of

$$PFD_{SIF} = PFD_{COMP} + PFD_{HUM} = 0.008 + 0.001 = 0.009$$

4.2.5 Sensitivity to baseline human error probability:

Obviously, if the baseline human error probability increases to 0.04 due to extra fatigue or extra stress due to schedule constraints, the PFD of the systematic human error will increase substantially and the SIL 2 target becomes even less attainable. Likewise, if suitable operational discipline is adopted to reduce the baseline human error with independent performance measurement to validate the results, the human error probability will be reduced (though it is likely not possible to reduce the baseline human error probability enough to achieve a SIL 2 target, if a SIF bypass is present).

4.3 Example Summary Table

Table 4 shows a summary of key parameters and results from the examples. The examples illustrate the effect of changes in the base human error probability for typical ways that human error can compromise an SIF. The table also shows some examples of error detection and correction that reduce the PFD.

5 Ways to Detect and Manage Human Error for SIFs

Here is a brief summary of techniques that PII has seen employed to detect and manage human error:

- Compare process values for redundant transmitters and alarm on deviation.
- Use logic solver features to detect shorts (bypasses) on de-energize to trip switches (level, pressure, temperature, etc.).
- Avoid blind switches except for tuning fork type level. A level switch at a fixed point can usually detect high level at the same fluid height even if the density changes.
- Provide position switch on bypass valves around final element valves with operator response to alarm to correct, or with shutdown.
- Staggered maintenance as discussed in prior sections.

- Strong maintenance procedures to ensure root valves are open, and all bypasses are removed from the logic solver. Provide alarms for any bypasses in the logic solver.
- Use transmitter diagnostics to alarm on frozen output as it will not only detect hardware failure, but will also detect closed root valves.
- Strong checklist procedures for operations to confirm root valves are open before startup. Requires auditing to determine the human error probabilities. If car seals or locks are used to confirm that root valves and bypass valves are in the correct position, provide a management system that includes auditing to determine the human error probability.
- Strong checklist procedures for operations to confirm sensors are responding correctly during startup. Requires auditing to determine the human error probabilities.
- Design startup bypasses that are automatically removed.
- Use alarm management to improve scenarios of operator response to alarms for extended periods of SIF bypass.

These techniques, if managed well, if documented, and if audited, can be evaluated quantitatively determine the human error probability effects that increase the PFD_{SIF} and the error recovery probability effects that reduce the PFD_{SIF}.

6 Conclusion

As can be seen from the quantitative examples, systematic errors have the potential to significantly impact a SIF in a negative manner. In addition, SIL verifications performed today often do not account for this contribution to probability of failure. In such cases, it becomes increasingly likely that the risk reduction assumed by analysts (who rely upon a SIL 2 to have a PFD of 0.01 or lower) is not sufficient to satisfy corporate risk criteria when the actual risk reduction estimated for the IPL is being counted on, such as an SIF replacing a relief valve, as opposed to analyses that are simply performed on a comparative basis where consistency is more important than the actual numbers.

The paper points to the need for companies to begin:

- Accounting for systematic error (and especially human error probability) in SIL Verifications; otherwise the risk reduction factor from this IPL will be unrealistically optimistic.
- Taking a more in-depth look at the management systems and work processes in place for operations and maintenance and their associated training and revalidation of performance.

Using the mathematics presented, companies can gain insight concerning the relative effectiveness of their practices and find areas where improvements can be made with minimal additional cost. Just as improved human factors improve safety, improved safety performance, if done properly with true operational discipline, should also improve reliability and plant availability.

Table 4: Summary Table for the Examples

Courtesy Process Improvement Institute, Inc., All Rights Reserved

	Example	4.1			4.1			4.2			4.2		
	Figure	3			3			4, 5, 6			4, 7		
		1st Pass	Detect & Correct the Error		1st Pass	Detect & Correct the Error		1st Pass	Detect & Correct the Error		1st Pass	Detect & Correct the Error	
	Maintenance							High Dependency		High Dependency	Low Dependency (Staggered)		Low Dependency (Staggered)
	Baseline Human Error Probability	0.02		0.02	0.04		0.04	0.02		0.02	0.02		0.02
	Target SIL	1		1	1		1	2		2	2		2
	PFD_{COMP}	0.039		0.039	0.039		0.039	0.008		0.008	0.008		0.008
PFD_{HUM} from Human Errors	$PFD_{ROOT\ VALVES}$	0.02	**Op Resp to ZSO (Not Open)**	**0**	0.04	**Op Resp to ZSO (Not Open)**	**0**	0.0077	**Op Resp to root valve Not Open**	**0**	0.00016		0.00016
	$PFD_{SIF\ BYPASS}$	0.02		0.02	0.04		0.04	0.02		0.02	0.02	**Improve P of bypass for SIF and Relay**	**0.0004**
	$PFD_{RELAY\ BYPASS}$							0.02		0.02	0.02		**0.0004**
	PFD_{MISCAL}	0.02		0.02	0.04		0.04	0.0077		0.0077	0.00016		0.00016
	Total $PFD_{SYS-HUM}$	0.06		0.04	0.12		0.08	0.055		0.048	0.040		0.001
	Achieved PFD_{SIF}	0.099		0.079	0.159		0.119	0.063		0.056	0.048		0.009
	Achieved SIL	1		1	1		1	1		1	1		2

7 Acronyms Used

1oo2	One out of two voting architecture
1oo3	One out of three voting architecture
2oo3	Two out of three voting architecture
3oo3	Three out of three voting architecture
λ	Failure Rate, units of time^{-1}
ANSI	American National Standards Institute
CD	Complete Dependence
COMP	Random hardware failure contributions to overall PFD
D	Dangerous
ETT	Energize To Trip
ETOPS	Extended-range Twin-engine Operational Performance Standards
F	Failure/error term
FE	Final Element
FAA	US Federal Aviation Administration
HD	High Dependence
HRA	Human Reliability Analysis
IPL	Independent Protection Layer
ISA	International Society of Automation
LD	Low Dependence
LS	Logic Solver
LOPA	Layer of Protection Analysis
MOC	Management of Change
P	Probability
PES	Programmable Electronic System
PFD	Probability of Failure (dangerous) on Demand
PII	Process Improvement Institute, Inc.
PS	Power supply
PSV	Pressure Safety Valve
S	Sensor
SS	Required Support Systems, such as a power supply in ETT
SIF	Safety Instrumented Function
SIL	Safety Integrity Level
SIS	Safety Instrumented System
SRS	Safety Requirements Specifications
SYS	Systematic failure contributions to overall PFD
HUM	Systematic errors and failures generated by human error
SYS-PROC	Systematic errors and failures generated randomly by the process
TI	Proof Test Interval, units of time
TR	Technical Report (such as for ISA technical reports)
ZD	Zero Dependence

8 References

8.1 References cited

[1] W.G. Bridges and H.W. Thomas, "Accounting for Human Error Probability in SIL Verification Calculations", 8th Global Congress on Process Safety, Houston, TX, April 1-4, 2012, American Institute of Chemical Engineers.

[2] IEC 61508, Functional Safety of Electrical / Electronic / Programmable Electronic Safety-Related Systems, International Electrotechnical Commission (IEC), 2010, Geneva, Switzerland.

[3] U.S. Chemical Safety and Hazard Investigation Board, *Investigation Report, Pesticide Chemical Runaway Reaction Pressure Vessel Explosion*, Report No. 2008-08-I-WV, January 2011.

[4] IEC 61511-1:2016+AMD1:2017 CSV, Consolidated version: Functional safety - Safety instrumented systems for the process industry sector - Part 1: Framework, definitions, system, hardware and application programming requirements, International Electrotechnical Commission (IEC), 2010, Geneva, Switzerland.

[5] ANSI/ISA-61511-1-2018 / IEC 61511-1:2016+AMD1:2017 CSV, Functional Safety – Safety Instrumented Systems for the Process Industry Sector – Part 1: Framework, definitions, system, hardware and application programming requirements (IEC 61511-1:2016+AMD1:2017 CSV, IDT), International Society of Automation, Research Triangle Park, North Carolina.

[6] ISA-TR84.00.02-2015, *Safety Integrity Level (SIL) Verification of Safety Instrumented Functions*, International Society of Automation, Research Triangle Park, North Carolina.

[7] ISA-TR84.00.02-2002, Safety Instrumented Functions (SIF)-Safety Integrity Level (SIL) Evaluation Techniques Part 2: Determining the SIL of a SIF via Simplified Equations, International Society of Automation, Research Triangle Park, North Carolina.

[8] CCPS, *Layer of Protection Analysis: Simplified Process Risk Assessment*, Center for Chemical Process Safety, American Institute of Chemical Engineers, New York, NY, 2001.

[9] A.M. Dowell, III, "Understanding IPL Boundaries", to be published in Process Saf Prog, March 2018. (Originally prepared for presentation at the American Institute of Chemical Engineers 2018 Spring Meeting, 14th Global Congress on Process Safety, Orlando, Florida, April 23–25, 2018).

[10] A. Swain and H. Guttmann, *Handbook of Human Reliability Analysis with Emphasis on Nuclear Power Plant Applications, Sandia National Laboratories*, 1983 [this document became NUREG/CR-1278– The Human Reliability Handbook, guidelines from the US NRC on Human Reliability Analysis].

[11] STTD. WSRC-TR-93-581, H.C. Benhardt et al, "Savanah River Site, Human Error Data Base Development for Nonreactor Nuclear Facilities (U)", Westinghouse Savannah River Company, Aiken, SC, February 28, 1994.

[12] W.G. Bridges and G.M. Collazo, "Human Factors and Their Optimization", 8th Global Congress on Process Safety, Houston, TX, April 1-4, 2012, American Institute of Chemical Engineers.

[13] U.S. Chemical Safety and Hazard Investigation Board, *Investigation Report, Refinery Explosion and Fire*, Report No. Report No. 2005-04-I-TX, March 2007.

[14] CCPS, *Guidelines for Independent Protection Layers and Initiating Events in Layer of Protection Analysis*, Center for Chemical Process Safety, American Institute of Chemical Engineers, New York, NY, 2012.

[15] J.V. Bukowski, Results of Statistical Analysis of Pressure Relief Valve Proof Test Data Designed to Validate a Mechanical Parts Failure Database, Technical Report, exida, Sellersville, PA, 2007.

[16] J.V. Bukowski and W.M. Goble, Villanova University, *Analysis of Pressure Relief Valve Proof Test Data: Findings and Implications*, 10th Plant Process Safety Symposium, American Institute of Chemical Engineers, 2008.

[17] J.V. Bukowski and W.M. Goble, Villanova University, *Analysis of Pressure Relief Valve Proof Test Data*, Process Saf Prog, American Institute of Chemical Engineers, March 2009.

[18] D. Gertman, H. Blackman, J. Marble, J. Byers, and C. Smith, *The SPAR-H Human Reliability Analysis Method*, NUREG/CR-6883, U.S. Nuclear Regulatory Commission, Office of Nuclear Regulatory Research, Washington, DC, August 2005.

[19] A. Swain, *Human Reliability Analysis*, Training Course, ABS Consulting (formerly JBF Associates), 1993.

[20] US Federal Aviation Administration, "AC 120-42B - Extended Operations (ETOPS and Polar Operations)", Washington, DC, June 13, 2008

8.2 Additional References

[1] A. Swain, Accident Sequence Evaluation Program (ASEP): Human Reliability Analysis Procedure, NUREG/CR-4772, U.S. Nuclear Regulatory Commission, Office of Nuclear Regulatory Research, Washington, DC, February 1987.

[2] Human Error Repository and Analysis (HERA) System, NUREG/CR-6903, U.S. Nuclear Regulatory Commission, Office of Nuclear Regulatory Research, Washington, DC, 2006.

Facility Siting, a Practical Approach

Presenter/ Yaneira Saud, CSP
ERM
840 West Sam Houston Parkway North, Suite 600, Houston, Texas
yaneira.saud@erm.com

Richard MacNguyen
ERM
Richard.MacNguyen@erm.com

Alejandro Jimenez, P.E.

Prepared for Presentation at
American Institute of Chemical Engineers
2019 Spring Meeting and 15th Global Congress on Process Safety
New Orleans, LA
March 31 – April 3, 2019

Facility Siting, a Practical Approach

Presenter/ Yaneira Saud, CSP
ERM
840 West Sam Houston Parkway North, Suite 600, Houston, Texas yaneira.saud@erm.com

Richard MacNguyen
ERM
Richard.MacNguyen@erm.com

Alejandro Jimenez, P.E

Keywords: facility siting, fire, explosion, toxic effect, risk, hazardous materials

Abstract

Under Process Safety Management Regulation, every facility is required to acknowledge or consider "facility siting" in the process hazard analysis (i.e. HAZOP), but does not provide a clear guidance on what is a proper approach to consideration and what is adequate. This requirement can be easily overlooked during the design, leading to decisions considered during late project stages, even after the intended permanently occupied buildings are already in place and mitigation measures are extemporaneous or budget prohibitive.

The American Petroleum Institute (API) and the Center for Chemical Process Safety (CCPS) has invested significant effort in develop diverse guidelines and literature, for example API developed RP-752/753 and CCPS develop the Guidelines for Facility Siting and Layout. Although these resources provide a good starting point and a sound process via practical flowcharts, the recommended practices leave a lot of room in the selection of adequate technical methods needed to complete the assessment; this is a gap particularly when required answers call for quantitative event simulation and frequency calculation. Some challenges medium and small operators face are for example, once the API screening flowchart is applied, and qualitative assessment is completed; but the building does not meet siting criteria for explosion or toxic release, what next? How detailed should the detailed analysis be? Can radiation and blast loads be determined without a full blown quantitative risk analysis (QRA) that is in most cases budget prohibitive? How much of sound analysis can be can be done with a constrained budget? In absence of benchmark public industrial development individual risk criteria, are there local/corporate siting decision criteria for decision making?

An overall practical quantitative method is discussed in this paper including four stages:

1. Stage 1: Defining building occupancy, building functionality, identifying buildings of concern and conducting spacing review

2. Stage 2: Review the hazard sources in more detail to identify potential impact to target buildings, by using recognized risk index methodologies to classify the sources in accordance with their risk: for example Dow's Fire & Explosion Index (F&EI). Screen out buildings outside Intermediate-Heavy-Severe Radius of Exposure
3. Stage 3: Perform optimized QRA only for the sources and targets affecting the buildings that were not screened-out during Stage II
4. Stage 4: Demonstrate Risk-based or As Low as Reasonably Practicable (ALARP)
    a. Tolerability verification: occupied buildings risk is below tolerable frequency for workers (as dictated by client or suggested by best practice)
    b. Risk-based/ALARP verification: All options to reduce risk to occupied buildings identified and assessed and every option not grossly disproportionate was discussed, assessed in terms of cost-benefit, and implemented

This 4-staged approach is flexible: Can be applied for a variety of process and multiple facilities and scope can be managed (i.e. when to stop and what to evaluate). This approach is cost-effective: A staged process allows screening-out what does not contribute to risk. It also strengthens qualitative assessment and reduces the scope of Stage 3.

1 Introduction

Under Process Safety Management Regulation (e.g., 29 CFR 1910.119), every regulated facility is required to acknowledge or consider "facility siting" during the process hazard analysis, but the regulation does not provide guidance on the proper approach and what level of risk is acceptable. This requirement can be easily overlooked during the design, leading to inadequate decisions considered during late project stages, even after the permanently occupied buildings are already in place and mitigation measures are poorly implemented or budget prohibitive.

The American Petroleum Institute (API) and the Center for Chemical Process Safety (CCPS) has invested significant effort in develop diverse guidelines and relevant literature. Although these resources provide a good starting point and a sound process via practical flowcharts, the recommended practices leave much room for interpretation, for example, around the selection of adequate technical methods needed to complete the assessment. This is a gap particularly when required answers call for quantitative event simulation and frequency calculation

1.1 Different approaches

As indicated previously, the process safety regulations do not provide a clear guidance on what approach is considered adequate, leaving an open-ended discussion on what method is adequate for a given facility. On that matter, the CCPS provides a clarification in their Guidelines for Risk Based Process Safety book indicating that the adequacy of the hazard identification and risk analysis approaches should depend on perceived risk, resources and organizational culture. In that sense, the general approach should comply with this premise. In order to determine the general approach, the current available approaches should be identified. Currently there are three main facility siting approaches, these are described in the following sections.

1.1.1 Spacing tables

Spacing tables provide minimum distances between different elements of a process facility, including occupied buildings and types of process units. Spacing tables show the recommended minimum separation distances between different components in a process plant. Table 1 shows the benefit and disadvantages of this method.

Table 1. Benefit and disadvantages of spacing tables

Benefit	Disadvantages

• Easy to apply
• Only requires layout and tables
• Distances considers past incidents
	• Minimum distances typically consider fire events only. Distances do not consider:
	    o Toxic clouds
	    o Flammable cloud
	    o Explosions
	    o BLEVE

1.1.2 Consequence-based

This method is based on calculating the potential effect produced by the worst-case accident to determine any potential impact to any location of concern, e.g., buildings or other process units. The calculations uses empirical and mathematical correlations to model these different potential accidents, e.g. determining the effects of hazardous material releases at the defined targets, which may include buildings, equipment, etc. Table 2 shows the benefit and disadvantages of this methodology

Table 2. Benefit and disadvantages of consequence-based approach

Benefit	Disadvantages

• Modeling is specific to the conditions in the plant
• Considers events not included in spacing tables:
    o Toxic/flammable cloud
    o Explosion
    o BLEVE
	• Requires a skilled analyst to perform consequence modeling
	• Identification of sources and specific accident events is required
	• Requires identification of worst-case events, which is often not clear and may be misidentified
	• May require specific software (e.g. PHAST, Effects, Canary, CFD, etc.)
	• Indicates only maximum impact distances for the worst-case scenario, but does not include the likelihood of occurrence, which is needed to determine the risk

1.1.3 Quantitative risk analysis (QRA)

The QRA is a structured analysis that considers the frequency and consequence of an event determining the risk of an operation or activity. As indicated, the QRA requires numeric calculations of the frequency of occurrence of all potential events, simulates the effects of all potential accident events (not only worst-case), evaluating the impact on defined receptors (e.g. buildings, equipment, etc.), and performing a risk calculations to determine the risk. Table 3 shows the benefit and disadvantages of this method.

Table 3. Benefit and disadvantages of QRA

Benefit	Disadvantages
• Assessment is specific to the process conditions and design • Gives a complete risk picture of defined targets • More detailed results to support decision making • Allows for ranking of targets in order of risk exposure • Ranking risk contribution of hazards, allowing focused risk control on targets or hazards with higher risk, producing a higher risk reduction	• Requires a skilled analyst to perform more complex consequence modeling and frequency estimation • Requires more information and resources • Requires specific software for consequence modeling (e.g. PHAST, Effects, Canary, CFD, etc.) • Requires calculation tools to perform the risk calculation (e.g. Safeti, Checkers, etc.) • Takes more time to perform the evaluation, which is typically more costly

2 The Challenge

Despite the best practices developed by API and CCPS, some companies, especially medium and small operators, face diverse challenges when implementing facility siting. The majority of these challenges involved the interpretation and implementation of the recommended practices. Other challenges involved a correct application of the risk management principles.

2.1 Recommended practices implementation

Many of the recommended practices are performance-based approaches rather than prescriptive, requiring that organizations define their objective. However, the correct performance objective is not clear for many companies, making it difficult to follow and fully implement the recommended practices. Some of the main challenges faced by these companies involved the following items:

- How do I determine which buildings need to be evaluated in the siting study?
- What is considered an occupied building?
- The first gate of API RP 752 is about whether the building is impacted by an event (fire, explosion, toxic). How do I determine this if all I have is a PHA, which is based on team discussion and not calculations?
- We did a review based on spacing tables, is that enough? What do we do if we did not meet the

distances?
- A qualitative assessment is completed but the building does not meet siting criteria for explosion or toxic, what is next?
- What are siting criteria? In the absence of benchmark public industrial development criteria, what siting decision criteria do I use for decision-making?
- API RP 752 includes technical terms, such as blast evaluation and thermal effects., How can we reduce the cost of these evaluations and still be confident that the appropriate events were considered?
- What is involved in the more detailed analysis mentioned in API RP 752?

2.2 Risk management principles

Risk management is the process to ensure that risk to employees, the public and the environment is consistently controlled within the organization's risk tolerance. In that sense, the facility siting assessment should answer the following questions:

- Hazard – what can go wrong?
- Consequence – How bad could it be?
- Likelihood – How often might it happen?
- Risk – is the risk within the risk tolerance

Many companies do not have a defined risk tolerance, and the resources to properly answer the questions defined above require an external party to complete the assessment. Furthermore, some organizations struggle to contract the facility siting study due to limited budget, which provides a challenge on performing a credible study without a comprehensive QRA, which isoften budget prohibitive. In these cases, the typical doubt includes:

- How much analysis can be done with a constrained budget?
- How detailed should the detailed analysis be? Can radiation and blast loads be determined without a comprehensive QRA?
- In absence of benchmark public industrial development individual risk criteria, are there local/corporate siting decision criteria for decision-making?

3 A Practical Approach

This paper proposes and discusses an optimized practical approach to manage the risk properly, considering the challenges identified in Section 2.

The general approach includes four stages:

1. Stage 1: Define building occupancy and functionality, identify buildings of concern and conduct a spacing review
2. Stage 2: Review the hazard sources in more detail to identify potential impact to target buildings, allowing to exclude from the assessment the process and buildings with no impact or low risk, based on a recognized risk index
3. Stage 3: Perform optimized QRA only for the sources and targets affecting the buildings that

were not screened-out during Stage 2,

4. Stage 4: Demonstrate Risk is As Low as Reasonably Practicable (ALARP) level
    a. Tolerability verification: occupied buildings risk is below tolerable frequency for workers (as dictated by client or suggested by best practice)
    b. ALARP verification: Identify and assess all options to reduce risk to buildings of concern. This discussion and assessments in terms of cost-benefit and implementation.

This approach allows eliminating buildings and process that do not contribute significantly to the risk, advancing to a more detailed analysis only considering the inventories and buildings with higher potential risk. Additionally, this approach utilizes each method identified in section 1. Table 4 shows a summary of the intention of each stage.

Table 4. Overall summary of the different stages of the approach

	Stage 1	Stage 2	Stage 3	Stage 4
Intention	Determine buildings that should be considered and the potential scenarios involved and review if there is a potential impact to the buildings	Evaluate events with a higher degree of assessment to determine initial risk controls and if further analysis is required	Determine Risk of buildings which were not excluded on previous stages	Demonstrate risk controls as adequate and ALARP
Events Considered	Fire (spacing tables) However, other events are identified, but no direct relation is establish between these and building impact	various	All	All
Basis	Spacing tables	Worst-Case event or risk index (e.g., Dow's Fire & Explosion index)	Quantitative Risk (consequence and frequency)	Workshop to demonstrate ALARP or develop risk control plan

3.1 Stage 1

The purpose of stage 1 is to close the gap on the scope of the assessment and perform a high-level review of the facility. This stage involves a review of all the buildings and hazards for the identifications of those buildings that require being included in the assessments; as well as identify the hazards present at the facility.

Table 5. Building inclusion criteria

Parameter	Criteria	Comments
Main Occupancy (man-hours/week)	40	Average numbers of man-hours per week inside the particular building. This value considers the hours spent by each individual.
Peak Occupancy (occupants)	5	Maximum number of occupants inside the building, at any given time during a year
Functional Significant	Yes	The building is key for the operations and production process. Destruction of the building and/or its content would lead to significant downtime
Emergency Essential	Yes	Building is required for emergency response and mitigation of the events, e.g., shelter, fire pump
High Value ($)	$250,000	Cost of building and content replacement as well as the indirect cost of losing the building is high. This value depends on the facility business risk tolerance

This stage also includes a hazard and event review, defining their potential consequence. Table 6 shows an example of a process equipment summary, developed to indicate operating conditions and potential hazards and events.

Table 6. Process equipment summary table example

No.	Inventory/Equipment	Process Conditions	Hazards	Major Accident Event
1	Equipment or process involved in the event, e.g., boiler, vessels, columns	capacity: xxx ft^3 Maximum Operating Pressure: xxx psig Temperature: xxx ^oF Operating thermodynamic state: e.g., Saturation Liquid at operated pressure	e.g., Pressurized Water, flammable, toxic	e.g., Steam Explosion, Vapor Cloud explosion, jet fire, pool fire

The final task during this stage is the review of the facility arrangement using accepted spacing tables to determine correct spacing; however, spacing tables only consider fire events

3.2 Stage 2

Stage 2 provides more detail and understanding of the hazards and risks. The intention of this stage is to identify low contributor to the risk and exclude them from further analysis (e.g., QRA). This stage considers two options, the use of an appropriate risk index and worst-case consequence modelling.

3.2.1 Risk Index

Risk indices provide a relative value of the risk of individual process units, allowing their classification according to their risks. In particular, Dow's Fire and Explosion Index (F&EI) classifies the risk into five categories and provides a radius of exposure based on the index value, which allows the identification of processes that could affect the buildings of concerns. Table 7 shows the hazard degree defined with the F&EI.

As the intention of this stage is to exclude process units with low risk, an exclusion criteria is required. The defined exclusion criteria considers exposure radius and hazard degree, as follows:

1. Any process unit that produce an F&EI greater than 97 (intermediate hazard degree) would be considered significant, and should be considered during stage 3 (e.g., QRA).
2. Any process unit with an F&EI below 96 (e.g., Moderate and Light) but that exposes a building of concern should also be considered during stage 3.
3. Any process unit with an F&E below 96 (e.g., Moderate and Light) without exposure to any building of concern will be excluded from the study in stage 3.

It is considered that the F&EI has some uncertainty regarding explosions, large jet fires and escalations effects, for that reason exclusion was not considered for the case in which a process unit with a F&EI greater than 96 (intermediate and higher hazard degree) does not show exposure of any building of concern.

Table 7. Dow's Fire and explosion index (F&E) levels

Fire and Explosion (F&E) index Range	Degree of Hazard	Exposure radius range (ft.)
1 – 60	Light	0.84 - 50
61 – 96	Moderate	51 - 81

Fire and Explosion (F&E) index Range	Degree of Hazard	Exposure radius range (ft.)
97 – 127	Intermediate	82 - 107
128 – 158	Heavy	108 - 133
> 159	Severe	> 134

3.2.2 Worst-Case consequence modeling

This option is based on identifying the worst-case scenario, and modelling them using a consequence analysis software (e.g., DNV-GL PHAST, TNO Effects, Canary, Shell FRED, BP Cirrus, CFD) to determine if the scenario could impair the buildings of concern. If the simulation show that the scenario could impair a building of concern, this particular scenario should be considered during Stage 3; otherwise, the scenario is considered adequate and no further analysis is performed. Figure 1 shows an example of the worst-case modeling output.

This option is based on scenarios, rather than overall process unit risk, allowing excluding scenarios that would not impair the buildings of concern, reducing the scope of stage 3.
This option requires a definition of impairment criteria and a greater analyst competence.

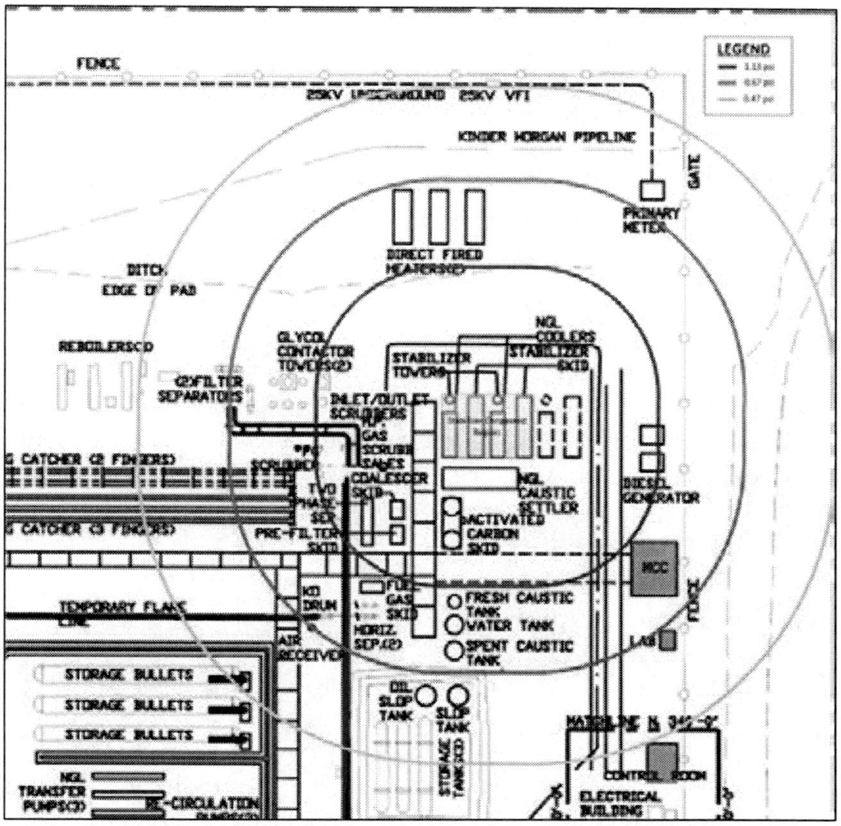

Figure 1. Example of worst-case explosion modelling

3.2.3 *Comparison between options*

Table 8 shows a comparison between the two methods in this stage

Table 8. Advantages and disadvantages of both stage two approaches

Option	Advantage	Disadvantage
Dow's F&E Index	Easy to apply Based on risk, intrinsically defined in the index, reducing the probability of errors in identifying and modelling accident scenarios	Exclusion is based on complete process unit, and potential for exclusion of scenarios of a particular unit is not possible, e.g. explosions, fire. Index basis is not known providing uncertainty of the method – it is assumed that the index was properly built and that risk is properly defined by process parameters

Option	Advantage	Disadvantage
Worst-Case Modeling	Potentially screen out scenarios that would not impact buildings of concern More confidence on result as the science behind the model is well understood	Minimum competence require to properly: • Identify all scenarios • Define Worst-Case Scenarios • Correctly model the events

3.3 Stage 3

Stage 3 is a quantitative risk assessment (QRA) considering the buildings and process units that require further analysis by the analysis performed during the previous stages. The assessment involved a frequency analysis, consequence modelling and risk calculations to determine diverse risk metrics. These risk metrics require a risk criteria to determine risk tolerability

Table 9. Used risk metrics and their typical risk criteria

Risk Metrics	Description	Typical Risk Criteria	Disadvantage
Individual risk per annum (IRPA)	Annual risk of fatality of a particular individual taking into account their exposure, throughout their working year, to all facility-induced hazards	1E-04/yr	Risk criteria involve all risks, including occupational risks and process risk when individual is outside the occupied buildings. Using this metric could be misleading if the total contribution is unknown or different groups are defined reducing the risk of a particular group.
Potential Loss of Life (PLL)	Annual risk of fatality of an individual taking into account the working group exposure, throughout their working year, to all facility-induced hazards	Not typically defined	Risk criteria involve all risks, including occupational risks and process risk when individual is outside the occupied buildings. Using this metric could be misleading if the total contribution is unknown or different groups are defined reducing the risk of a particular group
Building individual risk	Annual risk of fatality to an individual in a particular building, based on an hypothetical occupancy rate of 40 hours/week	1E-05/yr	Risk criteria does not take into consideration the number of occupants, buildings with higher occupancy represent higher safety risk

Risk Metrics	Description	Typical Risk Criteria	Disadvantage
Impairment Frequency	Frequency at which a particular building fail to provide protection from a hazardous event	1E-04/yr per type of event (e.g., fire, explosion, toxic)	Metric only consider frequency of impair the building, no credit is consider for occupants and time expend inside the buildings. This metric is adequate for functional significant, emergency essential and high value buildings.
Exceedance curves	This metrics is in graphic form, showing the cumulative frequency of exceeding a particular consequence parameter (e.g., overpressure, incident radiant flux) in the building.	1E-04/yr	Metric only considers frequency of building impairment. No credit is consider for occupants and time spent inside the buildings.
Expected loss	Predicted value of a loss, calculated as the sum of all possible loss multiplied by the probability of its occurrence	Not typically defined	Metric considers loss by an event in monetary ($) value. This metric is relevant for high value buildings

Implementation of all these risk metrics is not practical, potentially leading to confusion and misinterpretation of the results. In order to define the best metrics to be used, these criteria need to be reviewed in the context of the decision principles used to assess the building during stage 1. A building is assessed during this stage because the occupancy is greater than 40 man-hrs/week, the peak occupancy is greater than five, the building is functional significant, emergency essential or high value.

From an occupant's life safety perspective, the impairment frequency is considered the most adequate risk metric. Although this metric does not consider the occupant fraction, the occupancy ratio is not required for a proper risk comparison. The reason of this statement is explained as followed:
The building occupant risk can be evaluated as follows:

$$BOR = \frac{(\sum_i IF_i \cdot v_i) \cdot t_o \cdot n}{t_e}$$

Where
BOR – Building occupant Risk
IF_i – Impairment Frequency of a particular scenario (i)
v_i – Vulnerability of a particular scenario (i)
to – occupancy time
n – Occupants
te – Hazard exposure time

This equation means that the risk criterion will have the same relation with set values, particularly the vulnerability is considered 1 and the total IF is set, typically to 1E-04/yr per event type (e.g., fire, explosion toxic). The occupancy factors are not changed in the criterion

$$BOR\_Criterion = \frac{1 \cdot 10^{-04} \cdot t_o \cdot n}{t_e}$$

A risk assessment is a relative comparison between the calculated risk and the risk criteria. The comparison is still valid if the occupancy factor is removed from the BOR and the criterion. This approach allows estimating the building risk (BR) as the sum of the impairment frequency by the occupant's vulnerability, and compare it with simple impairment frequency criteria (typically, 1E-04/yr.), no matter the occupancy factor of the building.

$$BR = \sum_i IF_i \cdot v_i$$

$$BR\_Criteria = 1 \cdot 10^{-04}$$

This criterion is used in conjunction with an exceedance curve of the building to define a proper protection that ensure a BR of 1E-04/yr. Figure 2 shows an example of a building exceedance curve. This graph shows the frequency of exceed a particular effect parameter (e.g., overpressure, incident radiation heat flux), where the cumulative frequency is in the vertical axis and the particular effect is in the horizontal axis.

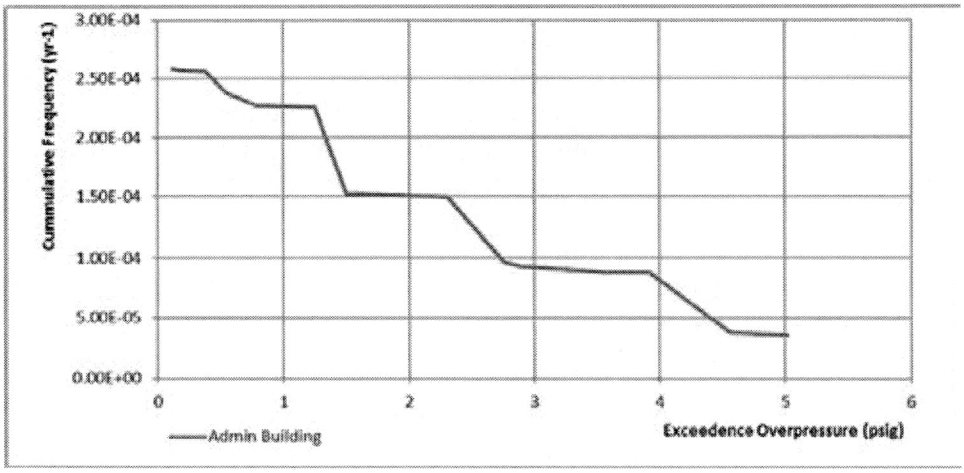

Figure 2. Exceedance curve example

For the emergency essential buildings, the most appropriate criterion is the fraction of events that impair these buildings, e.g. fire pump should not be impair by 85% of the incidents. For these buildings, a criterion per building of concern may need to be defined, e.g. fire pump, emergency shelter.

For the functional significant buildings the impairment frequency criterion still applies, but the criteria need to be defined in conjunction with the facility management as impairment would represent a

business impact.
For the high value buildings, the expected loss criterion is recommended; this criteria need to be defined with facility management as the insurance premium should be considered in defining an appropriate value.

3.4 Stage 4

Stage 4 comprises a final review of the results of the previous stages, and identifying actions and activities that would reduce the risk levels to a tolerable and acceptable level. This is considered to be the Risk Mitigation plan defined in API RP 752/753. The approach is based on the hierarchy of control, as shown in Figure 3, which defined actions effectiveness on reducing the risk level.

During stage 4, different control activities are proposed for each level in the hierarchy of controls, indicating the level of risk reduction achieved by each risk mitigation activity. The list is evaluated in a workshop with relevant decision making personnel to discuss practicability and potential cost of each proposed risk mitigation activity, defining the most appropriate activities to implement.

Often, in risk management, it is desired to reach to the "As Low As Reasonably Practicable" (ALARP) risk level, which is considered the optimum level of risk. ALARP means that the effort, time, and money needed to further control the risk is grossly disproportionate with the benefit gained by implementing the control. In that sense, the duty-holder must be able to show that implementing the control would be grossly disproportionate to the benefits of risk reduction that would be achieved. Thus, the process is not one of balancing the costs and benefits of measures but, rather, of adopting measures except where they are ruled out because they involve grossly disproportionate sacrifices. The workshop performed during this stage is intended to reach the ALARP level by discussing with the shareholders the proposed risk mitigation action defining which of those are grossly disproportionate with their benefit.

In order to understand the disproportionality aspect of ALARP, the following examples are typically provided:

- To spend $1MM to prevent five staff suffering bruised knees is obviously grossly disproportionate, but
- To spend $1MM to prevent a major explosion capable of killing 150 people is considered proportionate

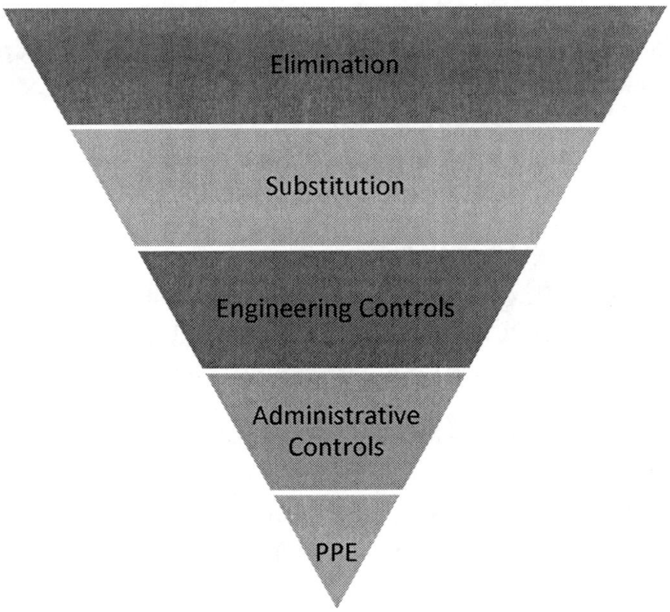

Figure 3. Hierarchy of controls

3.5 Benefit of the Approach

This practical approach follows the risk management principles, ensuring that risk to employees, the public and the environment is consistently controlled within the organization's risk tolerance. Table 10 summarizes how the approach complies with these principles.

Table 10. How the approach complies with risk management principles

Hazard – What can go wrong?	By going through all the stages, the different hazards are identified
Consequence – How bad could it be?	Every stage allows the consequence assessment of the different accidental hazards identified. The understanding of the potential impact of each accident event increases as the evaluation goes through the different stages. Stage 1 – Fire spacing table Stage 2 – potential exposure of accident events Stag 3 – impact of relevant accidental scenarios
Likelihood – How often might it happen	As the assessment progress from stage 2 to stage 3, the likelihood is considered in a frequency estimation of pertinent accident events. Additionally, stage 1 and 2 allows eliminating scenarios that would not impact to the receptors of concern.
Risk – Is the risk within the tolerance	As the evaluation progresses through the stages, the estimated facility risk will determine if the installation complies with the organization's risk tolerance. Additionally, a tolerability criteria is recommended if the organization has not defined one

Additionally, this approach is flexible enough allowing its implementation in a variety of operations and process units while being cost-efficient by optimizing the evaluation permitting to stop the assessment to close gaps found in each stage before moving to the next stage, which reduce the scope and cost of the next stage.

4 Case Study

A case study is presented as an example of this general approach. This case study examines a hypothetical process plant with diverse process equipment and buildings.

4.1 Stage 1

During this stage, the initial tasks include gathering all relevant facility information, identifying all the potential hazards and listing the buildings in the facility. All the buildings are evaluated against the defined criteria to identify which buildings will be considered in the evaluation. Then, the spacing table approach is used to identify buildings that do not comply with the recommended spacing distances. Table 5 provides the building selection criteria to identify the buildings that should be considered in the study.

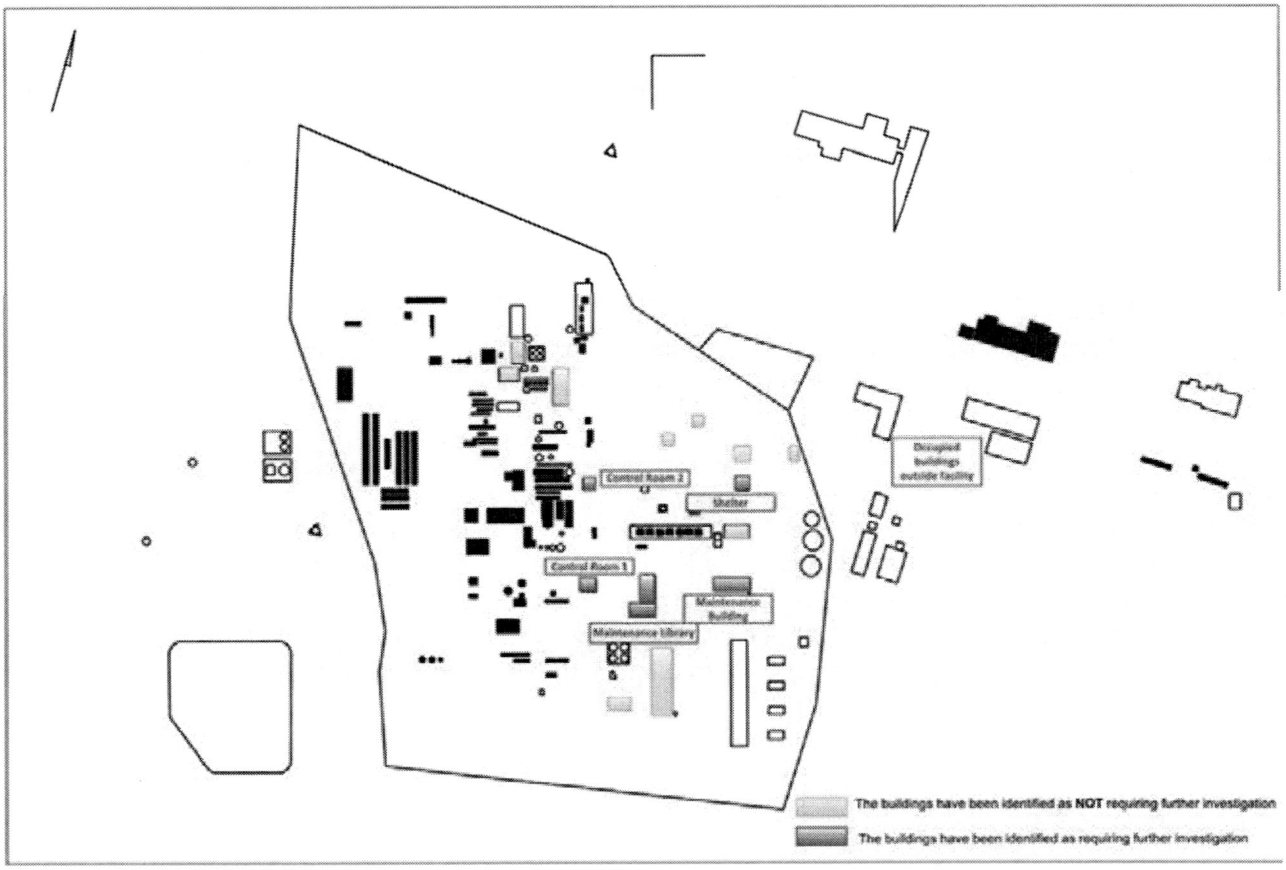

Figure 4. Example layout

The result of the building evaluation is summarized in Table 11, which shows the buildings considered for further analysis.

Table 11. Stage 1 - building summary example

Bldg. Ref	Building Description	Building Type	Area	peak Occupancy	Occupancy (manhr/wk)	Occupied	Functionally Significant	Further Analysis
CR1	Control Room 1	5	Piperack and Process	5	200	Yes	Yes	yes
CR2	Control Room 2	5	Process 2	6	240	Yes	Yes	yes
SH1	Shelter	5	Admin area	15	-	no	Yes	yes
ML	Maintenance Library	3	Admin area	6	22	yes	No	yes
MBD1	Maintenance Bldg	3	admin area	7	150	yes	No	yes
SR1	storage room 1	1	Pumping area	2	15	no	no	no
SR2	Storage room 2	1	Compression	2	10	no	no	no
MW1	Maintenance warehouse	1	Pumping area	3	12	no	no	no
SR2	Tools and equipment storage	1	Admin area	1	5	no	no	no
ISB	instrumentation a service building	1	Compression	1	2	no	no	no
ELS	Electrical substation	1	Pumping area	1	2	no	no	no
SB1	Service building 1	1	Admin area	2	4	no	no	no
SB2	Service building 2	1	Admin area	1	3	no	no	no
SPL	Sample Lab	2	Admin area	2	4	no	no	no
UTC	Utility control	1	Admin area	2	5	no	no	no

4.2 Stage 2

In this case study the F&EI approach is used. Dow's F&E Index is evaluated for each process area identifying the degree of hazards and radius of exposure. Additionally, based on the evaluation, the main hazards in each process units are identified. This information is summarized in a table, as shown in Table 12.

Table 12. Dow's F&E index summary example

Ref.	Process Description	Material	Pool Fire	Jet Fire	Flash Fire	VCE	BLEVE	Toxic	F&E Index	Degree of Hazard	R (ft.)
St-01	Propane Storage	Propane		✓	✓	✓	✓		85	Intermediate	71.4
Pm-01	Pump	LPG	✓	✓	✓	✓			92	Intermediate	77.28
PR-01	Pipe Rack/Process	LPG	✓	✓	✓	✓			110	Heavy	92.4
CP-01	Compressor 1	Propane		✓	✓	✓	✓		85	Intermediate	71.4
CP-02	Compressor 2	Propane			✓	✓	✓		95	Intermediate	79.8
PR-02	Pipe Rack	LPG	✓	✓	✓	✓	✓		100	Intermediate	84
St-02	Combustible Storage	Diesel	✓				✓		62	Moderate	52.08
Ct-01	Utility	Hot Oil	✓						41	Light	34.44

Based on the radius of exposure, the buildings exposed to process hazards can be identified.

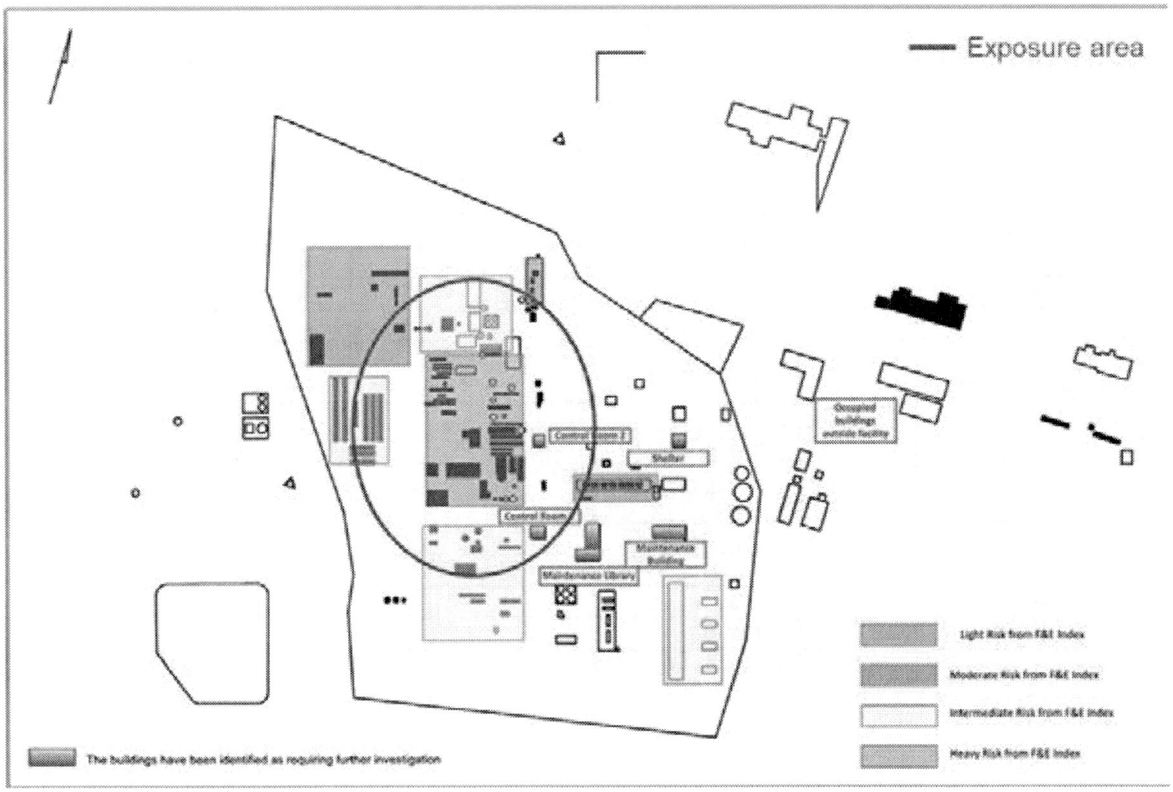

Figure 5. Dow's F&E index exposure area example

Table 13. Stage 2 building summary example

Bldg. Ref	Building Description	Building Type	Area	peak Occupancy	Occupancy (manhr/wk)	Occupied	Functionally Significant	Comply with F&EI spacing?	Stage III Required
CR1	Control Room 1	5	Piperack and Process	5	200	Yes	Yes	no	Yes
CR2	Control Room 2	5	Process 2	6	240	Yes	Yes	No	Yes
SH1	Shelter	5	Admin area	15	-	no	Yes	yes	No
ML	Maintenance Library	3	Admin area	6	22	yes	No	yes	No
MBD1	Maintenance Bldg	3	admin area	7	150	yes	No	yes	No

4.3 Stage 3

This stage involves a quantitative risk analysis (QRA) considering only relevant process units and buildings defined in the previous stages. The QRA involves a frequency analysis and consequence modeling of the relevant accident events.

The process units that present hazards to the buildings and those with a degree of hazard higher than intermediate were considered (F&E index greater than 81). Considering these process units, the standard QRA method is used to determine the impact on the buildings that were selected in the previous stage.

For each building, risk metrics were calculated. For example, an exceedance curve for radiant heat and overpressure was generated to determine the appropriate structural resistance.

Considering a risk criteria of 1E-04/yr, the main control building would require a structural resistance that withstand approximately 2.5 psig of overpressure and 25 kW/m^2 of radiant heat. In the case of overpressure, interaction with the structural engineers is required to provide adequate input to ensure the require resistance is implemented. Typically, the structural engineers would require in addition to overpressure, the impulse duration, which can be determined by reviewing the results of the explosion modeling.

The thermal resistance requires more definition because the building fire rating is typically defined in hours. Further heat transfer modeling is required to convert the 25 kW/m^2 to a heat load onto the building. Specification of the buildings construction is useful for performing the heat transfer evaluation that estimate the appropriate rate. The result can be used to discuss options to protect the building to the required thermal load with a fireproofing vendor.

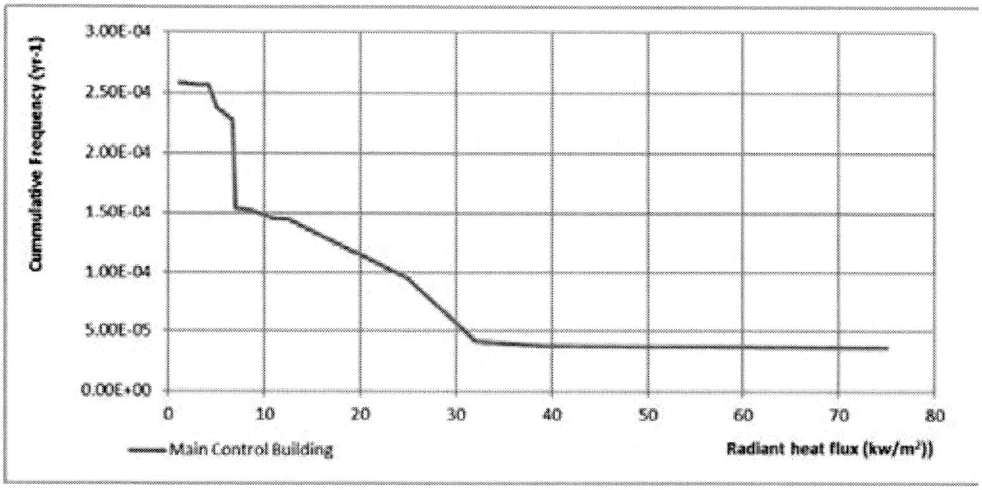

Figure 6. Thermal impact exceedance curve example

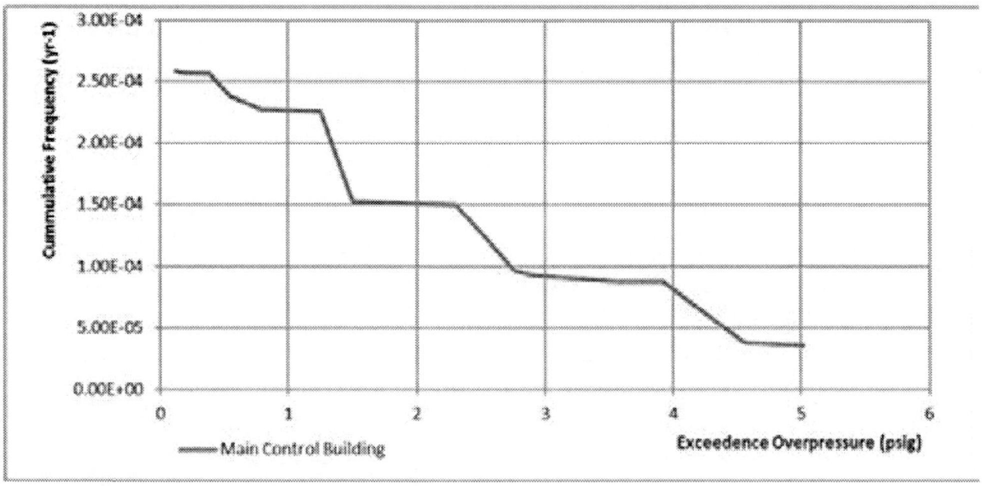

Figure 7. Overpressure exceedance curve example

4.4 Stage 4

Stage 4 involves the identification of risk reduction measures to reduce risk to a tolerable level. The obvious risk reduction would be to upgrade the buildings to resist the thermal and overpressure levels identified by the exceedance curves. However, the most appropriate approach is to identify risk reduction measures in each level of the hierarchy of controls. For example, the following risk control measures can be identified for our case study:

1. **Elimination**
   Elimination is the most effective risk control measure, which involves elimination of the hazard. However, it is typically not practical to achieve because it would involve the elimination of the process units that produce the undesired effect or the buildings of concern. In our case study, this risk reduction level is not practicable.

2. **Substitution**
   Substitution follows elimination in the effectiveness of control. This level involves the replacement of a circumstance with another with lower risk. For example, in our case we could move the building to a location where the worst-case event does not impact the building, where the cumulative frequency of impact is below the risk criteria (i.e., 1E-04).
   Another substitution control could include the substitution of buildings intention reducing the exposure time. For example, if the control room is moved to the building used as a Maintenance Warehouse, the new control room could be exposed to lower risk. In addition, the maintenance warehouse has only 12 man-hr/week representing a lower risk if it is moved to the building currently used as a control room.

3. **Engineering Control**
   The engineering controls involve the implementation of design features that will reduce the risk to tolerable levels. For example, in our case, the engineering controls would involve upgrading the affected buildings to a thermal and overpressure resistance level that provides a cumulative frequency of impact below the risk criteria (i.e., 1E-04/yr) shown on the exceedance curves.

4. **Administrative Controls**
   Administrative controls involve the modification of behaviors and organizational aspect to reduce the risk level. For example, a typical reduction measure includes reducing occupancy loads and improving the asset integrity program beyond the industry standards. In general, these controls are not fully effective, but need to be reviewed on a case-by-case basis. For example, in our case, the occupancy should be reduced to lower than 40 man-hrs/week and peak occupancy of 5 in order to have any impact to the risk (building inclusion criteria), which would be impractical for a control room.

5. **Personal Protective Equipment (PPE)**
   PPE is any equipment worn by personnel that would reduce the impact of the identified accidents. PPE represents the lowest effectiveness level in the hierarchy of controls. In general, due to the accidents considered in a facility siting study, PPE is generally ineffective in protecting personnel. For example, in our case, an explosion could produce a building collapse

endangering the occupants. The PPE worn by the occupants would not adequately protect them from injuries caused by a wall and structural members collapse. In some cases, for toxic and smoke ingress cases, a respirator can be provided to the occupants allowing the prompt escape through a clear escape route. However, these cases need to be assessed on a case-by-case bases, as it would depend on the accumulation rate and toxic incapacity level.

The final activity of the study is to perform the risk control workshop with the relevant stakeholders to evaluate ALARP. In our case, the relevant proposed risk control measure are presented with the considered level of risk reduction level, which can be estimated from the stage 3 results (QRA). During the workshop, effort and costs of risk controls are estimated and utilizing the ALARP principle, decisions are made on whether the implementation of these controls are practicable. It is important to understand that in order to achieve ALARP, the tolerable risk level needs to be initially achieved before decide that further risk reduction is not needed based on the ALARP principle.

5 Conclusions

A practical facility siting approach was presented. Using this approach, the facility siting is performed through different stages allowing for optimization of the assessment. Specific benefits of this approach include:

1. Optimized risk assessment by focusing on the events that significantly contribute to risk, while still achieving risk reduction to a tolerance level
2. Flexible approach that allows the assessment of diverse processes and facilities
3. Allows for definition and implementation of risk control measure before moving to the next stage, reducing overall assessment scope and cost
4. Reduced complexity and cost of quantitative risk analysis (QRA)
5. Interaction between different stakeholders to define optimum level of risk

6 References

[1] Guidelines for Risk Based Process Safety, Center for Chemical Process Safety, American Institute of Chemical Engineers, John Wiley & Sons, New Jersey, 2007

[2] API Recommended Practice 752, Management of Hazards Associated with Location of Process Plant Permanent Buildings, American Petroleum Institute, Third Edition, December 2009

[3] API Recommended Practice 753, Management of Hazards Associated with Location of Process Plant Portable Buildings, American Petroleum Institute, First Edition, June 2007

[4] Dow's Fire & Explosion Index Hazard Classification Guide, American Institute of Chemical Engineers, Seventh Edition, 1994

[5] Guidelines for Facility Siting and Layout, Center for Chemical Process Safety, American Institute of Chemical Engineers, New York, August, 2003

548758 Designing an Effective Detection and Mitigation System for a Plant – Build on Facility Siting Study and QRA Investments

AIChe GCPS Abstract – Murtaza Gandhi, David Black, Alex Fergusson, and Kristen Graham, BakerRisk

Designing an effective detection and mitigation system for a plant – build on Facility Siting Study and QRA investments.

Many plants worldwide are moving forward from conducting only consequence-based analyses to risk based siting analyses. Additionally, operators are relying more on instrumented systems to reduce outdoor personnel tasks, and to allow them to work further from process units. As these developments occur, plant operators are looking for ways to account for these changes when quantitatively determining the amount of risk reduction. One such approach on reducing risk is installation of a fire and gas detection system that can be used in conjunction with isolation to mitigate a hazard before it develops into a major gas release, explosion, or fire. By utilizing this approach, risk reduction credit can be taken for high consequence loss of containment scenarios which can dominate consequence based evaluations. However, the implementation of these systems is typically performed in an empirical manner. With the qualitative approach, sites are not able to implement such systems in a consistent manner and, as a result, they are not able to take credit for such systems in the calculations of their overall risk values.

Some of the questions being asked are: how many detectors are needed? Where to install them? Will the detection be effective? Most importantly, if the dispersion or fire is detected, can it be mitigated? All of these questions can be answered in a definitive manner using ISA's S84 Technical Report 7 on Fire & Gas systems. ISA84 TR7 provides guidance on designing fire and gas detection and mitigation system using a performance-based standard.

In order to determine the effectiveness of gas detection system, a series of controlled dispersions tests were performed to validate the various approaches in industry. This paper combines the results from those tests along with numerous fire and gas mitigation studies performed to test the different aspects of a detection and mitigation system's effectiveness. The paper proposes to discuss the numeric concepts of ISA84 TR7 pertinent to designing a fire and gas detection and mitigation system, and its impact to overall risk values for risk based siting studies. Methods of optimizing the number and types of detectors is also discussed.

The Good, the Bad and the Ugly – The Importance of Management of Change in Building Siting

Kris Smith
Risktec Solutions, Inc.
16055 Space Center Blvd, Houston, TX 77062
Kris.Smith@risktec.tuv.com

Jayant Kulkarni
Risktec Solutions, Inc.
15810 Park Ten Blvd, Houston, TX 77084
Jayant.Kulkarni@risktec.tuv.com

Prepared for Presentation at
American Institute of Chemical Engineers
2019 Spring Meeting and 15th Global Congress on Process Safety
New Orleans, LA
March 31 – April 3, 2019

The Good, the Bad and the Ugly – The Importance of Management of Change in Building Siting

Kris Smith
Risktec Solutions, Inc.
16055 Space Center Blvd, Houston, TX 77062
Kris.Smith@risktec.tuv.com

Jayant Kulkarni
Risktec Solutions

Keywords: Management of Change (MOC), Facility Siting

Abstract

According to the Occupational Safety and Health Administration's (OSHA) Process Safety Management (PSM) standard, Management of Change (MOC) is required by companies when any changes (e.g. equipment, procedures, staffing or buildings) are made to facilities containing highly hazardous and reactive chemicals. Several incident investigations conducted by the Chemical Safety Board (CSB) over recent years have identified that the root causes of catastrophic accidents usually include some element of improper or ineffective change management. Although many companies have mature MOC programs, there is still some uncertainty as to the effectiveness of their processes.

1 Introduction

Siting studies are required under OSHA's regulation, *Process Safety Management of Highly Hazardous and Reactive Chemicals.* The analysis is mandated, but the way in which the analysis is conducted and the way in which results of the study are managed are up to each company. Over many years of conducting siting studies while working for operating companies and since becoming consultants, we have had many opportunities to learn from what went well and what didn't. This paper focuses on five key lessons we've learned over the years and how to make sure the industry is better equipped to deal with siting issues in the future.

2 Lessons Learned from 20+ Years of Conducting Siting Studies

2.1 Siting is more than buildings and occupants

While most companies perform siting studies to comply with OSHA's PSM regulation and thus API RP-752, 753 and 756, siting goes well beyond just buildings and its occupants. Smart companies consider siting from a broader perspective and use information generated during a siting study to not only address buildings and occupants, but to also identify the best location for new units, new equipment, new support structures, potential offsite impact and future use. This requires that siting is considered during the early stages of a project. While most MOC programs capture the addition of new temporary or permanent buildings or the relocation of personnel, a true MOC program should consider factors associated with the location of any change.

For example, whilst installing a new atmospheric storage tank may or may not impact the results of an existing facility siting study, it can definitely exacerbate existing problems related to runoff, containment, smoke and emergency response. These should all be thoughtfully considered during the initial design to ensure runoff is to the safest location, containment isn't overloaded and smoke and firewater are manageable from an emergency response standpoint.

For new green field projects, a preliminary facility siting study should be done early in the select phase. This allows the location of process equipment, utilities and support buildings to be conscientiously chosen to provide the greatest distance possible between hazards and personnel for likely credible scenarios for fire, explosion and toxic or reactive releases. Many midstream companies are employing this approach when siting new plants. By reviewing siting issues early enough in the design phase, companies can build-in areas for future expansions and projects. In order to produce an early stage siting assessment, one would need a preliminary plant layout in 2D and some basic process data in order to determine the impacts of the new proposed facility.

For brown field and existing units, this becomes trickier because of the possible significant layout and materials already present in the facility, however it is still an extremely valuable analysis. Adding a train to an existing facility can be analyzed simply by identifying the location on the plot plan and using the data from the existing or similar available train to develop likely explosion overpressures, radiation levels and toxic release concentrations. These can be overlaid on the new unit's plot plan to determine if it exacerbates any existing facility siting issues or creates any new issues. This information can then be used to fine-tune the layout to provide optimum placement for preventing or mitigating explosion, fire and toxic/reactive release scenarios. Once the plot layout is finalized and the design is set, the existing siting study can be updated to include the new analysis generated by the change.

Companies considering adding whole new units to their complex can still conduct preliminary siting studies by using results from similar type units to estimate the impact to the site. For example, adding a new polymerization unit to an existing petrochemical complex can be evaluated using data from like facilities. If the company has existing polymerization units in other locations or the contractor conducting the study has experience with similar polymerization units, information gleaned from those units can be used to help site future units in the complex. Conducting a preliminary siting study using similar process information will help the company determine if adding the unit is feasible or if the risk of placing a new unit inside an existing complex is too high.

2.2 *Risk-based siting gives you greater control*

PSM requires that facilities covered under 29 CFR1019.119 conduct a facility siting analysis. By definition these facilities process flammable, explosive, toxic or reactive chemicals. Consequently, a release of any of these chemicals may have dire consequences. When addressing siting issues, the key is to prevent loss of containment. If chemicals remain contained within the process, then the facility, its staff and surrounding communities are safe.

Over the years, many companies have spent millions of dollars designing new buildings and relocating personnel. This is usually a result of a consequence-based facility siting study which will almost always tell you that buildings or people are too close to the process. This approach overlooks the fact that risk is comprised of both consequences and their likelihood. The worst case consequence used in the study for example is likely to occur very infrequently.

The most effective way of preventing undesirable consequences is by keeping hazardous chemicals in the "pipe." It is all about better design, strictly controlled operations, competent personnel and routine, predictive and preventive maintenance. Money spent on quality and reliability improvements not only results in better profitability, but also reduces the likelihood of loss of containment events.

Relocating personnel reduces the potential for fatalities should a loss of containment occur. It does not, however, reduce the frequency of a loss of containment event. The event still has the same likelihood of occurrence. If you are not only concerned about the people but also the asset, then resources should be focused on ensuring loss of containment events don't occur. MOC allows you the opportunity to do that. The purpose of the MOC program is then to, little by little, change by change, incrementally reduce the overall risk by improving performance and reliability.

2.3 *Use the "best tool" for the analysis*

There are several ways to conduct siting studies and many documents currently exist which discuss appropriate methods (e.g. Ref. 1, 2, 3, 4, 5 and 6). As a result, there are many "tools" in the market to assess siting and many experienced practitioners capable of applying these tools to conduct the analysis. The key to a good, cost-effective siting study, however, is to know when to use which tool.

A qualitative evaluation is a good first-step. It allows you to cost-effectively screen out buildings and situations that are of no or very little risk. Items that cannot be screened out require further thought and more in-depth analysis. In these situations, quantitative tools may be more appropriate. Siting studies can range in costs between a few thousand dollars to several hundred thousand dollars depending on the size of the facility, extent of quantification and level of detail. So, it is important to understand what each tool accomplishes and when it's time to use a different tool.

For example, it doesn't take a sophisticated computational model to determine that a concrete building built in the 1960s and sitting in the middle of a process unit is not the safest location for your operations personnel. Tools are available to calculate the probable overpressures at the building as well as its structural capability to withstand an explosion. Those analyses cost money and take time. A better approach for managing this situation is to look for opportunities to relocate control operations to a more remote location and, in the meantime, focus on the mechanical integrity of equipment in the nearby area. Perhaps the money might be better spent replacing old equipment, improving seal systems or adding detection and mitigation systems rather than calculating something that will only tell you what you already know. Again, this can be accomplished through the MOC system. Each time you make a change, evaluate what might be done to lower the risk or remove the hazard. This will build inherently safer systems while addressing existing risk issues.

As with all analyses, the more sophisticated the tool, the more complex and expensive the study. Starting with a qualitative approach allows you to use your money wisely, only using the sophisticated tools to evaluate the more difficult issues and, instead, spending money on actual prevention measures.

2.4 If you decommission it, it should go

Removing unused equipment and piping reduces risk. The energy released in an explosion is significantly higher if the release and ignition occur in confined or congested spaces. Therefore, as part of each change, consider "cleaning house."

If the change involves replacing equipment or adding equipment, consider how that effects the confinement and congestion of the area. Where possible, leave adequate space to prevent escalation or reduce the consequences of an explosion.

2.5 The siting report is not just a report

Companies have invested hundreds of thousands of dollars over the years conducting siting studies only to find out, after a short period of time, that the results are obsolete. Siting studies should be evergreen and should be considered process safety information. Individuals within the company who are responsible for the study should also be educated on what to do with the results of the study.

Frequently, progress on siting issues is hindered by personnel changes. Today, almost every company has performed siting studies, yet in many cases the facility risk remains unchanged because nothing was done with the results due to key people moving on or being assigned different priorities. Actions generated as part of a siting study should be managed the same way as actions generated by process hazards analyses, incident investigations and audits. People newly assigned to address these actions need to understand what the report is saying. Changes to processes need to be evaluated against the siting study as they may impact the results.

The only way to do this is to ensure the assumptions made during the study and the input data used for the study are preserved. To do this, companies should always retain this information in-house, which means asking consultants to provide this data along with their deliverables (make sure the contract requires all native files to be handed over). And then, companies need to ensure the report findings are translated into actionable and understandable recommendations which are tracked until closure.

Frequently, the people who understand the physics and chemistry involved in releases, fires and explosions are not the ones who need to fix the problem. It is in the best interest of the company that the information provided in the study can be understood and addressed by the workforce. This may involve providing additional education to key employees, working more closely with the analysts or asking more from those who conduct the study.

3 The Good, the Bad and the Ugly Examples

The following are examples of some good, bad and ugly practices we have seen over the years.

3.1. *The Good*

3.1.1. High Level Siting Assessment of New Units

Siting a process unit using a screening risk assessment makes good business sense. Issues around the layout of the new unit, its location within the site and existing and proposed buildings can be assessed without specific information on process flows, pressures and temperatures by using analyses from similar process units. Using experience developed over several years of performing siting studies for all different types of facilities, from oil and gas to petrochemical, to chemical manufacturers, a screening method can be developed to determine optimum design layouts.

For example, Company A planned to expand their existing operation using some of their available plant space. Assessing the impact of this new unit on their existing facility site helped identify that an existing Warehouse was vulnerable to projected blast overpressures. This allowed decisions to be made relative to the location of the existing warehouse and layout of the site's plot plan thus, reducing the risk to people housed in the Warehouse.

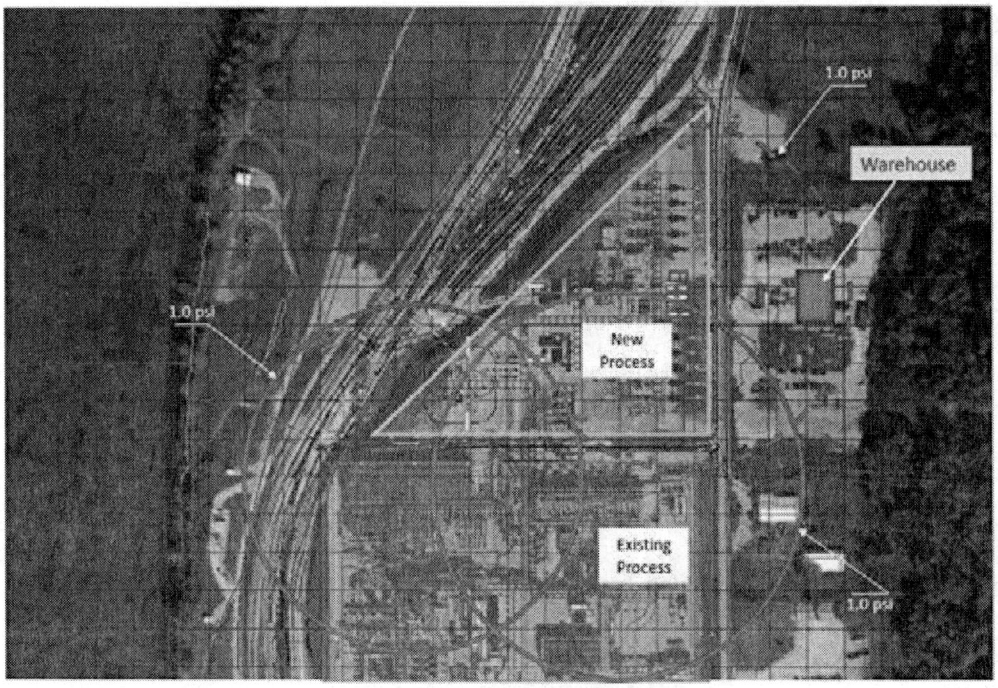

Figure 1: Selection of New Unit Location

This example shows that companies considering adding whole new units to their facility can still conduct a high level siting assessment by using results from similar type units to estimate the overall impact to the site.

Another example of using a screening method involved Company B who had plans to add a green field facility to their portfolio. Based on the screening method, it was determined that the proposed layout had an undesirable effect on the proposed control room that would require significant modifications, either structural or relocation. A small change to the layout was made while it was still in design which reduced the risk profile at the site at a fraction of the cost of structural upgrades.

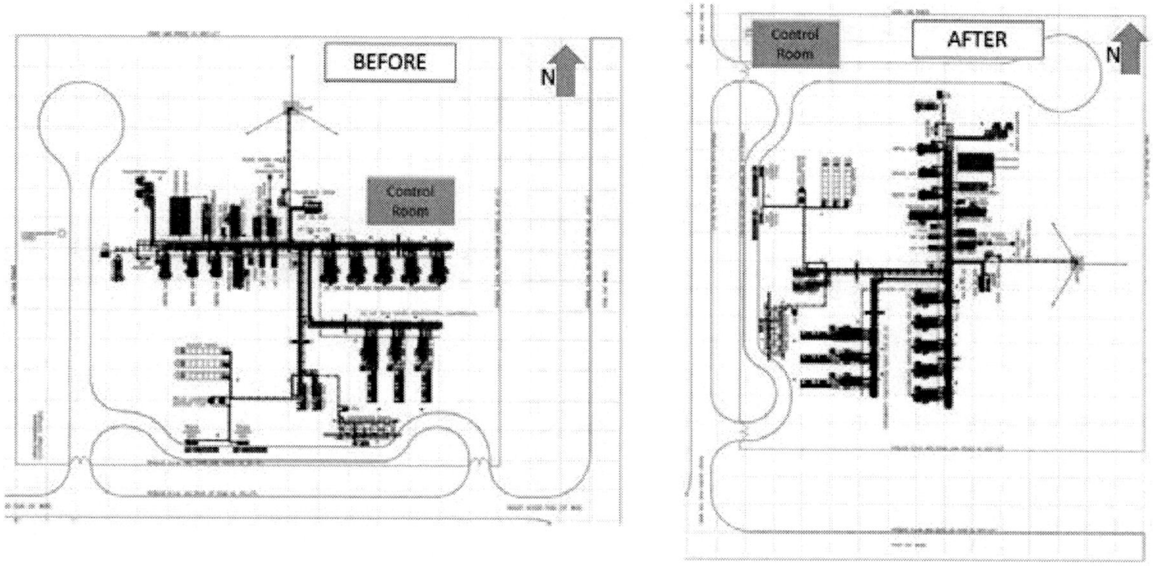

Figure 2: New Layout Before and After Siting Assessment

This early intervention allowed the location of process equipment, utilities and support buildings to be conscientiously chosen to provide the greatest distance possible between hazards and personnel for likely credible scenarios for fire, explosion and toxic or reactive releases. By reviewing siting issues early enough in the design phase, companies can build-in areas for future expansions and projects.

3.1.2. Changes to personnel

Management of personnel changes should also require an evaluation of the potential impact to the facility siting study. For example, Company C planned to reduce the number of operators due to budget cuts. Reducing the number of operators could impact the inspection frequency of equipment, training and competency of operators and response time to alarms which could lead to more frequent catastrophic events. However, Company C utilized a management of personnel change process which required a hazard analysis involving all concerned department representatives. In their analysis, these items were identified and appropriate safeguards, including reconsidering some staffing levels, were put in place to maintain the current risk level.

Personnel changes can have a profound impact to frequency of events and therefore, the risk associated with siting. It is as important to include an evaluation of the impact to risk from changes to personnel as it is to changes to equipment and processes.

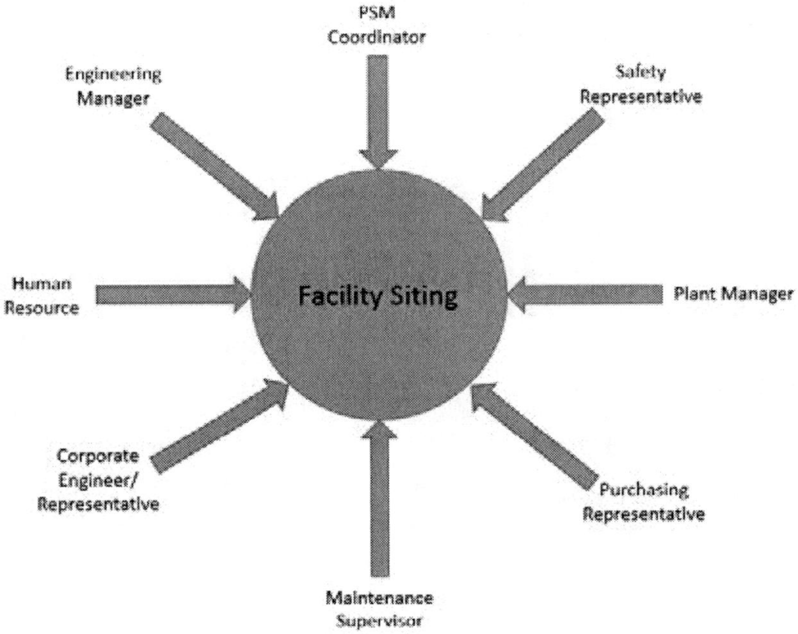

Figure 3: Facility Siting Authorized Personnel

3.1.3. Decommissioning

Decommissioned units, although not active, still pose a risk for overpressure if its equipment and structures remain onsite as explosion overpressures increase with confinement and congestion.

Company D was interested in building a new administration building near an abandoned unit. Performing a qualitative evaluation showed that by keeping the equipment in the decommissioned unit, the results from an explosion analysis would be unacceptable or cost-prohibitive. Once Company D considered removing the unused equipment, it was determined that the proposed location for the new administration building was within acceptable limits.

3.2 The Bad

The following are some examples of some bad and some down-right ugly situations resulting from either a lack of awareness or a general disregard of facility siting principles.

3.2.1. Structural modifications over time

Client E listed the building on their Siting Assessment as a Blast Resistant Building. Over time, an air intake vent was added after the building was located onsite. The manufacturer of the building provided documentation to support their claim. In the documentation, the manufacturer explicitly stated that any modifications to the building would require re-analysis to ensure the building maintained its blast resistance. This was not done at the time the vent was contemplated.

As part of a facility siting revalidation (required every 5 years per OSHA's PSM regulation), it was determined that the structural integrity of the building was lost when the vent was added and that the retrofit would need to be evaluated. Additionally, it was also noted during the revalidation that the building allowed egress from only one location, making it a concern with regard to escape and evacuation due to fire or toxic release. Ultimately, the building will require relocation or retrofit to address these risks – a costly mistake.

Figure 4: Building Integrity

Company F expanded their plant over time without considering the siting impact of the changes. Some of the major issues found during revalidation were:

- New hazards were added to the facility which had the ability to impact the Control Room;
- The Control Room was originally a blast rated building; however, over the years several modifications were made to the building to accommodate a new distributed control system and air conditioning. Both modifications affected the structural integrity of the building and it could no longer be considered blast rated;
- Pipe racks containing toxic materials were added above the building presenting a new hazard not previously considered. The building was not equipped with a toxic chemical detection system or an automated HVAC shutoff system.

Thus, the risk to personnel inside the building had significantly increased over time.

Figure 6: Control Room Revalidation

3.2.2. Siting portable buildings during turnaround.

In preparation for an upcoming turnaround, Company G required the addition of several light-wood portable trailers to house contractors during the maintenance activity. Company G placed the trailers outside the fence line as it was considered to be outside the process area and thus not at risk from explosion, fire or toxic releases. Unfortunately, explosions, fires and releases aren't stopped by fencing. Their location was within the potential blast overpressures and the contractors were at a higher risk due to the location of the trailers.

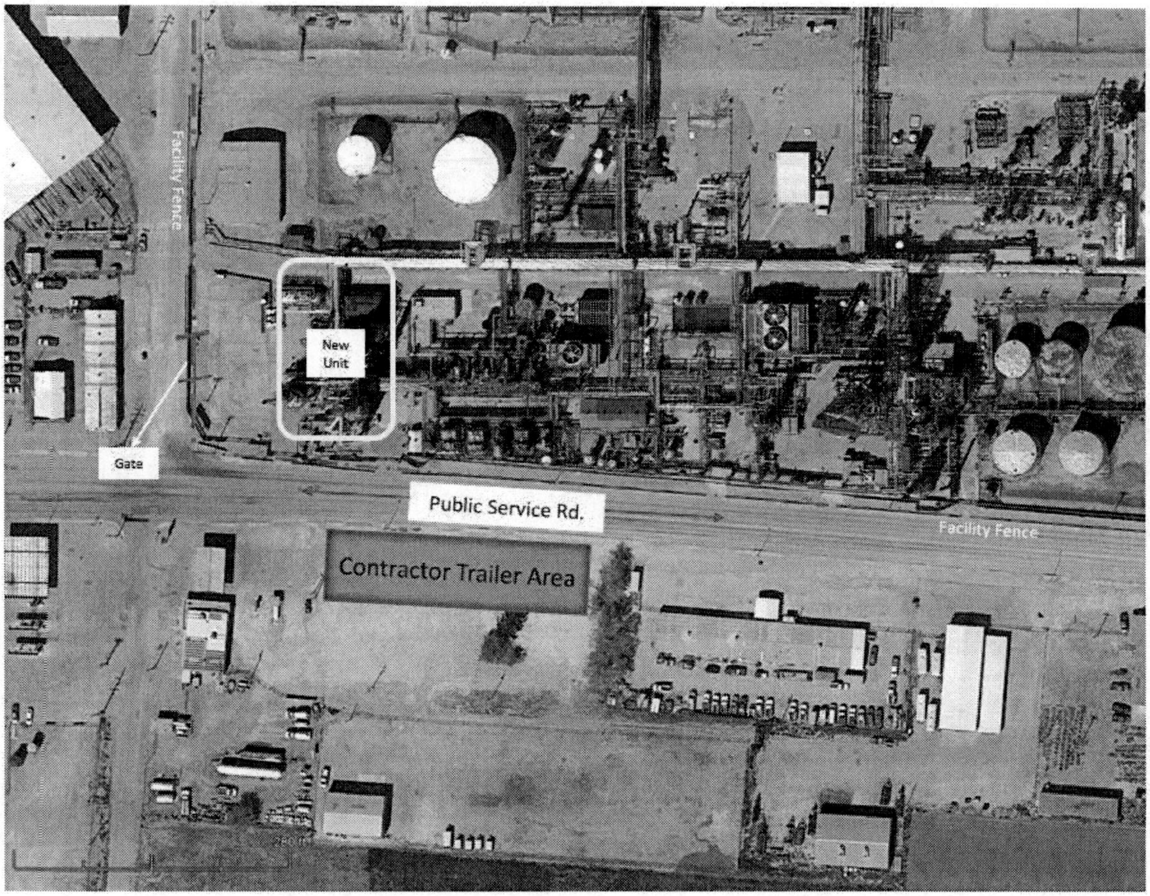

Figure 5: Siting Portable Buildings during Turn around

3.2.3. Acquisitions and facility siting

As part of due diligence for new acquisitions, it is appropriate and prudent to evaluate the findings from the other company's siting study before making final decisions. Costs for addressing facility siting issues can escalate into the millions if relocation and rebuilding are required.

For example, Client H purchased assets from another company. Client H's facility siting standard was based on consequences only and had resulted in relocation and rebuilding of many facility buildings over time. Their new acquisition had used a risk-based approach to facility siting and had managed risk by relocating non-essential personnel and improving reliability. In order to meet the acquiring company's acceptance criteria, a lot of capital investment would be necessary. Had facility siting been part of the due diligence process, decisions on the acquisition may have changed or at least been considered during the negotiations.

4 Conclusions

Conducting qualitative or quantitative siting studies when considering capital improvements allows you to make better informed decisions on layout and siting of new equipment or facilities. A preliminary siting study can be performed with limited information using similar units as the basis for the analysis.

Using the MOC process to encourage the use of inherently safer technologies is a cost-effective way to reduce risk and address legacy siting issues. Ensuring the chemical remains in the "pipe" reduces the likelihood of a catastrophic event and reduces the risk of siting-related fatalities.

Expect more from the siting experts. Use their knowledge to educate your workforce. Use their results to guide your decisions. Use the report as input to future changes. Manage the siting report as you would any key piece of process safety information and treat the recommendations as you would any other process safety study.

5 References

[1] American Petroleum Institute Recommended Practice 752, *Management of Hazards Associated with Location of Process Plant Buildings*, 3rd edition, 2009.

[2] American Petroleum Institute Recommended Practice 753, *Management of Hazards Associated with Location of Process Plant Portable Buildings*, 1st edition, 2007.

[3] American Petroleum Institute Recommended Practice 756, *Management of Hazards Associated with Location of Process Plant Tents*, 1st edition, 2014.

[4] CCPS. *Guidelines for Evaluating Process Plant Buildings for External Explosions, Fire and Toxic Releases.* Center for Chemical Process Safety, American Institute of Chemical Engineers, New York, NY, 2nd edition, 2012.

[5] CCPS. *Guidelines for Siting and Layout of Facilities.* Center for Chemical Process Safety, American Institute of Chemical Engineers, New York, NY, 2nd edition, 2018.

[6] CCPS. *Guidelines for Vapor Cloud Explosion, Pressure Vessel Burst, BLEVE and Flash Fire Hazards.* Center for Chemical Process Safety, American Institute of Chemical Engineers, New York, NY, 2nd edition, 2010.

Safe and Optimized Operations Through Smart Analytics & Big Data

Karl McManus
Chartered Chemical Engineer, IChemE
Wood, Galway Technology Park, Parkmore, Galway, Ireland
Karl.mcmanus@woodplc.com

Contributions by

Jason Payne
Analytics Manager, Digital Solutions
Wood, Galway Technology Park, Parkmore, Galway, Ireland
Jason.payne@woodplc.com

Ronan O'Malley
PhD, Senior Data Scientist,
Wood, Galway Technology Park, Parkmore, Galway, Ireland
ronan.omalley@woodplc.com

Prepared for Presentation at
American Institute of Chemical Engineers
2019 Spring Meeting and 15th Global Congress on Process Safety
New Orleans, LA
March 31 – April 3, 2019

Safe and Optimized Operations through Smart Analytics & Big Data

Keywords: Predictive Fault detection, asset integrity assurance, artificial intelligence, machine learning, sensor data

Abstract

Are offshore oil and gas operating companies using all their available data to demonstrate safe operations and optimise performance?

Wood have developed applications in the areas of Prognostics (Predictive Fault detection) and asset integrity assurance. Using supervised and unsupervised Artificial Intelligence (AI) learning techniques, automated pattern detection and classification, our engineering teams have reduced risk, proven process safety assurance and reduced OPEX costs by millions of dollars per year.

The following application examples (case studies) have been developed by operational chemical process engineers working closely with data analysts. The applications manipulate big data to support client operations and save OPEX costs.

- Prognostics Case Study: Valve Actuator Performance – The Wood valve actuator performance application uses sensor data and machine-based learning techniques to help predict failures of production critical subsea valve actuators.
- Asset Integrity Case Study: Water Ingress – Traditionally, major oil and gas companies spend significant amounts of revenue on Subsea Under Water Inspections (UWI). The Wood water ingress application utilises data driven condition monitoring to demonstrate asset integrity thus reducing the requirement for UWI and minimising OPEX costs.

1 Introduction

Digitization is not new, but it is transforming the oil and gas industry. Exploration and Production (E&P) companies are utilizing technology that captures and stores large amounts of data points from wellhead sensors, valves, pipelines and various equipment installed in the field to remote monitoring stations. With this data, operational processes can become much more efficient and safer. Digitization has allowed E&P companies to use technology, big data, and predictive analytics to extract resources more efficiently from each well, create a safer work environment, and do all of this with better labor utilization.

This paper outlines two case studies to provide Prognostics (Predictive Fault detection) and asset integrity assurance. The valve actuator performance application (case study 1) uses sensor data and machine-based learning techniques to help predict failures of production critical subsea valve actuators. Assurance to operations that no water ingress has occurred in deepwater facilities (case study 2) is achieved by event detection and event classification based on modelled and real data.

2 Case Study 1: Prognostics; Valve Actuator Performance

2.1 Problem definition

After approximately three years of production a number of subsea valve actuators started to fail unexpectedly. The root cause of the failure was thought to be hydrogen induced stress cracking of the actuator springs. These failures were undetected until the valve was hydraulically actuated from the control room. Spontaneous failures result in significant downtime (months), deferred production (e.g. 10mbd well shut-in), substantial engineering and intervention costs.

2.2 Solution

Condition monitoring of actuator performance became a high-profile project for an asset due to unrevealed valve failures. With hundreds of this type of actuator in the field, this became a large data capture, review and analysis project. Subsea gate valve signatures graphically illustrate changing hydraulic pressure and flow parameters at various points in the local valve actuator circuit during the stroking operation. By identifying the characteristics and data patterns of healthy valve signatures degrading trends can be identified. Six health indicators have been identified for a valve close signature and these are illustrated in Figure 1. Over time, specific rates of change can provide useful information on how a valve is performing or degrading. The six indicators are as follows:

1. Accumulated Volume – Has the piston performed a full stroke?
2. Valve Closing Time – Are there any signs of reduced spring force?
3. Clear Stop – Is there a clear stop point indicating that the actuator has reached the end of its stroke and stopped?
4. Initial Function Pressure Drop – Indication of initial spring movement?
5. Function Line Pressure Curvature – A damaged, cracked or plastically strained spring would show a change in the force extension characteristics.

6. Pressure Differential – An excessive DP between start to fully closed function pressure will indicate a reduction in spring force.

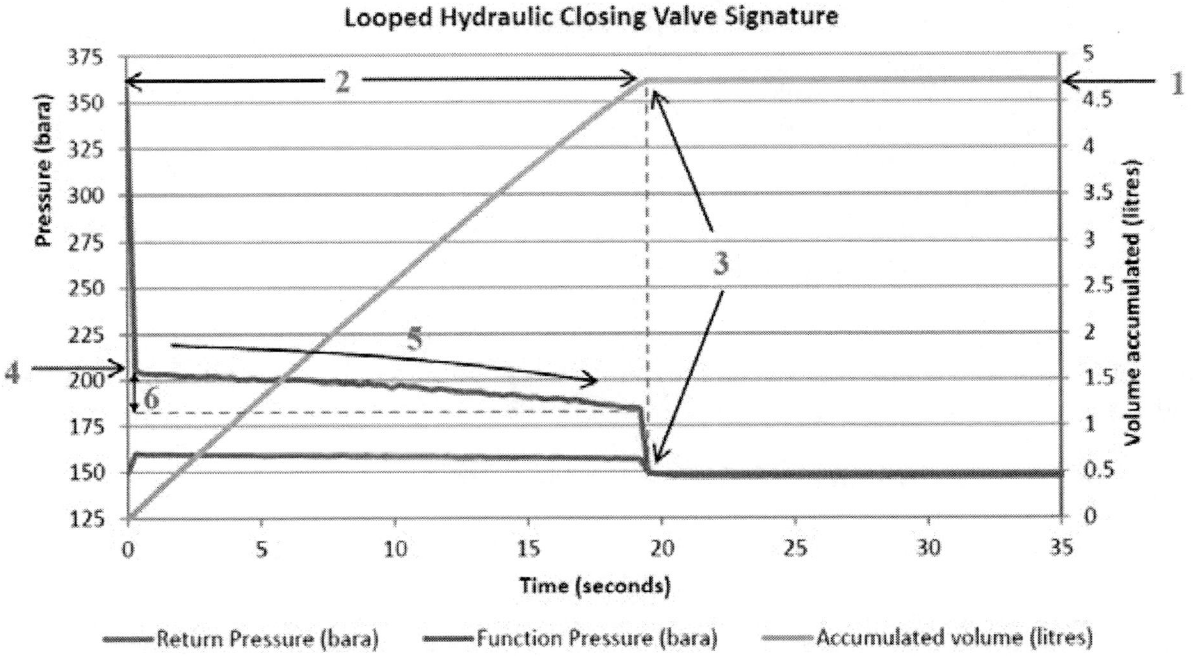

Figure 1. Healthy Closing Actuator

Figure 2 provides an example of one actuator with two signatures overlapped. The healthy signature is shown in solid lines. The same actuator in a degraded state is shown with dashed lines. By identifying early signs of degradation, the asset team can predict when a failure is likely to happen and put an intervention plan in place accordingly to minimise downtime and deferrals. An unexpected actuator failure can result in an oil well being shut-in for 2-3 weeks whilst a vessel is sourced, and intervention preparations (actuator lock out tool) are completed. This has resulted (depending on deferred cost rates) of > 1million USD lost/deferred production based on a well that produces 10mbd. By completing this activity during planned maintenance there are minimal deferral impacts.

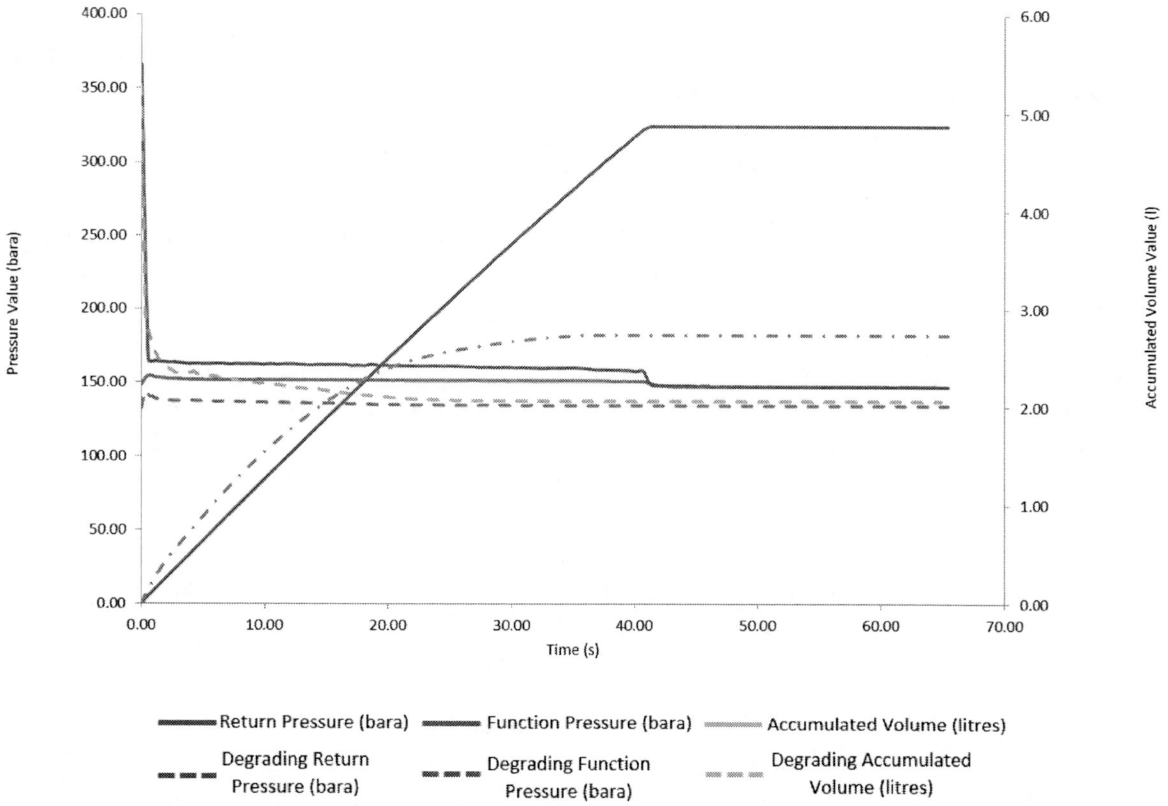

Figure 2. Example of Degradation from Healthy Signature

Currently there are more than 50,000 actuator signatures stored in the asset database. Two machine learning techniques were deployed to automate detection of well understood degradation patterns and track condition of equipment that has no recorded history of failure (or degradation):

- Supervised learning (similarity and information-based approaches) for well understood patterns
- Unsupervised learning (dimensional reduction, clustering) for equipment that has no recorded history of failure or degradation

Both techniques are deployed within human-in-the-loop applications. A human-in-the-loop application leverages machine learning or artificial intelligence to improve human performance, i.e. the human and machine work together to solve problems that can't be solved by human or machine working alone. For valve actuators, the digital process enables review of actuators at scale and at times that would be infeasible without the digital process.

Supervised learning currently employs a k-Nearest Neighbor (kNN) approach (where there are known and well-understood degradation patterns) and suitable for to the desired human-in-the-loop philosophy. The current algorithm has a greater than 90% accuracy at predicting both healthy and degrading actuator signals.

As actuators are imported from the field, the algorithm automatically performs a first pass classification of health and flags potential issues. When reviewing actuator signatures hardware engineers are presented with the machine learning classification and the basis for the classification, e.g. the three nearest neighbors from history. Nearest neighbors from history are signatures that have been classified in the past by hardware engineers and have been determined by the machine to be similar to the new signature. A majority voting system is applied by the machine to determine health classification, e.g. if 2 out of 3 neighbors are healthy then the signature is classified as healthy. Cases where the machine doesn't find similar neighbours requires engineering intervention. This manual classification is used by the machine in future assignments.

For assets with no recorded history of failure or degradation, unsupervised learning is employed to monitor the condition of actuators based on classification and association with existing human-labelled clusters. The algorithm forms a cluster, the hardware engineer examines the cluster and tags it based on domain expertise, e.g. healthy sensor with data errors or healthy operating. If a new cluster is developed or is developing this is automatically flagged by the machine and brought to the attention of the hardware engineer. This new or emerging cluster can then be efficiently reviewed within the application and tagged by the engineer, e.g., cluster to watch or potential degrading cluster formation. Each time a new signature is imported by the application it is plotted and the strength of its association to other known clusters is calculated.

By training and deploying machine learning models (supervised and unsupervised) to automatically identify these patterns results in giving approximately 18 months' notice before 'failure'. This is a significant step towards digitalization of workflow, significant reduction in manhours and improved inspection intervals and service to client. Digitalization makes it feasible for engineers to monitor, leverage and report on valve actuator data at regular (normally monthly) intervals. With a cloud hosted web application (workflow interface for engineer) this enables engineers based in all parts of the world access to database technology and connectors to historian/source data as required.

3 Case Study 2: Asset Integrity; Water Ingress Detection

3.1 Problem Definition:

Offshore oil and gas companies typically spend millions of USD per annum on inspections. A big risk to deepwater subsea assets is water ingress. At typical operating pressures (<100 bara) downstream of the choke valve, for deepwater oil-producing assets, a breach in equipment or pipework will result in water ingress due to relatively high hydrostatic pressure (~>130 bara). Water ingress has the potential to cause hydrate formation in the pipe which can lead to blockage. Inspections are performed at routine intervals and cannot detect water ingress.

3.2 Solution

The intent of this project was to use data analytics methodologies to strengthen data driven condition monitoring and reduce inspections and minimize OPEX of existing assets. Artificial intelligence can recognize patterns (neural networks) and identify healthy or abnormal behaviour operating conditions such as sea water ingress into the system. For Artificial Intelligence to recognize an ingress of water, it was fed pressure/temperature data of what this scenario may look like based on flow assurance modeling. A flow assurance model was created in OLGA to represent the ingress of sea water in the system. The intent was to pilot an approach of utilizing data generated from transient models developed and use this model to teach the machine what a water ingress event looked like. The model starts from the reservoir and ends at the manifold exit. Figure 3 outlines the leak location modelled on a manifold.

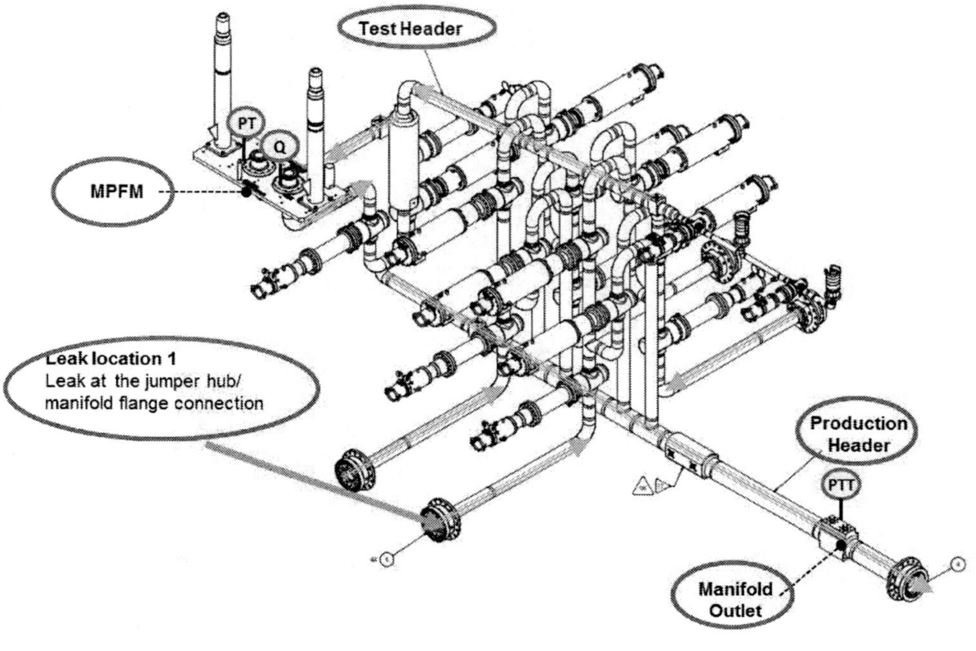

Figure 3. Manifold Isometric Pipework

Changes in temperature, pressure, choke position, liquid flow rate, liquid velocity and water-cut over time at the manifold exit due to sea water ingress were investigated. Using the modelling results software agents were developed to monitor the relationship between temperature upstream of the choke in each connected tree and production header temperature at the manifold exit. To initiate a temperature agent for a location, event detection and event classification takes place. The previous two years of real data from relevant instrumentation was queried from the historian at thirty-minute resolution. A linear model was fit between the independent variables (tree temperatures) and the dependent variable (manifold temperature). While the linear model was not an optimal representation of steady state operations, deviations from the model are a reasonable mix of ramp up / ramp down of production, explainable temperature anomalies (mainly due to sensor malfunction) or false alarms.

While detecting the water ingress simulations provided from the flow assurance model, a new variable called the harmonized signal is derived from the linear model. This harmonized variable is derived from the absolute value of the difference between successive residuals.

$$\text{harmonized} = \text{abs}(\text{diff}(\text{residuals}))$$

To extract events from this signal, a threshold is derived from its distribution.

$$\text{threshold} = \text{mean(harmonized)} + 3*\text{sd(harmonized))}$$

Any connected group of points was converted to an event at this location with an associated start and end time. This even boundary was widened by five hours each side and combined with any other events that result from the widening. As the event detection process was designed to be conservative, a second layer of classification was applied to label the event as normal or abnormal. This considers other instrument data around this location during the timeframe of the event to decide which label to apply. After experimentation it was found that the choke position data provided the most relevant information for automated labelling.

$$choke\ movement\ feature = \log(\max(choke\ position) - \min(choke\ position))$$

Figure 4 shows a 2D dimensionally reduced representation of events for a single manifold with highlighted normal events and anomalies.

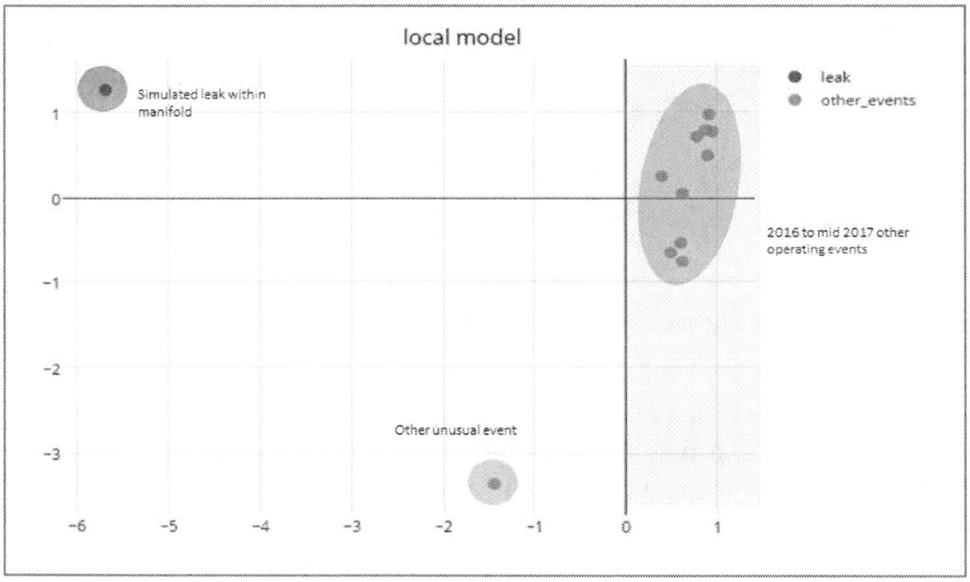

Figure 4. A 2D dimension reduction of events for a manifold.

Agent updates are performed on a regular basis ahead of scheduled engineering review, where new events are added to the cloud based web application. Event data can be analysed, reviewed and labelled by engineers during the regularly scheduled review process. By combining simulated leak data with historical field data and using advanced analytics, a strategy for deployment involved employing numerous autonomous agents each monitoring a section, rather than a single agent monitoring at a high level. Additional agents will be added to scans over time to increase sophistication.

4 Conclusion

When operating facilities and plants, engineers and managers need to be proactive in relation to process equipment performance. In the case studies discussed, with the utilization of historian data (process/hydraulic pressure, flow, temperature etc.) available it is possible to use technology, solutions and predictive analytics to make informed decisions more efficiently and create a safer work environment.

The case studies involve human-machine interaction, but the machine does not make the final decision. Competent process and operations engineers are still very much integral to the overall outcome. These case studies have provided significant positive impact on risk-based inspection (significant cost saving), data-driven innovation, automatic notification of potential anomalies or operational performance issues before they become significant. Companies can plan for the failure of equipment, minimize downtime through planned intervention/repair, reduce installation costs, have temporary/ standard operating procedures written (and approved) in advance of the change as part of the management of change process. Early identification of possible failure results in significant cost savings and ultimately minimize or eliminate production deferrals.

Data analytics is a continually evolving. To keep abreast of latest technology, the data analytics team in Wood are working on a research project (PREMSAM) with the Centre for Applied Data Analytics research (CeADAR, Ireland). These collaborations are envisaged to further develop methodologies for condition monitoring of process safety equipment performance. By developing innovative digital solutions to technical integrity management, operations and process safety management as outlined in the case studies above we hope to make industry safer, smarter and more cost effective.

5 References

[1] C. Aldrich, L. Auret, "Unsupervised Process Monitoring and Fault Diagnosis with Machine Learning Methods", Springer-Verlag, London, 2013

[2] Hastie, T., Tibshirani, R., Friedman, J. (2001). *The Elements of Statistical Learning*. New York, NY, USA: Springer New York Inc.

[3] Kelleher, J. D., MacNamee, B., D'Arcy, A. (2015). Fundamentals of Machine Learning for Predictive Data Analytics: Algorithms, Worked Examples, and Case Studies. Cambridge, MA: MIT Press. ISBN: 978-0-262-02944-5

Don't Worry!! Laboratory Safety Is Fun Do, Let's Play a Mobile Game ☺

Nishaben Desai Dholakiya
University of Pardubice, Faculty of Chemical Technology Czech Republic
E-mail: st53279@student.upce.cz

Milos Ferjencik
University of Pardubice, Faculty of Chemical Technology, Czech Republic
E-mail: Milos.Ferjencik@upce.cz

Josef Brozek
University of Pardubice, Faculty of Electrical Engineering and Informatics, Czech Republic, E-mail: Josef.Brozek@upce.cz

Zdenek Jalovy
University of Pardubice, Faculty of Chemical Technology Czech Republic
E-mail: Zdenek.Jalovy@upce.cz

Jan Kubik
University of Pardubice, Faculty of Electrical Engineering and Informatics, Czech Republic, E-mail: st52531@student.upce.cz

Prepared for Presentation at
American Institute of Chemical Engineers
2019 Spring Meeting and 15th Global Congress on Process Safety
New Orleans, LA
March 31 – April 3, 2019

Don't Worry!! Laboratory Safety Is Fun Do, Let's Play a Mobile Game ☺

Nishaben Desai Dholakiya
University of Pardubice, Faculty of Chemical Technology Czech Republic
E-mail: st53279@student.upce.cz
Milos Ferjencik
University of Pardubice, Faculty of Chemical Technology, Czech Republic
Josef Brozek
University of Pardubice, Faculty of Electrical Engineering and Informatics, Czech Republic
Zdenek Jalovy
University of Pardubice, Faculty of Chemical Technology Czech Republic
Jan Kubik
University of Pardubice, Faculty of Electrical Engineering and Informatics, Czech Republic

Keywords: Laboratory Safety, Spiral learning, Kolb's ELT,

Abstract

The overall aim of this article is to change the general perspective of laboratory safety from being boring to make it more interesting using mobile games. A type of Mobile Serious Game (MSG) is underway of designing for laboratory safety at the Institute of Energetic Materials at the Faculty of Chemical Technology, Pardubice University (CZ). This application is principally aiming a group of players who, after a period of time feel either over confidence or ignorant towards laboratory safety. It is believed that challenging nature of the game would make learning precisely interesting. Three different levels of game would be based on level of difficulty with respect to safety rules. i.e., level one is all about Personal Protective Equipment (PPE), level two would highlight one of the hazardous Nitration reactions, while level three would be designed for detailed learning of safety rules connected with the reaction. Second half of this paper would argue how constructivism and learning theories are believed to be suitable options while correlating safety with MSG. Whereas Kolb's Experiential Learning Theory (ELT) offers four stages of learning; a) concrete experience b) reflective observation c) Abstract conceptualization and d) active experimentation. Which means a learner/player would enter from any stage of learning cycle to complete learning process to gain knowledge of laboratory safety.

Introduction

A Mobile Serious Game (MSG), a combination of education and entertainment. has received significant concern in the field of education, and a mobile educational gaming community has been formed in recent years [1], Mobile applications are the biggest component of Apple services and helped that segment gross over $38billion in 2017 from Applications store. Furthermore, Applications Store has more than 65,000 educational applications, primarily intended for iPad, which cover a wide range of fields suitable for different levels and learning styles [2]. With a mobile phone in hand, you will be up-to-date on the latest news, regulations, and incidents. You can quickly communicate with anyone you need, regardless of whether you are on-site. You can even pull up any piece of information you need at any time [4]. Besides, mobile apps have proved to be valuable educational tools; laboratory instructors have been limited to using mobile devices only for virtual laboratories with simulated experiments. Which recently been changed, as researchers have developed a series of mobile applications that allow students to remotely interact with real data and equipment in real laboratories. Students reported deeper engagement levels using mobile apps and the virtual lab [5].

Chemical process safety by *OSHA* (Occupational Safety and Health Administration) *Safety Regulations, Checklists Audit Reports* [6] is a mobile application provides a compilation of laws and regulations regarding the handling of hazardous chemicals, as well as a database of checklists which can be downloaded and used for audits. Another OSHA-based app is *OSHA Safety*, which comes with all OSHA safety regulations and standards for hazards pre-downloaded so they can be accessed anywhere at any time without connection to internet or a cellular network [7]. When connected to a data source, the application can direct the user to the latest OSHA news. Both of these applications are free of charge.

Considering extensive ranges of mobile phone applications in the field of education and training, we came up with a concept LABSCAPE, a type of educational mobile game application which is underway of designing to improve laboratory safety. An overall aim of the LABSCAPE is to offer a new direction not only towards laboratory safety training; however it also aimed to make safety more interesting. It offers an online platform where a game player could potentially discuss his/her difficulties with other players from any part of the world, without associated risk. A player making mistake means losing a level rather than danger of himself/others while performing real laboratory experiments.

In order to contribute to reflect on the mobile game use in learning process for laboratory safety, This article will map out some guidelines of the pedagogical proposal using Kolb's Experiential Learning Theory (ELT). This identifies and details a complete four-stage learning cycle, showing how experiences can be translated depending on learner's choice to enter the learning cycle [8]. Whereas a spiral learning theory where a learner continuously constructing knowledge based on previous experience at the end of every cycle.

LABSCAPE for Nitration

The MSG is being designed for Nitration reaction for preparation of Adipic Acid by the masters or research postgraduate students at the University of Pardubice. Based on laboratory experiment data the Nitration reaction initiates with 10 ml of water, 20 ml of 65% nitric acid (14.3 g; 147.4 mmol) was added and the solution was warmed to 85°C. Which followed by 4 ml of Cyclohexanol (3.8 g; 38.4 mmol) gradually added such a way that temperature was maintained between 80-90°C. (Dinitrogen Tetraoxide and other nitrogen oxides are evolved during this period). After addition of Cyclohexanol, the reaction mixture was heated at 100°C for five minutes. Afterwards this mixture was cooled to 15°C, where the solid state of mixture was filtered, washed with cold water and crystallized form water. The yield of Adipic acid was measured at 2.3 g (42%).with 153-155°C of Melting Point. While the fact about hazards correlating this reaction are listed as; a) 65 % Nitric acid is corrosive to skin and eyes. The vapours can irritate respiratory system. b) Dinitrogen Tetraoxide (N_2O_4) released during reaction is toxic. It causes burns in contact with skin, eyes or of if inhaled. It also causes pulmonary edema if inhaled.

Screenshot in Figure 1 highlights a welcome screen of LABSCAPE, with list of options for game players. Game player needs to remember both categories of rules; a) game playing rules b) safety rule to clear the level and gain points. Colorful graphics and real like laboratory environment is likely to encourage players to continue playing for fun. For the major part of designing were used mainly the basic operations of 3D modeling using *Blender* software. All meshes were also made with the possible smallest amount of polygons due to high-poly models could have a negative impact on performance. While software UNITY would be used for coding part of the game and finally MATLAB software would provide some simulation part of the game. It is believed that launching LABSCAPE at an online play store would maximize possibilities to get maximum reviews from game users from around the globe which eventually would assist in data analysis for its usefulness for laboratory safety.

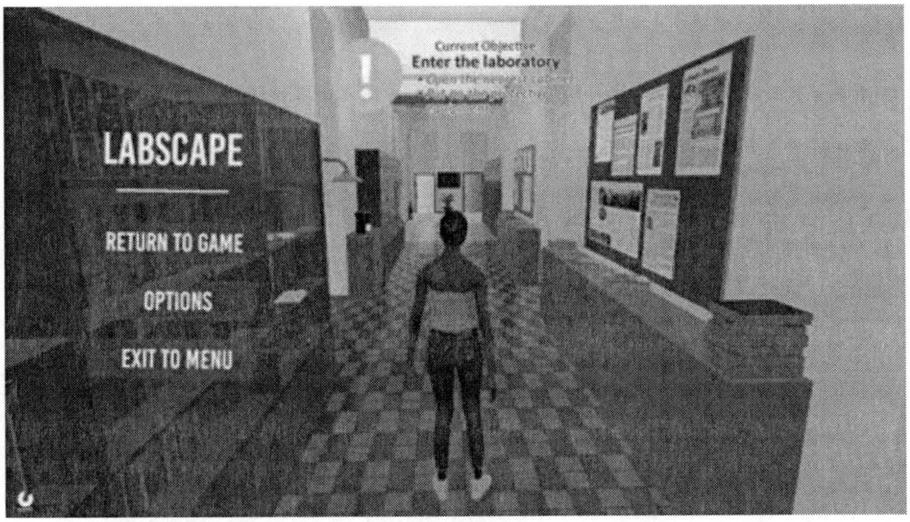

Figure 1: Screenshot of Welcome Screen (LABSCAPE)

Laboratory Safety (Level I)

While practicing engineering safety protocols in laboratory experimental procedure enhances students' commitment to improve their safety behavior and attitude toward carrying out operations in industrial process units. As students study the experimental procedure and identify hazards involved before the unit is run, potential undesirable incidents can be identified and hazard control measures can be processed within the system. This makes the laboratory environment more safe and resilient [9]. Table 1 highlights point's allocation pattern of LABSCAPE, with respect to consequences. Depending on type of mistakes and how often a player is making mistake, would decide a player to continue the same level or fail the level and start from same level. It isn't feasible to play with mobile in laboratory hours, it would be offered in free time before or after laboratory hours. It is believed that chatting with online players or coworkers relevant to LABSCAPE would make it as an open platform to correct mistakes, motivate to reentering the ELT cycle and challenge them.

Table 1: Correlating consequences

Steps	Causes	Consequences	LABSCAPE
Entering laboratory	No gloves No goggles No coat	HNO3 spill on hand HNO3 spill on body Eye irritation	Losing points/level
Pouring HNO3	Not reading label	HNO3 Spill on body HNO3 Inhale/ swallow	Losing points/level
Drinking coffee	Inside laboratory	Swallow chemical while drinking coffee	Losing level
Reentering laboratory	Without gloves	HNO3 spill on hand	Losing points/level

Figure 2: Screenshots of possible variation to enter laboratory using PPE

Figure 2 indicating the possibility of how a game player can make mistakes with PPEs before entering to laboratory. Depending on type of error, the player would lose points, while if mistakes are repetitive, he or she would likely to lose the level, which is indicated in Figure 3, with consequences of causes II and III of first step of entering laboratory. It was decided to let the player continue the level by entering laboratory even with missing PPE to help them remember the consequences of mistakes for further learning. For instant, player is dealing with HNO3 even without wearing gloves or goggles to make them analysis associated hazards as shown in Figure 3

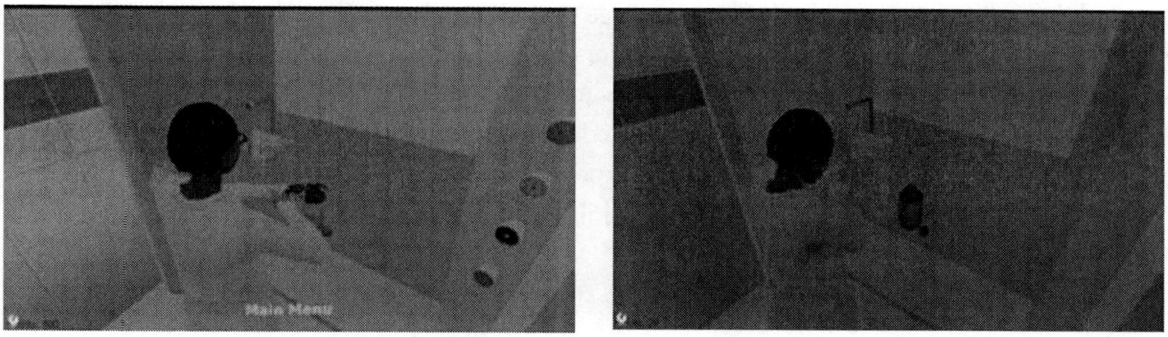

Figure 3: Screenshots of concequences of causes I and III

Kolb's Experiential Learning Theory (ELT) for LABSCAPE

Kolb defined learning as "the process whereby knowledge is created through the transformation of experience and knowledge results from the combination of grasping and transforming experience" [10]. The ELT model reveals two vertically related modes of grasping experience—Concrete Experience (CE) and Abstract Conceptualization (AC)—and two horizontally related modes of transforming experience—Reflective Observation (RO) and Active Experimentation (AE) [11]. Kolb's ELT cycle begins at any one of these four stages, but the following is one of the most often sequences suggested for the learning process [12]:

- Concrete Experience (CE): the learning process begins with the students actively carrying out an experience/activity
- Reflective Observation (RO): following the Concrete Experience stage, the students consciously reflect back on the experience;
- Abstract Conceptualization (AC): The students attempt to understand the general principles under the experience, trying to conceptualize a theory, model, or hypothesis of what is observed;
- Active Experimentation (AE): the students try to plan how to test a model, theory, and/or plan for new experiences, probably with different behaviors.

The impact and efficiency to achieve the complete navigation Kolb's learning cycle, it was believed to investigate methodological and pedagogical aspects of laboratory safety, using different levels of LABSCAPE. According to Kolb the ELT cycle is actually a learning spiral. When a concrete experience is enriched by reflection, given meaning by thinking and transformed by action the new experience created becomes richer, broader and deeper. Further iterations of the cycle continue the exploration and transfer to experiences in other contexts. John Dewey was the first to describe how the learning spiral leads to development; transforming the impulses, feelings and desires of concrete experience into higher order purposeful action [13]. A spiral ELT was introduced, where each cycle would represent each level of LABSCAPE. For instant, LABSCAPE level I defines four stages (CE, RO, AC and AE) of Kolb's ELT to learn introduction part of laboratory safety using PPE. While level two examine hazardous properties in preparation of Adipic Acid using Nitration reaction as shown in Figure 4. Level three is designed for student to understand how the changes in parameters influence consequences of reaction into which he was introduced during the level two. And finally, level four would actually be the implementation cycle, where learner would enter to laboratory to perform real experiments based on what he or she has learned in cycle one to three. Based on its application for laboratory experiment, it named as Introduce, understand, experiment and implement respectively as indicated in Figure 4.

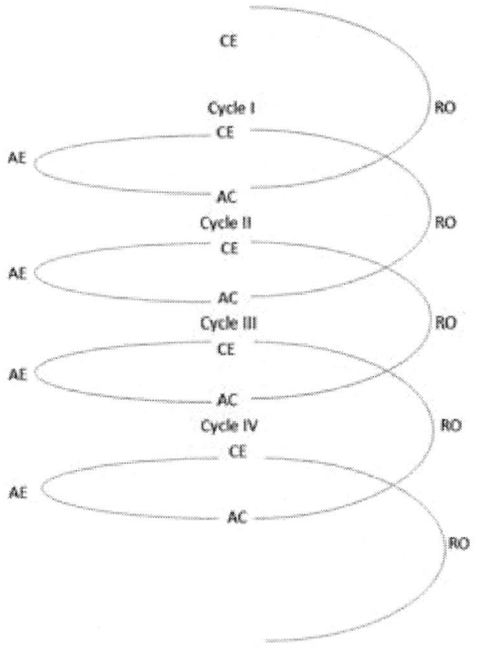

Figure 4: Spiral Learning using LABSCAPE

Table 2: Correlation of LABSCAPE with spiral ELT

Spiral	Concrete Experience	Reflective Observation	Abstract Conceptualization	Active Experimentation
Cycle I	Existing knowledge	Chatting	Developing	Playing
Cycle II	Reading manual	Brainstorming	Updating	Playing
Cycle III	Base of level II	Questioning	Developing	Playing
Cycle IV	Base of level I to III	Correlating	Correcting	Performing

Cycle I: It is designed to highlight the importance of PPE; CE would serve the gateway to enter the learning cycle using their existing knowledge. Discussing doubts or points with fellow players or chatting online with RO approach, would result in developing own concept for PPE as AC stage. This cycle would end with playing LABSCAPE (AE) one successfully, or losing the level would redirect to same level from scratch as indicated in Table 2.

Cycle II: This cycle highlights a Nitration reaction to prepare Adipic Acid which is detailed earlier in this article, would serve a CE stage, where a game player would recollect his experience for the reaction. Table 2 shows how Brainstorming season required to update existing knowledge of the reaction would offer RO with outcomes the cycle by constructing own ideas in AC stage to result in playing the game for AE.

Cycle III: Experience from cycle II would work as base knowledge to enter this cycle from CE stage. Questioning to elaborate those micro mistakes from past can serve the purpose of RO stage which results in development of some compact understanding towards the consequences in correlation of AC. Playing this final level of LABSCAPE precisely terms AE stage to end this cycle.

Cycle IV: Final cycle of ELT spiral where a game player enters to real laboratory to perform the Nitration reaction carrying the existing knowledge gained from LABSCAPE I to III levels for CE stage. Discussing game levels with other students in laboratory serves RO purpose with the output of AC to perform actual laboratory experiment (AE). This ELT spiral ends here with further learning cycles, where the base of the ELT would start from this level.

Conclusion

Mobile phone applications are increasingly in use for engineering education and chemical industry for different purposes. This paper was aimed to highlight the early stage progress of how we believe mobile game application could probably work on replacement of traditional safety training for senior staff or students who is working in same laboratory since long. Considering potential hazards of using mobile phone in laboratory, even if it has been suggested to play LABSCAPE before or after laboratory hours, it may be possible of continuity of LABSCAPE during laboratory hours. This would be conflicting towards the overall aim of this game with creating hazardous situation by playing LABSCAPE.

References

[1] J.Lu, Z.Meng, and J.B. Stav, "A new approach in improving operational efficiency of wireless response system," Computer and Information Technology (CIT), 2010 IEEE 10th International Conference on IEEE, (2010) 2676-2683

[2] C.Franzwa,Y. Tang, and A. Johnson, "Serious Game Design: Motivating Students through a Balance of Fun and Learning," Games and Virtual Worlds for Serious Applications (VS-GAMES), 5th International Conference on IEEE (2013), 1-7.

[3] D. Pano, C. Baulch, Smartphone use in process safety, Available web link with https://onlinelibrary.wiley.com/doi/10.1002/prs.11884, Accessed on February 13, 2019

[4] J. A. Frank and V. Kapila, Development of Mobile Interfaces to Interact with Automatic Control Experiments,. IEEE Control Systems, 2014

[5] Volodymyr Oliennikov, OSHA Safety Regulations, Checklists Audit Reports, 2016.

[6] W. Howard, OSHA Safety, 2013, Available web link at https://itunes.apple.com/us/app/osha-safety/id690989492?mt=8, access on January 15, 2019.

[7] W. T. Dotehlo, M. Marietto, and J. C. M. Ferreira, Kolb's experiential learning theory and Belhot's learning cycle guiding the use of computer simulation in engineering education: A pedagogical proposal to shift toward an experiential pedagogy, Computer Applications in Engineering Education 24 (2015), 79-81.

[8] M. Gunasekera, F. Khan and S. Ahmed, Learning Process Safety principles through practice, Process Saf Prog 37 (2018), 347- 354.

[9] N. D. Dholakiya, M. Ferjencik, D. Schofield and J. Kubik, Virtual learning for safety, why not a smartphone? , Available web link with https://onlinelibrary.wiley.com/doi/10.1002/prs.12005, Accessed on February 13, 2019.

[10] D.A. Kolb, Experiential learning: experience as the source of learning and development, Prentice-Hall, Englewood Cliffs, New Jersey, 1984.

[11] W. T. Dotehlo, M. Marietto, and J. C. M. Ferreira, Kolb's experiential learning theory and Belhot's learning cycle guiding the use of computer simulation in engineering

education: A pedagogical proposal to shift toward an experiential pedagogy, Computer Applications in Engineering Education 24 (2015), 79-81.

[12] D.A. Kolb, Experiential learning: experience as the source of learning and development, Prentice-Hall, Englewood Cliffs, New Jersey, 1984.

546961 Leveraging Advanced Data Analytics to Improve Process and Plant Safety

Abstract:

Digitalization is the next step in the industrial revolution to drive improvements in productivity within the Process Industries. This is made possible by the incredible speed at which new technological driver's the industry has available such as Increased computing power, improved communications, smarter sensors, virtualized architectures, cloud computing and process simulation. Historically, the responsibility to reduce process risk by reducing the likelihood of hazardous event was placed on the Safety Instrumented System (SIS). Today, plants are starting to look at areas to improve their overall plants operations to additionally reduce the initiating events that might have lead into these hazardous conditions.

What attendees will learn

In this session attendees will learn how advanced data analytics in their process plants can help reduce potentially hazardous scenarios, thus lowering their overall process risk. Reducing process risk through these methods will lessen the performance requirements on their SIS design which will help to lower costs and improved plant availability. The following areas to leverage data analytics for improving process and plant safety will be explored:

1. Optimize control loops to reduce potentially critical situations.
2. Reducing stress and work load of the operator by improving alarm and messaging systems.
3. Avoid unexpected plant downtimes by predicting failures before they happen to key process assets (valves, motors, drives, etc.)

551242 Alarm Management Using Data-Driven Methods

Alarm systems are a crucial part of modern control systems. They serve as a layer of protection and provide a prior indication to the operator about an abnormal process event. Alarms are designed and configured in the control systems and logged into historian for future references and reviews. With the progress in data mining and analysis methods, the alarm and event logs can be mined to generate diagnostic information about the healthiness and configuration of the alarm system. To demonstrate this, we propose a method to mine information from alarm & event log. This method is developed on an open-source programming platform. The input to the tool is an alarm and event log file from an asset management system or control system. This tool aids in (1) identifying and extracting alarm flood sequences, (2) providing graphs for visual analysis, (3) mining frequent alarm patterns and (4) bench marking the system against the Key Performance Indicators (KPI). The outputs can be used for offline analysis, rationalization of the alarm system, definition of suppression logics, and reduction in the alarm floods. The effectiveness of the proposed method and tool is established by application to a real industrial data set.

546373 Finding Unknown Anomalies before They Risk Process Safety

Four out of every five oil historical refinery accidents resulted from unexpected, unknown process and mechanical anomalies rather than known, expected failure modes. These unknown anomalies caused unexpected expenses of more than $100 million and ruined excellent company safety records. The oil refinery industry works diligently to address findings from all accidents to avoid repeat incidents. However, the next unknown anomaly which may cause a large property damage loss lurks obscured in a fog of maintenance and operations variance. Rather than wait fearfully for the next unknown anomaly to surface, AI Factory detects all known and unknowns as early as possible. This paper will explain how the latest Artificial Intelligence and Deep Learning techniques from BHGE, a GE company, help eliminate operational surprises. Because refineries generate a wealth of rich data, they need a fundamentally different approach to use information to detect anomalies which lead to 9 digit accidents. Using AI Factory, refiners will find unknown anomalies early and avoid catastrophic surprises. In addition, they can now rapidly refine data driven models and visualize unique, industrial scale operations.

546723 Digitalization in Process Safety: Fire Risk Prediction Analytics

As PETRONAS move towards digitalization, we envision to converge HSSE and operational excellence and realized the value of our extensive process data. One of our focus is to identify emerging process safety risk utilizing process safety data. With that, we embark on a pilot analytics project with the aim to enable timely interventions to prevent process safety incidents.

The analytic model is build based on Data-driven model and SME-input model. Data-driven model learns the pattern of past incidents while the SME input model detects abnormality based on available process safety information. Weightage is assigned to process indicators and alarm excursions to obtain a risk score and provide an early warning for intervention.

Furthermore, skeptical analysis was conducted against process safety leading indicators determine if there are any early warning indicators prior to an incident.

The pilot model was run on 5-years data in one of our facility with a selected recall accuracy and prediction accuracy. This set the basis of minimum viable product. Real time monitoring is currently in place for the pilot model. The results will serve as a basis for future risk analytic projects.

When Emergency Response Goes Fatally Wrong: Lessons Learned from Investigating Two Major Incidents

Scott Davis
Gexcon US
4833 Rugby Avenue Suite 100
Bethesda, MD
sgdavis@gexcon.com

Prepared for Presentation at
American Institute of Chemical Engineers
2019 Spring Meeting and 15th Global Congress on Process Safety
New Orleans, LA
March 31 – April 3, 2019

When Emergency Response Goes Fatally Wrong: Lessons Learned from Investigating Two Major Incidents

Scott Davis
Gexcon US
4833 Rugby Avenue Suite 100
Bethesda, MD
sgdavis@gexcon.com

Keywords: emergency response planning, lessons learned, case studies, major explosion incidents

Abstract

Emergency response plans are an essential, yet oftentimes overlooked, layer of protection for facilities where all other safeguards failed to prevent an incident. While catastrophic accidents such as large releases of chemicals, fires or explosions are devastating for the process industry, experience investigating numerous incidents has shown that a lack of an effective emergency response plan can lead to an unnecessary and tragic escalation of the initial incident. More specifically, investigation of two recent incidents: (1) 2013 West Texas Explosion and (2) 2015 explosion on the FPSO Cidade São Mateus; will demonstrate how fatal consequences resulted from the lack of emergency planning in the first case study and from deficiencies in the emergency response plan or implementation of the plan in the second case study. More specifically, the FPSO explosion resulted in 9 fatalities, eight of which were responding to a leak of a light-end hydrocarbon, and the West Texas explosion that resulted 15 fatalities, most of whom were responding to the fire that preceding the devastating explosion. Lessons will show that the fatal consequences could have been avoided with the proper emergency planning. In addition, while emergency response plans consider "maximum credible" scenarios, past events have shown that low probability high consequences should also be at least considered.

1 Introduction

Emergency response plans are an essential, yet oftentimes overlooked, layer of protection for facilities where all other layers of protection failed to prevent an incident. An emergency plan specifies procedures for handling sudden or unexpected situations, where the ultimate objectives are to be prepared to: (1) prevent fatalities and injuries; (2) reduce damage to buildings, stock, and equipment; (3) protect the environment and the community; and (4) accelerate the resumption of normal operations [1]. Developing the emergency response plan consists of five major steps (see Figure 1). The first step is to assess the vulnerability of the site and determine what hazards pose a threat to the installation. Such an exercise will help determine not only what "credible" events can occur at a given facility, but also what catastrophic or major events can occur. It should be noted that major emergencies are generally rare events.

Next, it is necessary to develop your emergency response plan and determine what means are currently available to execute the plan and what is still needed. At the planning stage, it is important that several groups be asked to participate, such as: subject matter experts and process engineers; the health and safety group; and even appropriate municipal officials should also be consulted since control may be exercised by the local government in major emergencies and additional resources may be required. Once the emergency response plan is developed, the necessary facilities and equipment need to not only be provided but tested and maintained. Communication, training and periodic drills will ensure adequate performance if the plan must be carried out. Lastly, there should be a periodic review of the emergency response plan.

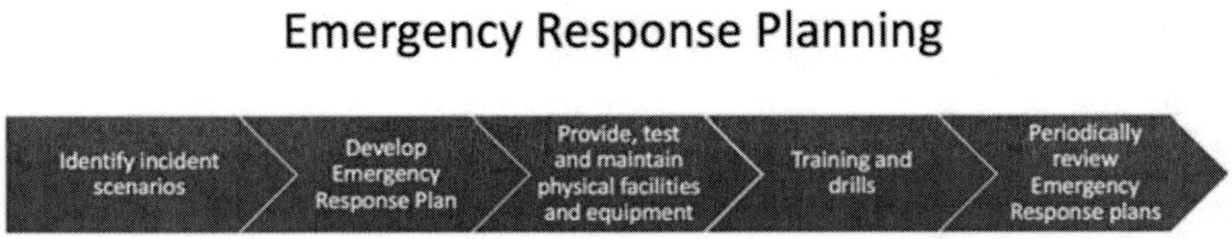

Figure 1: Steps of Emergency Response Planning

While the major benefit is to provide guidance during an emergency, developing the plan has other advantages. For example, previously unrecognized hazardous conditions, which would aggravate an emergency situation, could be discovered during the emergency planning and means put in place to mitigate or eliminate these hazards. The planning process may also identify deficiencies, such as the lack of resources (equipment, trained personnel, supplies), which can be easily corrected before an emergency occurs.

While catastrophic accidents such as large releases of chemicals, fires or explosions are devastating for the process industry, experience investigating numerous incidents has shown that a lack of an effective emergency response plan can lead to an unnecessary and tragic escalation of the incident. Chaos can quickly occur during an emergency if there are lack of resources and trained personnel coupled with an urgent need for rapid decisions and a very short period of time. The stress of the situation can lead to poor judgment resulting in severe losses, and normal channels of authority and communication cannot be relied upon to function routinely. Therefore, it is imperative to eliminate the "it can't happen here" mentality, as emergencies can occur and preplanning is necessary.

The importance of emergency planning will be discussed within the context of two recent incidents: (1) 2013 West Texas Explosion and (2) 2015 explosion on the FPSO Cidade São Mateus. These two case studies will demonstrate how the lack of emergency planning resulted in devastating consequences in the first case study and from deficiencies in the emergency response plan or implementation of the plan in the second case study.

2 Background of Incidents

2013 West Texas Explosion

On April 17th, 2013, an explosion occurred at the West Fertilizer Company storage and distribution facility in West, Texas. The explosion at West Fertilizer resulted from an intense fire in the seed and fertilizer area of the facility that led to the detonation of approximately 30 tons of ammonium nitrate stored inside a wooden receiving bin. The explosion occurred while emergency services personnel were responding to a fire at the facility. Fifteen people were killed, more than 250 were injured, and numerous buildings were damaged or destroyed.

The seed and fertilizer building (Figure 2) was built in 1958 and stored fertilizer and seed, which were distributed to farmers. Figure 3 shows the layout of the seed and fertilizer building, which included an ammonium nitrate (AN) storage bin neighboring the seed room, as well as an empty bin used for ammonium nitrate storage, and bins containing ammonium nitrate, K-Mag (fertilizer mix of potassium, magnesium and sulfur), Potash, ammonium sulfate, DAP (diammonium phosphate), and a bin for potassium chloride along the west wall. This building was capable of receiving rail shipments directly from the rail car by means of a rail spur junction and an underground conveyor belt and bucket elevator; the bucket elevator dumped directly into the northern AN storage bin, or onto a conveyor belt feeding the bins along the west wall.

Figure 2: West Fertilizer Facility – viewed from the west (upper) and east (lower)

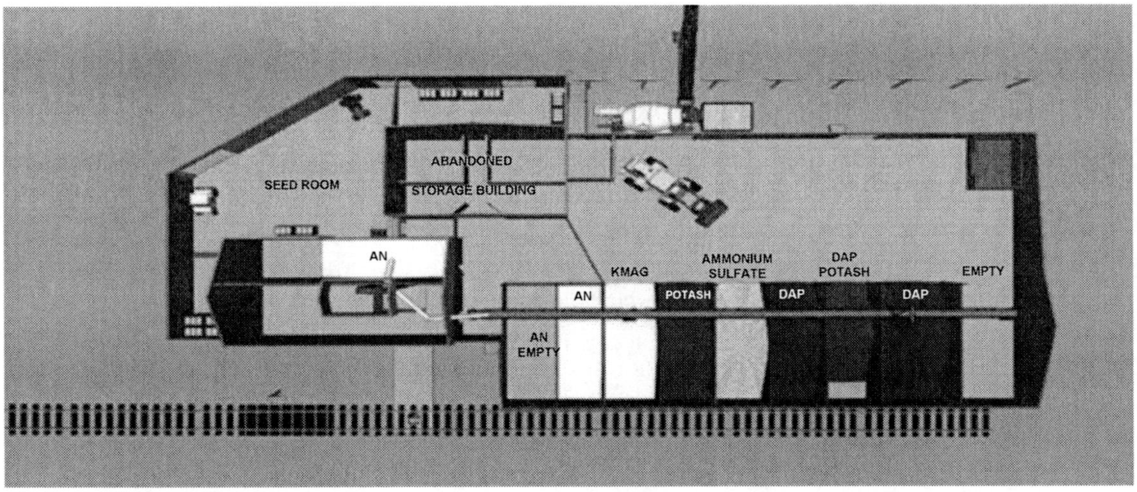

Figure 3: Layout of the Seed and Fertilizer Building.

On April 17th, 2013, at approximately 7:29 pm, a fire was reported, and fire fighters responded to the fertilizer plant [1]. All operations were already stopped for the day and employees had left the plant at about 5:30-6:00 pm. GexCon obtained several videos that recorded the progression of the initial fire [3]. Figure 4 shows the initial fire as viewed from the east and west; note that the highest point on the building represents the location of the bucket elevator by the northern AN storage bin. The image shows that the fire is isolated to the northern section of the building, which includes the seed room and portions of the northern AN storage bin containing about 30 tons of ammonium nitrate.

At approximately 7:50 pm, a massive explosion occurred at the fertilizer plant. Figure 5 shows images from a video taken from the high school parking lot to the southeast and Figure 6 the ensuing blast damage and crater from the detonation. Based on the videos from the day of the incident, there were actually two explosions: one smaller blast followed by a larger blast roughly 300 msec later.

Figure 4: Close up of the Seed and Fertilizer Building during fire as viewed from east (left image) and viewed from the west (right image)

Figure 5 – View of the explosion (viewed from southeast)

Fifteen people were killed as a result of the explosion, which included voluntary firefighters, and over 250 people were injured [3]. Damage was extensive throughout the neighboring community, which included homes, schools, an apartment building and a nursing home.

Figure 6: Blast damage and resulting crater

2015 explosion on the FPSO Cidade São Mateus

Due to the confidentiality of matters in this case, all references are made to two publicly available reports regarding the incident [5], [6]. The Cidade São Mateus was a Floating, Production, Storage and Offloading of Oil and Gas Export (FPSO), with a capacity to produce 4,000m³/day of oil and 10,000,000m³/day of gas as shown in Figure 7.

Figure 7: 2015 FPSO Cidade São Mateus.

On February 11, 2015 there was a leak of condensate (light-end hydrocarbon liquid) discovered in the pump room at a spade blind flange connection. Figure 8 shows that the pump room is situated just forward of the engine room and aft of the storage tanks. The layout and levels of the pump room is given in the left image of Figure 9, while the leaking spade blind flange is in the right image.

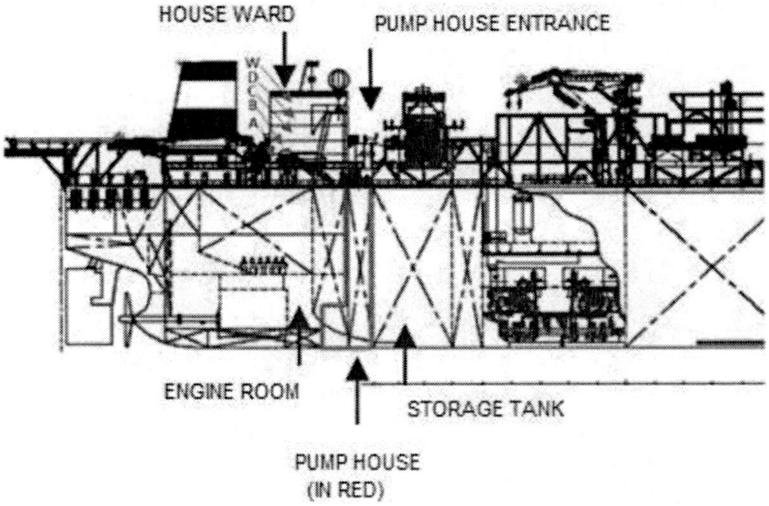

Figure 8: Cutaway side view of the pump room on the FPSO Cidade São Mateus.

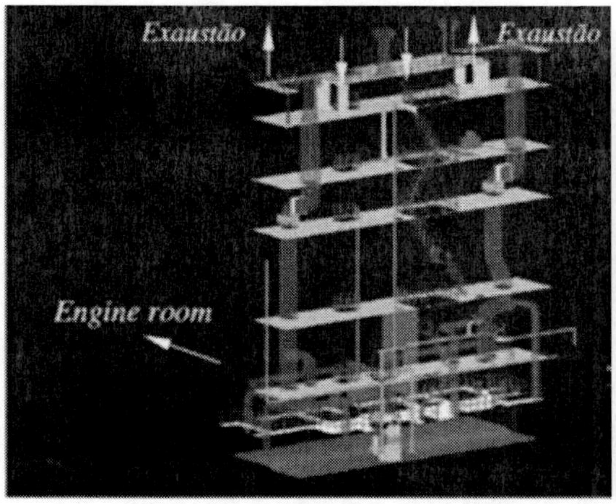

Figure 9: Model of the pump room on the FPSO Cidade São Mateus (left) and spade blind flange (right).

Just prior to the leak, the operations crew was attempting to drain condensate center cargo tank 6 (6C) into an alternate tank and they used an alternate pump alignment (stripping pump). However, the valve on discharge side of stripping pump was closed and they were pumping against a spade blind that was inadequate for such pressures, which ultimately failed the blind and seals. The gas detectors in the pump room began to alarm, and ultimately all three were alarming.

The ventilation was interlocked to shutdown per "confirmed gas" in the pump room. Emergency response teams entered the pump room to attempt clean up and repair the flange. There were three separate entries, and on during the third entry the explosion occurred. Figure 10 through Figure 12 show the level of blast damage to the pump room. Nine people were killed, of which eight were first responders.

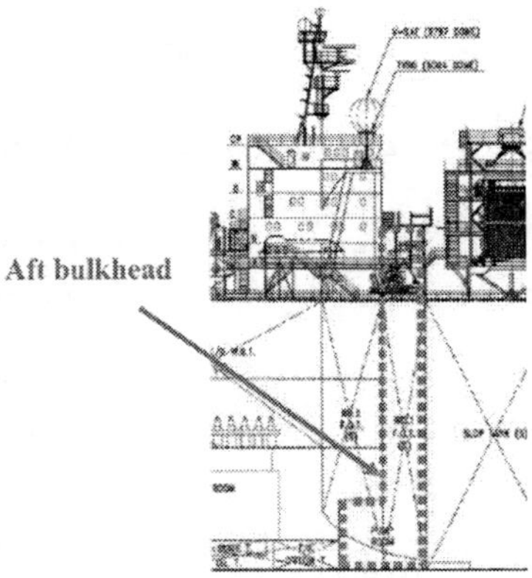

Figure 10: Damage to level above the pumps – complete bulkhead failure.

Figure 11: Damage to the engine room.

Figure 12: Damage to upper level of pump room.

3 Emergency Response

Response during the 2013 West Texas Explosion

As the fire occurred 20 minutes prior to the explosion, fire personnel were deployed to scene. Of the 15 people killed, 12 were first responders (8 volunteer firefighters and emergency services), and the other individuals were good Samaritans who supported the emergency response at the fertilizer plant. One of the volunteer fire fighters, who died in the blast, was employee of West Fertilizer. In fact, he was the plant foreman and knew what was stored in the facility. The emergency response is outlined in Figure 13.

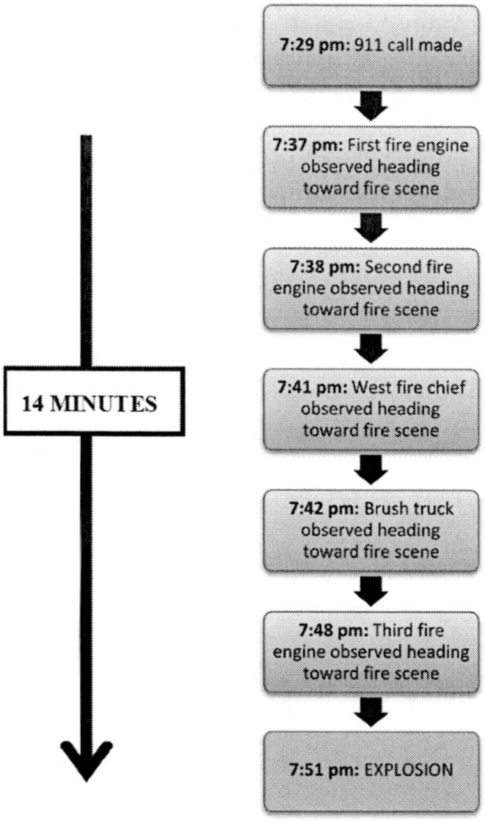

Figure 13: Emergency Response Plan at West Texas

Plant personnel and management did not understand the unique explosive nature of ammonium nitrate. More specifically, while the conditions necessary for detonation are not precisely known, the data suggest that detonations are promoted by severe shock or by heating with confinement, such as the massive fire at the plant. In fact, pools of molten ammonium nitrate may form in the event of fire, and if the molten mass becomes confined (e.g., in the elevator pit) it could explode. These hazards were not known nor understood by plant personnel and emergency responders. Therefore, responders were unaware of the potential of ammonium nitrate to detonate, had no knowledge of how long the fire had been burning, no robust incident pre-planning process in place, and no previous ammonium nitrate-related fire emergency training or drills. Hence, the strategy and tactics utilized by the West Volunteer Fire Department were not appropriate for the rapidly developing situation and they were exposed to unnecessary and extreme risk that proved fatal.

Response during 2015 explosion on the FPSO Cidade São Mateus

The first detector alarmed at 11:31, and by 11:38 all three detectors reached the over limit levels. The 1st team entered the pump room between 11:49-11:54 to assess the gas alarms, and they reported a leak at the flange and a 2-meter diameter pool under the flange. A 2nd team entered the pump room around 12:07 to assess the damage and whether repairs could be made. Despite the confirmed gas on all three detectors, the 2nd team still entered, and a personal gas detector alarmed at 100% LEL. The 2nd team exited around 12:15 and the ship was partially demobilized around 12:20 for lunch. The 3rd team enters around 12:27 to clean and repair, with ladders, wrenches, fire hose and absorbent blankets. The explosion occurred around 12:38 and killed four members in the pump room and four members at the top of the pump room on watch. Due to the failure of the wall into the engine room, the utilities operator was fatally injured in the engine control room.

There was a detailed Emergency Response Plan on the FPSO, along with the muster points for each person involved in the emergency response as shown Figure 14.

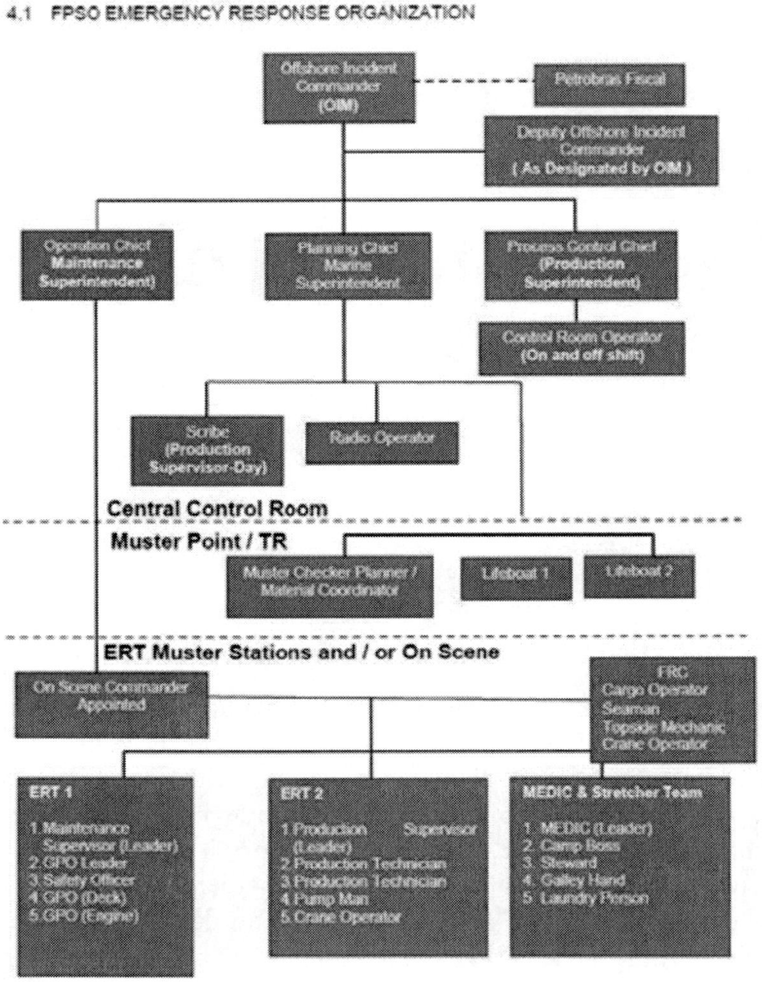

Figure 14: Organization Chart and muster points of the emergency response team

The Offshore Incident Commander is the Offshore Installation Manager (OIM) and would have the roles of implementing a command structure that would satisfy the organizational needs to protect life, environment and the installation, as well as deploying quick and efficiency actions to control incidents, as well as managing response actions to ensure that the escalating potential, risks of injury to personnel and damages to the installation are minimized. In the event of a leak in potentially explosive atmospheres, the emergency response must consider the possibility of the event escalating to a fire or explosion scenario. Thus, systems and procedures should be provided for the crew to control the situation and mitigate the risks of escalation from explosive atmospheres.

Investigation showed that the vessel was not designed for condensate, but for crude and gas. Yet despite the fact that there was confirmed gas detection within the pump room and on personal gas detectors, the crews entered three times. On the third entry, the explosion occurred. It was also reported that an explosive atmosphere in the pump or engine room was not included as an accident scenario in the emergency response plan, and there were no clear instructions of this particular scenario. Despite the knowledge that there was confirmed gas detection on the fixed sensors installed at the bottom of the pump room and also on personal detectors, Emergency Response Teams were sent into the pump room to evaluate the scenario in an explosive atmosphere.

The actions of Command essentially played down the risk of the condensate leak and allowed personnel to enter the pump room on multiple occasions to clean the spill without first performing a risk analysis, and without additional monitoring the room's atmospheric conditions. This noncompliance with the applicable emergency procedure was not only observed during entry operations, but also through the partial or total demobilization of the people irrespective of their muster points with an ongoing confirmed gas detection in the pump room. This demobilization resulted in two members being injured as they were found in a damaged elevator shaft after the blast.

4 Analysis

While deficiencies in the emergency response plans resulted in unnecessary and tragic consequences, the reasons each were lacking are very different and distinct.

For the West Texas Explosion, the Texas State Fire Marshal incident investigation [2] concluded that neither West Fertilizer nor West Fire Department had pre-fire and pre-incident plan at the facility. -Therefore, the recommendations provided were to ensure that local fire departments implement an Emergency Response Plan, with a particular focus on high-hazard occupancies. It was identified that fire departments visit and pre-incident plan responses to commercial structures in their jurisdiction. Further, they concluded that it is imperative that high-risk occupancies containing hazardous materials are identified and appropriate emergency response strategies and tactics are developed through pre-plan documents and SOP directives. Despite the extremely low likelihood of ammonia nitrate detonations, the recommendations state that high-hazard and high-risk occupancies containing heavy fire loads and hazardous materials should be identified by Fire Departments and all members made aware of their potential dangers. This will better prepare them for not only credible hazards, but also those high consequence hazards which may be found in the event of a fire and also provide them with insight as to when it is prudent to attack a fire or evacuate the area. More specific recommendations regarding the emergency response planning included:

- Develop a detailed Emergency Response Plan and Evacuation Plan for the area within 1 mile of existing ammonia nitrate buildings
- Develop standard operating procedures/guidelines to train and educate members on firefighting and emergency scene strategies and tactics for high-hazard occupancies
- Develop pre-incident plans by visiting the commercial and high-risk properties within the response jurisdiction.
- Utilize risk management systems on all emergency responses to determine when to employ alternative tactics.

- Incorporate incident management systems in every emergency incident, training exercise and basic operational process of the fire department to provide appropriate command and control for conducting operations.
- Implement a strategic incident action plan (IAP) for all emergency operations through a coordinated application of appropriate emergency scene tactics for the presented dangers
- Ensure that an incident safety officer (ISO), a person who is independent from the incident commander, is appointed and effectively utilized at every incident that would require an ISO.
- Establish a hazardous materials program and actively train members on the tactics to use when an incident involving hazardous materials occurs

In contrast, for the 2015 FPSO event, there was a detailed Emergency Response Plan for the vessel. Despite this plan, investigations concluded that there were significant deficiencies in the plan that needed to be remedied, and largely included reference to personnel exposure, failure to identify the accident scenarios for the emergency plan, training and mobilization. The execution of the plan failed to properly identify the potential high consequence and catastrophic result of the actions of the emergency responders. More specifically:

- Exposure to personnel
    - In the event of a leak in potentially explosive atmospheres indoors, the plan must consider the possibility of event escalation to a fire and/or explosion. Thus, systems and procedures should be provided once there is the formation of an explosive atmosphere. To ensure the success of mitigation actions, emergency procedures and resources for emergency response should be pre-planned, drawing on the use of all available resources, including the time of scenario assessment.
    - To determine that all functions involved in emergency response, including land-based support, have periodic theoretical training in emergency response procedures.
    - Set in the emergency response plan the available material and human resources to the response of each accidental
- Failure to identify accident scenarios or not included from the risk scenarios
    - Establish criteria and procedures for the migration of risk scenarios identified in the risk analysis and safety studies for emergency response plans. In addition to the selected scenarios, the procedures should contain scenarios that represent challenges to response actions, such as fatalities, and multiple injuries, among others.
    - Establish periodically simulated emergency covering all identified accident scenarios, including scenarios with multiple injuries and fatalities. Such simulations should include the activation of external resources to inform platform that response structure is effectively tested and that corrective measures arising from simulated ensure continuous improvement of the response actions of all entities involved.
- Demobilization of meeting points
    - Periodically coordinate all staff regarding the hazards involved in the assumptions made in the response procedure to the emergency unit.

  - o Prohibit the release of people to normal activities before the full control of the emergency.
- Ignition of the explosive environment
  - o As the ignition was introduced by the action of the emergency responders within explosive atmospheres, review the procedures related to emergency response and safe work practices considering the possible ignition sources identified in existing rules

5 Conclusion

The lessons from two specific incidents show that critical deficiencies in the emergency response plan or implementation of the plan can result in fatal consequences that could have been avoided. The 2013 West Texas explosion shows how important it is to not only recognize high risk events, but also have the appropriate plans in place to avoid unnecessary escalation of an already tragic incident, where 12 of the 15 people fatally injured were first responders. By preparing first responders for not only credible hazards, but also those high consequence hazards, can provide them with crucial insight as to when it is prudent to attack a fire or evacuate the area. In addition, not properly following or playing down the risk of a high consequence event was observed in the 2015 FPSO event, where eight of the nine crew members killed by the blast were first responders. The failure to recognize the potentially high consequence and catastrophic event, led to an emergency response that unnecessarily exposed responders and ultimately led to an exacerbated event. This study confirms that emergency response plans are an essential layer of protection for facilities where all other safeguards failed to prevent an incident, which could have prevented unnecessary and tragic escalation of the incident.

6 References

[1] Preparedness for Emergency Response - FEMA Training - FEMA.gov

[2] Texas State Fire Marshal, "Firefighter Fatality Investigation FFF FY 13-06," West, TX, 2013

[3] Davis, SG, DeBold, TF, & Marsegan, C (2017). Investigation Findings and Lessons Learned in the West Fertilizer Explosion. Journal of Fire Science. Volume 35, Issue 5, September 5, 2017: pp. 379-395.

[4] "A Public Health Report on Injuries Related to the West (Texas) Fertilizer Plant Explosion," June 24, 2014.

[5] Brazilian Navy Directorate of Ports and Coasts, Maritime Accident Safety Investigation Report, FPSO Cidade São Mateus - Explosion followed by flooding and casualties, February 11th, 2015.

[6] Investigation report of the explosion incident occurred on 11/02/2015 in the FPSO Cidade São Mateus, Brazilian National Agency of Petroleum, Natural Gas and Biofuels, Superintendence of Operational Safety and the Environment, August 2015.

Major Explosion at a Chemical Company in China: the Causes and Lessons Learned

Wei Wang
China University of Petroleum (East China)
No. 66, West Changjiang Road, Huangdao District, Qingdao, China
15621097585@163.com

Yi Liu
China University of Petroleum (East China)
No. 66, West Changjiang Road, Huangdao District, Qingdao, China
liuyi@upc.edu.cn

Dongfeng Zhao
China University of Petroleum (East China)
No. 66, West Changjiang Road, Huangdao District, Qingdao, China
zhaodf@vip.sina.com

Yanchun Wang
China University of Petroleum (East China)
No. 66, West Changjiang Road, Huangdao District, Qingdao, China
Wychun25@163.com

Jian Chen
China University of Petroleum (East China)
No. 66, West Changjiang Road, Huangdao District, Qingdao, China
836771528@qq.com

Hanhai Dong
China University of Petroleum (East China)
No. 66, West Changjiang Road, Huangdao District, Qingdao, China
donghanhaiupc@126.com

Prepared for Presentation at
American Institute of Chemical Engineers
2019 Spring Meeting and 15th Global Congress on Process Safety
New Orleans, LA
March 31 – April 3, 2019

Major Explosion at a Chemical Company in China: the Causes and Lessons Learned

Wei Wang
China University of Petroleum (East China)
No. 66, West Changjiang Road, Huangdao District, Qingdao, China
15621097585@163.com

Yi Liu
China University of Petroleum (East China)

Dongfeng Zhao
China University of Petroleum (East China)

Jian Chen
China University of Petroleum (East China)

Yanchun Wang
China University of Petroleum (East China)

Hanhai Dong
China University of Petroleum (East China)

Keywords: Fishbone Diagram; Analytic Hierarchy Process; Process Safety Management; Incident Investigation

Abstract

On July 12, 2018, a devastating explosion incident occurred in Yibin City Hengda Technology Co., Ltd. in Sichuan Province, resulting in 19 deaths and 12 injuries, and impacting an area of about 1,800 square meters. This was one of the most disastrous explosions that occurred in a chemical plant in China in the past five years. The incident reflects the general deficiencies of process safety management in small and medium-scale companies. Therefore, it is significant to analyze the causes of the incident and put forward countermeasures to avoid such kind of incidents.

According to the characteristics of complexity, diversity, and uncertainty about the influencing factors of the explosion, this paper presented a systematic analysis from qualitative and quantitative perspectives. The fishbone diagram model was used to identify influence factors from six aspects of people, design, management, environment, safety prevention, and materials. Furthermore, an assessment index including 26 factors was determined. The analytic hierarchy process method was used to analyze the importance degree of each factor quantitatively, and the key factors leading to the incident were obtained. The results showed that the sum of weights of three main aspects, including people, design, and safety prevention is nearly80%, so they were the dominant causes leading to this incident. Lack of chemical industry experience, non-standard process design, and incomplete automatic control system were the focus of prevention among 26 factors.

Process safety management is the application of management systems to the identification, understanding, and control of process hazards to prevent incidents from occurring. This analysis considered process hazard analysis, pre-startup safety review, operation procedures, training, and emergency plan and response, which were the main elements in process safety management framework. Finally, we proposed safety precautions for the identification, storage, production, and use of hazardous chemicals, as well as recommendations for process equipment management.

1 Introduction to the event

At 18:42:33 on July 12, 2018, an explosion occurred in Yibin Hengda Technology Co., Ltd. in the industrial park of Jiang'an County, Sichuan Province, causing 19 fatalities and injuring 12 people. The explosion caused the second and third workshops to catch fire, and the combustion area is $1800m^2$ [1]. The first workshop, warehouse, drying room, analysis room, office building, and other buildings had been damaged to some extent. The fire department dispatched 42 fire engines and more than 200 firefights. It took more than two hours to extinguish the fire on the spot as shown in Figure 1. It was estimated that the total energy of the explosion in this incident was equivalent to 235 kg of TNT. The direct economic loss was 41.42 million yuan.

At the time of the incident, the company was producing imazamox and triazole. Triazole is a pesticide, and imazamox is an herbicide. The process of the incident was the dehydration of butyramide in the process of synthesis of imazamox. At about 11:30 on July 12, the company collected 2 tons of chemicals labeled as raw materials (actually the explosive chemical sodium chlorate, not the raw material required for the production of butyramide). At around 14:00, the deputy director of the company's second workshop received raw materials from the warehouse. At around 15:30, the workers in the second workshop raised the raw materials for production through the lift to the third floor of the second workshop. At around 16:00, the 2R301 kettle for dehydration of butyramide was completed, and the feed was in an empty state. Before 17:20, the 2R301 kettle was completed. At about 6 pm, the dehydration completed. At 18:42:33, it was due to the on-site handover time. A chemical explosion occurred in the 2R301 kettle on the third floor of the second workshop. The direct cause of the incident was that the operator replaced the unlabeled sodium chlorate with butyramide and put it into the 2R301 kettle for dehydration, which caused the explosion and fire. Sodium chlorate is a kind of explosive chemical, which is easy to form an explosive mixture when mixed with butyramide. The explosion led to the disintegration of 2R301 kettle. The high-temperature toluene vapor rushed out rapidly formed an explosive mixture with the external air and produced a secondary explosion, which also caused the explosion and combustion of the materials such as sodium chlorate, toluene, and methanol stored at the workshop site.

Figure 1. Extinguishing at the scene.

According to the investigation, in this incident, the second workshop was the core area of the explosion, and the overall structure was severely damaged as shown in Figure 2. The platform on the third floor collapsed. The reactor with a height of nearly two meters and a diameter of about one meter fell from the upper level.

Figure 2. Severely damage to the second workshop.

2 Incident evaluation index system based on the fishbone diagram

Fishbone Diagram is also named as cause and branches figure analysis. Kaoru Ishikawa, a Japanese management master, invented it for finding out the root cause of incident [2]. It indicates the relationship between a problem and its underlying causes. It is a qualitative analysis tool to help identify incident causes [3]. With step by step in-depth study to analyze its influencing factors,

Fishbone Diagrams can systematize complicated incident causes. When starting the specific report, we first analyze the factors which affect the safety most and then followed by departure from the overall reason to find the basic reason and to detect and determine the main reason [4,5,6].

After the incident, the government set up a team of experts to identify risk factors from six aspects: human, design, management, environment, safety prevention, and material and establish an incident evaluation index system with 26 elements. The fishbone diagram that induced the explosion is shown in Figure 3.

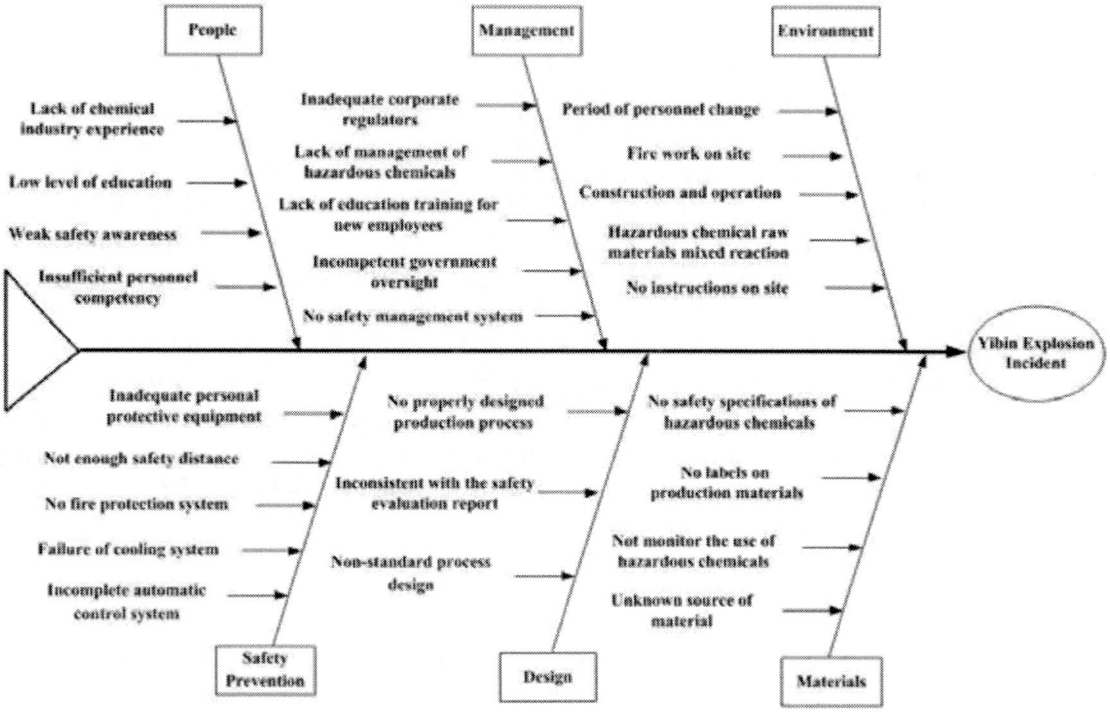

Figure 3. Fishbone diagram of Yibin explosion incident [7].

2.1 People

Neither the safety managers nor the operators demonstrated proficiency in the requirements of process safety competency. There were two main aspects below:

- Corporate management - The general manager of the enterprise had responsibility for the mechanical processing and production contract. However, he did not have appropriate experience in the chemical industry. The technical director did not have experience with management of process safety and production systems. The corporate legal person responsible for finance and logistics also had no experience in chemical engineering.

- Operators - Among the 19 people who died in the incident, 16 were operators, and most of these operators were local farmers. They lacked the basic knowledge of chemical process safety production, and they did not understand the process safety risks in the production process.

2.2 Design

The second workshop where the incident occurred was not constructed and installed strictly by the design drawings, and it was originally planned to have nine sets of 5000L tanks and one set of 3000L tanks and produce 5-nitroisophthalic acid. As built, the workshop had three sets of 5000L tanks, seven sets of 3000L tanks, and eight sets of 2000L tanks were set up. The product was also different from the original, but imazamox and triazole. A similar situation existed in the third workshop. The production technology of imazamox and triazole used by the company had no formal technical source, and the enterprise also did not entrust professional institutions for process calculation and construction drawing design. The technical director of the enterprise only planned the required production equipment and operation process according to his experience, but he knew nothing about the process safety risks of related technologies. It was during the production of imazamox that the distillation kettle used for the dehydration of butyramide by toluene first exploded. According to the *Notice on Further Strengthening the Safety Design Management of Hazardous Chemicals Construction Project*, Hengda company needed to start construction in April 2018, but the company had already started construction in July 2017. Also, the fire separation design of each building did not meet the requirements of the *Petrochemical Enterprise Design Fire Code (GB50160).*

2.3 Management

The lack of process safety culture is the main cause of management failure. Safety culture is developed through the interactions of factors such as safety policies and safety goals, the discipline of the organization, the behavior and attitude of the company staff, and the competence of workers. Improper operating discipline and procedures can lead to a negative process safety culture and an increased likelihood of process safety incidents. The enterprise had not formulated the post safety responsibility system, safety management rules, and regulations. Furthermore, the enterprise manager and the person in charge of each department only had the oral authorization, and there was no formal appointment document to clarify the post responsibility. The safety management organization was not complete. The company's safety and environmental protection department had only one safety officer who did not have safety management capabilities. There were also no full-time or part-time safety officers in each workshop. There was a lack of management of hazardous chemicals in the process safety management. The technology provider concealed some chemical name marks for confidentiality, and there was no safety specification. Required knowledge and competence connected with safety for workers were the pivotal content of safety training. New employees did not have security education directly work or internship. The government had not strictly regulated the safety departments of local key hazardous chemicals companies.

2.4 *Environment*

The factory built the workshop while producing, and used the polypropylene pipe which was easy to produce static electricity to transport flammable and explosive liquid materials. The building materials, hardware, and chemicals in the warehouse were mixed and piled with hidden dangers. The incident occurred during shift time, resulting in a large number of casualties. In the on-site operation workshop, there was neither operation instructions nor warning signs for hazardous chemicals. At the time of the incident, the second workshop produced two products at the same time. This practice of producing multiple produced at the same time in the same workshop has excellent security risks.

2.5 *Safety prevention*

Compared with the 20 elements of CCPS, China drew on the process safety management experience of the Occupational Safety and Health Administration (OSHA) and further optimized the formation of 12 elements for the production of Chinese chemical companies. After that, a series of process safety technologies were developed and promoted for the key elements of risk management and equipment integrity. In 2013, the SAWS issued *Guidance for enhancing chemical process safety management*. The Ministry of Emergency Management, which was integrated by the State Administration of Work Safety(SAWS) in 2018, attaches great importance to the supervision of hazardous chemicals, hazardous chemical processes, and major hazardous sources, referred to as "Two Priorities and One Major." If an enterprise violates The relevant provisions, it shall be punished according to the upper limit of economic punishment according to law.

In the production process of imazamox and triazole, the nitrification, oxidation, diazotization, hydrazine hydrate, hydrogen peroxide, methanol, and other dangerous chemicals and processes involved were all "Two Priorities and One Major" projects, which needed strict supervision. However, the company's automatic control system, flammable and toxic gas alarm system and fire water system and other safety facilities were not installed. Once there was a danger, the enterprise itself could not deal with an emergency. The lack of an automated control system led to more than ten operators performing manual operations around the reactor, which was an important cause of major casualties.

2.6 *Materials*

According to the investigation, the safety evaluation report made by the enterprise before the incident did not fully consider the process safety information, such as the lack of description of dangerous chemicals and the risk identification of the process, so the enterprise had great potential risks. Workers didn't know the source of raw materials, and many chemical production raw materials were not marked. Workers knew nothing about the risks of these dangerous chemicals, and company officials failed to communicate information about the hazards of the process to employees. The workers compiled codes to replace the names of products and raw materials, and these numbers were not abbreviated as specifications. The raw materials used by Hengda Company, such as hydrazine hydrate, nitric acid, toluene, hydrochloric acid were dangerous chemicals or precursor chemicals, but the company still purchased them illegally without relevant licenses. In the production process, regulators did not supervise the management of hazardous chemicals.

3 Determination of the weights of the incident incentive index based on the analytic hierarchy process

The Analytical Hierarchy Process (AHP) is a decision-making method proposed and used by American operations researcher T.L.Saaty in the mid-1970s [8]. AHP is a process that breaks down a problem in a hierarchical matrix with increasing level of detail to determine the most significant causes of an incident. According to the obtained fishbone diagram of the Yibin explosion incident, an incremental layer structure for analytic hierarchy analysis was constructed, as shown in Fig. 4, where the A, B, and C layers illustrated the primary effect and hierarchy of causes.

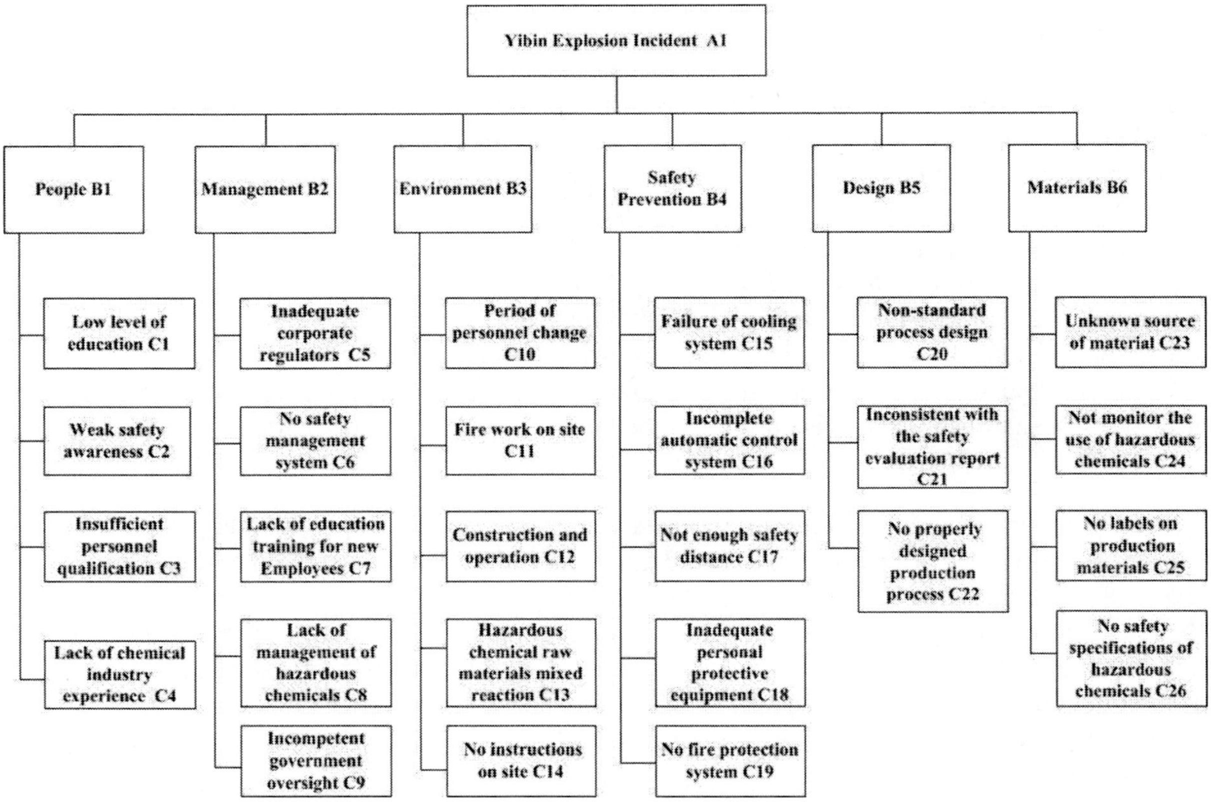

Figure 4. Hierarchical structures of safety evaluation system for Yibin explosion incident.

3.1 The single-layer index weights of the influencing factors

According to the actual data collected and expert opinions, we used 1-9 scale to construct the judgment matrix (A) of the first-level evaluation index, as shown in table 1, and obtained the influence of various factors in layer B on layer A. λ_{max} is the maximum eigenvalue, CI is the consistency index, RI is the average random consistency index, and CR is the random consistency ratio; When CR<0.1, it passed the consistency test [9]. According to table 1, the weights of design is the largest, followed by people and safety prevention, with the sum of the weights of the three as high as 0.8. It can be seen that the leading cause of this incident is due to the low level of process design, operators' insufficient qualification of chemical safety production and non-compliant protection measures. Therefore, enterprises should pay attention to intrinsic safety design, improve the safety awareness of staff by popularizing safety knowledge and operational guidelines, and finally achieve the goal of safe production.

Table 1. Judgment matrix of primary evaluation indicators and weights results

A	B1	B2	B3	B4	B5	B6	Weights	Consistency	Ranking
B1	1	2	3	1/2	1/3	7	0.1645	λ_{max}=6.4698 CI=0.094 RI=1.26 CR=0.0746<0.10 Satisfy consistency	3
B2	1/2	1	2	1/3	1/4	6	0.1060		4
B3	1/3	1/2	1	1/4	1/5	5	0.0700		5
B4	2	3	4	1	1/2	8	0.2545		2
B5	3	4	5	2	1	9	0.3810		1
B6	1/7	1/6	1/5	1/8	1/9	1	0.0239		6

3.2 The comprehensive weights of the influencing factors

We built each judgment matrix of layer C to layer B and calculated the single layer weights of layer C, and finally, we obtained the comprehensive weights of incidents caused by bottom influencing factors. We considered the product of these two layers as the comprehensive weights of the sub-factor. For example, the comprehensive weights of C1 is equal to the single-layer weights of C1 to B1 multiplied by the single-layer weights of B1 to A. The comprehensive weights result and ranking of each sub-factor were shown in table 2.

Table 2. Single and synthetic weights of C layer indexes

		Single weights	Consistency	Synthetic	Ranking
B1	C1	0.0843	λ_{max}=4.1375 CI=0.0458 RI=0.9 CR=0.0515<0.1 Satisfy consistency	0.0139	16
	C2	0.2120		0.0349	8
	C3	0.1544		0.0254	12
	C4	0.5492		0.0904	3
B2	C5	0.1556	λ_{max}=5.2550 CI=0.0638 RI=1.12 CR=0.0569<0.1	0.0165	15
	C6	0.0566		0.0060	21

	C7	0.4735	Satisfy consistency	0.0502	5
	C8	0.2578		0.0273	11
	C9	0.0566		0.0060	22
B3	C10	0.6253	λmax=5.3598 CI=0.08995 RI=1.12 CR=0.0803<0.10 Satisfy consistency	0.0438	7
	C11	0.0901		0.0063	19
	C12	0.0435		0.0030	23
	C13	0.1976		0.0138	17
	C14	0.0435		0.0030	24
B4	C15	0.0781	λ_{max}=5.1717 CI=0.042925 RI=1.12 CR=0.0383<0.1 Satisfy consistency	0.0199	13
	C16	0.5895		0.1501	2
	C17	0.1271		0.0324	9
	C18	0.0781		0.0199	14
	C19	0.1271		0.0324	10
B5	C20	0.6833	λ_{max}=3.0246 CI=0.0123 RI=0.58 CR=0.0236<0.1 Satisfy consistency	0.2604	1
	C21	0.1168		0.0445	6
	C22	0.1998		0.0761	4
B6	C23	0.1242	λ_{max}=4.1891 CI=0.0630 RI=0.9 CR=0.0708<0.1 Satisfy consistency	0.0030	25
	C24	0.2626		0.0063	20
	C25	0.5465		0.0131	18
	C26	0.0667		0.0016	26

As can be seen from table 2, the full weights of C16 (Incomplete automatic control system) and C20 (Non-standard process design) was greater than 0.1. The full weights of C4 (Lack of chemical industry experience) and C22 (No properly designed production process) were very close to 0.1. The combined weights of these four factors reached 0.577, which was more than half of the total. Therefore, these four influencing factors were the primary causes of the explosion incident in Yibin, Sichuan, as strict control and prevention were given priority.

Among them, both C20 and C22 belong to the design aspect, and enterprises should entrust qualified formal companies to design the process and production plan. In the method of safety design, the operation and design should be analyzed in real time to check the status of dangerous materials, effectively reduce the operating temperature and pressure, and control the influence of hazards factors in the process on chemical equipment. The equipment used in production shall meet the strength and anti-corrosion performance required by the process, and shall be designed and manufactured by qualified factories.C4 is a human problem. Therefore, it is necessary to conduct chemical safety education and operation training for operators before entering the plant. Many of Hengda's workers were the relatives and friends of management personnel. They were generally poorly educated and skilled in operation, and had not received any training in chemical safety. The workers did not go through systematic learning, but the workshop director directly taught them operating skills. C16 is a problem of safety prevention. The automatic control system can not only improve the operation efficiency of production equipment but more importantly, it can protect the safety of chemical personnel and reduce the loss caused by incidents. Enterprises should follow the safety production regulations, install automatic control systems, toxic and harmful gas alarm systems, and fire protection systems to effectively prevent incidents.

The weights of C7 (Lack of education training for new employees), C10 (Period of personnel change) and C21 (Inconsistent with the safety evaluation report) were all around 0.05. The weights of C2 (Weak safety awareness), C19 (No fire protection system) and C17 (Not enough safety distance) also exceeded 0.03. These five factors were important causes of incidents and should be effectively controlled.

C7 is a new employee training management problem. Enterprises should pay attention to the educational training of employees before entering the plant and implement the three-level safety training education of the plant, workshop, and team. C10 involves change management. At the time of the incident, it was during the shift time of employees, the change of personnel and the change of materials and equipment, which had great potential risks. Enterprises should establish a change management system, conduct comprehensive training for the staff who play a vital role in the change management process and clarify the position and responsibility of each person in the change system.

4 Lessons learned from the causes of the incident

The Yibin explosion was a pharmaceutical chemical incident. Through the investigation and analysis of historical disasters in the chemical industry in China, it was found that the proportion of pharmaceutical chemical incidents in the chemical industry was huge [10]. It was characterized by a random change of chemical process, small scale, low level of process control, involving dangerous materials and complex chemical reaction. Especially in small and medium-sized chemical enterprises, the lack of process safety management is an essential cause of such incidents. From the perspective of process safety management, this paper summarized and put forward suggestions for preventing such incidents.

- Process Hazard Analysis - In the design stage, the risk identification and assessment tool system should be applied to identify the risks of materials in the process of technology. In the review stage, HAZOP, SIL are used to review the process, layout and material flow. Companies should use scientific models to calculate the explosion-proof grade and safety distance of buildings. By adding the explosion-proof wall and the explosion-proof film, the building close to the installation area can reach the required safety factor. After the device is put into operation, it needs to be compared with the hazard analysis in the design stage, and the hazard that is inconsistent with the design should be analyzed again. Hengda's actual operating equipment, raw materials, and products were inconsistent with the original plan, and the company had not discovered and controlled the real hazards [11].

- Pre-startup Safety Review - The company shall organize a safety inspection team, whose task is to confirm whether the improvement measures and safety guarantee measures in the process hazard analysis report have been implemented as required; Whether adequate training of personnel in operation, maintenance, and management of process equipment has been completed; Whether necessary safety information of process equipment is completed and updated; Whether necessary equipment testing and checking have been completed and confirmed; Whether the process equipment is constructed and installed according to the design requirements [12].

- Operation Procedures - The incident occurred during the change management phase, and there was no review or other risk assessment during the shift. Staff should review the operating procedures as often as possible to ensure that the operating procedures reflect current operating conditions, including changes in chemicals, process equipment, and facilities. The enterprise shall confirm the adaptability and effectiveness of the operating procedures and ensure that the operators understand and master the operating systems [13].

- Training - A complete process safety management is inseparable from the participation of operators, and the human factor is the most unstable factor in process safety management. Therefore, Enterprises should establish strict production operation disciplines and require operators to master process safety information, equipment integrity management points, and emergency response policies and finally improve workers' ability to control equipment safety operations [14].

- Emergency Plan and Response - Enterprises need to establish a set of overall emergency plans, which usually stipulate in written form how the factory should deal with abnormal or urgent situations. Also, enterprises should establish emergency response teams, traditionally composed of enterprise personnel, but also include external staff; The responsibilities of each team member should be clear, ensuring that members have no questions about accountability and delegation. In case of an emergency, the person in charge may determine the safe area according to the emergency response manual and direct the person to evacuate to a safe place [15].

5 Conclusion

The circumstances of the incident described in this article are typical in China, and the incident reflects the common safety problems in small and medium-sized chemical enterprises. Managers should understand these lessons we learned from this incident. By using fish-bone diagram method, the causes of the explosion incident were obtained through qualitative analysis, including six aspects, such as people, design, management, environment, safety prevention and materials, and 26 factors influencing the incident were further analyzed. Then the analytic hierarchy process was used for quantitative analysis, and the results showed that lack of chemical industry experience, non-standard process design, and incomplete automatic control system were the main factors affecting the incident, so the enterprise should strictly control and prevent these factors. Finally, we summarized the causes of such incidents and put forward the measures from five aspects of process safety management, including process hazard analysis, pre-production safety inspection, operation procedures, training, and emergency plan and response plan. An incident brings not an only economic loss but also bad social influence to the factory. It is our hope that this paper will help prevent future accidents.

6 References

[1] Sichuan Provincial Emergency Management Office, Yibin Hengda Technology Co., LTD. "7.12" Major Explosion Fire Accident Investigation Report, Available at http://www.chinasafety.gov.cn/gk/jsts/201902/t20190214_224397.shtml, Accessed on February 14, 2019.

[2] Tongyuan Luo, Chao Wu and Lixiang Duan, Fishbone diagram and risk matrix analysis method and its application in safety assessment of natural gas spherical tank, Ind. Eng. J. 25(02), 2007, 138-141.

[3] Xing, W., QC Team Based Materials, China Society Press, Beijing, China, 2004.

[4] T Deng, X., Chen, Q. and Jiang, D., Application of fishbone chart analysis method in prevention for steam turbine overspeed incident, Guangdong Electr. Power, 20(02), 2013, 73-77.

[5] Changlong Zhu, Jiayong Zhu, Li Wang and M. Sam Mannan. Lessons learned from analyzing a VCE incident at a chemical plant, Journal of Loss Prevention in the Process Industries,50,2017, 397-402.

[6] Dennis P. Nolan, Chapter 5: Specialized Reviews—CHAZOP, EHAZOP, Bow-Tie Analysis, Layers of Protection Analysis, Safety Integrity Level, FishboneDiagram, and Cyber Security Vulnerability Analysis, Safety and Security Review for the Process Industries (Fourth Edition), Gulf Publishing Company,2015,Pages 17-27.

[7] Xinhua News Agency, Emergency management department: 7·12 incident enterprises have no law and discipline, and government supervision is lax, Available at http://www.gov.cn/xinwen/2018-07/17/content_5307196.htm?_zbs_baidu_bk, Accessed on July 17, 2018.

[8] Seong KonLee, GentoMogi and Jong WookKim. Decision support for prioritizing energy technologies against high oil prices: A fuzzy analytic hierarchy process approach, Journal of Loss Prevention in the Process Industries,22(6), 2009, 915-920.

[9] Saaty T.L., A scaling method for priorities in hierarchical structures, Journal of Mathematical Psychology, 15 (3),1977, 234-281.

[10] Heda Zhang and Xiaoping Zheng, Characteristics of hazardous chemical incidents in China: A statistical investigation, Journal of Loss Prevention in the Process Industries, 25(4), 2012, 686-693.

[11] Hao WU, Igor PEÑARRUBIA, Lin CUI and Jinsong ZHAO, Process safety management considerations for biofuel production, Higher Education Press, 4(3), 2017, 357-367.

[12] Yuan Gao, Yunxiao Fan, Jing Wang, Xi Li and Jingjing Pei, The mediating role of safety management practices in process safety culture in the Chinese oil industry, Journal of Loss Prevention in the Process Industries, 57, 2019, 223-230.

[13] Paul R.Amyotte and Cathleen S.Lupien, Chapter Three - Elements of Process Safety Management, Methods in Chemical Process Safety, 1, 2017,87-148.

[14] T. Kontogiannis, M.C. Leva and N. Balfe, Total Safety Management: Principles, processes and methods, Safety Science, 100(B), 2017, 128-142.

[15] Hao WU, Igor PEÑARRUBIA, Lin CUI and Jinsong ZHAO, Process safety management considerations for biofuel production, Front. Eng. Manag, 4(3), 2017, 357- 367.

Taming the Wild River Rapids – How Process Safety Can Apply Outside the Process Industries

Trish Kerin
IChemE Safety Centre
Level 7, 455 Bourke St, Melbourne, Victoria, Australia
tkerin@icheme.org

Prepared for Presentation at
American Institute of Chemical Engineers
2019 Spring Meeting and 15th Global Congress on Process Safety
New Orleans, LA
March 31 – April 3, 2019

—

Taming the Wild River Rapids – How Process Safety Can Apply Outside the Process Industries

Trish Kerin
IChemE Safety Centre
Level 7, 455 Bourke St, Melbourne, Victoria, Australia
tkerin@icheme.org

Keywords: lessons learned, process safety, design, maintenance

Abstract

On the 25th October 2016 the Thunder River Rapids ride at Dreamworld on Australia's Gold Coast suffered a pump failure, resulting in the tragic deaths of four riders. This incident was investigated by the police and later a coronial inquest to determine how a tragedy like this could occur. As a process safety professional, it was interesting to follow the coronial inquest findings and then think about how process safety management may have helped to prevent this incident. Assuming that a process safety event can be defined as a high consequence or potential consequence low likelihood loss of control event [1], this incident had all the hallmarks. Therefore, it is possible to apply the IChemE Safety Centre (ISC) framework [2] for process safety to identify where the system failed and understand how it could be averted in future. This paper will explore various elements of the incident and how process safety management techniques may have helped, including that the ride had been modified to be different from all other similar rides, that the staff did not know how to stop the ride in an emergency and that all means to stop the ride required manual detection and intervention, with no alarms or shut down systems installed.

1 The Incident

The Thunder River Rapids ride at Dreamworld in Queensland, Australia, was a water ride, simulating white water rafting. Up to six riders sit in a circular raft, which flows along a turbulent river, including travelling up a conveyor to ride down a waterfall. The water flow is driven by pumps and made turbulent by the installation of underwater obstacles. It was in full operation on 25th October 2016, when the pump managing the water flow tripped out. This resulted in the water level on the ride dropping. This drop in water level caused a raft to become stuck on the rails at the conveyor leading to the disembarkation platform. This raft remained stationary for some 57 seconds while the ride continued to run before it was impacted by the next raft in the sequence. This following raft contained 6 passengers, four adults and two children. When this raft struck the stationary one, it flipped over, killing the four adult passengers. The two children managed to escape the raft [3]. This incident is

—

currently under a Coronial Inquest. As such this paper is written from several media outlets that have been reporting the proceedings and based on the evidence reported, not the findings of the Coroner. The Coronial Report is due to be published in mid 2019.

2 History of the Ride

The Thunder River Rapids ride was completed in 1986 and was likely an in-house design and construction [4]. Sometime after construction, the ride was modified to remove some of the timber slats that form the base of the conveyor [4]. Two in every three slats were removed, though is unclear why this was done. This did create significant gaps in the base of the conveyor belt. The ride was also refurbished some months prior to the incident. [5].

Over the history of the ride there had been a number of incidents that had occurred with the rafts and the pumps. This in itself is not unusual, in that pumps can trip out from time to time, however there had been an incident in 2001 where the pump had tripped, and a raft became stuck [6]. This was then struck by a second raft, which flipped over. In this incident there were no passengers in the raft, and therefore no injuries. The investigation found that the ride operator had failed to adequately monitor the ride and he was dismissed. Details of this incident were not shared broadly and the operators and maintainers of the ride in 2017 had no knowledge of it. At the time an email was written by a park engineer discussing the incident and stated "I shudder when I think if there had been guests on the ride" [3]. This demonstrates that someone at some time recognized the significance of this incident, but it was then lost in history.

3 Technical and Organizational Causes of the Incident

There is one key immediate cause of this incident, however there are many different and wide-ranging causal factors. The immediate cause has been determined to be the tripping of the water pump, which in turn allowed the water level to drop, causing the first raft to become stuck. This provided the obstacle that the second raft, with six passengers, struck, causing it to then flip over. But let us examine the broader aspects of this incident, from the pump reliability to the safety systems and organizational capability. The ISC framework [2] provides a useful template to examine the causes. This framework in represented by Figure 1. Fundamentally leadership underpins all aspects of process safety. There are six functional elements that must be robust in an organization to manage major hazard safety. These are; knowledge & competence; engineering & design; systems & procedures; assurance; human factors and culture. In this incident there were clear failures in each of these elements.

—

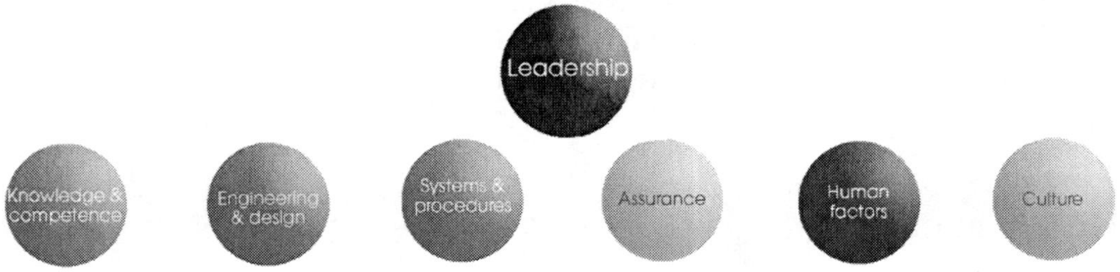

Figure 1. ISC Framework for Process Safety

3.1 Knowledge & Competence

The training of the ride operators took only 90 minutes and was unstructured with no evidence of competence consistently recorded [7]. There was a procedure manual however it was thought to be confusing by the ride operators [8]. The operator running the ride on 25th October 2016 had been trained for 90 minutes that morning, as it was her first day running the Thunder River Rapids. The junior operator stated that the trainer "showed me the front page of the manual and just flipped through it. I had to go through it myself. She said just sign off on it when you are happy." [9] The trainer stated that she sometimes had to "invent" training techniques because of the complexity of the control panel [8]. There was no clear guidance on how and when to use the emergency stop for the ride. The new operator testified that she was told "don't worry about it, you don't need to use it" with regard to the emergency stop button [10]. This is disputed by the person who delivered the training. The new operator believed it was the role of the senior operator to shut the ride down. It should be noted that the emergency stop would shut down the moving equipment with in two seconds, where the normal shutdown took around eight seconds. The ride operator that day stated that she "would have waited for the senior operator to tell her to hit the e stop" [10]. The senior operator that day was "surprised" to learn there was an emergency stop that would have shut the ride down within two seconds [7].

3.2 Engineering & Design

Intamin Amusement Rides designed and built the first river rapids rides in the world, and following this tragedy released a statement stating that the ride was "neither built by Intamin nor supplied by Intamin". The ride at Dreamworld was a custom build, and this was said to be common for rides built in the 1980's in the Gold Coast[4].

3.2.1 The Timber Slats

—

A few years after the ride was opened, a modification was made to the timber slats in the conveyor system [4]. The slats appear to serve at least two safety purposes. The first was to provide a solid platform for a passenger if they fell from the raft, essentially isolating them for the moving machinery and providing a platform to stand on [11]. The second was so that the rafts did not get caught in gaps and become stuck or indeed flip over. The Thunder River Rapids ride had two of every three slats removed. This provided large gaps between the remaining slats. There does not appear to be a theme park anywhere else that has made this change. There also appears to be no evidence as to why the change was made, nor was there a considered Management of Change (MoC) review undertaken. It is not clear if the organization had a structured and systematic MoC process. The safety regulator did not approve the modification and believed it added in additional hazards [4, 12]. In this incident, the raft became lodged in the gap, and as the conveyor continued to move forward, the raft was driven up vertically. When the passengers then fell from the raft, there was no solid base and they landed in the mechanism of the conveyor. There had also been conveyor framework modifications made to the ride in early 2016 in an attempt to stop rafts flipping over [13]. It is unclear what this modification was or indeed why it was made.

Figure 2. Aerial view of Thunder River Rapids showing the large gaps in the timber slat conveyor after the incident [10].

Figure 3. Aerial view of Infinity Falls, Orlando, showing close spacing of the boards on the conveyor [14].

3.2.2 The Pump

The water pumps had failed three times in the days leading up to the incident, and then twice on 25th October 2016, before the final failure that lead to the tragedy [15]. The pump's purpose was to keep the water flowing through the ride to keep the rafts moving. If the pumps stopped the water not only stopped flowing, it also drained back, and the water level dropped such that rafts could become stuck on the underwater obstacles that produced the rapids [16]. The reason for the pumps tripping out was not determined, the protocol seemed to be to just reset them to get the ride operating again [3]. There also is some confusion as to whether an alarm was sounded when the pumps tripped out, with an old procedure having the comment about an alarm crossed out [17]. At the time of the incident

—

the pumps were approximately 10 years old, and the manufacturer stated that they had a life expectancy of about a decade. There did not appear to be any plans to replace the pumps, however there was a maintenance request into the pump supplier in the week prior, but this visit did not eventuate [18].

3.2.3 Low Water

Low water in the ride was a critical hazard, as the rafts could become stuck or indeed flip. There had been previous incidents where low water level hazard led to rafts flipping, notably in 2001. There was no low water level switch that would shut down the ride if the water level became too low [3]. The only indication the operators had was to notice the water stain level on the wall of the ride canal to determine if the water was too low [19]. When the water level dropped on 25th October, one empty raft became stuck on the rails near the base of the conveyor. As the second raft containing four adults and two children engaged with the conveyor it was then driven into the stuck raft. A safety review conducted after the event stated that the primary cause was a lack of a safety rated water level detection system. It was believed that such a device could have been fitted for as little as AUD 3,000. The ride had been refurbished some seven months prior to the incident, but a device was not fitted. The nearby log ride at the park had such a safety device fitted [18].

3.2.4 Emergency Stop

Around two years prior to the incident an external safety report was written suggesting the emergency stop system should be simplified, however this was rejected by the company as the emergency system in place was seen to be an "acceptable risk". The stopping procedure for the ride contained several individual steps. There was a single button which would stop all components of the ride, however this was located at the unloading area and not the control panel where the operators worked. [20]

There was a great amount of confusion regarding the emergency stop button and what it actually did. The senior operator stated that they were "surprised" to learn there was an emergency stop that would stop all components and the junior operator on her first day was unsure what the button did [7]. She stated she was told to not worry about the button because she did not need to use it and testified, she "would have waited for my senior operation to tell me to press it". She also did not know the button would stop the conveyor from moving. [10] The maintenance workers who performed the daily checks on the ride also stated they did not know what the emergency stop button did as they never tested it [21].

3.2.5 Maintenance

Seven months prior to the incident a directive to cut all repairs and maintenance spending was ordered by the owners. At the time of the incident the facility was approximately AUD125,000 over maintenance budget for the year to date. Capital expenditure could continue to be spent during this time [16]. One maintainer stated that he performed comprehensive daily, monthly and annual checks on the ride, which include consumable

—

replacements such as nuts and bolts [22]. When the regulator inspected the ride after the incident, he noted it was in "generally poor condition" with excessive corrosion and rust on the electrical cables [23].

The maintenance planner noted that if spending could be deferred it would be. He also noted that the ride was shut annually for review, but on the nine occasions he witnessed a shutdown, he never saw an engineer inspect the ride nor were any risk assessment carried out [24].

3.2.6 Seatbelts

The Thunder River Rapids rafts had seat belts fitted for the safety of the riders. It should be noted that in May 2017, some seven months after the Dreamworld incident, that an 11 year old girl died on a similar ride, Splash Canyon, at Drayton Manor in the United Kingdom. The Splash Canyon ride rafts had no seatbelts installed at all, and as the child stood to swap seats with a fellow rider she fell overboard and struck her head. She died days later in hospital. This was not the first time a rider had fallen from the Splash Canyon ride [25]. This incident highlights the importance of proper restraints on these types of rides.

The seatbelts provided on the Thunder River Rapids rafts consisted of Velcro straps. It was vital that these were engaged correctly to provide the necessary restraint. Unfortunately, the Velcro was worn, and this made correct engagement more difficult. It is thought that at least one of the four riders may have survived the incident if the rafts had a better quality seatbelt, such as a mechanical buckle type that are on aircraft. This is because as the raft was shaken by the mechanism the belt failed to hold and the rider dropped to the conveyor system below [23, 26].

3.3 Systems & Procedures

Dreamworld has a series of operating procedures for the Thunder River Rapids ride however they were thought to be confusing [8]. During the investigation into the 2001 incident where empty rafts flipped, it was determined that the operator failed to follow the startup and emergency procedures. The investigation also reviewed the operating procedures and determined that the chance of a similar incident occurring with passengers on the rafts "is nil" [6]. A senior engineer believed that the cause of the 2016 incident was the operator not following procedures and that if procedures were followed the "system is a safe system" [27, 28].

There was also a lack of adequate emergency response procedures. Operators had discussed wanting better procedures and undertaking emergency drills, however, even two years after the incident, these have not been conducted at all [29].

3.4 Assurance

—

Over the history of the ride there had been a number of third-party reviews, inspections or audits carried out. Three years prior the incident a safety audit was conducted, and it recommended replacing the complicated emergency shut down process with a single emergency stop button, but this recommendation was never actioned [30].

There was also a legal requirement for the rides to have a "certificate to operate" and the Thunder River Rapids ride certificate was due to expire in January 2016. The company sought at extension as they were unable to find a "competent person" as defined under the legislation to inspect the ride [16]. Eight day before the incident the ride was inspected by a third-party engineer. He declared the ride as "structurally sound and properly functioning" even though he did not complete the full safety audit and merely did a visual inspection. He also did not inspect the log book as was required by law in a push to have the ride certified before it expired, though he admitted that he could have refused to certify the ride because of a "lack of documentation" [31, 32].

3.5 Human Factors

The control panel for the ride was confusing [3] and the operators needed to perform 36 different tasks in a one minute period, consistently [10]. This made for a system that encouraged human error, as it was not possible for the operators to be effective under these conditions. In addition to this, the engineering team believed the operators needed to simply follow procedures for the ride to be safe, not taking into account how difficult the procedures were or the actual workload. A senior engineer stated "In my view if procedures are followed the system is a safe system, you may as well not open the park. Procedures are there to be followed, if we don't follow procedures in everyday life, things happen." This blind focus on procedures does not take into account how the human interfaces with the machine and how the system advantages or disadvantages human performance.

3.6 Culture

A maintenance worker stated that he believed that the "safety culture at Dreamworld is very sound" [22] however this is in contrast to other workers who testified that familiarity had bred contempt with respect to the Thunder River Rapids ride [16], where new attractions were given more focus that existing ones [17]. Junior operators believed there was a culture of fear and had engaged with their union over their working conditions and the demands put on them regarding lack of rest breaks [33].

The culture also appeared to not learn from incidents, with the 2001 incident of flipping rafts said to be a one off and having "nil" chance of occurring with riders on board [6]. even though the consequence appeared to be understood at the time with an engineer writing in an email "I shudder when I think if there had been guests on the ride." [3].

The park relied on its apparent good safety history rather than challenging how and when an incident could occur [11]. They did not have an effective incident tracking system to

—

help with the learning process. They introduced a database but there was "general reluctance to use what was a fairly complex database" by the operators [34].

Safety did not appear to be a value in the company culture. The former safety manager stated that "everyone thought it was someone else's responsibility". He believed that deficiencies in the corporate structure led to a lot of problems. In 2018 his corporate safety role was made redundant.

Around five months prior to the incident, a memo was issued to staff to 'remain vigilant when monitoring conveyor movements" for items that might interfere with the rafts. This suggests there was some understanding of hazards and consequences, but it does to appear to have driven any improvement activities or modifications to prevent an incident [24].

4 Ramifications

The full ramifications of this incident are not yet clear. The Queensland government has initiated some legislative improvements to manage theme parks, but there is likely to be more recommendations coming from the Coronial Inquest when it releases its findings. Aside from the tragic loss of four lives, from the Dreamworld perspective, their guest numbers have remained low since the incident occurred, and it is struggling to attract patrons. This has led to their parent company suffering financial losses for the past two years and 50% decreases in the share price over the two year period [35].

The incident also provided an interesting lesson in crisis management. The company was scheduled to hold their annual general meeting in the days following the incident. The meeting went ahead and in it the CEO was awarded a performance bonus of $840,000 including both shares and a cash component despite the four deaths days before [36]. From a public relations perspective this was not a successful event. The CEO also stated they had been in contact with the victim's families, but this was publicly disputed by the families [37]. This further damaged the company reputation at a difficult time.

This incident will not only have ramifications for the families of the deceased, but for the survivors and the staff who responded due to the nature of injuries sustained by the victims. Four former staff members are suing Dreamworld for psychological damage [38].

These ramifications show how a process safety type incident can have significant impact on a business and its ability to continue operations in a positive way.

5 Could the Principles of Process Safety Have Helped?

As had been shown, there were clear and systematic failures across the whole organization that led to this tragedy. Starting with knowledge and competence, which relates to

—

organizational capability, had the operators been trained better in the safe operation of the ride as well as in emergency response, their performance that day may have been improved. From an engineering and design perspective, the design and installation of the ride was inadequate, it underwent undocumented changes which introduced hazards, and it was not adequately maintained. The systems and procedures for the operation and maintenance the ride was confusing and did not appear to contain necessary information. The assurance processes conducted did not seem to be thorough, rather they focused on keeping the ride operating and recommendations were not actioned. The ride was not designed with human factors in mind, from the high workload to the confusing operating panel. The organization culture did not seem to place a value on safety, it only focused on cost reduction. It also did not appear to be a learning culture, where lessons from the past are understood and embedded. Lastly at the time of crisis, the overall leadership displayed by the parent company was publicly found to be lacking.

There are transferrable learnings for the process industries from this incident, with some clear questions to ask. How are you managing your competency program to ensure you have the right capability? Are your designs adequate, maintained and any changes managed? Are your procedures thorough, identifying hazards, consequences and controls and easy to comprehend? How are you monitoring the operation of your facility? Does the design and operation of your facility enhance human performance or make it more difficult? Does your culture support safe operations?

6 Conclusion

A focus on the elements of process safety can help identify what is important in a business. In this instance there had been recommendations for a low-cost instrumented safety system to be installed that would have prevented the incident from occurring. The importance of this was not understood by the organization, and the opportunity to upgrade the ride during its recent refurbishment was ignored. It is clear that the hazard and potential consequences and controls were not understood. As is often the case, when the hazard is not understood and therefore not managed, people suffer tragic consequences.

7 References

[1] Kerin, T. *Managing Process Safety. In the Core Body of Knowledge for Generalist OHS Professionals*. Safety Institute of Australia, Tullamarine, VIC. 2017

[2] ISC (IChemE Safety Centre). Safety Centre Framework, Available at https://www.icheme.org/knowledge/safety-centre/framework/, Accessed on January 2, 2019

—

[3] ABC News, Available at https://www.abc.net.au/news/2018-06-18/dreamworld-fatal-accident-ride-had-broken-down-inquest-told/9871898, Accessed on January 2, 2019

[4] Australian Financial Review, Available at https://www.afr.com/news/dreamworlds-rapids-ride-different-from-others-around-the-world-20161101-gsfd07, Accessed on January 2, 2019

[5] 9 News, Video 11, Available at https://www.9news.com.au/2018/12/06/03/33/dreamworld-inquest-enters-final-days, Accessed on January 2, 2019

[6] news.com.au, Available at https://www.news.com.au/national/queensland/courts-law/coroner-james-mcdougall-requests-proof-all-dreamworld-rides-were-certified/news-story/e77275b97b0bbee9c1c541bda6df0035, Accessed on January 2, 2019

[7] ABC News, Available at https://www.abc.net.au/news/2018-06-19/dreamworld-operator-on-first-day-when-four-people-died-on-ride/9885214, Accessed on January 2, 2019

[8] ABC News, Available at https://www.abc.net.au/news/2018-06-27/dreamworld-trainer-promoted-since-four-killed-on-ride/9914188, Accessed on January 2, 2019

[9] 9 News, Video 5, Available at https://www.9news.com.au/2018/12/06/03/33/dreamworld-inquest-enters-final-days, Accessed on January 2, 2019

[10] ABC News, Available at https://www.abc.net.au/news/2018-06-20/dreamworld-operator-impossible-load-responsibilities-manage-ride/9889776, Accessed on January 2, 2019

[11] 9 News, Video 1, Available at https://www.9news.com.au/2018/12/06/03/33/dreamworld-inquest-enters-final-days, Accessed on January 2, 2019

[12] 9News, Available at https://www.9news.com.au/2018/11/20/03/32/dreamworld-ride-passed-wphs-safety-audit, Accessed on January 2, 2019

[13] ABC News, Available at https://www.abc.net.au/news/2018-10-08/dreamworld-inquest-hears-total-failure-to-ensure-safety/10350034, Accessed on January 2, 2019

[14] Intamin Worldwide, Available at https://www.intaminworldwide.com/infinity-falls-orlando/ , Accessed on December 12, 2018

—

[15] 9 News, Video 8, Available at https://www.9news.com.au/2018/12/06/03/33/dreamworld-inquest-enters-final-days, Accessed on January 2, 2019

[16] ABC News, Available at https://www.abc.net.au/news/2018-06-25/dreamworld-inquest-told-of-ride-maintenance-cutbacks/9906214, Accessed on January 2, 2019

[17] Australian Financial Review, Available at https://www.afr.com/real-estate/commercial/hotels-and-leisure/dreamworld-prioritised-new-attractions-over-maintenance-of-old-rides-inquest-20180626-h11urb, Accessed on January 2, 2019

[18] 9 News, Video 10, Available at https://www.9news.com.au/2018/12/06/03/33/dreamworld-inquest-enters-final-days, Accessed on January 2, 2019

[19] ABC News, Available at https://www.abc.net.au/news/2018-06-21/dreamworld-ride-operator-highly-distressed-give-evidence-inquest/9893308, Accessed on January 2, 2019

[20] news.com.au, Available at https://www.news.com.au/national/queensland/courts-law/dreamworld-engineer-tells-families-i-apologise-for-not-being-to-make-eye-contact-with-you/news-story/40545dc747c95eb96b0293c888d77d96, Accessed on January 2, 2019

[21] Sunshine Coast Daily, Available at https://www.sunshinecoastdaily.com.au/news/dreamworld-thunder-river-rapids-button-wasnt-regul/3543997/, Accessed on January 2, 2019

[22] ABC News, Available at https://www.abc.net.au/news/2018-06-26/dreamworld-ride-operator-says-conveyor-was-never-stopped/9910192, Accessed on January 2, 2019

[23] news.com.au, Available at https://www.news.com.au/national/queensland/courts-law/safety-report-shows-simple-change-on-dreamworld-ride-could-have-saved-lives/news-story/9c1fa2c195efd12f86a3e8ad516ea1b8, Accessed on January 2, 2019

[24] 9 News, Video 20, Available at https://www.9news.com.au/2018/12/06/03/33/dreamworld-inquest-enters-final-days, Accessed on January 2, 2019

[25] news.com.au, Available at https://www.news.com.au/travel/travel-updates/incidents/girl-11-dies-after-falling-into-the-water-on-splash-canyon-ride-during-school-trip-to-uk-theme-park/news-story/f83330b0ce12ea11df2d089ece676519, Accessed on January 2, 2019

—

[26] 9News, Available at https://www.9news.com.au/2018/12/06/03/33/dreamworld-inquest-enters-final-days, Accessed on January 2, 2019

[27] 9News, Available at https://www.9news.com.au/2018/10/16/14/43/dreamworld-inquest-thunder-river-rapids-ride-engineers-give-evidence, Accessed on January 2, 2019

[28] 9News, Available at https://www.9news.com.au/2018/10/17/10/49/dreamworld-inquest-engineer-to-appear-again, Accessed on January 2, 2019

[29] Australian Financial Review, Available at https://www.afr.com/news/dreamworld-ride-trainer-promoted-after-tragedy-staff-wanted-emergency-training-20180628-h11yke, Accessed on January 2, 2019

[30] 9 News, Video 12, Available at https://www.9news.com.au/2018/12/06/03/33/dreamworld-inquest-enters-final-days, Accessed on January 2, 2019

[31] ABC news, Available at https://www.abc.net.au/news/2018-10-09/dreamworld-ride-declared-safe-days-before-tragedy/10352164, Accessed on January 2, 2019

[32] ABC News, Available at https://www.abc.net.au/news/2018-10-17/dreamworld-supervisor-sympathy-has-families-in-tears-inquest/10387186, Accessed on January 2, 2019

[33] 9 News, Video 9, Available at https://www.9news.com.au/2018/12/06/03/33/dreamworld-inquest-enters-final-days, Accessed on January 2, 2019

[34] 9 News, Video 16, Available at https://www.9news.com.au/2018/12/06/03/33/dreamworld-inquest-enters-final-days, Accessed on January 2, 2019

[35] Australian Financial Review, Available at https://www.afr.com/research-tools/AAD/share-prices?days=1095, Accessed on November 26, 2018

[36] The Guardian, Available at https://www.theguardian.com/australia-news/2016/oct/27/deborah-thomas-set-to-receive-840000-in-bonuses-despite-dreamworld-deaths, Accessed on December 19, 2018

[37] ABC News, Available at https://www.abc.net.au/news/2016-10-27/dreamworlds-shareholders-ask-parent-company-to-support-victims/7970104, Accessed on January 2, 2019

—

[38] news.com.au, Available at https://www.news.com.au/lifestyle/real-life/news-life/dreamworld-first-responders-to-sue-theme-park-for-psychological-damage/news-story/e8fd7399eec9d165c657dbdb3ac35968, Accessed on January 2, 2019

Fatal Toxic Chemical Release at DuPont

Tamara Qureshi
U.S. Chemical Safety and Hazard Investigation Board
1750 Pennsylvania Ave, NW Suite 910 Washington, DC 20006
tamara.qureshi@csb.gov

Dan Tillema, Steve Cutchen, Lauren Grim, and Chris Lyon
U.S. Chemical Safety and Hazard Investigation Board
1750 Pennsylvania Ave, NW Suite 910 Washington, DC 20006

Prepared for Presentation at
American Institute of Chemical Engineers
2019 Spring Meeting and 15th Global Congress on Process Safety
New Orleans, LA
March 31 – April 3, 2019

Fatal Toxic Chemical Release at DuPont

Tamara Qureshi
U.S. Chemical Safety and Hazard Investigation Board
1750 Pennsylvania Ave, NW Suite 910 Washington, DC 20006
tamara.qureshi@csb.gov

Dan Tillema, Steve Cutchen, Lauren Grim, and Chris Lyon
U.S. Chemical Safety and Hazard Investigation Board

Keywords: DuPont, toxic release, gas detection, alarm

Abstract

On November 15, 2014, four workers were killed after approximately 24,000 pounds of highly toxic methyl mercaptan released from an insecticide production unit (Lannate® Unit) at the E.I. du Pont de Nemours chemical manufacturing facility in La Porte, Texas. This paper will discuss the events leading to the methyl mercaptan release, the deficiencies in DuPont La Porte's toxic gas detection system, and offer lessons on how to prevent some of the deficiencies in the toxic gas detection system from occurring. An effective toxic gas detection system can protect workers by alerting them of a potentially dangerous situation and preventing them from unknowingly entering a potential toxic chemical release.

1 Introduction

On November 15, 2014, approximately 24,000 pounds of highly toxic methyl mercaptan released from an insecticide production[1] unit (Lannate® Unit) at the E.I. du Pont de Nemours chemical manufacturing facility in La Porte, Texas. Four workers at the DuPont La Porte facility entered the Lannate® Unit's manufacturing building (manufacturing building) and the highly toxic methyl mercaptan release. Three of these workers were responding to a coworker's distress call from inside the manufacturing building. The fourth worker entered after all the workers inside the manufacturing building stopped communicating with workers inside the control room. The Lannate® Unit's methyl mercaptan toxic gas detection system did not effectively warn workers of the large chemical release occurring within the manufacturing building. Additionally, this toxic gas detection system did not provide emergency responders with sufficient information to effectively evaluate the potential harm to the public.

[1] The Insecticide Business Unit (IBU) was divided into two primary insecticide product lines: Lannate®, which is DuPont's trade name for methomyl; and Agricultural Products Intermediates (API), which produced insecticides and nematicides known as Vydate®, the trade name for oxamyl [3, p. 11].

The DuPont La Porte methyl mercaptan toxic gas detection system had several deficiencies. First, the methyl mercaptan detector system did not meet its design intent. The alarm set point for methyl mercaptan concentration and the method of conveying alarms did not forewarn workers of the immediately dangerous to life and health (IDLH) conditions they would encounter by entering the manufacturing building. The methyl mercaptan gas detectors only alarmed on the Lannate® control room operator's computer screen. There was no audible or visible alarm provided in the field for outside personnel. Second, the placement of the methyl mercaptan toxic gas detectors did not fully portray the extent of the November 15, 2014 methyl mercaptan release. Third, the DuPont La Porte site normalized the hazards associated with methyl mercaptan. Finally, the alarms did not prompt protective action or investigations into the sources of alarms. During the hours prior to the November 15, 2014 incident, multiple highly toxic chemical gas detectors alarmed (sounded). However, the DuPont emergency response team was not notified nor was the area cleared of personnel. In addition, methyl mercaptan releases on November 13 and 14, 2014, were identified by methyl mercaptan detectors, but were never reported as releases nor investigated as serious process safety incidents. Personnel believed they knew the source of the releases and normalized their responses to these alarms. Consequently, there was a delayed awareness of the release by operations, and workers who entered the manufacturing building after the release began were not alerted of nor prepared for the lethal concentrations of methyl mercaptan being released, resulting in the death of two of these workers.

The DuPont La Porte incident is the CSB's third investigation of a fatal DuPont incident in five years.[2] On September 30, 2015, the CSB issued a presentation, safety video, and an Interim Recommendations report [1],[3] which detailed needed safety improvements at the facility and issued recommendations to the DuPont La Porte facility to be implemented prior to resuming operations of the Lannate® Unit.[4] Due to the amount of methyl mercaptan stored at the DuPont La Porte site, the Lannate® process was covered by Occupational Safety Health Administration's (OSHA) Process Safety Management (PSM) program and the Environmental Protection Agency's (EPA) Risk Management Plan (RMP) Rule [2].

An effective toxic gas detection system can make the difference between personnel being prepared for a potentially dangerous situation or entering a potential toxic chemical release. To be effective, a toxic gas detection system needs to consider the chemical's hazards, the gas detector's

[2] The CSB also investigated an incident at a DuPont plant in Belle, West Virginia (three incidents that occurred in a 33-hour period over January 22-23, 2010, resulting in one fatality), and another in Buffalo, New York (a single incident that occurred on November 9, 2010, resulting in one fatality and one serious injury) [16], [17].

[3] In the Interim Recommendations, the CSB identified significant process safety management deficiencies at the DuPont La Porte facility, including delayed maintenance of safety-critical equipment, lack of written procedures, poor hazard analysis practices, lack of implementation of important inherently safer design concepts, and a lack of hazard recognition.

[4] In response to the CSB findings and recommendations, DuPont committed to making changes at the DuPont La Porte facility. DuPont La Porte, however, was not able to address all of these findings and recommendations, because in the spring of 2016 DuPont announced that it would not reopen the Lannate® Unit. The company had determined that "significant changes in market conditions during the period of the shutdown [would] persist over the long term [14]" and the cost required for the restart "[was] not a long-term viable and cost-efficient option for the DuPont Crop Protection business [15]." In 2017, DuPont dismantled and removed the buildings and equipment from the DuPont La Porte facility that were associated with its insecticide and herbicide business units, which included the Lannate® Unit.

capabilities, the location of the gas detectors, and the alarms associated with the gas detectors. This paper will discuss some of the events leading to the methyl mercaptan release, the deficiencies in DuPont La Porte's toxic gas detection system and offer lessons on how to prevent some of the deficiencies in the toxic gas detection system from occurring.

2 Events Leading to the Incident

On Monday, November 10, 2014, an inadvertent chemical dilution caused operating difficulties that forced a shutdown of DuPont La Porte's Lannate® Unit [3, pp. 20-21]. On Wednesday, November 12, 2014, operators tried to restart the Lannate® process. During the shutdown, the methyl mercaptan piping to the reaction section of the process plugged due to the formation of a clathrate hydrate (hydrate),[5] halting restart. At that time, operators did not know the source of the plugging and began troubleshooting the process to try to clear the piping, unaware that the solid hydrate had formed.

Two days later, on Friday, November 14, 2014, the troubleshooting efforts to clear the plugging were still ongoing. That morning, a technical team composed of engineers and other employees experienced in the Lannate® process, met with operations personnel to discuss troubleshooting options to clear the plugging. This meeting identified the likely scenario that water had entered the methyl mercaptan system and a solid hydrate formed as a result. Based on their understanding of the hydrate,[6] the technical team asked operators to put hot water on the outside of the methyl mercaptan piping, under the insulation, to warm the piping and its contents, to break-up (dissociate[7]) the plugging. The team also realized that when heated, methyl mercaptan would expand, requiring a safe place to vent to avoid over-pressuring the piping. To address this concern, operations personnel opened valves between the methyl mercaptan piping and a waste gas vent header, not recognizing that this alignment created a pathway for liquid methyl mercaptan to eventually release from the piping inside the manufacturing building when drain valves were opened. In addition, when operations personnel opened other valves while troubleshooting, such as drain valves, methyl mercaptan released to the atmosphere causing a strong methyl mercaptan odor[8] that could be smelled by site personnel.

[5] At low temperatures (≤ 52 °F) water and methyl mercaptan form a solid, ice-like material called methyl mercaptan clathrate hydrate [10]. Temperatures in the Houston area for the 24 hours preceding the incident averaged approximately 40 °F and had been consistently below 55 °F since Tuesday, November 11, 2014. Although the potential for methyl mercaptan hydrate formation had been identified in DuPont's methyl mercaptan technical standard and in a process hazard analysis years earlier, DuPont never implemented safeguards, such as heat tracing, or developed a procedure to dissociate the hydrate safely [3, p. 22].

[6] DuPont's methyl mercaptan technical standard identifies the potential to form a hydrate at low temperatures. The technical standard states, "[methyl mercaptan] will form a hydrate with water, which is a solid below 40 deg F [degrees Fahrenheit] per information provided by a [methyl mercaptan] supplier."

[7] Methyl mercaptan clathrate hydrates dissociate back to the liquid phases of methyl mercaptan and water at approximately 52 °F [10].

[8] Methyl mercaptan has a low odor threshold. It gives off a smell like rotten cabbage. It can be smelled at only two parts per billion (ppb) [9, p. 1]. OSHA's permissible exposure limit (PEL), however, is 10 parts per million (ppm), which is 5,000 times the minimum concentration it can be smelled. The low odor threshold of mercaptans makes them useful as fuel gas odorants [9, p. 2].

During this troubleshooting process, methyl mercaptan released both outside and inside the manufacturing building, triggering 32 methyl mercaptan gas alarms on the control panel throughout the 17 hours preceding the incident. While methyl mercaptan is a toxic and flammable chemical, and at high enough concentrations lethal, operations personnel did not respond in accordance with the nature of the emergency circumstances, because they associated the alarms with ongoing troubleshooting efforts. They did not perceive the methyl mercaptan alarms as signifying a serious hazard because they normalized[9] both the methyl mercaptan odor within the Lannate® Unit as well as the detector alarms associated with them.

Early in the morning on Saturday, November 15, 2014, the hot water from the hoses warmed the hydrate plugging, causing it to dissociate, clearing the plugging. Due to the valve alignment, liquid methyl mercaptan flowed into the waste gas vent header, located inside the manufacturing building [3, pp. 25-26]. This vent header, however, was not intended for liquid methyl mercaptan.

At 2:51 am alarms began to sound on the control system indicating high pressure in equipment located inside the manufacturing building. Operations personnel did not realize that liquid methyl mercaptan was causing the high pressure. Instead, they attributed the high pressure to a common, long-standing problem with process condensate (that DuPont La Porte personnel believed to be mostly water) accumulating in the vent header piping [3, p. 28]. The Lannate® control room operator (Board Operator) contacted the night shift supervisor (Shift Supervisor) and another operator (Operator 1) separately, asking them to help troubleshoot the high-pressure situation. The Shift Supervisor and Operator 1 each went separately to the manufacturing building, likely with the intent to drain liquid from the vent header piping manually.[10] Sometime between 3:01 am and 3:13 am, two sets of drain valves on the vent header piping, located on the third floor of the manufacturing building were opened (Figure 1). But instead of being comprised mostly of water as workers believed, liquid methyl mercaptan that had filled the piping escaped out of the valves, vaporized, and killed the Shift Supervisor.

Between 3:24 am and 3:26 am, three methyl mercaptan detectors inside the manufacturing building sensed at least 25 parts per million (ppm) methyl mercaptan[11] and triggered alarms at the control panel. The Board Operator—still focused on reducing the high pressure—did not realize a major chemical release inside the manufacturing building was occurring.

[9] Normalization of deviance is a phenomenon in which individuals or work teams gradually accept a lower standard of performance until the lower standard becomes the norm. It is typically the result of conditions slowly changing and eroding over time [11, p. 4].

[10] In 2011, DuPont installed a NOx (nitrogen oxides) reducing scrubbed (NRS) incinerator system. The waste gas vent header piping to the NRS Incinerator was designed without sufficient consideration of liquid accumulation. It contained low points where liquid could accumulate, and there was no engineered equipment provided, such as a knock out drum, to remove liquid safely from the waste gas vent header piping system. The liquid accumulation in the piping would create pressure in process equipment. Pressure buildup within the equipment was resolved by manually draining liquid from the waste gas vent header. See DuPont La Porte, Texas Chemical Facility Toxic Chemical Release: Interim Recommendations [3, p. 18].

[11] The methyl mercaptan detectors alarm at 25 ppm–well above the National Institute of Occupational Safety and Health (NIOSH) recommended exposure limit (REL) of 0.5 ppm and the OSHA enforceable (PEL) ceiling limit of 10 ppm. OSHA recommends that employers consider using the alternative occupational exposure limits because the agency believes that exposures above some of these alternative occupational exposure limits may be hazardous to workers, even when the exposure levels are in compliance with the relevant PELs [5]. The Cal-OSHA permissible exposure limit (PEL) is an 8-hour time-weighted average (TWA) of 0.5 ppm and NIOSH recommended exposure limit (REL) is a 15- minute ceiling limit of 0.5 ppm [7], [6].

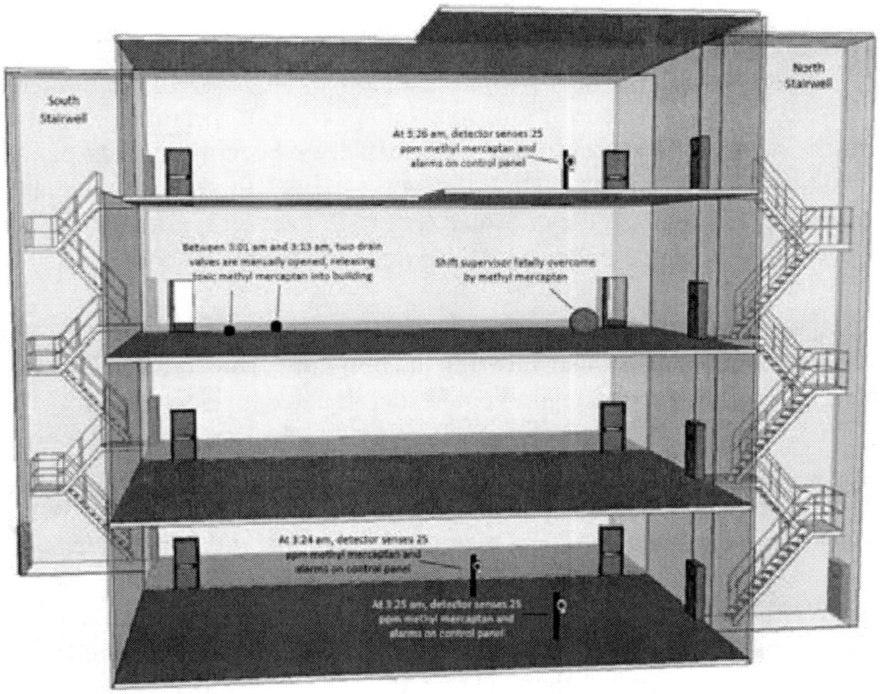

Figure 1. Depiction Showing Location Where Drain Valves Were Opened. These drain valves released toxic methyl mercaptan into the manufacturing building. Methyl mercaptan detectors on the first and fourth floors detected high concentrations of methyl mercaptan shortly after the release began. This depiction is a simplified graphic of the manufacturing building and does not show the location of Operator 1. Source: CSB

At about 3:30 am, Operator 1 made an urgent call for help over the radio. Various personnel interpreted her communication in different ways, generally as either "We need help!" or "I need help on the fourth floor!" The Board Operator tried to get more information through radio communication, but neither the Shift Supervisor nor Operator 1 ever responded [3, p. 29].
Two operators (Operator 2 and Operator 3) who were in the control room and heard the distress call ran to the manufacturing building to help. Another operator (Operator 4) saw Operator 2 and Operator 3 running to the manufacturing building, and he followed them. None of these operators knew there was a major release of toxic methyl mercaptan inside of the manufacturing building, and therefore they did not wear any respiratory protection when they ran into the manufacturing building. The manufacturing building lacked automatic visual or audible alarms to alert field workers or prevent them from entering a potentially toxic atmosphere.

Operator 2, Operator 3, and Operator 4 each entered the south stairway and took different routes inside the manufacturing building to try to find Operator 1. Operator 2 went to the third floor, where he was fatally overcome by methyl mercaptan [3, p. 29]. Operator 4 went to the second floor. He walked about ten feet and hit what he later described as a "wall" of methyl mercaptan, but he managed to retreat to the stairwell [3, p. 29]. Operator 3 went to the fourth floor but did not find anyone. He then announced on the Lannate® Unit's public-address system that he did not see anyone on the fourth floor. The Board Operator responded that the Shift Supervisor and Operator 1 may be on the third floor. Operator 3 then began to feel light headed. He made his way to the stairwell and lost consciousness while descending from the fourth floor [3, p. 29].

When Board Operator could not reach people inside the manufacturing building, an operator in the control room, Operator 6[12] (the brother of Operator 2), then grabbed three 5-minute escape respirators. Other operators in the control room warned him not to enter the manufacturing building because they did not know what was going on or where the non-responsive operators were. Operator 6, however, took the escape respirators and rushed to the manufacturing building [3, p. 29].

At about 3:40 am, Operator 6 encountered Operator 4 in the south stairway of the manufacturing building and put an escape respirator on him. The breathing air helped Operator 4 recover, and he exited the manufacturing building safely. A worker—likely Operator 6—then manually activated the manufacturing building fume release alarm, an alarm intended to alert area workers of a toxic chemical release in the building. Operator 6 then went to the third floor of the manufacturing building. At some point inside the manufacturing building after helping Operator 4, Operator 6 retrieved a self-contained breathing apparatus (SCBA). He tried to rescue his brother (Operator 2) by putting one of the escape respirators on his face. Operator 6 was then fatally overcome by methyl mercaptan before he could connect his own respirator mask to his SCBA tank.

3 Lessons from the DuPont La Porte Incident

3.1 Ensure Toxic Gas Detector Systems Operate as Intended

DuPont La Porte's methyl mercaptan technical standard stated, "[a]utomatic or continuous specific chemical detection systems *must* be installed where a [methyl mercaptan] release could have any off-site impact […] [Emphasis in the original]." DuPont La Porte's methyl mercaptan toxic gas detection system consists of methyl mercaptan detectors, alarms, distributed control system (DCS), control panel, and a control room operator. The manufacturing building is equipped with gas detectors for methyl mercaptan in various locations. The third floor, where the release occurred, however, did not have any methyl mercaptan detectors. The detectors inside the manufacturing building were also set to alarm on the control panel when they detected a 25 ppm concentration of methyl mercaptan and were intended to protect workers from exposure to methyl mercaptan.[13]

DuPont used hydrogen sulfide (H_2S) detectors to identify methyl mercaptan. DuPont records indicate that the manufacturer has stated the detector should be capable of detecting methyl mercaptan within 1 ppm.[14] The alarms activated when the detectors sensed the hydrogen sulfide equivalent to 25 ppm methyl mercaptan.[15] The implementation of the methyl mercaptan toxic gas

[12] Operator 6 was a board operator for another process in the Lannate® Unit.

[13] The 25 ppm alarm set point is 50 times the NIOSH's REL of 0.5 ppm and more than double OSHA's ceiling limit of 10 ppm [7]. Therefore, alarming at this detection level could not protect workers from exposure to methyl mercaptan.

[14] DuPont La Porte's detectors for methyl mercaptan are calibrated for 0-50 ppm using a hydrogen sulfide (H_2S) calibration gas.

[15] The methyl mercaptan detectors were evaluated post-incident. Although the detectors inside the building have the ability to span 0-50 ppm, these detectors were set to span 0-25 ppm. Post-incident, the span was increased to 50 ppm. This configuration of the alarms limited the amount of information communicated to DuPont La Porte personnel.

detection system inside the manufacturing building, however, did not meet the design intent, federal safety regulations, or DuPont corporate standards.

The methyl mercaptan detectors outside and inside the manufacturing building were installed for two different purposes—the outside detectors were geared towards an offsite release and the inside detectors were meant to protect workers from a hazardous environment. DuPont La Porte documentation stated that the basis for the alarms outside the manufacturing building was to provide an early warning mechanism for a potential offsite methyl mercaptan release. The alarm setpoint was based on the ERPG (Emergency Response Planning Guidelines)-2[16] concentration[17] for methyl mercaptan which is 25 ppm. DuPont La Porte installed the methyl mercaptan detectors inside the manufacturing building after the methyl mercaptan detectors outside the building. Unlike the outside detectors, there was no documentation for the basis for setting the alarm setpoint to 25 ppm for the methyl mercaptan detectors inside the manufacturing building nor was there evidence that these setpoints were set at 25 ppm to prevent methyl mercaptan alarms from becoming a nuisance. It appears that the setpoints were copied from the outside methyl mercaptan detectors even though the inside detectors served a different purpose—worker protection. The methyl mercaptan detectors alarmed at 25 ppm—well above the NIOSH recommended exposure limit (REL) of 0.5 ppm and the OSHA enforceable PEL ceiling limit of 10 ppm.

Depending on the purpose of a toxic gas detection system alarm, there are different sets of values for facilities to consider. For instance, ERPG values are designed for the protection of the public during an emergency and not intended for the everyday routine protection of workers [4].[18] For worker protection, OSHA recommends that employers consider using the lower alternative occupational exposure limits because the agency believes that exposures above some of these alternative occupational exposure limits may be hazardous to workers, even when the exposure levels are in compliance with the relevant PELs [5]. For example, the Cal-OSHA permissible exposure limit (PEL) is an 8-hour time-weighted average (TWA) of 0.5 ppm and NIOSH's REL is a 15- minute ceiling limit of 0.5 ppm [6], [7].

Additionally, information from the methyl mercaptan detectors inside the manufacturing building only displayed on the DCS monitor and was not recorded by the site's DCS and data historian [3, p. 37]. CSB investigators, however, were able to obtain information from alarm logs about the detectors.

DuPont La Porte's management of change documentation for installation of methyl mercaptan detectors inside the manufacturing building stated the condition prompting the change was the

[16] ERPG-2 is the maximum airborne concentration below which nearly all individuals could be exposed for up to 1 hour without experiencing or developing irreversible or other serious health effects or symptoms which could impair an individual's ability to take protective action. At concentrations above ERPG-2, some members of the community may experience significant health effects which could impair them from taking protective action. Emergency responders often alert the community to take protective action in areas susceptible to the ERPG-2 concentration and above [19].

[17] The ERPG values for methyl mercaptan in parts per million (ppm) are: ERPG-1, 0.005; ERPG-2, 25; ERPG-3, 100 [18, p. 6].

[18] Emergency Response Planning Guidelines (ERPGs) are chemical concentration levels that are used during emergency response and when preparing for emergencies. They are designed to anticipate public health effects not occupational health effects from exposure to airborne chemicals [4].

possibility that personnel could enter a potentially hazardous location without knowing methyl mercaptan had been released:

> *Present Conditions and Description of Proposed Change:*
>
> *Personnel can enter a potentially hazardous location without knowing MeSH [methyl mercaptan] has been released to the environment.*

The methyl mercaptan gas detectors, however, only alarmed on the Lannate® control room operator's computer screen.

DuPont La Porte site did not have audible or visual alarms on or in the manufacturing building to alert field personnel of a methyl mercaptan release automatically. DuPont relied on verbal communication of alarms that automatically displayed on a continuously manned control board. Without being made aware of the major chemical release inside the building, Operator 2, Operator 3, and Operator 4 entered without respiratory protection. As a result, Operator 2 was killed from methyl mercaptan exposure, and Operator 3 and Operator 4 were exposed to dangerous concentrations of methyl mercaptan.

Facilities should make strategic decisions when choosing a toxic gas detector to ensure that a toxic gas detection system operates as intended. Considerations should include regulatory requirements, NIOSH recommendations, chemical interferences, and utility (worker protection, release modeling, etc.). The methyl mercaptan detectors at the DuPont La Porte facility had a span of 0-50 ppm. This range may be appropriate for early detection of a toxic leak when coupled with alarms and clear instructions to evacuate the premises at lower concentrations. If the goal is to understand other aspects of a release, like potential lethal concentration (150 ppm is methyl mercaptan's IDLH concentration) or modeling an offsite release, then a wider concentration range should be considered.

Additionally, facilities should also consider the use of automatic alarms designed to alert personnel of hazardous conditions—such as a release of a toxic chemical or explosive atmosphere—which can relay critical safety information immediately to personnel, without potential delays inherent in a system that solely relies on a control room operator to verbally communicate safety alarms to operators in the field.

3.2 Ensure Toxic Gas Detector Placement is Strategic

Although the DuPont La Porte site had 13[19] fixed methyl mercaptan detectors inside and outside the manufacturing building, there were no fixed methyl mercaptan detectors located on the third floor inside the manufacturing building where a majority of the November 15, 2014 methyl mercaptan release occurred. Additionally, no workers were provided with either portable or personal methyl mercaptan monitors. Part of the Lannate® process occurred inside the closed manufacturing building, a four-story building containing a large amount of piping and process equipment, dividing walls, multiple stairwells, and an elevator. The Lannate® Unit had methyl

[19] The site had a total of 15 methyl mercaptan detectors—there were two additional methyl mercaptan detectors installed in another process building adjacent to the manufacturing building.

mercaptan process piping both inside and outside the manufacturing building. A facility siting study and a process hazards analysis had identified the necessity of having methyl mercaptan detectors in the Lannate® manufacturing building. These studies identified the need for protecting workers from toxicity and explosivity hazards related to potential methyl mercaptan releases.

DuPont La Porte installed 10 methyl mercaptan detectors near process piping outside the manufacturing building. The design basis for the outside methyl mercaptan detectors was to provide an early warning of a methyl mercaptan release. DuPont La Porte placed these detectors around its methyl mercaptan storage tank and the area where methyl mercaptan could be unloaded. There were, however, no fixed methyl mercaptan detectors along the fence line that could relay whether there was an offsite release.

DuPont La Porte installed three methyl mercaptan detectors inside the manufacturing building. The stated design purpose for the methyl mercaptan detectors inside the building was worker protection. The inside detectors were located within the building on the first and fourth floors. These detector locations were determined based on potential release scenarios from a 2001 process hazard analysis (PHA).

The methyl mercaptan release occurred on the third floor where there were no methyl mercaptan detectors. Although the 2001 PHA did not identify methyl mercaptan detector locations on the third floor, DuPont La Porte routed the waste gas vent header piping along the third floor and this piping was directly connected to the top of the methyl mercaptan storage tank. During abnormal or upset conditions, the waste gas vent header could have been a potential source for a methyl mercaptan release. For example, ineffective hazard management while troubleshooting the plugged methyl mercaptan feed piping during the November 15th incident allowed liquid methyl mercaptan to flow into the waste gas vent header piping towards the NRS incinerator—a location where it was never intended to go.

Additionally, the routine draining of liquid from the waste gas vent header should have triggered a management of change review and considered whether additional safeguards like methyl mercaptan detectors were necessary. As discussed earlier, DuPont La Porte never fully resolved liquid accumulation in the waste gas vent header piping to the NRS incinerator with an engineered solution. Instead, operators were directed to drain the system by area management through written instructions in a daily instruction log book to operations staff [8, p. 18].[20] This routine draining presented potential dangers because DuPont La Porte personnel never sampled or analyzed the liquid workers routinely drained from this system [8, p. 18]. Daily instructions before the incident requested the operators to sample the uncharacterized liquid. DuPont La Porte management wanted samples of the liquid for a subsequent laboratory analysis to identify its chemical composition. Workers reported that the liquid had a strong, noxious odor of sulfur compounds [8, p. 18]. Even though DuPont La Porte was aware that the liquid likely contained more than just water and was unaware of the actual chemical composition [8, p. 18], instructions to workers did

[20] Beginning January 14, 2014, and continuing until the incident, the daily instructions directed operators to drain the methyl mercaptan waste gas vent header's "low point" on the 3rd floor once per shift. After the installation of a flexible hose from the waste gas vent header drain valve to the floor, the instructions changed to include directing the liquid through the hose under a running safety shower until the liquid stops flowing from the hose [8, p. 18].

not specify that additional precautions—like wearing additional PPE—were needed when draining the liquid from the waste gas vent header to the floor inside the manufacturing building.

Consequently, DuPont La Porte workers dealt with the common problem of liquid accumulation in the waste gas vent header on a routine basis by draining the liquid (line breaking) without engineered solutions or ensuring the use of safety procedures or personal protective equipment. On November 15, 2014, when the liquid drain valves were opened this time, flammable and highly toxic methyl mercaptan flowed onto the floor and filled the manufacturing building with toxic vapor and there no methyl mercaptan detectors on the third floor to alert personnel of the hazardous conditions that existed there.

As with toxic detector selection, the placement of toxic gas detectors must be strategic in order to capture the potential areas where a release could occur and the extent of the release. The methyl mercaptan detector placement both externally and internally at the DuPont La Porte facility did not operate as intended in part because the detectors did not fully cover the areas where the majority of the methyl mercaptan was released. For example, there was no real time data for emergency responders to use to evaluate the potential for offsite concentrations during the methyl mercaptan release. Although emergency responders conducted offsite monitoring hours after the initial release, the information obtained from air monitoring could not give an accurate portrayal of the methyl mercaptan release. Furthermore, the methyl mercaptan monitors inside the building did not provide accurate information about the release because they were not located near the open drain valves.

Facilities can adopt a variety of methods to select the location of toxic gas monitors with the understanding that they should be prepared for scenarios where the monitoring location may not be effective. One technique that a facility could use is to develop a series of release scenarios and based on those scenarios place monitors in the path of those releases (including by the potential sources of a leak). This technique, however, may not cover all potential release areas. Additionally, facilities should evaluate additional locations for toxic gas detectors if there is a change in process that creates new potential release scenarios. Another technique, if focusing on worker protection, is to give workers personal monitors when they enter toxic gas hazardous areas, especially if these areas are not normally occupied or congested with equipment. At DuPont La Porte, commercially available personal hydrogen sulfide monitors could have been issued to workers. Additionally, if using fixed toxic gas monitors for personnel protection, they should be installed near the breathing zone.

3.3 Ensure Personnel Understand Toxic Gas Hazards

DuPont La Porte personnel did not fully appreciate the effects of methyl mercaptan's physical characteristics. Although methyl mercaptan[21] is a highly toxic, flammable chemical, its low odor

[21] Personnel at DuPont La Porte would refer to methyl mercaptan as "MeSH." MeSH is shorthand based upon methyl mercaptan's chemical formula: CH_3SH. "Me" represents the methyl group (CH_3) and "SH" represents the mercaptan group (SH)—and when combined, these initialisms forms MeSH.

threshold[22] and ability to cause olfactory fatigue in personnel can generate a false sense of security when working with the chemical.

Although DuPont's methyl mercaptan technical standard states, "[a]ll DuPont sites *will* protect the site and neighboring businesses and communities from the escape of a highly toxic material [,]" methyl mercaptan detectors provided essentially no value to DuPont personnel [Emphasis in the original]. DuPont operations staff became accustomed to the odor of methyl mercaptan being present in their work environment. The low odor threshold of methyl mercaptan resulted in a false perception of safety. Consequently, the workers did not associate detecting the odor of methyl mercaptan as representing an immediate hazard.

> *"And I smell MeSH [methyl mercaptan] all the time ... but it never ... I smelled MeSH but I never felt I couldn't breathe. Like whether it's, there's -- whether it's a small leak or anything, it just never -- I didn't think MeSH was that dangerous. I thought it, you know, but it is."*

As seen by this comment from a DuPont La Porte worker, smelling methyl mercaptan at low concentrations without having any observable, adverse effects can potentially normalize the danger of a highly toxic chemical.

Furthermore, extended exposure to methyl mercaptan could potentially eliminate the perception of a methyl mercaptan release because olfactory fatigue will prevent workers from being able to smell the chemical. The odor threshold for methyl mercaptan is approximately two parts per billion (2 ppb). However, olfactory fatigue is known to occur with methyl mercaptan, resulting in the exposed person to no longer being able to detect the chemical by smell. As a result, using odor as an indicator of a release may not provide adequate warning of hazardous concentrations [9].

In the events leading to the major methyl mercaptan release, there were many smaller releases, filling the area with methyl mercaptan odor. In part because personnel believed they understood the source of releases and normalized the risk associated with these smaller releases, personnel did not apply additional safeguards like clearing the area or wearing the required personal protective equipment required by the site when dealing with a methyl mercaptan release.

Facilities should ensure that employees understand the hazards of chemicals that they are working with and may be potentially exposed to if there is a release. In cases where there are multiple hazardous chemicals, such as methyl isocyanate[23] (MIC) and methyl mercaptan (as was the case at DuPont La Porte), it is important that companies effectively train workers on the hazards of each chemical at their facility. Furthermore, companies should train workers that handle multiple highly toxic chemicals, such as methyl mercaptan and MIC, to treat these chemicals with equal

[22] Odor threshold, in general, is the lowest concentration of a gas or a material's vapor that can be detected by odor. Odor threshold values are not fixed physiological parameters or physical constants but are statistical points representing the best estimate value from a group of individual responses [12, p. 2].

[23] Methyl isocyanate (MIC) is a raw material used in the production of insecticides at La Porte including the Lannate® process. In December 3, 1984, a methyl isocyanate release at the Union Carbide insecticide plant in Bhopal, India resulted in an estimated 3,800 people that died within days, and tens of thousands that were injured. Eventually, the release killed tens of thousands of people. See CSB Safety Message on 30th anniversary of fatal chemical release that killed thousands in Bhopal, India [13].

importance. This training is important for workers to help ensure they have a working knowledge of those chemical hazards.

3.4 Ensure Response to a Toxic Gas Detector Alarm Is Appropriate

The toxic gas detection system at DuPont La Porte had a history of false readings. Because of these false readings, personnel supplanted their ability to observe and detect a leak over the toxic gas detection system. Personnel adapted their responses to a potential toxic gas release to include determining whether the toxic gas detectors were functioning properly by using their perception of smell before implementing the site's leak response procedures.

For example, the Lannate® Unit used chlorine in addition to methyl mercaptan. Workers could observe from a distance a chlorine leak by the green color of the gas or the immediate adverse effects (i.e., difficulty breathing) of the gas. Workers, similarly, would also use their sense of smell to detect methyl mercaptan. Methyl mercaptan gas, however, is colorless and the odor is not a reliable indicator of a release (for the reasons noted above). In both situations where workers used their physical observations to determine where there was a leak, workers were at risk for being overcome by a toxic chemical release because they were not using additional safeguards like personal protective equipment (PPE).

DuPont's highly toxic material standards require that toxic gas detection systems be treated the same as safety interlocks. DuPont defines a safety interlock as a system or function that detects an out-of-limits (abnormal) condition or improper sequence and brings it to a safe condition. In this case, an alarm should trigger a predetermined response from operators to bring the system to a safe condition. In practice, the methyl mercaptan detectors were not treated as a safety interlock.

Multiple DuPont employees informed the CSB that odor was the primary method used to locate a methyl mercaptan leak. DuPont personnel routinely responded to methyl mercaptan detector alarms without any additional respiratory protection. When they responded, if a significant leak was identified, these personnel consistently stated they would leave to get a SCBA and then return.

On the night of the incident, multiple methyl mercaptan detector alarms activated, but because the workers were aware that methyl mercaptan was being released from their activities associated with clearing the hydrate, these alarms were normalized and not treated as a safety interlock. The alarms were not treated as a near miss and there was no actual investigation into the source or how to prevent these releases. No additional personal protective equipment, such as air-purifying respirators or SCBAs was used. In addition, DuPont did not order a unit evacuation and did not notify the emergency response team until hours later, after Operator 1 made the distress call.

When a toxic detection system has a history of false readings it creates an opportunity for workers to normalize toxic gas detector readings and their response, making the toxic gas detection system ineffective. Facilities should take steps to address the potential for false readings and reduce opportunities for workers to normalize false readings and their alarms. One method is to treat toxic gas detector alarms as a near miss and investigate them even if there are no injuries or the readings are determined to be false. This method should trigger an effort to understand alarm management and safety management system issues and address them. Additionally, facilities should ensure that workers are using leak response procedures to prevent personnel from entering into potential hazardous situations without proper safeguards, like PPE.

4 Conclusions

The Lannate® Unit's toxic gas detection system did not effectively warn workers of the November 15, 2014 methyl mercaptan release occurring inside the manufacturing building. The manufacturing building, including the building's overall construction, detectors, and alarms, was not designed such that personnel could be readily notified of a major chemical release in the manufacturing building. Additionally, the methyl mercaptan toxic gas system did not provide real-time, accurate information to warn of potential offsite consequences.

There are several lessons that facilities can learn from the incident to help improve their toxic gas detection systems. Gas detectors must operate as intended to prevent workers from normalizing the hazard the detectors are protecting workers from. For toxic gas detection system to be effective, the detector selection and placement have to be strategic to identify and quantify releases. Finally, personnel have to understand and appreciate the hazards associated with the toxic gases they work with to help ensure they respond appropriately to potential toxic gas releases.

5 References

[1] U.S. Chemical Safety and Hazard Investigation Board (USCSB), "DuPont La Porte Facility Toxic Chemical Release," [Online]. Available: https://www.csb.gov/dupont-la-porte-facility-toxic-chemical-release-/. [Accessed 1 March 2019].

[2] U.S. Environmental Protection Agency (USEPA), "List of Regulated Substances under the Risk Management Plan (RMP) Program," [Online]. Available: https://www.epa.gov/rmp/list-regulated-substances-under-risk-management-plan-rmp-program. [Accessed 2017 August 28].

[3] U.S. Chemical Safety and Hazard Investigation Board (USCSB), "DuPont La Porte, Texas Chemical Facility Toxic Chemical Release: Interim Recommendations," 30 September 2015. [Online]. Available: https://www.csb.gov/assets/1/20/dupont_la_porte_interim_recommendations_2015-09-30_final1.pdf?15526. [Accessed 28 February 2019].

[4] National Oceanic and Atmospheric Administration (NOAA) Office of Response and Restoration, "Emergency Response Planning Guidelines (ERPGs)," [Online]. Available: https://response.restoration.noaa.gov/oil-and-chemical-spills/chemical-spills/resources/emergency-response-planning-guidelines-erpgs.html. [Accessed 10 March 2019].

[5] United States Department of Labor (USDOL) Occupational Safety and Health Administration (OSHA), "Permissible Exposure Limits - Annotated Tables," [Online]. Available: https://www.osha.gov/dsg/annotated-pels/. [Accessed 28 February 2019].

[6] United States Department of Labor (USDOL) Occupational Safety and Health Administration (OSHA), "Annotated Table Z-1," [Online]. Available: https://www.osha.gov/dsg/annotated-pels/tablez-1.html. [Accessed 28 February 2019].

[7] Centers for Diseases Control and Prevention National Institute of Occupational Safety and Health (NIOSH), "NIOSH Pocket Guide to Chemical Hazards - Methyl Mercaptan," 11 April 2016. [Online]. Available: https://www.cdc.gov/niosh/npg/npgd0425.html. [Accessed 28 February 2019].

[8] U.S. Chemical Safety and Hazard Investigation Board (USCSB), "DuPont La Porte, Texas Chemical Facility Toxic Chemical Release: Interim Recommendations," 30 September 2015. [Online]. Available: https://www.csb.gov/assets/1/20/dupont_la_porte_interim_recommendations_2015-09-30_final1.pdf?15526.

[9] Agency for Toxic Substances and Disease Registry (ATSDR), "Methyl Mercaptan," [Online]. Available: https://www.atsdr.cdc.gov/MHMI/mmg139.pdf. [Accessed 28 February 2019].

[10] A. H. Mohammadi and D. Richon, "Equilibrium Data of Sulfur Dioxide and Methyl Mercaptan Clathrate Hydrates," *Journal of Chemical & Engineering Data,* vol. 56, pp. 1666-1668, 2011.

[11] Center for Chemical Process Safety (CCPS), Recognizing Catastrophic Incident Warning Signs in the Process Industries, Hoboken: Wiley, 2012.

[12] A. H. Lehocky, S. S. Murnane and D. Patrick, Odor Thresholds for Chemicals with Established Occupational Health Standards, Falls Church: AIHA, 2013.

[13] USCSB, "On 30th Anniversary of Fatal Chemical Release that Killed Thousands in Bhopal, India, CSB Safety Message Warns it Could Happen Again," [Online]. Available: http://www.csb.gov/on-30th-anniversary-of-fatal-chemical-release-that-killed-thousands-in-bhopal-india-csb-safety-message-warns-it-could-happen-again-/?pg=4.

[14] K. Baker, "DuPont Closing Texas Plant Where 4 Died," The News Journal, 1 April 2016. [Online]. Available: https://www.usatoday.com/story/money/2016/04/01/dupont-closing-texas-plant-where-4-died/82512270/. [Accessed 29 June 2017].

[15] DuPont, "Form 8-K," 31 March 2016. [Online]. Available: https://www.sec.gov/Archives/edgar/data/30554/000003055416000115/dd33116form8-k.htm.

[16] U.S. Chemical Safety and Hazard Investigation Board (USCSB), "E.I. DuPont de Nemours & Co., Inc. Belle, West Virgina Investigative Report," September 2011. [Online]. Available: https://www.csb.gov/assets/1/20/csb%20final%20report.pdf?13966. [Accessed 1 March 2019].

[17] U.S. Chemical Safety and Hazard Investigation Board (USCSB), "E. I. DuPont De Nemours Co. Fatal Hotwork Explosion," [Online]. Available: https://www.csb.gov/e-i-dupont-de-nemours-co-fatal-hotwork-explosion/. [Accessed 1 March 2019].

[18] American Industrial Hygiene Association (AIHA), "Current ERPG Values," 2014. [Online]. Available: https://www.aiha.org/get-involved/AIHAGuidelineFoundation/EmergencyResponsePlanningGuidelines/Documents/2014%20ERPG%20Values.pdf.

[19] F. Cavender, S. Phillips and M. Holland, "Development of Emergency Response Planning Guidelines (ERPGs)," *Journal of Medical Toxicology,* vol. 4, no. 2, pp. 127-131, 2008.

553063 Hurricane Harvey 20 Months Later - Insights on the Arkema Crosby Flooding

On August 24, 2017, Hurricane Harvey, a Category 4 hurricane, made landfall in southeast Texas producing an unprecedented amount of rainfall. Hurricane Harvey killed 68 people and flooded over 300,000 structures, forcing roughly 40,000 people to evacuate their homes. Hurricane Harvey required a large emergency response effort supported by local, state, and Federal officials.

At Arkema's Crosby site, Harvey delivered an unprecedented amount of rainfall. The National Oceanic and Atmospheric Administration estimates show that the rainfall during Harvey had less than a one-in-one-thousand-year likelihood of occurring. Harris County has estimated that at Crosby, Harvey was a 5,000 year to 20,000 year event. The plant's hurricane preparedness procedure was followed and resulted in a safe shutdown and hurricane preparedness of the manufacturing units. However, finished goods inventory remaining in the warehouses was overwhelmed by high water and overcame multiple layers of protection used to safely store organic peroxide finished goods.

This presentation will explain the sites preparedness ahead of time, the events leading up to the flooding, the incident itself, and the emergency response.

The presentation will focus on the impacts, the main lessons learned and the key recommendations from the CSB report.

AUTHOR INDEX

AUTHOR INDEX

AIChE
120 Wall Street, FL 23
New York, NY 10005-4020

CURRAN ASSOCIATES INC.
proceedings
.com

ISBN 978-1-5108-8691-9

9 781510 886919